AF830832

Lecture Notes in Mechanical Engineering

Series Editors

Francisco Cavas-Martínez, Departamento de Estructuras, Universidad Politécnica de Cartagena, Cartagena, Murcia, Spain

Fakher Chaari, National School of Engineers, University of Sfax, Sfax, Tunisia

Francesco Gherardini, Dipartimento di Ingegneria, Università di Modena e Reggio Emilia, Modena, Italy

Mohamed Haddar, National School of Engineers of Sfax (ENIS), Sfax, Tunisia

Vitalii Ivanov, Department of Manufacturing Engineering Machine and Tools, Sumy State University, Sumy, Ukraine

Young W. Kwon, Department of Manufacturing Engineering and Aerospace Engineering, Graduate School of Engineering and Applied Science, Monterey, CA, USA

Justyna Trojanowska, Poznan University of Technology, Poznan, Poland

Lecture Notes in Mechanical Engineering (LNME) publishes the latest developments in Mechanical Engineering—quickly, informally and with high quality. Original research reported in proceedings and post-proceedings represents the core of LNME. Volumes published in LNME embrace all aspects, subfields and new challenges of mechanical engineering. Topics in the series include:

- Engineering Design
- Machinery and Machine Elements
- Mechanical Structures and Stress Analysis
- Automotive Engineering
- Engine Technology
- Aerospace Technology and Astronautics
- Nanotechnology and Microengineering
- Control, Robotics, Mechatronics
- MEMS
- Theoretical and Applied Mechanics
- Dynamical Systems, Control
- Fluid Mechanics
- Engineering Thermodynamics, Heat and Mass Transfer
- Manufacturing
- Precision Engineering, Instrumentation, Measurement
- Materials Engineering
- Tribology and Surface Technology

To submit a proposal or request further information, please contact the Springer Editor of your location:

China: Dr. Mengchu Huang at mengchu.huang@springer.com
India: Priya Vyas at priya.vyas@springer.com
Rest of Asia, Australia, New Zealand: Swati Meherishi at swati.meherishi@springer.com
All other countries: Dr. Leontina Di Cecco at Leontina.dicecco@springer.com

To submit a proposal for a monograph, please check our Springer Tracts in Mechanical Engineering at http://www.springer.com/series/11693 or contact Leontina.dicecco@springer.com

Indexed by SCOPUS. The books of the series are submitted for indexing to Web of Science.

More information about this series at http://www.springer.com/series/11236

P. M. Pandey · Pradeep Kumar ·
Vikas Sharma
Editors

Advances in Production and Industrial Engineering

Select Proceedings of ICETMIE 2019

Springer

Editors
P. M. Pandey
Department of Mechanical Engineering
Indian Institute of Technology Delhi
New Delhi, Delhi, India

Pradeep Kumar
Department of Mechanical Engineering
Indian Institute of Technology Roorkee
Roorkee, India

Vikas Sharma
Department of Mechanical Engineering
The NorthCap University
Gurgaon, India

ISSN 2195-4356 ISSN 2195-4364 (electronic)
Lecture Notes in Mechanical Engineering
ISBN 978-981-15-5518-3 ISBN 978-981-15-5519-0 (eBook)
https://doi.org/10.1007/978-981-15-5519-0

© Springer Nature Singapore Pte Ltd. 2021
This work is subject to copyright. All rights are reserved by the Publisher, whether the whole or part of the material is concerned, specifically the rights of translation, reprinting, reuse of illustrations, recitation, broadcasting, reproduction on microfilms or in any other physical way, and transmission or information storage and retrieval, electronic adaptation, computer software, or by similar or dissimilar methodology now known or hereafter developed.
The use of general descriptive names, registered names, trademarks, service marks, etc. in this publication does not imply, even in the absence of a specific statement, that such names are exempt from the relevant protective laws and regulations and therefore free for general use.
The publisher, the authors and the editors are safe to assume that the advice and information in this book are believed to be true and accurate at the date of publication. Neither the publisher nor the authors or the editors give a warranty, expressed or implied, with respect to the material contained herein or for any errors or omissions that may have been made. The publisher remains neutral with regard to jurisdictional claims in published maps and institutional affiliations.

This Springer imprint is published by the registered company Springer Nature Singapore Pte Ltd.
The registered company address is: 152 Beach Road, #21-01/04 Gateway East, Singapore 189721, Singapore

Preface

We would like to thank the authors for taking their precious time to submit manuscript as per the quality parameters required in this book, Advances in Production and Industrial Engineering.

All the accepted manuscripts were presented in ICETMIE 2019. ICETMIE is a biennial International conference which aims to provide a platform for academicians and practitioners to explore emerging technologies in the field of Mechanical and Industrial Engineering and further to contribute and disseminate their experience and research work for the purpose of exploring solutions to the global challenges. This conference provided an opportunity for researchers to learn about the latest developments and emerging trends in mechanical and industrial engineering through scientific information interchange between researchers, developers, engineers, students and practitioners in this field.

The purpose of this book is to provide the details of the latest advancements in research and developments of various advanced machining processes such as additive manufacturing processes, application of alloys/composite techniques, composites, ceramics, and polymers/processing. This book will be useful for industrial experts, entrepreneurs, university professors, and research scholars.

New Delhi, India — P. M. Pandey
Roorkee, India — Pradeep Kumar
Gurgaon, India — Vikas Sharma

Contents

About the Editors

Dr. P. M. Pandey is currently professor at the Department of Mechanical Engineering, IIT Delhi. He obtained his B.Tech. from HBTI, Kanpur and M.Tech. and Ph.D. from Indian Institute of Technology Kanpur. His major areas of research interests include Additive Manufacturing/3D Printing and Tooling, CAD/CAM, Non-traditional Machining and Finishing, FEA of Manufacturing Processes, Biomedical application of 3D Printing. He has published 149 research papers in respected international journals and more than 45 papers in International and National Conferences. Dr. Pandey is currently editor-in-chief and authors of many well renowned books. He has more than 4008 citations with h-index: 30.

Dr. Pradeep Kumar is currently a professor at the Department of Mechanical & Industrial Engineering, Indian Institute of Technology Roorkee. He obtained Bachelor of Engineering in Industrial Engineering and Master of Engineering in Production Engineering (Gold Medalist) and Ph.D. Manufacturing and Production Engineering from University of Roorkee (Now, IIT Roorkee). He also served the Delhi Technological University as Vice Chancellor. He has been honored as eminent mechanical engineer for his contribution in the field of mechanical engineering by The Institution of Engineers (India). Dr. Kumar has supervised more than 38 Ph.D. in the field of industrial and mechanical engineering. He has published several research papers in reputed international journals and conferences.

Dr. Vikas Sharma is currently assistant professor at the Mechanical Engineering Department, The NorthCap University, Gurugram, Haryana, India. He graduated in 2007 with a bachelor of mechanical engineering from Kurukshetra University. He post-graduated in CAD/CAM & Robotics from Thapar University, Patiala 2010. Dr. Sharma earned a Ph.D., in Laser Machining/Composite Materials from Thapar University, Patiala in 2017. His general area of research includes Laser Beam Machining, Optimization Techniques, Composite materials, FEM, Advanced

Casting Methods, Additive Manufacturing. He has several peer-reviewed international and national journal publications and conference proceedings. Dr. Sharma is currently reviewer of many renowned journals like Elsevier, Springer, Sage, etc. He has also published two patents in field of mechanical and bio-mechanical engineering.

Effect of Angular Orientation of Continuous Fibers on the Extensional Properties of Carbon Fiber Composites

Joginder Singh and M. R. Tyagi

Abstract The strength of the unidirectional composites depends on factors like types of the fibers, types of the matrix, the volume fraction of fibers (V_f), volume fraction of matrix (V_m), angle of fibers to the horizontal axis, etc. If the direction of fiber and the applied force is the same then the angular orientation is 0° and the composite behaves like an isotropic material. In this condition, if extensional stress is applied to the composite then predominantly extensional strains are generated. In the same condition, if shear stress is applied to the composite then predominantly shear strains are generated. If the direction of fiber and applied stress is not the same then the composite behaves like an anisotropic material. In this condition, if extensional stress is applied to the composite then both extensional strains and shear strain are generated in the same component. In the same condition, if shear stress is applied to the composite then both shear strains and extensional strains are generated. So, as the angle of orientation increases, the behavior of composite moves from isotropic material to anisotropic material. The maximum value of the angle of orientation can be 90° when the fiber is oriented in the lateral direction. At such orientation, the tensile strength of the composite depends on matrix and the fibers act as stress concentration factor (SCF). Therefore, fibers have a negative influence on the tensile strength of the composite. The elastic behavior of Carbon Fiber Reinforced Composites (CFRC) was studied as a function of the angular orientation of the carbon fiber in the composite. The change in modulus of elasticity, as well as variation of stress and strain with an increase in the angle of orientation the carbon fiber, were calculated by theoretical approach. The results of the same are presented here.

Keywords Carbon fibers · Matrix · Volume fraction · Elastic modulus · Strain · Stress · Orientation · CFRC

J. Singh (✉) · M. R. Tyagi
Department of Mechanical Engineering, Manav Rachna University, Faridabad, Haryana, India
e-mail: joginder@mru.edu.in

© Springer Nature Singapore Pte Ltd. 2021

P. M. Pandey et al. (eds.), *Advances in Production and Industrial Engineering*, Lecture Notes in Mechanical Engineering,
https://doi.org/10.1007/978-981-15-5519-0_1

1 Introduction

In every field of engineering, people are working on various methods to save energy, so that global warming can be reduced. Materials to be used in engineering parts play a very crucial role in that sense. Initially, parts were developed from steel and other heavy metal materials. Researches were conducted in material science to reduce density. Then, Aluminum was discovered and no doubt density was reduced very much. And its uses increased everywhere like aerospace, automobile, heating, ventilation, and air conditioning (HVAC), etc. Even then further reduction was required for human welfare. Aluminum Metal Foam came into existence. By this material, density was reduced further without compromising the strength and other requirements [1]. Research is a journey, it's not a destination. High technology materials came into existence i.e., composites. Composites are not new because it had been used by many countries for bow and arrow years ago. But composites like glass fiber, carbon fibers, etc. were new for engineering application point of view. Initially, its application was limited in aerospace and biomedical. Now optimization is also going on in this field also [2]. Yang et al. [3] studied how composite strength and Elastic Modulus effect if fiber alignment was not proper. Similarly, the angular orientation of fiber with material axes also affects the composite mechanical properties but that was not discussed in the paper. S. Kumar et al. [4] conducted an experiment on a single composite but using different fibers at the same time and varying the fiber orientation for the evaluation of bending and tensile properties. According to them, elongation was maximum at 45° and lowest at 30°. Elastic modulus was maximum at 30°. Hang et al. [5] studied the composite's specimen with various widths ranging from 10 to 40 mm. They observed the delamination in the specimen after the thickness of 20 mm. Naresh et al. [6] worked on shear strength and flexural properties for carbon epoxy and glass epoxy. They observed that flexural strength and modulus are highest at fiber orientation of 0° for both the composites. Sahin et al. [7] found the observation between fiber orientation and dry behavior. The weight of the composite was affected by the factors of speed and load and not by the fiber orientation. Routray et al. [8] performed the experiment with basalt fibers by varying the fiber orientation and thickness to check the Modulus of Elasticity, load, and stress. They got that composite gave the better result at 0° than 90° for the discussed properties. They also found that the production technology of Basalt is much simpler than Glass fibers. Hossain et al. [9] developed the jute fiber with epoxy resin using the technique of vacuum bagging and characterized by tensile tests at different fiber orientations i.e., 0°, 45°, and 90°. They found that the tensile strength of the composite is very sensitive to the tensile strength of jute fibers and defects in jute fibers. Wu et al. [10] studied the hybrid composite by the combination of carbon and glass fiber. They revealed that tensile strength increases as carbon fibers increases but there is very little change in compressive strength. Tensile fracture strain and compressive fracture strain decreases as the carbon fiber increases. Rahmani et al. [11] confirmed the relation between elastic modulus and fiber orientation of multiple layers of carbon fiber composites. They confirmed that for tensile strength of composite fiber orientation is the most enhancing parameters. Defects like bond line defects, microvoids and porosity minimize the mechanical properties. Cordin et al. [12] developed the biobased lyocell-reinforced polypropylene composite at various fiber orientations. They got a pattern between the elastic modulus and fiber orientation with the help of the rule of mixtures and the result was as per the expectation. But they modified the rule of mixtures of elastic modulus by improving the contribution of reinforcing fibers.

In this study, the calculation of the Unidirectional Composite was conducted to know the effect of the angular orientation of continuous fibers on the Elastic properties of carbon fiber in unidirectional composites. As the fiber orientation changes, there is a change in Elastic properties like Elastic Modulus, Shear Modulus, Poisson's Ratio, etc. Here only Elastic Modulus, Stress, and Strain at different angles with material are discussed.

2 Basic Properties of Carbon Fiber and Epoxy Resin

Materials can be of different types like isotropic, special orthotropic, generally orthotropic, and anisotropic. As the forces acting on these materials, they show different behavior. In isotropic materials, the extensional stresses result in predominantly extensional strains whereas shear stresses induce shear strains. The nature of stress–strain relationships for various materials is summarized in Table 1 [13].

Table 1 Types of stress and strain developed in different materials

Type of materials	Stresses induced	Strains developed
Isotropic	Extensional stresses	Extensional strains
Isotropic	Shear stresses	Shear strains
Special Orthotropic Lamina	Extensional stresses	Extensional strains
Special Orthotropic Lamina	Shear stresses	Shear strains
Generally Orthotropic Lamina	Extensional stresses	Extensional Strains And Shear Strains
Generally Orthotropic Lamina	Shear stresses	Extensional strains and shear strains
Anisotropic	Extensional STRESSES	Extensional strains and shear strains
Anisotropic	Shear stresses	Extensional strains and shear strains

Table 2 Basic properties of carbon fiber and cast epoxy resin

S. no.	Type of properties	Symbols	Values
1	Elastic modulus of fiber	E_f	350,000 N/mm^2
2	Tensile strength of fiber	σ_{uf}	2100 N/mm^2
3	Shear modulus of fiber	G_f	13,000 N/mm^2
4	Shear strength of fiber	τ_f	23 N/mm^2
5	Poisson's ratio of fiber	ν_f	0.20
6	Elastic modulus of epoxy	E_m	3500 N/mm^2
7	Tensile strength of cast epoxy resin	σ_{um}	130 N/mm^2
8	Shear modulus of epoxy	G_m	2000 N/mm^2
9	Shear strength of epoxy	τ_m	6.33 N/mm^2
10	Poisson's ratio of epoxy	ν_m	0.35

In composite, fiber orientations to x-axis decide the type of material and the stresses and strains will be developed in the composite. A Unidirectional Composite has two main constituents i.e., Carbon Fiber and Epoxy. These constituents have so many different properties. Every engineering or non-engineering application has different requirements. Some applications even required thermal and electrical properties. For the model being used here in this study, the required properties are shown in Table 2 [14].

3 Specimen Specification

Orthotropic materials can be of two types i.e., Specially Orthotropic Material and Generally Orthotropic Material. In Special Orthotropic Condition, load on the composite is aligned with fibers. It means there will be no angle between the fibers and the load. In Generally Orthotropic Condition, load is not aligned with the fibers. It means there will be an angle between the fibers and the load. This is also known as Off-Axis Loading Condition. Here the focus is on Off-Axis Loading Condition [15].

3.1 X-Section Area of the Specimen of Unidirectional Composite

The geometry of the specimens was taken as per the ASTM E8M standard. The same design is also selected while tested for the specimen of steel. Accordingly, the X-Section area of the Unidirectional Composite specimen, with dimensions in mm, is shown in Fig. 1.

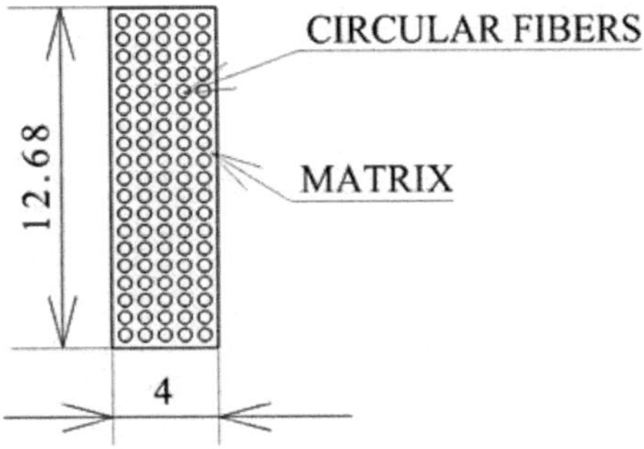

Fig. 1 X-Section of the unidirectional composite

So, X-Sectional Area of Unidirectional Composite

$$A_c = 12.68 \times 4 = 50.72\,\text{mm}^2$$

3.2 Volume Fraction of Carbon Fiber and Epoxy

The volume fraction of Carbon Fiber and Epoxy is an important factor for the strength of the composite. It has been found that almost the entire load on a Unidirectional Composite is carried by carbon fibers if the volume fraction of the fiber is 0.5. Therefore, the volume fraction of fiber is taken i.e., 0.5 neglecting the presence of voids. Volume fractions can be calculated as

$$V_f + V_m = 1 \tag{1}$$

where, V_f—Volume Fraction of Fiber and V_m—Volume Fraction of Matrix

So, $V_f = V_m = 0.5$.

X-Sectional Area of the Carbon Fiber and the Matrix was calculated as

$$A_f = V_f \times A_c \tag{2}$$

$$A_m = V_m \times A_c \tag{3}$$

Thus,

$$A_f = 25.36\,\text{mm}^2 \text{ and } A_m = 25.36\,\text{mm}^2$$

Since the values of the volume Fraction of the Fiber and the Matrix are the same. The values for X-Sectional Areas of Carbon Fiber and Matrix are also the same.

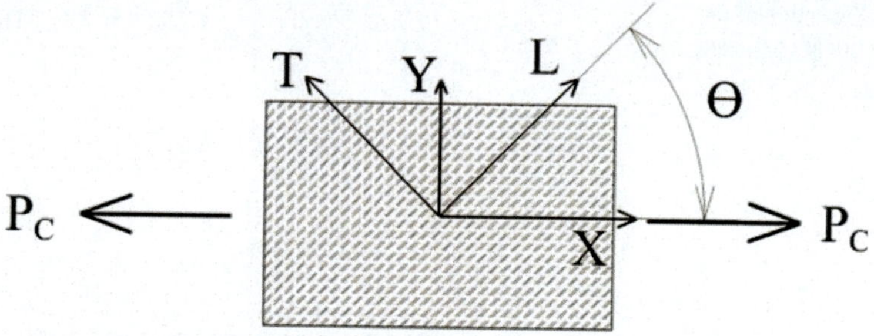

Fig. 2 Load carried by the unidirectional composite

3.3 *Load (Force) Applied on the Unidirectional Composite and Orientation of the Fibers*

The load is applied on Unidirectional Composite only in X-direction as shown in Fig. 2. In the present study, 45400 N loads have been considered. This is the maximum load before failure on a test specimen of steel in the tensile test on the Universal Testing Machine (UTM) [16]. The orientation of the fibers is denoted by 'θ,' the angle between the line of action of the applied force and the direction of the orientation of the carbon fiber in the composite. In the present study, the orientation varied from 0 to 90 degrees in the steps of 5°.

Load carried by the Unidirectional Composite in X-Direction = $\mathbf{P_c}$ = 45400 N.

4 Basic Calculations

4.1 *Load Carried by the Fibers and Matrix (P_f and P_m)*

The applied load on the composite is shared by the fibers and matrix. The relation between the load shared by the Composite, Fibers, and Matrix can be expressed as

$$P_c = P_f + P_m \tag{4}$$

The ratio of loads is given by the following relation in terms of Elastic Moduli and Volume Fractions.

$$\frac{P_f}{P_c} = \frac{\frac{E_f}{E_m}}{\frac{E_f}{E_m} + \frac{V_m}{V_f}} \tag{5}$$

So, from Eqs. (4) and (5),
$P_f = 44950.495$ N and, $P_m = 449.505$ N as all other parameter values are known.

4.2 Stresses in Fibers and Matrix (σ_f and σ_m)

Stresses in Fibers and Matrix can be calculated with the formula of stress, force, and area.

$$\begin{aligned}\text{Stress in Fibers} = \sigma_f &= \text{Load Carried by Fibers/Area of Fibers}\\ &= P_f/A_f\\ &= 4495.495/25.36\\ \sigma_f &= 1772.496\,\text{N/mm}^2\end{aligned}$$

$$\begin{aligned}\text{Stress in Matrix} = \sigma_m &= \text{Load Carried by Matrix/Area of Matrix}\\ &= P_m/A_m\\ &= 449.505/25.36\\ \sigma_m &= 17.725\,\text{N/mm}^2\end{aligned}$$

4.3 Stresses in the Longitudinal Direction of the Unidirectional Composite (σ_L)

This is the stress when the direction of load and fibers are same. Stress in the Longitudinal Direction of the Composite can be calculated as

$$\sigma_L = \sigma_f V_f + \sigma_m V_m \tag{6}$$

So, $\sigma_L = 895.11$ N/mm^2.

4.4 Elastic Modulus of Unidirectional Composite in Longitudinal Direction (E_L)

As per the rule of mixture, Elastic Modulus of Unidirectional Composite in Longitudinal Direction can be calculated as

$$E_L = E_f V_f + E_m V_m \tag{7}$$

So, $E_L = 350000 \times 0.5 + 3500 \times 0.5 = 176750$ N/mm^2.

4.5 Halpin–Tsai Equations for Elastic Modulus of Unidirectional Composite in Transverse Direction (e_T)

Elastic Modulus of Unidirectional Composite in Transverse Direction can be calculated with the help of Halpin–Tsai Equations [17]. It can be calculated as

$$\frac{E_T}{E_m} = \frac{1 + \xi\,\eta\,V_f}{1 - \eta\,V_f} \tag{8}$$

where

$$\eta = \frac{\frac{E_f}{E_m} - 1}{\frac{E_f}{E_m} + \xi} \tag{9}$$

where

ξ Zeta
η Iota

ξ (Zeta) is a measure of reinforcement and depends on the loading condition, packing geometry, and fiber geometry. The value of Zeta 'ξ' can be taken 2 for fiber with circular or square x-sections as per the Tsai and Halpin. Fibers used here are circular as shown in Fig. 1. The value of 'η' (Iota) can be obtained from Eq. (9) as all other values are known.

$$\eta = 0.97$$

So, from Eq. (3), Elastic Modulus of Composite in Transverse Direction was obtained as

$$E_T = 13400\,\text{N/mm}^2$$

4.6 Stresses in the Transverse Direction of the Unidirectional Composite (σ_T)

This is the stress when the load is perpendicular to the direction of fibers. Stress in the Transverse Direction of the Unidirectional Composite can be calculated as

$$\frac{\sigma_T}{E_T} = \frac{\sigma_f}{E_f}V_f + \frac{\sigma_m}{E_m}V_m \tag{10}$$

So, $\sigma_T = 67.86$ N/mm^2.

4.7 Halpin–Tsai Equations for in-Plane Shear Modulus of a Unidirectional Composite (G_{LT})

As the direction of the load and the fibers are not aligned, shear stress will generate. Therefore, In-plane Shear Modulus of a Unidirectional Composite can be calculated as

$$\frac{G_{LT}}{G_m} = \frac{1 + \xi\,\eta\,V_f}{1 - \eta\,V_f} \tag{11}$$

where

$$\eta = \frac{\frac{G_f}{G_m} - 1}{\frac{G_f}{G_m} + \xi} \tag{12}$$

The value of Zeta 'ξ' can be taken 1 for In-Plane Shear Modulus of a Unidirectional Composite as per the Tsai–Halpin [18]. The value of 'η' (Iota) can be obtained from Eq. (12) as all other values are known.

$$\eta = 0.73$$

So, from Eq. (5), In-plane shear modulus of a Unidirectional Composite was obtained as

$$G_{LT} = 4315.79\,\mathrm{N/mm^2}$$

4.8 Major and Minor Poisson's Ratio of a Unidirectional Composite (ν_{LT} and ν_{LT})

For the In-plane loading of a Unidirectional Composite, two Poisson Ratios can be defined i.e., Major Poisson's Ratio and Minor Poisson's Ratio. As per the rule of mixture, Major Poisson's Ratio can be calculated as

$$\nu_{LT} = \nu_f V_f + \nu_m V_m \tag{13}$$

So, $\nu_{LT} = 0.275$

Minor Poisson Ratio can be calculated from the relation given below

$$\frac{\nu_{LT}}{E_L} = \frac{\nu_{TL}}{E_T} \tag{14}$$

So, $\nu_{TL} = 0.021$.

4.9 Strain in L and T Direction of a Unidirectional Composite (ϵ_L and ϵ_T)

The strain in L Direction of a Unidirectional Composite can be calculated as

$$\epsilon_L = \frac{\sigma_L}{E_L} - \nu_{TL}\frac{\sigma_T}{E_T} \tag{15}$$

So, $\epsilon_L = 0.0051$.
The strain in T Direction of a Unidirectional Composite can be calculated as

$$\epsilon_T = \frac{\sigma_T}{E_T} - \nu_{LT}\frac{\sigma_L}{E_L} \tag{16}$$

So, $\epsilon_T = 0.0037$.

4.10 In-plane Shear Stress and Shear Strain of a Unidirectional Composite (τ_{LT} and γ_{LT})

In-plane Shear Stress of a Unidirectional Composite can be calculated as

$$\frac{\tau_{LT}}{G_{LT}} = \frac{\tau_f}{G_f}V_f + \frac{\tau_m}{G_m}V_m \tag{17}$$

So, $\tau_{LT} = 10.648\ \text{N/mm}^2$.
In-plane Shear Strain of a Unidirectional Composite can be calculated as

$$\gamma_{LT} = \frac{\tau_{LT}}{G_{LT}} \tag{18}$$

So, $\gamma_{LT} = 0.00247$.

5 Results and Discussion

Off-Axis load applied on Unidirectional Composite. Therefore, extensional stress developed the extensional strain and shear strain. Similarly, shear stress developed the shear strain and extensional strain also. Here only extensional properties considered i.e., '$E_{x,}$' '$\epsilon_{x,}$' and '$\sigma_{x.}$'

5.1 *Elastic Modulus of the Unidirectional Composite in X-Direction as Fiber Orientation Varies from 0° to 90° (E_x)*

Elastic Modulus of a Unidirectional Composite in X-Direction can be calculated as

$$\frac{1}{E_x} = \frac{\cos^4\theta}{E_L} + \frac{\sin^4\theta}{E_T} + \frac{1}{4}\left(\frac{1}{G_{LT}} - \frac{2\nu_{LT}}{E_L}\right)\sin^2 2\theta \quad (19)$$

The calculated value of 'E_x' is shown in Table 3. The graph between the Elastic Modulus and Fiber Orientation of a Unidirectional Composite is shown in Fig. 3.

Table 3 Elastic modulus, strains, and stresses of unidirectional composite at various angles

Fiber orientation (θ, Degree)	Elastic modulus in X-direction (E_x, N/mm^2)	Strain in X-direction (ϵ_x)	Stress in X-direction (σ_x, N/mm^2)
0	176,750 (HV)	0.00496 (HV)	876.4 (HV)
5	136,993	0.00473	648.6
10	82,818	0.00450	372.5
15	51,158	0.00426	217.7
20	34,429	0.00402	138.2
25	25,167	0.00378	95.2
30	19,721	0.00357	70.4
35	16,370	0.00338	55.3
40	14,263	0.00321	45.8
45	12,950	0.00308	39.9
50	12,182	0.00299	36.4
55	11,810	0.00294	34.7
60	11,738 (LV)	0.00293 (LV)	34.3 (LV)
65	11,895	0.00296	35.2
70	12,213	0.00303	37.0
75	12,614	0.00314	39.6
80	13,007	0.00329	42.8
85	13,294	0.00347	46.1
90	13,400	0.00367	49.2

HV—Highest Value, LV—Lowest Value

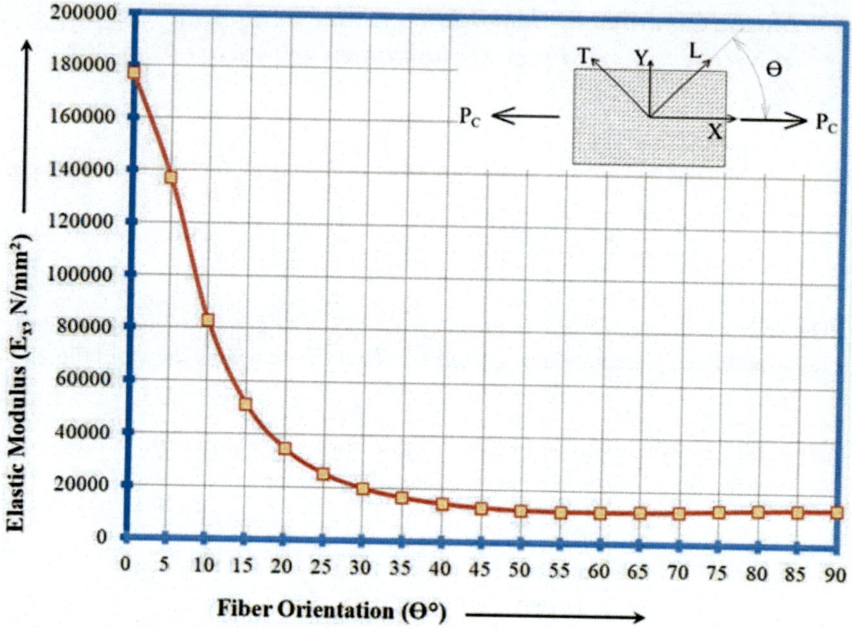

Fig. 3 Graph between the elastic modulus and fiber orientation of a unidirectional composite

5.2 Strains of the Unidirectional Composite in X-Direction as Fiber Orientation Varies from 0° to 90° (ϵ_x)

Strain is nothing but change in length upon original length. So it is a unit less entity. Strain of the Unidirectional Composite in X-Direction is dependent on 'ϵ_L,' 'ϵ_T,' 'γ_{LT},' and fiber orientation [19]. It can be calculated as

$$\epsilon_x = \epsilon_L \cos^2\theta + \varepsilon_T \sin^2\theta - \gamma_{LT} \sin\theta \cos\theta \tag{20}$$

The calculated values of 'ϵ_x' are shown in Table 3. The graph between the Strain and Fiber Orientation of the Unidirectional Composite is shown in Fig. 4.

5.3 Stresses of the Unidirectional Composite in X-Direction as Fiber Orientation Varies from 0° to 90° (σ_x)

Stresses of the Unidirectional Composite in X-Direction can be calculated as

$$E_x = \frac{\sigma_x}{\epsilon_x} \tag{21}$$

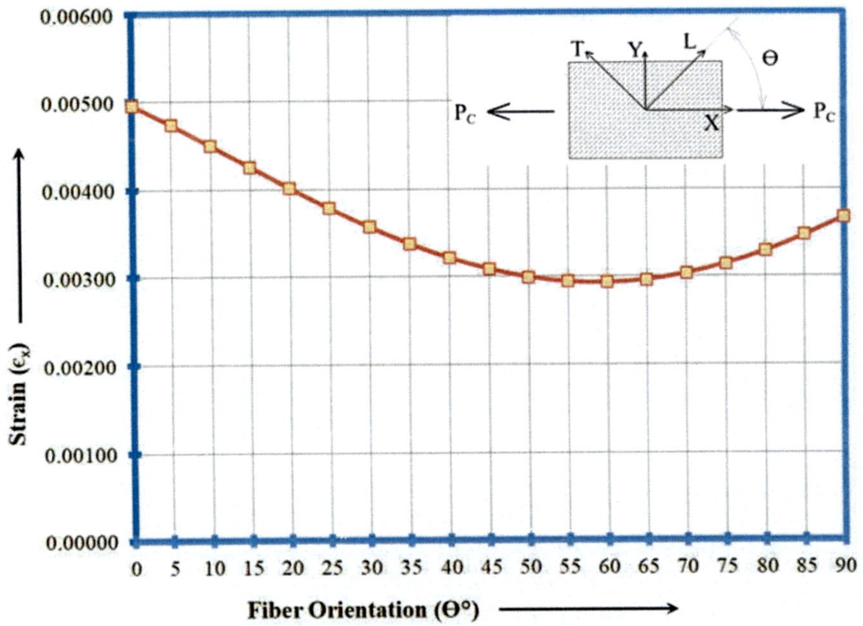

Fig. 4 Graph between the strain and fiber orientation of the unidirectional composite

The value of 'σ_x' can be obtained from Eq. (21) as Elastic Modulus and Strain values are calculated in Sects. 5.1 and 5.2. The calculated values of 'σ_x' calculated are shown in Table 3 [20]. The graph between the Stress and Fiber Orientation of the Unidirectional Composite is shown in Fig. 5.

5.4 Discussion

- As the fiber orientation increases from 0° to 90°, the Elastic Modulus of Unidirectional Composite decreases rapidly from 0° to 20° and then it decreases slowly up to 60°. After that, it becomes stable from 60° to 90°. Its highest value is at 0° i.e., 176,750 N/mm^2 and the lowest value is at 60° i.e., 11,738 N/mm^2.
- As the fiber orientation increases from 0° to 90°, strain induces in Unidirectional Composite. Strain decreases slowly from 0° to 60° and again increases up to 90°. Its highest value is at 0° i.e., 0.00496 and the lowest value is at 60° i.e., 0.00293.
- As the fiber orientation increases from 0° to 90°, Stress induces in the Unidirectional Composite. Stresses of Unidirectional Composite decrease rapidly from 0° to 20° and then it decreases slowly up to 60°. After that, it becomes stable from 60° to 90°. Its highest value is at 0° i.e., 876.4 N/mm^2 and the lowest value is at 60° i.e., 34.3 N/mm^2.

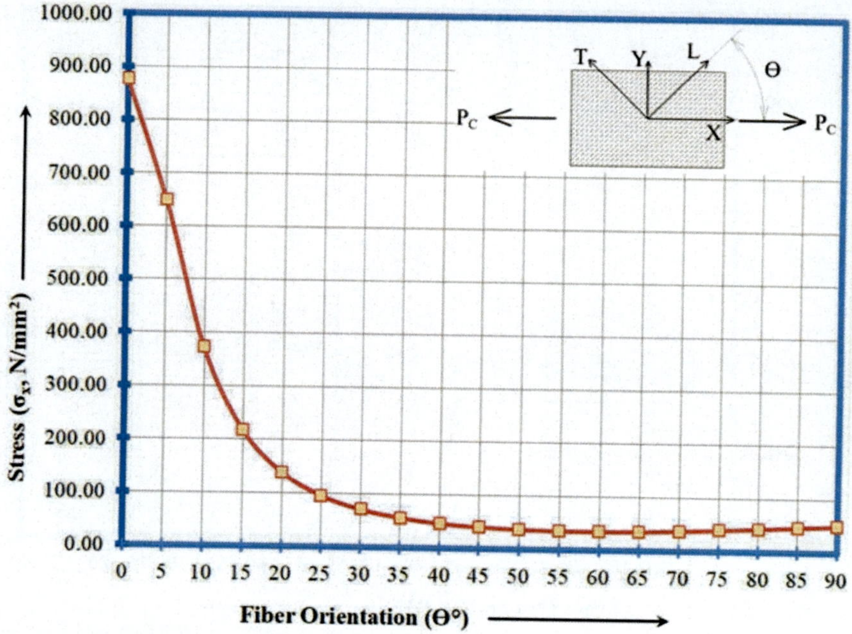

Fig. 5 Graph between the stress and fiber orientation of the unidirectional composite

6 Conclusion

- Elastic Modulus, Strain, and Stress have the highest values at 90° of Fiber Orientation i.e., at Longitudinal Direction. That is why Special Orthotropic Materials are more preferred for engineering applications Generally Orthotropic Materials.
- Elastic Modulus, Strain, and Stress have the lowest values at 60° of Fiber Orientation, not at 90°.
- Elastic Modulus, Strain, and Stress decreases from 0° to 60° and again starts increases from 60° to 90° of Fiber Orientation.
- This property can be utilized to optimize the carbon fiber reinforced composites for obtaining desired properties for specific engineering applications.

References

1. Andure MW, Jirapure SC, Dha Mande LP (2012) Advance automobile material for light weight future—a review. In: International conference on benchmarks in engineering science and technology ICBEST 2012. Proceedings published by International Journal of Computer Applications® (IJCA)

2. O'Toole Brendan J (2005) Introduction to composite materials. Department of Mechanical Engineering, Howard R. Hughes College of Engineering, University of Nevada Las Vegas. www.egr.unlv.edu/~bj
3. Yang X, Nanni A, Haug S, Sun C (2002) Strength and modulus degradation of CFRP laminates from fiber misalignment. Accepted for publication in Journal of Materials in Civil Engineering, ASCE
4. Kumar S, Krishna S, Rajanna S (2014) Study on effect of thickness and fiber orientation on a tensile and flexural properties of a hybrid composite. Int J Eng Res Appl 4(8):56–66 (Version 6). www.ijera.com. ISSN: 2248-9622
5. Hang L, Xue W, Terutake M, Jia X, Dongmei L (2015) Effects of specimen width on the tensile strength of aligned short-carbon-fiber reinforced epoxy composite laminates. In: 20th international conference on composite materials, Copenhagen, 19–24th July 2015
6. Naresh K, Krishnapillai S, Ramachandran V (2017) Effect of fiber orientation on carbon/epoxy and glass/epoxy composites subjected to shear and bending. Solid State Phenomena 267:103–108. ISSN: 1662-9779. www.scientific.net, Trans Tech Publications, Switzerland
7. Sahin Y, Tokdede A, Yankas M (2017) The effect of fiber orientation on the dry wear behaviour of carbon fibre-reinforced epoxy composites. Mater Methods Technol 11. ISSN: 1314-7269, J Int Sci Publ. www.scientific-publications.net
8. Routray S, Biswal KC, Barik MR (2015) The effect of fiber orientation on the dry wear behavior of carbon fibre reinforced epoxy composites. Res J Recent Sci 4(ISC-2014):202–208. ISSN: 2277-2502
9. Hossain MR, Islam MA, Vuurea AV, Verpoest I (2013) Effect of fiber orientation on the tensile properties of jute epoxy laminated composite. J Sci Res 5(1):43–54. www.banglajol.info/index.php/JSR
10. Wu W, Wang Q, Li W (2018) Comparison of tensile and compressive properties of carbon/glass interlayer and intralayer hybrid composites. Materials. www.mdpi.com/journal/materials
11. Rahmani H, Najafi S, Ashori A, Golriz M (2015) Elastic properties of carbon fibre-reinforced epoxy composites. Polym Polym Compos 23(7) (Smithers Information Ltd.)
12. Cordin M, Bechtold T, Pham T (2018) Effect of fibre orientation on the mechanical properties of polypropylene–lyocell composites. University of Innsbruck, Austria: Member of EPNOE—European Polysaccharide Network of Excellence, www.epnoe.eu, Cellulose
13. Broutman LJ (1967) Fiber reinforced plastics. In: Broutman LJ, Krock RH (eds) Modern composite materials. Addison-Wesley, Reading, MA, Chap. 13
14. Tsai SW, Azzi VD (1966) Strength of laminated composite materials. A/AA J 4(2):296–301
15. WilsonHB, Hill JL (1965) Mathematical studies of composite materials. Rohm and Haas Special Report No 5–50, AD 468569
16. Chen PE, Lin JM (1969) Transverse properties of fibrous composites. Mate Res Stand MTRSA 9(8):29–33
17. Hill R (1965) Theory of mechanical properties of fiber-strengthened materials: III. Self-consistent model. J Mech Phys Solids 13:189
18. Lempriere BM (1968) Poisson's ratio in orthotropic materials. AIM J 6(11):2226–2277
19. Agarwal BD, Narang JN (1977) Strength and failure mechanism of anisotropic composites. Fiber Sci Tec/Mol 10(1):37–52
20. Jones BH (1969) Determination of design allowable for composite materials. In: Composite materials: testing and design, STP 460. American Society for Testing and Materials, Philadelphia, PA, pp 307–320

Simulation of Nanowires on c-Si Surface for Reflectance Reduction

Arka Bera, Sourav Nag, and Arjyajyoti Goswami

Abstract Presence of surface structures has been found to reduce surface reflectance. Structures like Nanoholes, Nanopyramids, AR coating, etc. have been found useful for the purpose. In this work, the reflectance of c-Si with nanowires on its surface is simulated with the aim of reducing its surface reflectance. It is observed that the presence of Si nanowires on the surface of c-Si reduced the surface reflectance to 23% as compared to a reflectance of 38% in the case of bare c-Si. The simulations are based on Fast Fourier Transform and implemented through an online simulation tool called Stanford Stratified Structure Solver (S4). Results obtained from simulation are thereafter utilized in development of a mathematical model.

Keywords Surface structuring · Fast Fourier transform simulations · Reflectance

1 Introduction

Between 1992 and 2018, there has been significant growth in the field of Photovoltaics. During this era of time, photovoltaics (PV), additionally referred to as solar PV, has evolved from a distinct segment market of small- scale applications to a mainstream electricity supply [1]. For several years, Japan and pioneering European countries progressed in the research. With the improvement in technology, the cost of solar energy decreased significantly. Programs in several countries were involved in promoting PV deployment, such as the Energiewende in Germany, the Million Solar Roofs project in the United States, and China's 2011 five-year plan for solar energy production [2]. Since then, the deployment of photovoltaics has gained momentum on a worldwide scale, with various industries aiming to utilize solar energy as an alternative to conventional sources of energy. In the early twenty-first century, a

A. Bera (✉) · A. Goswami
Department of Mechanical Engineering, National Institute of Technology, Durgapur, India
e-mail: ab.17u10648@btech.nitdgp.ac.in

S. Nag
Department of Chemical Engineering, National Institute of Technology, Durgapur, India

© Springer Nature Singapore Pte Ltd. 2021

P. M. Pandey et al. (eds.), *Advances in Production and Industrial Engineering*, Lecture Notes in Mechanical Engineering,
https://doi.org/10.1007/978-981-15-5519-0_2

market for utility scale plants emerged in addition to rooftop solar panels and other distributed applications [3]. By 2015, about 30 nations had reached grid parity [4].

Crystalline Silicon (C-Si) is widely used by the photovoltaic industry for making traditional, conventional, and water-based solar cells. In 2013, conventional crystalline silicon technology dominated worldwide photovoltaic production, with multi-Si finding wider application as opposed to mono-Si, accounting for 54% and 36%t, respectively. 121 Gigawatts (GW) of crystalline silicon (87%) is included in the all-time deployed PV capacity (cumulative as of 2013) [5].

Crystalline Silicon generally reduces the reflection by 38%. Here we will be using Crystalline Silicon nanowires by growing them over the surface of C-Si so that the reflection reduces more. Silicon is chosen as it is cheap, easily available, and non-reactive. Alternatively, Indium Tellurium Gallide or Indium Arsenide could be used for the reduction of surface reflectance.

There are various nano-structures that have been implemented to reduce the surface reflectance of crystalline silicon. These include structures such as nanopyramidal arrays [6], nanohole arrays [7], and nanogratings [8]. There is a need to study the effect of nanowire parameters on the reflectance reduction from the surface of c-Si, which is being proposed in this paper. Studies by Van Trinh Pham et al. [9] have shown an increase in the length of nanowire decreases surface reflectance.

The work studies the influence of Si nanowire parameters in reducing the surface reflectance of C-Si by means of simulation of Si nanowires on a silicon wafer using S4 method. The work also aims at establishing a co- relation between surface reflectance and geometrical parameters of nanowires such as diameter and length using Response Surface Methodology. The rest of the paper is organized as follows: Sect. 2 introduces S4 method, including simulation parameters and explains the experimental methodology. Section 3 shows the results obtained from simulation and determines the error obtained between the simulation results and the mathematical model.

2 Simulation and Experimental Methodology

The nanowire simulations are geometrically well-defined surface structures consisting of equidistant, straight and parallel projections on the surface of c-Si. A nanowire is defined by two main geometrical parameters, namely, radius and length. Radius of nanowire refers to the distance between center and outer surface of a single nanowire. The length is the height of the nanowire above the c-Si surface. The schematic of a nanowire and its different geometrical features are shown in Fig. 1.

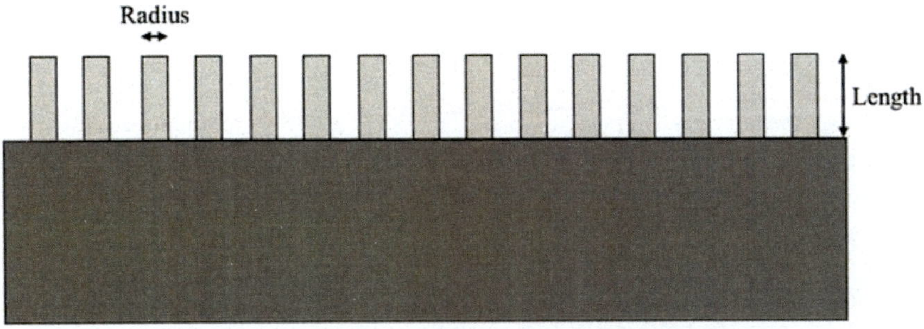

Fig. 1 Schematic of nanowire geometry

2.1 Simulation Tool

The simulation method used here is the Fourier Modal Method that is implemented through Stanford Stratified Structure Solver (S4). S4 is a code that analyzes mathematical functions or signals in terms of frequency, and solves problems of layered periodic structures. It internally uses Rigorous Coupled Wave Analysis (RCWA) and the S-matrix algorithmic program. RCWA (alternatively called the Fourier Modal Method (FMM)) is a semi-analytical methodology in studies of computational electromagnetics that is usually applied to resolve scattering from structures having periodic dielectric parts [10]. It is a Fourier-space approach. Devices and fields are considered as a sum of spatial harmonics [11]. FMM is the most widespread method for modeling diffraction gratings. The method is characterized by expanding the electromagnetic fields into Floquet–Fourier series, and medium permittivity (and probably medium permeability as well) into Fourier series, and determination of the ensuing Maxwell equations as a problem involving eigenvalues of a matrix [12]. The S4 tool was developed to carry out the analysis of optical proliferation in the generalized 3D structure. Development of S4 is accredited to Victor Liu, a member of the Fan Group in the Stanford Electrical Engineering Department [13–15].

The simulation space is divided into 4 regions. The top and bottom surfaces are chosen as semi-infinite layers of vacuum (made transparent in the schematic). The c-Si is further divided into 2 layers. The thickness of the upper layer is equal to the length of the nanowires to be simulated. This is a heterogeneous layer with its permittivity periodically changing between permittivity of c-Si and vacuum. The lower layer of c-Si is a homogeneous medium having constant properties of c-Si (which of course varies with the wavelength of the incident radiation). The schematic of the simulation setup is shown in Fig. 2.

An initial trial was made by comparing the surface reflectance of bare c-Si and c-Si with nanowires on its surface. The dimensions of the nanowires were taken as length: 1000 nm and radius: 75 nm. Qualitatively it can be seen directly that the presence of nanowires reduced the surface reflectance as compared to bare c-Si as can be seen in Fig. 3.

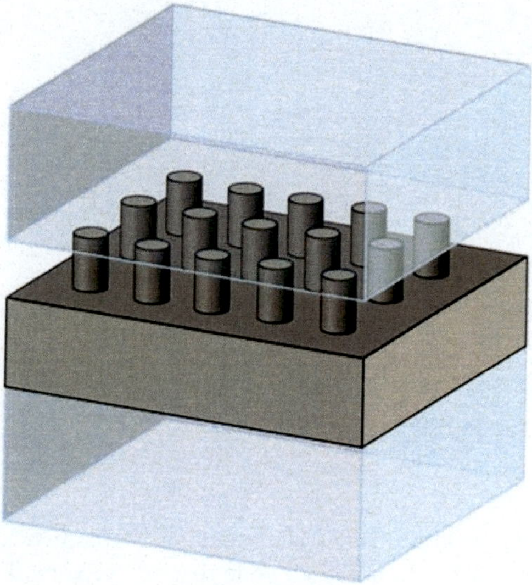

Fig. 2 Schematic of the simulation setup

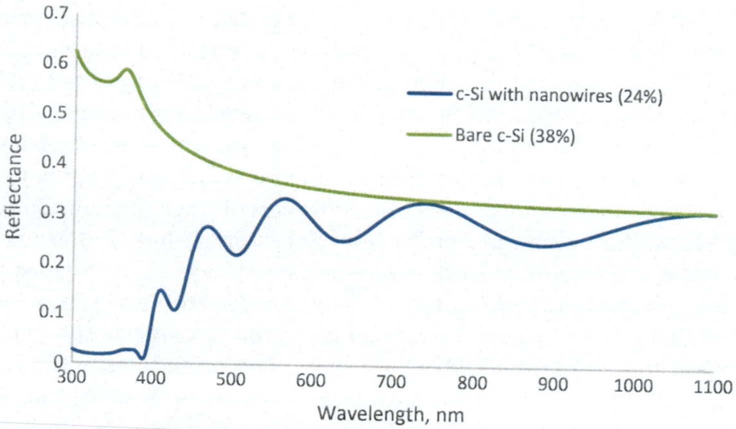

Fig. 3 Initial trial for reflectance reduction

More detailed experimentations were carried out through Central Composite Design as described in the next section.

Table 1 Simulation parameters at different levels of the design

Parameters	−1	0	+1
Radius	25 nm	55 nm	85 nm
Length	500 nm	800 nm	1100 nm

2.2 *Experimental Methodology*

The silicon nanowires were simulated on a base of crystalline silicon having a thickness of 30,000 nm. The wafer was of dimensions 1000 nm × 1000 nm. The mediums for incidence, reflectance, and transmittance of light rays were taken to be semi-infinite layers of vacuum. Vacuum has relative permittivity with real part = 1 and imaginary part = 0 for all wavelengths.

The reference point for a single nanowire was taken at (500 nm, 500 nm) with a circular cross-section. Thus, a matrix of nanowires was obtained about this central reference point. While simulating the silicon nanowires on the surface of the silicon surface, the radius of the nanowires was varied from 25 to 85 nm in steps of 10 nm, and length of the nanowires was varied from 500 to 1100 nm, in steps of 100 nm. These values were selected based on a literature search. In this case, 2 factor 3 level CCD is being used for Design of Experiments (DoE). The simulation parameters and their values at different levels are given in Table 1.

The simulation is conducted over a wavelength range of 300–1100 nm in steps of 1. The maximum Fourier expansion order is set to 1.

The simulated surface reflectance at various combinations of design parameters are shown in Table 2.

Table 2 Simulated values of surface reflectance at different design parameters

Standard order	Factor 1 A: R (nm)	Factor 2 B: L (nm)	Simulated surface reflectance (%)
1	25	500	32.58
2	85	500	24.62
3	25	1100	30.66
4	85	1100	22.45
5	25	800	31.31
6	85	800	23.95
7	55	500	28.10
8	55	1100	26.25
9	55	800	26.99
10	55	800	26.99
11	55	800	26.99
12	55	800	26.99
13	55	800	26.99

The mathematical model co-relating surface reflectance and the nanowire geometrical parameters is obtained as:

$$\text{Reflectance} = +38.81290 - 0.18938 * R - 4.16863E{-}003 * L \\ - 6.94444E{-}006 * R * L + 5.83716E{-}004 * R^2 + 7.81609E{-}007 * L^2 \quad (1)$$

where R = radius of nanowire.

L = length of nanowire.

3 Results and Discussions

The obtained equation has an R-squared value of: 0.9977. It shows very good correlation.

From the ANOVA analysis, it is found that the Model F-value of 612.21 implies the model is significant. There is only a 0.01% chance that a "Model F-Value" this large could occur due to noise. The "Predicted R-Squared" of 0.9781 is in reasonable agreement with the "Adjusted R-Squared" of 0.9961.

"Adeq Precision" measures the signal to noise ratio. A ratio greater than 4 is desirable. Ratio of 80.334 indicates an adequate signal. This model can be used to navigate the design space.

It is observed that for minimum surface reflectance the length and radius of the nanowires should be at the highest level as can be seen from Fig. 4. But there are fabrication limitations beyond which the length of a nanowire cannot be manufactured otherwise it will not be able to sustain itself and also the handling of the wafer with such long nanowires will be very difficult and the sample will be prone to damage.

The developed model is used to mathematically predict the surface reflectance for all combinations of the design parameters and the results are shown in Table 3.

The model is also tested for a few combinations of design parameters other than those in the design matrix and the results are tabulated in Table 4.

The comparative graphs in Fig. 5 shows the reflectance values obtained from simulation and from the mathematical model. As can be seen, there is an appreciable overlap between the two that indicates that the mathematical model is sufficient for predicting the surface reflectance of c-Si with nanowires.

After 5 trials, it is seen that there is an average error of 0.33.

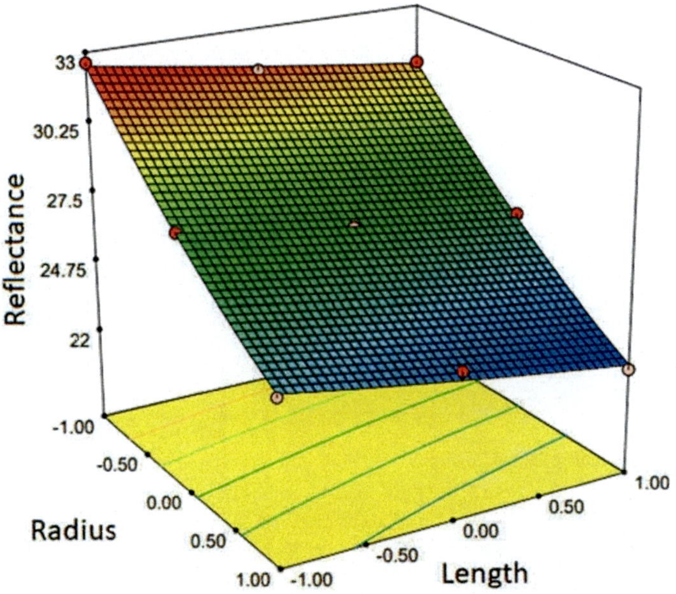

Fig. 4 Interaction plot of reflectance versus radius and length of the nanowires

Table 3 Calculated values of surface reflectance for various combinations of nanowire design parameters

Radius (nm)	Length (nm)	Mathematically calculated reflectance (%)	Error = |Mathematical – Experimental|
25	500	32.47	0.11
85	500	24.75	0.13
25	1100	30.61	0.05
85	1100	22.64	0.19
25	800	31.47	0.16
85	800	23.63	0.32
55	500	28.08	0.02
55	1100	26.10	0.15
55	800	27.02	0.03

Table 4 Calculation of error in the mathematical model obtained

Radius (nm)	Length (nm)	Simulated reflectance (%)	Calculated reflectance (%)	Error
45	700	28.28	28.72	0.44
45	1000	27.95	27.77	0.18
65	900	25.87	25.44	0.43
65	1000	25.46	25.13	0.33
75	1000	24.24	23.98	0.26

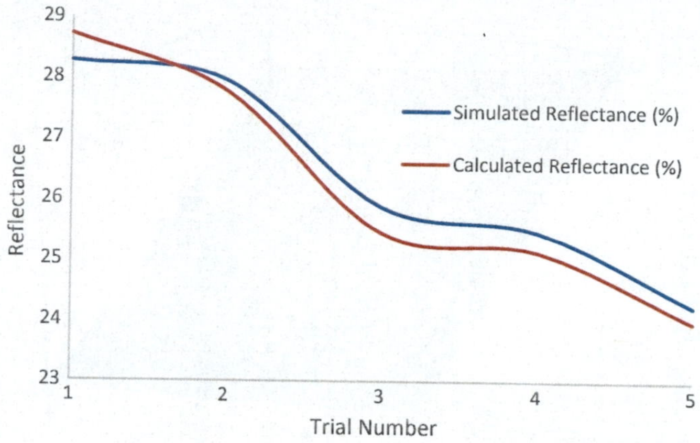

Fig. 5 Comparison of reflectance values from simulation and from mathematical model

4 Conclusion

The surface reflectance of crystalline silicon can be reduced by the growth of nanowires on its surface. Simulations performed with the help of the Fourier Modal Method show that reflectance reduces from 38 to ~23% a reduction of 40%. The mathematical expression developed can predict the value of surface reflectance, given the radius and length parameters, within an error margin of 33%. The "Pred R-squared" value of 0.9781 is in reasonable agreement with the "Adj R-Squared" value of 0.9961. This shows that the mathematical model is fairly accurate. The mathematical model obtained can be used to obtain optimal anti-reflectance nanowire dimensions, but that would be part of separate work. A difficulty that would be faced while performing real-life experimentation would be varied structuring of the nanowires. The simulation provides a starting point from which dimensions for experimentation can be proposed.

References

1. Accessed from https://en.wikipedia.org/wiki/Growth_of_photovoltaics on 24th May, 2019
2. Hu A-G (2017) The Five-Year Plan: a new tool for energy saving and emissions reduction in China. Institute for Contemporary China Studies, Tsinghua University, Beijing 100084, China
3. Wolfe P (2012) SolarPhotovoltaicProjectsinthemainstreampowermarket.Routledge, p 225. ISBN 9780415520485
4. Deutsche Bank Markets Research, 27 February 2015
5. Accessed from https://en.wikipedia.org/wiki/Crystalline_silicon#Market_share on 24th May, 2019
6. Sivasubramaniam S, Alkasai MM (2014) Inverted nanopyramid texturing for silicon solar cells using interference lithography. Microelectron Eng 119:146

7. WuL, Bai P, Zhou X, Li EP (2012) Reflection and transmission modes in nanohole-array-based plasmonic sensors. IEEE Photon J 4(1):26
8. Shimotsuma Y, Kazansky P, Qiu J, Hirao K (2003) Self-Organized nanogratings in glass irradiated by ultrashort light pulses. Phys Rev Lett 91(24):1–4
9. Pham VT et al (2014) Effect of nanowire length on the performance of silicon nanowires based solar cell. Adv Nat Sci Nanosci Nanotechnol 5:045014
10. Accessed from https://nanohub.org/resources/s4sim on 24th May, 2019
11. Heider F, Roberts J, Huang J, Lam J, Forouhi AR (2013) Effects of measured spectral range on accuracy and repeatability of OCD analysis. Solid State Technol 56(5):21
12. Li L (2014) Fourier modal method. In: Popov E (ed) Gratings: theory and numeric applications, second revisited edition. Aix Marseille Université, CNRS, Centrale Marseille, Institut Fresnel, pp 13.1–13.40. 978-2-85399-943-4. ffhal-00985928f
13. Liu V, Fan S (2012) S4: A free electromagnetic solver for layered periodic structures. Comput Phys Commun 183:2233–2244. https://doi.org/10.1016/j.cpc.2012.04.026
14. Ghebrebrhan M, Bermel P, Avniel Y, Joannopoulos JD, Johnson SG (2009) Global optimization of silicon photovoltaic cell front coatings. Opt Express 17:7505
15. Kang J, Wang X, Bermel P, Liu C (2014) S4: Stanford stratified structure solver. https://nanohub.org/resources/s4sim. https://doi.org/10.21981/D35T3G11T

Investigation of Surface Roughness of Miniature Spur Gears Fabricated Using WEDM by RSM Approach

Vyom Singh, Abhishek Patel, Ashish Goyal, Anand Pandey, Ravikant Gupta, and Rahul Goyal

Abstract The present investigation is based on micro-geometry of miniature spur gears manufactured by wire electrical discharge machining (WEDM). Effects of pulse-on time, pulse-off time and feed rate on surface roughness were analysed using response surface methodology and ANOVA analysis. The miniature gears have an outside diameter of 9.889 mm, face width 3.105 mm and nine miniature spur gears were fabricated. It was revealed from the observation that errors were seen during the fabrication of miniature spur gear at a high level of discharge energy. Scanning electron microscopy is also performed to analyse the geometry of the gears. This study found optimum surface roughness for miniature spur gear is 1.48 μs.

Keywords WEDM · Miniature spur gear · RSM · Surface roughness · SEM

1 Introduction

Miniature gear is widely used in the industrial and scientific fields. Mostly such gears are used in robotics, scientific instruments, miniature pumps and mechanical watches. Miniature gear can be further categorized as micro and meso gear depending upon the outer diameter. Micro gears have an outer diameter of less than 1 mm whereas the meso gears have an outer diameter ranging from 1 to 10 mm, i.e. $1\ \text{mm} < D_o < 10\ \text{mm}$. The gear manufacturing process is mainly divided into two parts, i.e. Forming and Machining. These conventional methods come with some limitations while manufacturing miniature spur gear, due to such limitations there has been research work done on alternative methods to eradicate the issues encountered during manufacturing via the conventional methods. Wire Electric Discharge Machining (WEDM) is widely acknowledged for the manufacturing of miniaturized components. It is a non-traditional machining technique which uses a thin wire as the tool electrode, usually brass in which the electrical energy is transformed to thermal energy for

V. Singh · A. Patel · A. Goyal (✉) · A. Pandey · R. Gupta · R. Goyal
Department of Mechanical Engineering, Manipal University Jaipur, Dehmi Kalan, Jaipur 303007, Rajasthan, India
e-mail: ashish.goyal@jaipur.manipal.edu

© Springer Nature Singapore Pte Ltd. 2021

P. M. Pandey et al. (eds.), *Advances in Production and Industrial Engineering*, Lecture Notes in Mechanical Engineering,
https://doi.org/10.1007/978-981-15-5519-0_3

cutting of the materials. Wire EDM produces a precise, fine, wear-resistance and corrosion-resistance surface.

Gupta et al. [5] concluded that WEDM can be used to obtain the better surface roughness for the fabrication of miniature gears. A comparative analysis was also proposed for the machining of spur gear by hobbing and WEDM process. Ali et al. [1] proposed the model to increases the machining flexibility. Higher efficiency and dimensional accuracy was also achieved by the proposed method. The mentioned model can be used for the fabrication of micro parts. Mahapatra et al. [8] presented WEDM parameters were optimized using the Taguchi loss function. Moreover, the single pitch deviation is significantly affected by the interaction between pulse-off time and wire feed rate.

Han et al. [6] concluded that the surface roughness was mainly depended on the selection of the precise process parameter of WEDM. It was observed that by a decrease in discharge energy and current, SR could be improved. Rao et al. [9] reported that SR was influenced by wire tension and spark gap voltage. It was proposed that machining of advanced material by WEDM provides a better surface from the conventional process. Raju et al. [10] proposed a model for the selection of process parameters during the machining by WEDM process. The better surface finish can be achieved by an increase in servo voltage ab wire tension.

Shandilya et al. [11] presented an experimental study by the WEDM process using the advanced optimization methodology. It was reported that responses are depended on the lower level of the voltage and wire feed. Gupta et al. [2] performed the experiments by the design of experiment methodology for the fabrication of miniature gears. The gears were fabricated by the brass materials. The total profile deviation and was also measured for the miniature spur gears. Somashekhar et al. [7] used an artificial neural network to optimize micro-EDM input parameters for MRR. The experiments were performed by using the design of experiments methodology and modelling of material removal is performed by using the ANN approach. Goyal [4] has performed experiments on Inconel 625 material by the WEDM process. The design of experiments methodology was used to analyse the outcome of input parameters on the output parameters. Goyal et al. [3] have done experiments on WEDM machine on Inconel 625 material. The machining is done by the DoE approach. It was concluded that pulse-on time and current influence the response. Present work reports about the preliminary investigations on the behaviour of surface roughness of miniature gears made of brass, one of the most commonly used miniature gear materials, with three process parameters of WEDM i.e. pulse-on time, pulse-off time and wire feed rate. The SEM analysis was also performed to analyses the surface integrity of the machined miniature spur gear.

2 Experimental Detail's

In the current research, a Brass rectangular plat of 6 mm thickness was used as a specimen to perform the experiments. The experiments were performed on WEDM (Electronica Maxicut) machine using brass wire of 0.25 mm dia. as a tool electrode and distilled mineral water as dielectric medium. Three process parameters i.e. pulse-on-time (Ton), pulse-off-time (Toff) and wire feed rate (Wf) were selected and the ranges for these were set according to machine limitations and upon the basis of preliminary experiments. The variation of these parameters was used to investigate response surface roughness. A square workpiece of 10 cm * 15 cm * 6 mm was taken for machining. The surface roughness is measured in micrometre. Figure 1 shows the WEDM machine set up. Table 2 shows different parameters and their range used in the experimentation.

Surface roughness of machined surface was measured by Mitutoyo's Surftest (SJ-210). The RSM (Response Surface Methodology) approach for the three variables with each varying at three levels requires total of 9 experimental runs. Figure 2 presents the image of surface roughness tester.

Fig. 1 WEDM machine set up

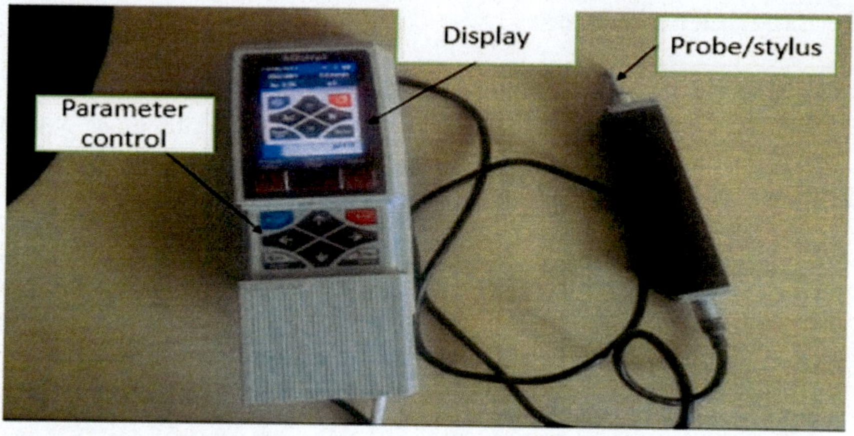

Fig. 2 Surface roughness tester

Table 1 Different parameters and ranges used for experimentation

Symbol	Control factor	Unit	Level 1	Level 2	Level 3
A	Pulse-On-Time	μs	1	3	4
B	Pulse-Off-Time	μs	6	8	10
C	Wire Feed Rate	mm/min	5	6	7

Fig. 3 Machined sample miniature gear

Therefore, in total, 9 gears were manufactured corresponding to the input values. Table 1 shows the process parameters and their range. Figure 3 shows the machined miniature spur gear.

3 Response Surface Methodology

Response surface methodology approach is the procedure used for determining the conditional behaviour of various process parameters and the output parameters in order to find the problem and to rectify the problem. RSM is a collection of mathematical and statistical methods useful for modelling and analysis of engineering problems. The main utilization of RSM is for the optimization of the responses which are of the key interest of influence for other responses. Based on the experimental investigation results and the model of response, an optimal point is then deduced

$$y = f^{'}(x)\beta + \grave{e} \tag{1}$$

where y is the output of the interest and the input variables are denoted by $x_{1,}\ x_{2\ldots}\ x_{n,}$ f(x) is a vector function of p elements that consists of power and cross-products of power of $x_{1,}\ x_{2\ldots}\ x_n$ up to a certain degree, β is a vector of p constant coefficients referred to as parameters and è is a random experimental error.

There are two important models used in RSM. This is the first-degree model and the second-degree model

$$y = \beta_o + \sum_{i=1}^{k} \beta_i x_i + \grave{e} \tag{2}$$

$$y = \beta_0 + \sum_{i=1}^{k} \beta_i x_i + \sum_{i<j} \sum \beta_{ij} x_i x_j + \sum_{i=1}^{k} \beta_{ii} x_i^2 + \grave{e} \tag{3}$$

Figure 4 shows factors at the central point and the middle point of the edge Box-Behnken designs are a class of rotatable or nearly rotatable second-order designs based on 3-level incomplete factorial designs. This cube consists of the central point and the middle point of the edge. Table 2 shows the orthogonal array obtained for the input and output parameters.

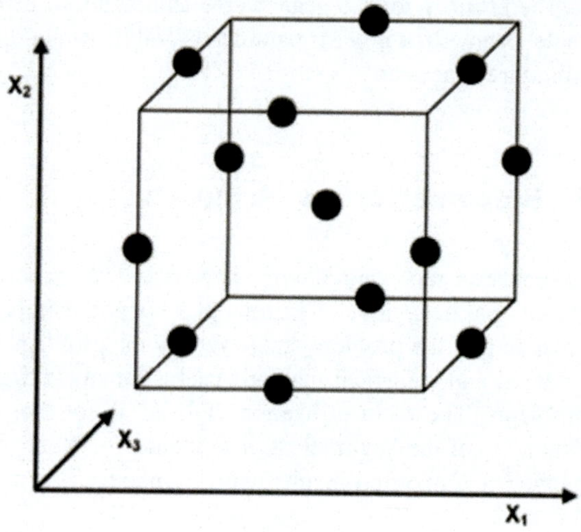

Fig. 4 Centre point and middle points of the design

Table 2 Orthogonal array obtained for different input and output responses

		Factor 1	Factor 2	Factor 3	Response 2
Std	Run	A:TON	B:TOFF	C:WF	RA
		μs	μs	m/min	μm
3	1	0.4	80	7	1.558
9	2	0.4	60	6	1.964
6	3	0.4	100	5	2.741
2	4	0.1	80	6	1.789
4	5	0.1	60	5	2.236
8	6	0.1	100	7	1.486
5	7	0.3	80	5	2.374
7	8	0.3	60	7	1.786
1	9	0.3	100	6	2.246

4 Result and Discussion

Figure 5 depicts a 3-D surface plot between input parameters and response parameter. The three-dimensional response phase diagram shows that when there is increase pulse-off time and pulse-on time, there is a significant increase in surface roughness. Figure 6 shows normal plot of residuals. The obtained results show that increasing pulse-on time also increases surface roughness. This is because when pulse-on time is increased it will make discharge last longer which will eventually lead to high-discharge energy.

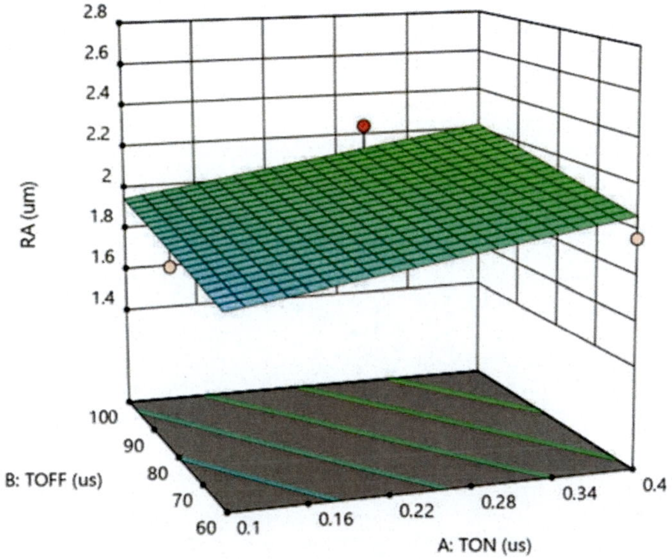

Fig. 5 3-D surface plot

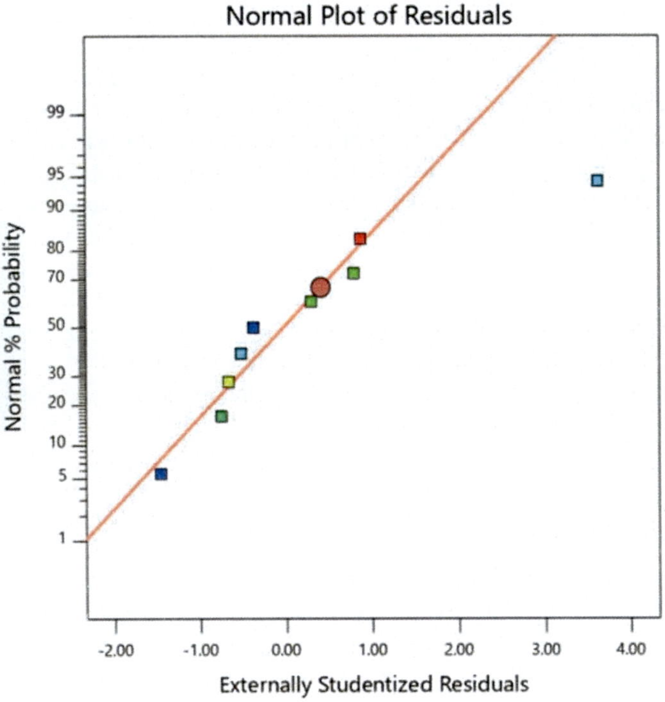

Fig. 6 Normal plot of residuals

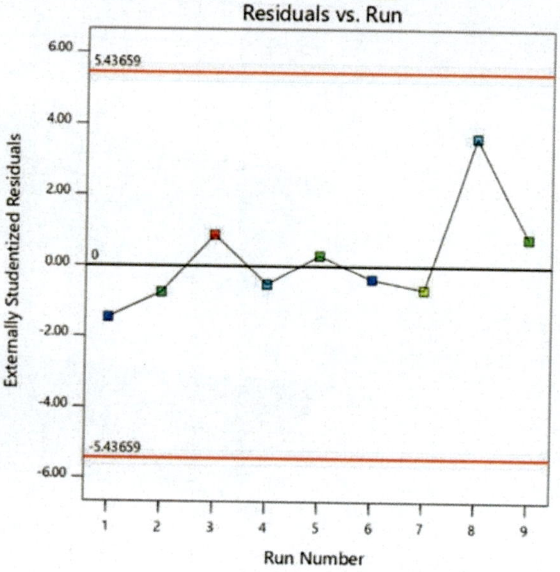

Fig. 7 Graph of run versus residuals

For the constant value of pulse-on time, increasing pulse-off time also increases surface roughness because it will discharge more number of spark. The poor flushing done in the machining zone by the SR is increased. Figure 7 shows that all the test runs are within the limit.

Figure 8 shows the effect of different factors on surface roughness, A(Ton), B(Toff) and (Wf). The surface plot is measure in colour coding, for this given figure it shows as minimum < blue < green < red < maximum. Red shows Design points above-predicated value and pink shows design points below predicated values.

Data Points should be approximately linear. A non-linear indicates non-normality in the error term, which may be corrected by the transformation. All factors show the individual relation with each parameter. In residual versus run graph, the run differs more than it predicated value and it is also in red control limits.

It was observed that lower feed rate provided the chances of wire breakage and the time of machining is also increase.

4.1 ANOVA Table

ANOVA is used to find the significant input parameters which are affecting the response.

If the P-value is less 0.05 then that parameter and model are significant. Table 3 presents the ANOVA table for the SR.

ANOVA indicates the parameters that are significant in WEDM process. The significant parameter is the feed rate. The R^2 (coefficient of determination) value

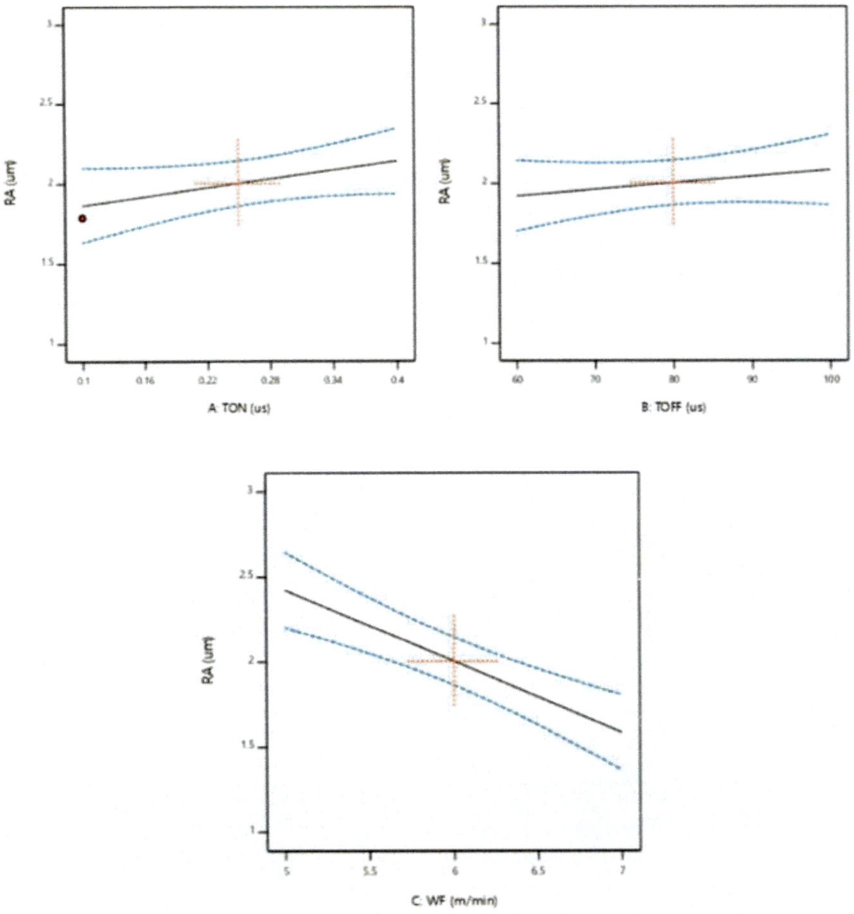

Fig. 8 Graph of different factors with Surface roughness. **a** (T_{on}), **b** (T_{off}) and **c** (WF)

Table 3 ANOVA table for SR

Source	Sum of squares	df	Mean square	F-value	p-value
Model	1.22	3	0.4066	15.28	0.006
A-TON	0.1209	1	0.1209	4.54	0.0863
B-TOFF	0.0395	1	0.0395	1.49	0.2773
C-WF	1.06	1	1.06	39.8	0.0015
Residual	0.1331	5	0.0266		
Cor Total	1.35	8			

is 90.16% represents a good percentage of variation. After optimization, the 47 solutions of input parameters and the solution with minimum SR are selected. All the 3 variables Ton, Toff, WF are 3 major factors, interaction form a linear equation which affect the quality of brass. The Developed mathematical model is used to predict surface roughness using Design expert 11. The Regression equation of Surface roughness developed by the design expert 11 considering input and the output parameters.

$$Y_{SR} = 2.00 + 0.1394 * A + 0.0812 * B - 0.4202 * C \tag{4}$$

4.2 Material Characterization of Miniature Spur Gear

The materials are scanned under a scanning electron microscope (SEM) after the machining operation to study and analyse the different characteristics, properties and behaviour of the material. Figure 9 shows the flank surface of the miniature gear. Figure 10 presents the SEM image shows a top view of the miniature gears. The above image shows the use of proper parameters that can make miniature gear by a uniform distribution of crates. This gear is machined from the following parameters: $T_{on} = 0.1$, Toff = 60, WF = 6. Figure 11 depicts that the miniature gear has a diameter of 9.889 mm.

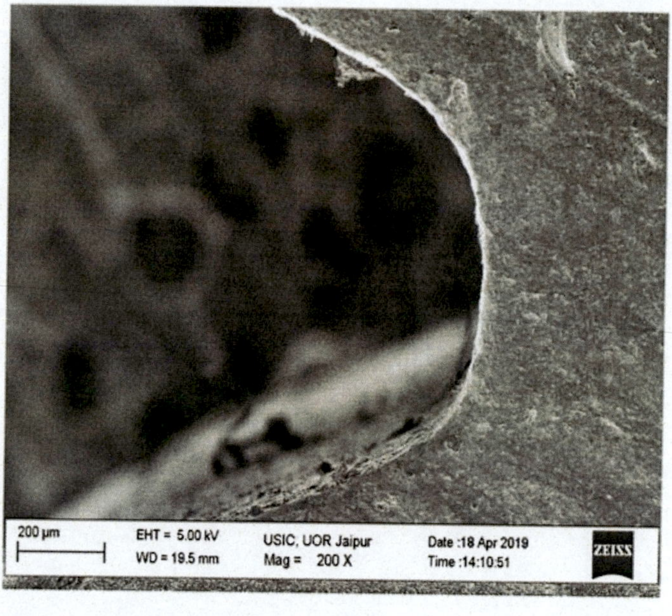

Fig. 9 SEM image of miniature gear at 200 × zoom

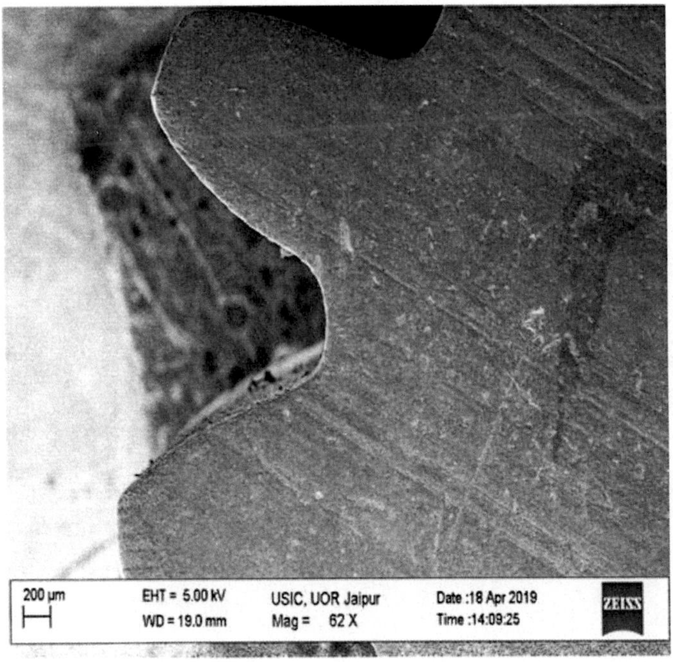

Fig. 10 Top view of spur gear

Fig. 11 Fabricated miniature spur gear

5 Conclusions

The effect of WEDM machining parameters on the surface roughness was investigated in the present experimental work. The following conclusions were made:

1. Increases of pulse-on time increase discharge energy which increases melting and evaporation of the material and creation of uneven shaped deeper craters which increases surface roughness
2. Feed rate is found to be the most significant parameter for the optimization of the process. Consumption of the tool electrode is more at the higher discharge energy.
3. The SEM image proves the capability of WEDM to machined miniature gear with uniform distribution of regular-shaped craters and defect-free flank surfaces.
4. The optimal machining parameters can be achieved at the pulse-on time of 0.4 μs, pulse-off time of 60 μs and feed rate of 6.861 mm/min to get surface roughness of 1.4861 μm.

References

1. Ali MY, Mohammad AS (2008) Experimental study of conventional WEDM for micro-fabrication. Mater Manuf Process 23(7):641–645
2. Gupta K, Jain NK (2013) On micro-geometry of miniature gears manufactured by WEDM. Mater Manuf Process 28(10):1153–1159
3. Goyal A (2017) Investigation of material removal rate and surface roughness during wire electrical discharge machining (WEDM) of Inconel 625 super alloy by cryogenic treated tool electrode. J King Saud Univ Sci: 0–7
4. Goyal A, Pandey A, Sharma P (2017) Machinability of Inconel 625 aerospace material using cryogenically treated WEDM. Solid State Phenom 266:38–42
5. Gupta K, Jain NK (2014) Analysis and optimization of surface finish of wire electrical discharge machined miniature gears. Proc IMechE Part-B J Eng Manuf 228(5):673–681
6. Han F, Jiang J, Yu D (2007) Influence of machining parameters on surface roughness in finish cut of wedm. Int J Adv Manuf Technol 34:538–546
7. Somashekhar KP, Ramachandran N, Mathew J (2010) Optimization of material removal rate in micro-EDM using artificial neural network and genetic algorithms. Mater Manuf Process 25:467–475
8. Mahapatra SS, Patnaik A (2007) Optimization of WEDM process parameters using Taguchi method. Int J Adv Manuf Technol 34(9–10):911–925
9. Rao PS, Ramji K, Satyanarayana B (2014) Experimental investigation and optimization of wire EDM parameters for surface roughness, MRR and white layer in machining of aluminum alloy. Procedia Mater Sci 5:2197–2206
10. Raju P, Sarcar M, Satyanarayana B (2014) Optimization of wire electric discharge machining parameters for surface roughness on 316L stainless steel using full factorial experimental design. Procedia Mater Sci 5:1670–1676
11. Shandilya P, Jain P, Jain N (2012) Parametric optimization during wire electrical discharge machining using response surface methodology. Procedia Eng 38:2371–2377

Multi-objective Analysis of Nd-YAG Laser Welding on Dissimilar Metals

Dhruv Bhatt, Ashish Goyal, and Vyom Singh

Abstract Welding is the science of metal joining, usually performed on similar metals. In the current study, Nd-YAG laser welding system was used for joining two different categories of steel (Mild Steel and SS-304), expected to have different properties. Dissimilar joints play a crucial role in the current industrial situation where multiple materials are used for everyday tasks. DOE was used to optimize the system input parameters. Quality of a weld is determined by the aspect ratio (depth of weld/width of weld), which was determined using metallography. Microhardness test was also inducted to study the hardness of these welds. A relationship was determined between microhardness, quality of weld and input parameters using Grey Relational Analysis (GRA) methodology.

Keywords GRA · Nd-YAG · Laser welding · Microhardness · Aspect ratio

1 Introduction

Laser Beam Welding (LBW) is a fusion joining process that produces coalescence of materials with the heat obtained from a concentrated beam of coherent, monochromatic light impinging on the joint to be welded. It is a non-contact process, requiring no pressure to be applied. Inert gas shielding is generally employed to prevent oxidation of the molten puddle and filler metals may be occasionally used. Dissimilar metal welding, compared to traditional laser welding, is the joining of two distinct metals which won't usually weld together as they have different mechanical and chemical properties, and are from different alloy groups. This benefits various industries, such as the automotive and electronics industry, heavy industry, oil and gas industry, etc.

Tadamalle et al. [11] studied the effect of welding speed on different process efficiencies in SS-304L. Bead on SS-304L was created by varying welding speed. The semi-empirical method was used to predict the melting efficiency of Nd-YAG laser

D. Bhatt · A. Goyal (✉) · V. Singh
Department of Mechanical Engineering, Manipal University Jaipur, Dehmi Kalan, Jaipur 303007, Rajasthan, India
e-mail: ashish.goyal@jaipur.manipal.edu

© Springer Nature Singapore Pte Ltd. 2021

P. M. Pandey et al. (eds.), *Advances in Production and Industrial Engineering*, Lecture Notes in Mechanical Engineering,
https://doi.org/10.1007/978-981-15-5519-0_4

welding process. The study concluded that process efficiency depends on operating parameters, chemical and thermomechanical properties of the material and laser power surface. Bagchi et al. [2] used SYSWELD to optimize and analyse the laser welding on Hastelloy C-276. The observation shows that the predicting aspect ratio is inversely proportional to heat input and that the location of peak temperature depended on overlapping pulse. Dong et al. [6] studied the impact of micro laser weld on microstructure. The work was aimed to reveal phase content within various microstructural zones of laser micro-welded crossed Nitinol wires by transmission electron microscope (TEM) with the assistance of FIB (focussed ion beam) technique.

Casalino et al. [4, 5] used Yb Fiber laser to carry out the work on 6 mm thick aluminium allot sheets in butt configuration. ANN (Artificial Neutral Network) was implemented to investigate the main effects of process parameters in laser welding process quality. Hu et al. [9] used Inconel 625 specimens for this study on laser welding. Inconel 625 specimens are made using selective laser melting (SLM) process. It was observed that fractures after the process were found on base alloy and not in welding zone also columnar grains that exhibited great resistance to tensile fractures were a major constituent of welding zone.

Wang et al. [12, 13] used laser welding to join two dissimilar commonly used materials, molybdenum and copper. The study aimed to device a method, which would prove to be successful in microwelding of the two materials. Energy and dispersive spectroscopy (EDS) showed no trace of diffusion of elements at the welded interface. Goyal et al. [7] optimize the process parameter of the WEDM machine by using the design of experiments methodology. The ANOVA analysis was also performed to show the relationship between the significant and non-significant parameters. Hekmatjou et al. [8] used Nd-YAG laser welding technique to weld 5456 aluminium alloy plates. The study focussed on investigating weld penetration and tendency of liquid cracking in heat-affected zone. Casalino et al. [4] simulated laser welding for lightweight metal sheets. Ansys parametric design language was used to device the programme. Butt joints of both homogenous (Titanium alloys plates-Ti6Al4V) and dissimilar (Al alloy and T40) metal plates were made. Ai et al. [1] investigated the weld bead integrity of welded joints of dissimilar metals. An optimization model was proposed by taking the integrity of the weld bead and weld area into consideration. Relationships between process parameters and weld bead integrity were developed by genetic algorithm optimized back proportion neutral network (GA-BPNN). Lei et al. [10] used ultrasonic-assisted laser welding on AZ31B Mg alloy to predict its effect of ultrasonic vibration on microstructure, mechanical property and welding characteristics of welded joints. It was observed found that the ultrasonic-assisted laser welding was successful in bringing weld porosity down to 0.9 from 4.3%. Bahtt et al. [3] performed experiment by using the ND: YAG laser welding process. The comparative study was also presented on the stainless steel and brass materials. In this study, two metals that possess dissimilar chemical or mechanical properties, and so aren't necessarily a natural fit for each other are being joined using Nd-YAG laser beam welding technique. There are certain factors that must be considered while joining dissimilar metals like solubility, intermetallic compounds, weldability, thermal expansion corrosion, melting rate and end service conditions. The laser beam

Table 1 Chemical composition of SS-304

Element	Fe	Mo	Ni	Cr	P	S	C	Si	Mn
%age	Remaining	1.24	7.93	19.04	0.018	0.019	0.081	0.368	1.74

Table 2 Chemical composition of mild steel (low carbon steel)

Element	C	Ag	Mn	S	P
%age	0.17	0.4	0.8	0.04	0.04

which is precisely focused on the metals, melt each of the metals together until they form one connected joint. If the two metals are not two wildly dissimilar, they will often join seamlessly together.

2 Materials and Methodology

Nd-YAG Laser Welding System's parameters were taken into account for designing the Taguchi L_{27} orthogonal array on Minitab-18. This method tests pairs of combinations and allows for the collection of data to determine factors affecting the output most. Therefore, this is preferred over the conventional way of testing possible combinations. The process helps in dealing with the study in a more professional and systematic manner, thereby saving resources and time. The experiment was conducted using Nd-YAG Laser welding system at BARC, Mumbai, India. The composition of SS-304 and mild steel are given in Tables 1 and 2, respectively. In the present experimental work, Helium (He) was used as a shielding gas.

In this study, five machine parameters were selected of the Nd-YAG laser welding system. Those five parameters were then optimized using L-27 Taguchi array. Based on this experiment design experiment were performed and determined the quality of weld using the aspect ratio (depth of weld/width of weld) of welds made. Figure 1 shows the Nd-YAG Laser Welding System setup.

3 Metallography

To analyse the quality of welds, the aspect ratio was considered as a reliable factor. For the successful calculation of aspect ratio, metallography was adopted. This method includes cutting the welded sample in two parts and later on etchants are used to smoothen the machined surface, later with help of optical microscope the weld region is analysed.

Fig. 1 Nd-YAG laser welding setup

Table 3 Etchant and concentration

Metal	Etchant	Concentration
Stainless Steel and Mild Steel	Aqua Regia with HCl and HNO_3	2 parts of HCl and 1 part of HNO_3

The width and depth of weld are calculated, resulting in a successful calculation of aspect ratio. The more the aspect ratio is the better is the weld quality [12]. The etchants are selected based on the materials, for this study we selected these etchants listed in Table 3.

4 Grey Relational Analysis (GRA)

In GRA the data (aspect ratio) is normalized in the range of 0–1. The process is termed as normalization. Based on this, grey relational coefficients are calculated, which helps in representing a relationship between the actual data (normalized) and the ideal data. Later the grey relation grade is determined by averaging grey relation coefficient to selected responses. Overall quality characteristics depend on the calculated grey relation grade.

4.1 Normalization

Raw data is prepared using normalization of signal to noise ratio (S/N Ratio) for analysis to transform the original sequence to comparable sequence. Normalization (Linear) is required as the unit and range in single data sequence may differ from others. Types of data normalizations are based on the requirement of Higher the Better (HB), Nominal the Best (NB), Lower the Better (LB) criteria. If target value of original sequence is infinite, then it has a characteristic for "higher is better". The original sequence can be normalized as follows:

$$x_i^* = \frac{x_i^o(k) - \min x_i^o(k)}{max x_i^o(k) - min x_i^o(k)} \tag{1}$$

where $i = 1 \ldots$ m; $k = 1 \ldots$ n, m is the number of experimental data items and n is the number of parameters. $x_i^o(k)$ denotes the original sequence, x_i^* the sequence after the data pre-processing, max $x_i^o(k)$ the largest value of$x_i^o(k)$, min $x_i^o(k)$ the smallest value of $x_i^o(k)$ and x^o the anticipated value.

4.2 Determination of Deviation Sequences Δ0i(k)

The deviation sequence, Δ0i(k) is the absolute difference between the reference sequence $x_0^*(k)$ and the comparability sequence $x_i^*(k)$ after normalization. It is determined using equation:

$$\Delta 0\mathrm{i(k)} = |x_o(k) - x_i^*(k)| \tag{2}$$

4.3 Grey Relational Coefficient (GRC)

GRC for all the sequences express the relationship between the ideal (best) and actual normalized S/N ratio. If the two sequences agree at all points, then their grey relational coefficient stands 1. The grey relational coefficient $\zeta_i(k)$ for the k-th performance characteristics in the ith experiment can be expressed as:

$$\zeta_i(k) = \frac{x_i^o(k\Delta min + \tau \Delta \mathrm{max})}{\Delta o_i(k) + \tau \Delta max} \tag{3}$$

ΔO_i is the deviation sequence of the reference sequence and $x_o^*(k)$ is the comparability sequence. ζ is the identification coefficient: ζ ϵ[0, 1] (the value can be adjusted based on actual system requirements). A value of ζ is the smaller and the distinguished ability is the larger. $\tau = 0.5$ is generally used. Grey relational coefficient for 27 comparability sequences.

4.4 Grey Relational Grade (GRG)

After the grey relational coefficient is derived, GRG is usual to take the average value of the grey relational coefficients as the grey relational grade. The grey relational grade is expressed as:

$$\gamma_i = \frac{1}{n}\sum_{k=1}^{n} \epsilon_i(k) \tag{4}$$

In a practical situation, the importance of factors varies, hence GRG was expressed as above.

The grey relational grade γ_i represents the level of correlation between the comparability sequence and the reference sequence. If two sequences are identical, then the value of GRG is 1. The GRG also indicates the degree of influence that the comparability sequence could exert over the reference sequence. Therefore, if a comparability sequence is more important than the other comparability sequences to the reference sequence, then GRG for that comparability sequence and reference sequence will be higher than GRG of the other.

5 Experimentation

Table 4, shows the resultant of Taguchi's L_{27} orthogonal array, which was designed using Minitab 18.

5.1 Aspect Ratio

Aspect Ratio can simply be calculated using a basic equation:

$$\frac{depth}{width} = aspectratio \tag{5}$$

For calculating it we need to first determine the exact depth and width of the weld. This is done using the metallography of the welded job. After this aspect ratio for each of the 27 samples is calculated we use GRA to determine the best set. Figure 2

Table 4 L_{27} Orthogonal array table

Sr. no.	PP (KW)	PD (ms)	Frequency (Hz)	Defocussing (mm)	Coff	Aspect ratio	Microhardness (HV)
1	1.0	4	6	1	1	0.500	232
2	1.0	4	6	1	0	0.312	219
3	1.0	4	6	1	−1	0.750	171
4	1.0	10	12	2	1	0.666	246
5	1.0	10	12	2	0	0.300	229
6	1.0	10	12	2	−1	0.727	160
7	1.0	16	18	3	1	0.416	248
8	1.0	16	18	3	0	0.357	230
9	1.0	16	18	3	−1	0.428	175
10	1.5	4	12	3	1	0.142	251
11	1.5	4	12	3	0	0.222	222
12	1.5	4	12	3	−1	0.250	177
13	1.5	10	18	1	1	0.769	245
14	1.5	10	18	1	0	1.100	239
15	1.5	10	18	1	−1	1.200	167
16	1.5	16	6	2	1	0.636	238
17	1.5	16	6	2	0	0.727	234
18	1.5	16	6	2	−1	0.818	177
19	3.0	4	18	2	1	0.833	231
20	3.0	4	18	2	0	0.750	237
21	3.0	4	18	2	−1	1.200	168
22	3.0	10	6	3	1	0.526	233
23	3.0	10	6	3	0	0.666	234
24	3.0	10	6	3	−1	0.600	170
25	3.0	16	12	1	1	1.285	248
26	3.0	16	12	1	0	1.666	238
27	3.0	16	12	1	−1	1.285	166

is the metallography result of one of the 27 samples. The image is obtained using a digital microscope.

The joint between two metals (denoted by A) in the above figure can clearly be seen, The C section (denoted by B) is the welded region. The depth and Width of the C is measured using the optical microscope.

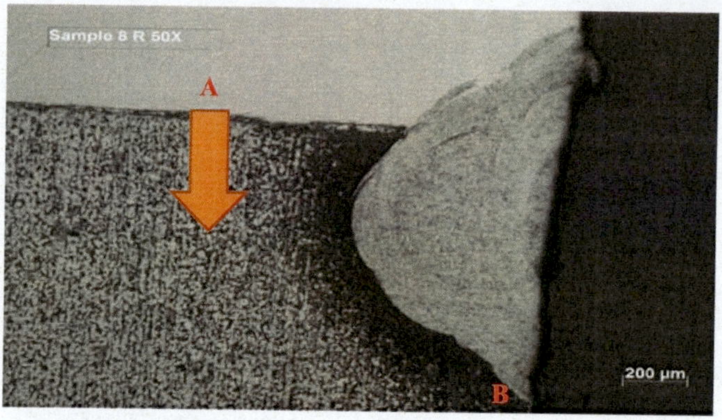

Fig. 2 Etched metallography result

5.2 *Microhardness*

Microhardness test was carried out on each of the 27 samples to determine the hardness of the welds made. The microhardness should ideally be equal to the base metal, but in the current scenario we have two metals. Hence, the microhardness close to level, where the welds are made (on the stainless-steel side or in the centre or the low carbon steel end of the joint. The +1 value in Coff means that the weld was made on the high carbon steel end of the joint, 0 means it was welded in the centre and -1 determines that it was welded on the low carbon steel end of the joint. So, if the high carbon end weld is analysed, we notice that the microhardness is more due to more carbon content in the joint, this is since more carbon is being fused in to prepare a weld.

6 Result and Discussion

Aspect Ratio for all the 27 samples was calculated and the result was analysed using Grey Relational Analysis (GRA) technique, the results were calculated for "higher is better". On the same principle the microhardness was calculated for each of the 27 samples. The microhardness of a welded region must be equal to the base metal (ideally) but in the case of dissimilar metal joining it would prefer it near the range harder metal, to avoid the defects. Later the ranks were determined to select the best outcome of the experiment. The study model was aimed to give a full proof solution to the issues related to dissimilar welding.

6.1 Normalization and Deviation Sequence

Using Eqs. 1 and 2, the normalized values and deviation sequence were calculated. The calculated normalized values and deviation sequence are mentioned in Table 5.

Table 5 Normalization and deviation sequence

S. no.	Normalization		Deviation sequence	
	Aspect ratio	Microhardness	Aspect ration	Microhardness
1	0.235	1	0.765	0.082
2	0.112	0.944	0.888	0.138
3	0399	0.737	0.601	0.345
4	0.344	1.060	0.656	0.022
5	0.104	0.987	0.896	0.095
6	0.384	0.690	0.616	0.392
7	0.180	1.069	0.820	0.013
8	0.141	0.991	0.859	0.091
9	0.188	0.754	0.812	0.328
10	0	1.082	1.0	0
11	0.052	0.957	0948	0.125
12	0.071	0.763	0.929	0.319
13	0.411	1.056	0.589	0.026
14	0.629	1.030	0.371	0.052
15	0.694	0.720	0.306	0.362
16	0.324	1.026	0.676	0.056
17	0.384	1.009	0.616	0.073
18	0.444	0.763	0.556	0.319
19	0.453	0.996	0.547	0.086
20	0.399	1.022	0.601	0.060
21	0.694	0.724	0.306	0.358
22	0.252	1.004	0.748	0.078
23	0.344	1.009	0.656	0.073
24	0.301	0.733	0.699	0.349
25	0.750	1.069	0.250	0.13
26	1	1.026	0	0.056
27	0.750	0.716	0.250	0.366

6.2 Grey Relational Coefficient and Grade with Rank

Based on the deviation Table 6, Grey Relational Coefficient and Grey Relational Grade, these are calculated on the lines of equations and, respectively.

According to GRA rank, the 26th sample was pretty close to the ideal weld. Hence the input characteristics on the first three ranks can be considered apt.

Table 6 Grey relational coefficient, grade and rank

S. no.	Grey relational coefficient		Grey relational grade	Rank
	Aspect ratio	Microhardness		
1	0.395	0.705	0.550	14
2	0.360	0.587	0.474	21
3	0.454	0.363	0.408	23
4	0.432	0.901	0.667	5
5	0.358	0.674	0.516	16
6	0.448	0.333	0.391	24
7	0.379	0.938	0.658	7
8	0.368	0.684	0.526	15
9	0.381	0.374	0.378	26
10	0.333	1.0	0.667	6
11	0.345	0.611	0.478	20
12	0.350	0.381	0.365	27
13	0.454	0.883	0.671	4
14	0.574	0.791	0.683	3
15	0.621	0.351	0.486	19
16	0.425	0.778	0.602	9
17	0.448	0.728	0.588	10
18	0.473	0.381	0.427	22
19	0.478	0.695	0.586	11
20	0.454	0.765	0.609	8
21	0.621	0.354	0.487	18
22	0.401	0.717	0.559	13
23	0.432	0.728	0.580	12
24	0.417	0.360	0.388	25
25	0.667	0.938	0.802	2
26	1.0	0.778	0.889	1
27	0.667	0.349	0.502	17

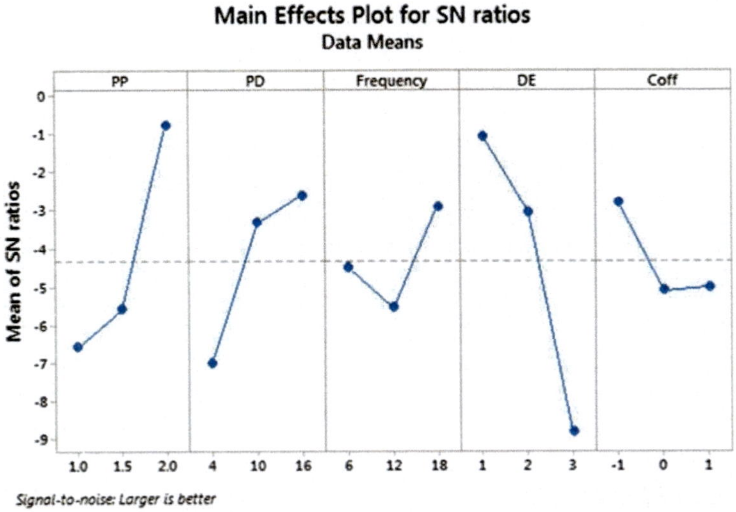

Fig. 3 Main effects plot for S/N ratios (Aspect ratio)

6.3 Aspect Ratio

As discussed earlier, aspect ratio is to be more for a good quality weld hence by taking "higher is better" criteria into consideration we have plotted mean of S/N ratio graph and according to Fig. 3 the aspect ratio was affected significantly by all five parameters. Using it is observed that the defocussing played the main role in influencing the aspect ratio.

6.4 Micro-hardness

Microhardness of the weld at three different locations (based on defocussing) varied a lot. The welds which were made on the Stainless-steel side (+1 D) of the joint had higher microhardness compared to the once made on the low carbon steel side (−1) and between the range of 220–240 HV. The microhardness according to the mean of S/N ratio graph (Fig. 4) was mainly dependent on defocussing. "Larger was Greater" criteria were selected even in this plotting as the number of defects was lower when the welded region had higher microhardness (towards the higher end of stainless steel).

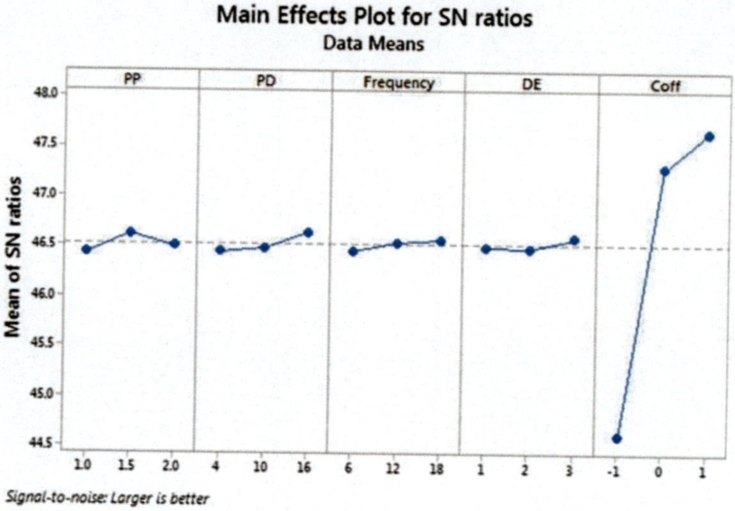

Fig. 4 Main effects plot for S/N ratios (Microhardness)

7 Conclusion

Based on the present experimental work, the following conclusion has been made.

1. The optimization of process parameters of an Nd-YAG Laser Welding system on dissimilar metals, was performed successfully using L27 Taguchi orthogonal array. The results were analysed using Grey Relational Analysis and mean for S/N ratio.
2. Pulse Power and Defocussing played a major role in influencing the aspect ratio of the welds, thereby directly effecting the quality of welds.
3. Centre Off, which played the least role in affecting the aspect ratio, surprisingly showed more influence on microhardness.
4. In future, the level of research can be extended by including new materials and by introducing hybrid optimization techniques.

References

1. Ai Y, Shao X, Jiang P, Li P, Liu Y, Liu W (2014) Welded joints integrity analysis and optimization for fiber laser welding of dissimilar materials. Opt Lasers Eng 86:62–74
2. Bagchi A, Saravanan S, Kumar GS, Murugan G, Raghukandan K (2017) Numerical simulation and optimization in pulsed Nd: YAG laser welding of Hastelloy C-276 through Taguchi method and artificial neural network. Optik 146:80–89
3. Bhatt D, Goyal A (2018) Effect of parameters of Nd YAG laser welding on AISI 316 Stainless steel and Brass. IOP Conf Ser Mater Sci Eng 455(1): 012118

4. Casalino G, Facchini F, Mortello M, Mummolo G (2016) ANN modelling to optimize manufacturing processes: the case of laser welding. IFAC- 49(12):378–383
5. Casalino G, Mortello M, Peyre P (2016) FEM analysis of fiber laser welding of titanium and aluminum. Procedia CIRP 41:992–997
6. Dong P, Li H, Wang W, Zhou J (2018) Microstructural characterization of laser micro-welded Nitinol wires. Mater Charact 135:40–45
7. Goyal A, Pandey A, Sharma P (2017) Machinability of Inconel 625 aerospace material using cryogenically treated WEDM. Solid State Phenom 266:38–42
8. Hekmatjou H, Naffakh-Moosavy H (2018) Hot cracking in pulsed Nd: YAG laser welding of AA5456. Opt Laser Technol 103:22–32
9. Hu X, Xue Z, Zhao G, Yun J, Shi D, Yang X (2019) Laser welding of a selective laser melted Ni-base superalloy: microstructure and high temperature mechanical property. Mater Sci Eng A 745:335–345
10. Lei Z, Bi J, Li P, Guo T, Zhao Y, Zhang D (2018) Analysis on welding characteristics of ultrasonic assisted laser welding of AZ31B magnesium alloy. Opt Laser Technol 105:15–22
11. Tadamalle AP, Reddy YP, Ramjee E, Reddy V (2014) Evaluation of Nd: YAG laser welding efficiencies for 304L stainless steel. Procedia Mater Sci 6:1731–1739
12. Wang X, Tang H, Shao M, Jin H, Liu H (2019) Laser impact welding: Investigation on microstructure and mechanical properties of molybdenum-copper welding joint. Int J Refract Metal Hard Mater 80:1–10
13. Wang X, Shao M, Jin H, Tang H, Liu H (2019) Laser impact welding of aluminum to brass. J Mater Process Technol 269:190–199

Fabrication and Tensile Testing of DHAK Fiber Reinforced Polyester Composites

Partha Pratim Das, Gorrepati Srinivasa Rao, Eswara Krishna Mussada, Gadudasu Babu Rao, Bhupendra Prakash Sharma, and Umesh Kumar Vates

Abstract In this Study, DHAK is identified as a potential fiber for reinforcement in making the composites. DHAK fiber reinforced polyester composites were manufactured up to a highest volume fraction of about 0.18 using hand layup method. The density of DHAK is found to be 1098.5 kg/m^3. The tensile behavior of the composites was examined as a function of fiber content. It is found that the mean tensile strength and modulus of the composites increases with an increase in fiber content. It has been observed that the highest mean tensile strength of the composite is 114.97 MPa for the maximum volume fraction, which is about 2.356 times higher than that of the matrix. The maximum mean tensile modulus of the composite is calculated as 16.14 GPa for highest volume fraction, which is about 2.07 times higher than that of a matrix. It is found that a firm bond in existed between the DHAK fiber and matrix at the highest volume fraction 0.18. It is found from the results that using DHAK fiber placing in the matrix plays an important part in terms of maximum tensile strength and modulus of a new manufactured composite.

P. P. Das · G. Srinivasa Rao (✉) · B. P. Sharma · U. K. Vates
Department of Mechanical Engineering, ASET, Amity University, Noida, U.P., India
e-mail: srgorrepati@amity.edu

P. P. Das
e-mail: das1.parthapratim@gmail.com

B. P. Sharma
e-mail: bpsharma@amity.edu

U. K. Vates
e-mail: ukvates@amity.edu

E. K. Mussada
Department of Mechanical Engineering, AIT Amity University, Noida, U.P., India

E. K. Mussada · G. Babu Rao
Department of Mechanical Engineering, Karunya Institute of Technology & Science, Coimbatore, India
e-mail: ekmussada@amity.edu

G. Babu Rao
e-mail: baburao@karunya.edu

© Springer Nature Singapore Pte Ltd. 2021

P. M. Pandey et al. (eds.), *Advances in Production and Industrial Engineering*,
Lecture Notes in Mechanical Engineering,
https://doi.org/10.1007/978-981-15-5519-0_5

Keywords Natural fibers · Fibers · Tensile properties

1 Introduction

The natural fibers have lot of benefits over synthetic fibers owing to their light weight, recyclable, availability, and ecofriendly. These fibers can be extracted either from plants or the stems of the trees. From the last two decades, a huge number of plants have been identified such as rice straw, banana, vakka, elephant grass, sisal, bamboo, jowar, etc. The composites made using these fibers exhibited maximum tensile, flexural, and impact strength when the appropriate adding agents are used for providing the solid strong connection between the fibers and the matrix. The impact strength and tensile properties of a rice straw reinforced in epoxy resin composites were investigated by Ratna Prasad et al. [1]. It has been observed that the composites fabricated by rice straw revealed maximum impact and tensile strength compared to that of the plain resin matrix. Murali Mohan Rao et al. [2] has reported on the tensile strength of the composites manufactured by chemically extracted elephant grass fibers with $KMnO_4$ treated fibers. It has been found that the specimens made by chemically treated fibers using $KMnO_4$ have showed a substantial growth in the ultimate tensile strength of the fibers taken out by general manual process. The composites prepared from various fibers such as jute, sisal, hemp, coir, kenaf, and its tensile properties were published by Wambua et al. [3] and reported that coir specimens exhibited lower tensile strength values and whereas hemp specimens showed the highest. A systematic study was conducted by Mishra et al. [4] on various mechanical properties of banana, hemp, and sisal specimens fabricated by novolac resin, with and without chemical treatments. Their studies revealed that the mechanical properties of the specimens were greatly affected by the chemical treatment process of the fibers as compared to the untreated fibers. Symington et al. [5] conducted a study on the influence of water content on mechanical behavior of abaca, kenaf, coir, flax, sisal, and jute. It is found that the specimens fabricated by jute exhibited higher mechanical properties than abaca, kenaf, coir, flax, and sisal. The mechanical behavior of thermosetting and thermoplastics composites made from recently identified fibers like sisal, hemp, banana, and coir using various methods was reported by several investigators [6–11]. The results show that the chemical treated fibers with polyester composites have showed higher impact and flexural strength as compared to the fibers taken out by manual process. Rajesh et al. [12] conducted a systematic study on mechanical and degradation properties of successive alkali treated completely biodegradable sisal fiber reinforced PLA composites. The results show that the alkali treated poly lactic acid composites exhibited higher flexural strength compared to that of plain poly lactic acid composites. Prasad [13] made an attempt to study the mechanical characteristics of various natural FRP. The investigations revealed that the jowar specimens prevailed superior mechanical characteristics than bamboo and sisal composites.

Fig. 1 DHAK plant and its dried stems

Though, enormous work is published on different natural fibers and their counterparts but no work was published on tensile testing of composites made by DHAK as fiber in the pure resin. Therefore, the main objective of this current work is to fabricate a new composite material for suitable applications and to evaluate the tensile behavior of DHAK composites.

2 Experimental Procedure

2.1 Materials

The fiber used in this study was a DHAK and collected from DHAK tree, which grows in western part of Uttar Pradesh, India. The DHAK plant and its dried stems are displayed in Fig. 1. Polyester resin, catalyst, and accelerator were acquired from ICA (Pvt) Ltd. (Delhi, India). The molds were cut as per ASTM standards in the medium-hard rubber. The sizes of the tensile specimens are 160 mm * 12.5 mm * 3 mm. The OHP transparent plastic sheet was used for covering the top side of the molds in longitudinal direction of the fiber in order to prevent the air. Four sets of specimens for individual volume fraction (0.06, 0.09, 0.11, 0.16, and 0.18) were manufactured by followed by ASTM D638-89.

2.2 Preparation of Fiber

Here, the fibers were extracted by retting and manual process. For doing that, initially the dried stems of DHAK plant were collected from the fields of the western parts of Uttar Pradesh. The long stems were cut at the knots of the DHAK and these stems were kept in water for 40–45 days to get them wet. Later, these stems were beaten smoothly by mallet in order for separating the fibers from DHAK. Moreover, separated fibers were combed further to get the required size and shape of individual fibers. The obtained fibers were allowed to dry under the sun light for about 2–3 h and successively dried in the oven. Subsequently, the fibers were cut to the necessary length as per ASTM standard.

2.3 Composite Preparation

Hand layup procedure was followed for fabricating the composites in the current work. The equal amount, i.e., 1.5 ml of catalyst and accelerator were mixed in plain resin of 100 ml to form a solid bond. The molds were filled with the fibers and resin simultaneously in unidirectional by layer this would help to remove the air present in the mold. Later, a transparent OHP sheet was allowed to stricken on the mold from one side by gently pressing by the steel rule subsequently with a heavy load. Moreover, the mold was remained for one day for curing and removes the specimens from the mold without breakage.

2.4 Tensile Testing of Specimens

The sizes of the above test composites are 160 * 12.5 * 3 mm. The four sets of each weight fraction were tested as per ASTM D638-89 standard. Prepared composites were examined at a cross speed of 0.5 mm/min by an UTM at ITS College, UP.

3 Result and Discussion

3.1 Tensile Testing of DHAK Composites

The influence of fiber amount on mean tensile strength and modulus of DHAK fiber reinforced specimens are represented in Figs. 5 and 6. It can be seen from Figs. 5 and 6 that the ultimate tensile strength and modulus of DHAK fiber reinforced composites gradually increases as the fiber quantity was increased. Highest mean ultimate strength and modulus were found as 114.97 MPa for the maximum volume fraction 0.18 that would be 2.356 times higher than that of a matrix. These results show that, making the composites by reinforcing less amount of DHAK fiber in the resin matrix advances the ultimate tensile strength of the composites. This was happened due to fixed bond between the DHAK and the resin matrix at the maximum volume fraction.

The same trend has been observed for the mean tensile modulus of DHAK fiber reinforced composites, as the mean tensile modulus increases with increase of volume fraction represented in Fig. 6. The highest mean tensile modulus of the composite is calculated as 16.14 GPa for highest volume fraction, which is about 2.07 times higher than that of a resin matrix.

The bare and tensile test coupons for different VF (0.06, 0.09, 0.11, 0.16, and 0.18) are represented in Fig. 2.

Failed bare and tensile tested composites are also shown in Figs. 3 and 4.

It can be seen clearly from the results that the DHAK fiber composites exhibited highest mean tensile strength of 114.91 MPa for the highest VF 0.18 which is much higher compared to that of the composites made by natural fibers like rice straw, sisal, banana, vakka, and elephant grass at 0.37 volume fractions. This indicates

Fig. 2 Prepared tensile specimens of bare and DHAK reinforced composites at various volume fractions

Fig. 3 Failed bare specimens of DHAK

Fig. 4 Failed tensile specimens of DHAK

that DHAK fiber reinforced composite will be a good replacement over the existing natural fiber composites in the near future (Figs. 5 and 6).

The density and tensile properties of existed fibers along with DHAK fiber reinforced composites have been represented in Table 1 for better comparison in the literature [11–16]. It is found that the density of DHAK is less compared to the many existed natural fibers like rice straw, sisal, banana, coir, hemp, and slightly higher than vakka, bamboo, and jowar which is a pretty property for fabricating light weight components.

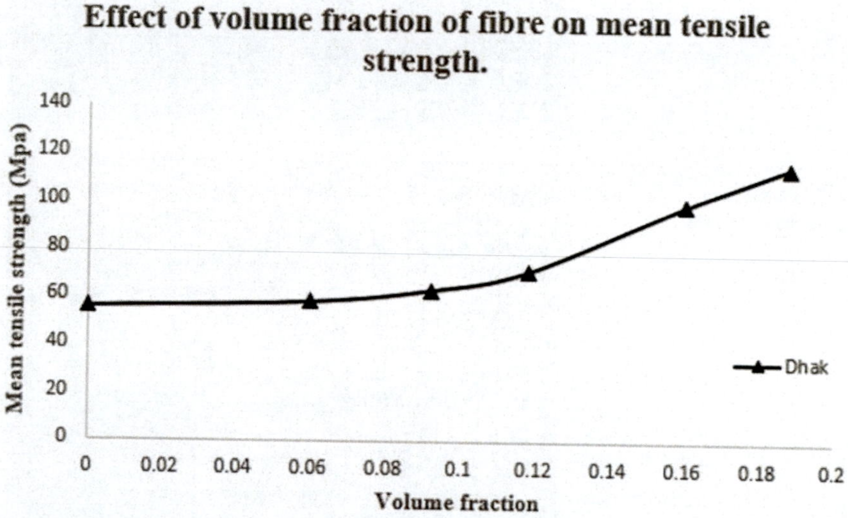

Fig. 5 Influence of VF on mean tensile strength of DHAK composites

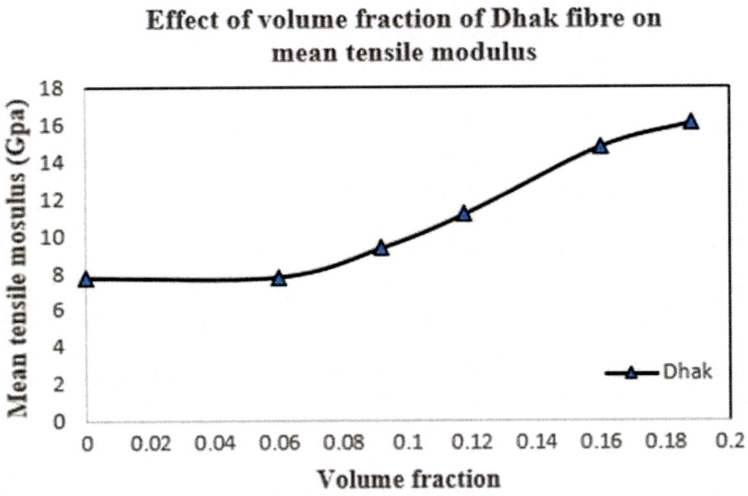

Fig. 6 Influence of VF on mean tensile modulus of DHAK composites

Table 1 Tensile properties and density of DHAK fiber reinforced composites along with other natural fiber reinforced composites [11–16]

S. no.	Name of the composite material	Density (Kg/m^3)	VF of fiber	Ultimate tensile strength (MPa)	Mean tensile modulus (GPa)
1	Matrix	1074.5	0.00	15.1	0.44
2	Rice straw	1258	0.41	46.1	1.04
3	Sisal	1450	0.37	50.0	1.6
4	Banana	1350	0.374	60.9	1.08
5	Vakka	810	0.372	66.0	1.79
6	Elephant grass	817	0.31	80.5	1.52
7	Bamboo	910	0.371	121.5	2.23
8	Jowar	922	0.40	124	2.75
9	DHAK	1098.5	0.18	114.91	1.614

4 Conclusions

In this study, the fabrication of DHAK fiber has been carried out and tensile testing is performed on the developed composites. The key points regarding the DHAK fiber reinforced composites are concluded as follows:

1. DHAK is identified as a potential fiber for reinforcement in making the composites.

2. The density of DHAK is found to be 1098.5 kg/m^3, which is less compared to the many existed fibers like rice straw, sisal, banana, coir, hemp, and slightly higher than vakka, bamboo, and jowar. This is a pretty property for manufacturing light weight components.
3. It is observed that the mean tensile strength and modulus of the composites increases with an increase in fiber content.
4. DHAK composites revealed highest tensile strength and modulus values at higher volume fraction 0.18 as 114.97 MPa and 16.14 GPa.
5. It is found that using DHAK in the resin plays an important role in terms of maximum tensile strength and modulus of a new manufactured composite.

References

1. Ratna Prasad AV, Mohan Rao K, Gupta AVSSKS (2007) Tensile and impact behavior of rice straw-polyester composites. Indian J Fiber Text Res 32:399–403
2. Murali Mohan Rao K, Ratna Prasad AV, Ranga Babu MNV, Mohan Rao K, Gupta AVSSKS (2007) The tensile behavior of elephant grass fiber reinforced composites fabricated from retting, chemical extraction and treated by $KMnO_4$. J Mater Sci 42:3266–3272. https://doi.org/10.1007/s10853-006-0657-8
3. Wambua P, Ivens J, Verpoest I (2003) Natural fibers: can they replace glass in fibre reinforced plastics? Compos Sci Technol 63:1259–1264
4. Mishra S, Naik JB, Patil YP (2000) Com Sci Tech 60:1729
5. Symington MC, Banks WM, West Opukuro David, Pethrick RA (2009) Tensile testing of cellulose-based natural fibers for structural composite applications. J Compos Mater 43:1083–1086
6. Li Y, Mai YW, Lin Y (2000) Sisal fibre and its composites: a review of recent developments. Compos Sci Tech 60:2037–2055
7. Singh B, Gupta M, Verma A (1996) Influence of fiber surface treatment on the properties of sisal–polyester composites. Polym Compos 17:910–918
8. Manikandan Nair KC, Diwan SM, Thomas S (1996) Tensile properties of short sisal fiber reinforced polystyrene composites. J Appl Polym Sci 60:1483–1497
9. Geethamma VG, Joseph R, Thomas S (1995) Short coir fibre-reinforced natural rubber composites: effects of fibre length, orientation and alkali treatment. J Appl Polym Sci 55:583–594
10. Mishra S, Naik JB, Patel YP (2000) The compatibilising effect of maleic anhydride on swelling and mechanical properties of plant-fibre-reinforced novolac composites. Compos Sci Tech 60:1729–1735
11. Ratna Prasad AV, Mohan RK (2011) Mechanical properties of natural fiber reinforced polyester composites: jowar, sisal, and bamboo. Mater Des 32(8–9):508–513
12. Rajesh G, Ratana Prasad AV, Gupta AVSSKS (20115) Mechanical and degradation properties of successive alkali treated completely biodegradable sisal fiber reinforced PLA composites. J Reinf Plast Compos 34(12):951–961
13. Murali Mohan Rao K, Mohan Rao K, Ratna Prasad AV (2010) Fabrication and testing of natural fiber reinforced composites: vakka, sisal, bamboo, banana. Mater Des 31:508–513

14. Murali Mohan Rao K, Mohan Rao K (2007) Extraction and tensile properties of natural fibres: Vakka, date and bamboo. Compos Struct 77:288–295
15. Romildo DTF, Karen S, George LE, Ghavami K (2000) Durability of alkali sensitive sisal and coconut fibres in cement mortar composites. Cem Concr Compos 22:127–143
16. Geethamma VG, Mathew KT, Lakshminarayana R, Thomas S (1998) Composite of short coir fibers and natural rubber: effect of chemical modification, loading and orientation of fiber. Polymer 39:1483–1491

Determination of Safety Stock in Divergent Supply Chains with Non-stationary Demand Process

A. M. Ranjith and V. Madhusudanan Pillai

Abstract In the present world, most of the firms are facing uncertain demand due to different market conditions. This type of demand can be modelled as non-stationary demand. In this paper, different multi-echelon divergent supply chain scenarios with non-stationary demand process are considered. Stock out plays an important role in supply chains; it reduces the service level of each member in a supply chain and causes high supply chain inventory cost. In order to avoid stock out in a supply chain, safety stock is to be maintained by each member. Developed simulation models for studying the operation of divergent supply chains under non-stationary demand processes. It is found that the stock out which leads to lost sales depends on two factors, i.e. the inertia of the non-stationary demand process and the structure of the divergent supply chain. In detail, the structure of the divergent supply chain is, basically, described in terms of the divergence factor. In this situation, we propose a mathematical relation to estimate safety stock by incorporating the divergence factor of the supply chain with non-stationary demand process. The performance of the supply chain against the divergence is analysed and the analysis shows that estimation of safety stock based on divergence factor can give a good estimate of the safety stock for divergent supply chain.

Keywords Supply chain · Non-stationary demand · Safety stock · Divergence factor

1 Introduction

A supply chain is an integrated network of multiple members and echelons consisting of retailers, wholesalers, distributors, manufacturers, and suppliers. The function of a supply chain is to meet customer demand in a timely manner. On the raw materials supplied by the suppliers to the manufacture of an end product, a series

A. M. Ranjith · V. M. Pillai (✉)
Department of Mechanical Engineering, National Institute of Technology Calicut, Kozhikode 673601, India
e-mail: vmp@nitc.ac.in

© Springer Nature Singapore Pte Ltd. 2021

P. M. Pandey et al. (eds.), *Advances in Production and Industrial Engineering*, Lecture Notes in Mechanical Engineering,
https://doi.org/10.1007/978-981-15-5519-0_6

of processes and value additions are involved and the product should be transported to the end customer through different members of the supply chain. In addition, apart from all these processes, each member of the network has the responsibility to do their operations without any delay. Meeting the demand in a timely manner is a difficult task for each member in the supply chain. Stock out is one of the main factors which make the replenishment of products to members in the supply chain tedious. To overcome this difficulty, people follow certain inventory policies and keep safety stocks in their inventory. Furthermore, stock outs occur in supply chain with stationary demand models, many researchers have reported solution for those stock outs [1, 2].

What would be the effect of non-stationary demand process in the level of stock out in a supply chain? This question has already been answered by researchers, but not fully. Most of the research reported in supply chain with non-stationary demand emphasised only on problems related to serial supply chains [1, 2 and 3]. Literature on non-stationary demand processes can be classified as whether the works focussed on optimisation or performance evaluation. Morton and Pentico [4] focused on setting of inventory policies in a single-stage supply chain with non-stationary demand process incorporating holding cost and backorder cost. Ettl et al. [5] reported a model that minimises the total inventory cost in a multi-stage supply chain system with non-stationary demand process. The non-stationary demand process was broken into stationary phases and adopted a rolling-horizon approach where the optimisation was performed for each demand phase. Non-stationary demand process was also modelled as a Markov-modulated Poisson demand process. A base-stock policy with state-dependent order-up-to level is optimal for a serial supply chain with non-stationary demand and the demand is modelled as a Markov-modulated Poisson demand process [6]. Abhyankar and Grave [7] developed an optimisation model to determine the best position to place an order in a two-stage serial supply chain with non-stationary demand. A state-dependent base-stock policy is optimal for a single-stage supply chain with Markov-modulated demand [8]. Neal and Willems [1] modified the existing algorithms of stationary demand models to optimise safety stock across the supply chain with non-stationary demand. Janssen et al. [9] proposed an age-based inventory policy for perishable product having constant shelf life with non-stationary demand. Tunc et al. [10] investigate the supply chain cost of employing stationary inventory policies when demand is non-stationary.

Based on the performance evaluation of supply chain, we consider the literature focused mainly on the bullwhip of supply chain. These papers generally assumed that each stage follows an adaptive base-stock policy. Grave [2] used an autoregressive integrated moving average (ARIMA) as demand process and describes the order amplification from one stage to another. In addition, the study shows an expression for estimating the safety stock in different echelons of a serial supply chain, accounting the parameters of non-stationary demand. Inventory management of a perishable product with fixed shelf life and non-stationary demand was investigated by PaulWorms et al. [11]. Grave and Willems [2] considered a constant service time policy and extended the optimisation algorithms of stationary demand to non-stationary demand to determine the safety stock. The sharing of demand faced information by retailer could reduce the variance of its forecast in upstream members of the supply chain with non-stationary demand [12]. Another study [13] presented dynamic programming approaches to solve for the reordering policies of supply chain with uncertain demand distribution.

This paper proposes a mathematical model for the estimation of safety stock in a multi-echelon divergent supply chain which faces non-stationary demand. The literature review shows that the literature is mainly concerned with safety stock estimation and the optimal inventory policy identification for a single item in serial supply chains. However, no works are reported on safety stock determination for divergent supply chain with non-stationary demand.

The remainder of this paper is organised as follows. Section 2 provides a brief description of demand models. Furthermore, the section gives an idea about non-stationary demand processes. Section 3 discusses the performance analysis of multi-echelon divergent supply chain. Safety stock estimation for the divergent supply chain is described in Sect. 4 and results are discussed in Sect. 5.

2 Demand Models

There are mainly two demand models available, stationary and non-stationary demand models. Stationary demand models are stochastic, but the process remains in statistical equilibrium with probabilistic properties, i.e. it does not change its parameters overtime. In particular, demand is varying about the fixed mean level and with constant variance. That is if the demand is normally distributed then parameter of the distribution will not change with time. Non-stationary demand models are also stochastic, but the demand histories behave like a random walk. In detail, the process does not remain in statistical equilibrium with probabilistic properties.

2.1 Non-stationary Demand

Non-stationary demand processes are modelled as autoregressive integrated moving average by Grave [7]. The model is given as follows:

$$d_1 = \mu + \varepsilon_1 \text{ and}$$

$$d_{t+1} = d_t - (1 - \alpha)\varepsilon_t + \varepsilon_{t+1} \text{ for t} = 1, 2, \ldots \quad (1)$$

In the given equations, d_t is the observed demand in period t,α is the measure of inertia in the non-stationary demand process, which means for larger α value there is less inertia in the demand process. Less inertia causes more fluctuation in the demand process. We have taken the value of α between 0 to 1, i.e. $0 \leq \alpha \leq 1$. If the value of α is one, the demand process would be a random walk on a continuous state space.μ is the mean value considered for the non-stationary demand process. The white noise (ε_t) of the non-stationary demand process is normally distributed random noise with $E[\varepsilon_t] = 0$ and $Var[\varepsilon_t] = \sigma^2$. The demand process becomes a stationary process when the value of $\alpha = 0$.

3 Multi-Stage Divergent Supply Chain: Divergent Factor

Supply chain structure can be differentiated based on the number of members available in each echelon. The differentiation is purely based on the degree of divergence of the supply chain which is given in Dominguez et al. [14]. The characterisation of divergent supply chain is known as divergence factor (*divF*).

$$divF = \sqrt{\sum_{i=1}^{m_s} \left(\frac{\left(m_i - m_M/m_s\right)^2}{m_s - 1} \right)} \tag{2}$$

where, i is the echelon indexing of the divergent supply chain and m_i represents the number of members available in echelon i. Number of echelons is given as m_s and the total number of members available in the supply chain is expressed as m_M.

3.1 Illustration of a Divergent Supply Chain Structure and Divergence Factor

A three-echelon divergent supply chain as shown in Fig. 1 is considered for the illustration and it has three retailers ($i = 1$), one distributor ($i = 2$) and one factory ($i = 3$). The notations used in Fig. 1 are, $D_i(t)$ is the demand of the customer, $S_i(t)$ is the amount of shipping quantity from the upstream members to their immediate downstream member and $O_i(t)$ is the order placed by the downstream members to their immediate upstream members.

Here, $m_M = 5$, $m_s = 3$ and the available members in echelons 1, 2 and 3 are 3, 1, 1 respectively, i.e. $m_1 = 3$, $m_2 = 1$ and $m_3 = 1$. The divergence factor can be calculated as follows,

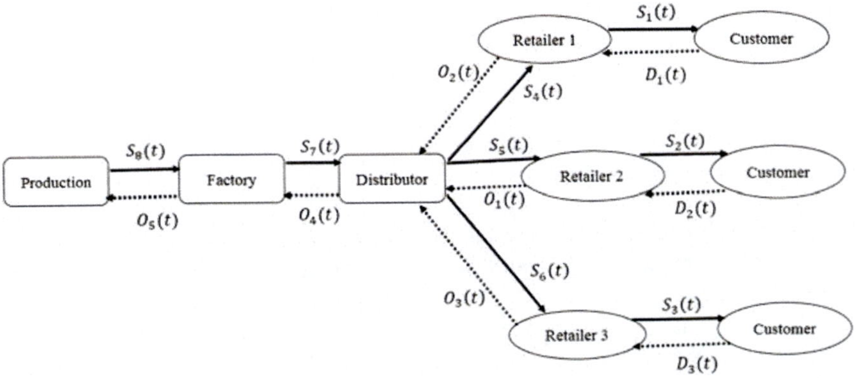

Fig. 1 Illustration of three-echelon divergent supply chain structure

$$divF = \sqrt{\sum_{i=1}^{3}\left(\frac{\left(m_i - 5/3\right)^2}{3-1}\right)} = \sqrt{\left(\frac{\left(3 - 5/3\right)^2}{3-1}\right) + \left(\frac{\left(1 - 5/3\right)^2}{3-1}\right) + \left(\frac{\left(1 - 5/3\right)^2}{3-1}\right)}$$

$$= \sqrt{\left(\frac{8}{9}\right) + \left(\frac{2}{9}\right) + \left(\frac{2}{9}\right)} = 1.154$$

4 Simulation Model and Assumptions

Three-echelon divergent supply chains are considered for the study and Fig. 1 shows such a supply chain with shipment and order flow among the members. In this study the non-stationary demand distribution for customer is considered. The supply chain consists of retailers ($i = 1$), distributors ($i = 2$) and one factory ($i = 3$). The retailers face customer demand ($D_i(t)$) and the factory has an unlimited production capacity and whatever quantity demanded by the factory manager is released after production. An order ($O_i(t)$) is placed at the end of a time period by each member to its immediate upstream member in the supply chain based on Order-Up-To (OUT) level inventory policy, which is a periodic review type inventory system [15]. In this inventory policy, a review takes place after a fixed order interval. At the review point, order-up-to level and inventory position are determined to decide about the order quantity. Order-Up-To level is the sum of forecasted demand of risk period and safety stock. At a review point, if the inventory position is less than the base-stock-level (Order-Up-To level), an order is placed. Otherwise, no order would happen. The review period considered for every member of the supply chain is one period. An order to reach an upstream member is considered as zero and delivery lead time is one period.

Table 1 Initial inventory and cost details

	Retailer	Distributor	Factory
Holding cost per unit	5	3	1
Lost sales cost per unit	10	6	2
Initial Inventory	190	190	190

The allocated quantity or quantity to be shipped ($S_i(t)$) is transported from each member to its immediate downstream member and is received at the beginning of time period t. Every member in the supply chain will decide the shipping quantity based on the on-hand inventory available on their shelf and allocation rationing rule. The exponentially weighted moving average is considered as the forecasting method of each member in the supply chain.

Excel-based simulation model of divergent supply chain is developed for the analysis. Initially, the replication length is assumed as 30 and after execution of initial simulation model actual replication is calculated for 95% confidence interval of total supply chain cost with 5% error. Demand data is generated using the Eq. 1 with a mean of 80 and the measure of inertia of the process is taken from the range "0 to 0.9". White noise is generated using a normal distribution with mean as zero and variance as 100. The simulation model is executed for a time period of 104 weeks and first 52 weeks are the warm-up period for the simulation analysis. Initial forecast value is assumed as the mean value used for the demand generation. The cost details and the initial inventory details are given in Table. 1 [16]. A detailed description of the model considered for determination of safety stock for divergent supply chain with non-stationary demand is given in the next section. It is assumed that all the retailers of the supply chain have the same inventory-related cost as given in Table 1, and in distribution and factory echelons of the supply chain, there is only one member in each stage.

5 Safety Stock Estimation for Multi-Stage Divergent Supply Chain

Safety stock is an additional quantity of the item kept in inventory in order to reduce the risk that the item will be out of stock. Safety stock is needed to cover the demand during the risk period, which includes ordering lead time, replenishment lead time and review period, in case of actual demand exceeds expected demand and/or actual lead time exceeds the expected lead time. The quantity of an items maintained by each members in a supply chain is based on the demand of those items. Demands would be not same at all time. It will vary depends on the market conditions and customers taste [6]. Emphasise by the members the supply chain of an item is to maintain sufficient inventory to meet downstream member demand/customer demand in a timely manner. Sometimes the upstream members are not able to meet those demands and it will lead to back order/lost sales. The stock outs in the member of supply chain can be managed with safety stock. Thus, the managers need to keep sufficient safety stock to avoid stock out and to meet the demand of downstream member/customer in a timely manner. Safety stock estimation in serial supply chains with non-stationary demand has already been considered in previous research works [2]. In [2], the risk period as well as the measure of inertia (α) of the non-stationary demand process are considered for safety stock estimation. Supply chain networks are considered by previous researchers [5, 14]. They [5] did not consider the structural aspects of the supply chain. Dominguez et al. [14] proposed a mathematical model to differentiate the supply chain structure in terms of divergence aspects and they conducted a comparative for analysing bullwhip effect in serial and divergent supply chain with stationary demand. The expression for the safety stock to cover uncertain demand over the lead time is described by accommodating the measure inertia of the non-stationary demand process.

This paper focuses on divergent supply chain with non-stationary demand and the estimation of safety stock for its members. The safety stock determination model suggested by Grave [2] for a serial supply chain is given in Eq. 3.

$$Safety\ \ Stock\ = \ Z\ \times\ \sigma_D \tag{3}$$

where, Z is the service level factor which can be determined from the standard normal table corresponding to the desired service level. The standard deviation of risk period demand is expressed as σ_D and is different for members belongs to different echelons.

Equation 4 is the proposed model for the safety stock calculation of divergent supply chain with non-stationary demand process. This relation is suitable for the extreme downstream members in a supply chain, i.e. this is only suitable for members in stage ($i = 1$). The standard deviation of demand during risk period can be estimated using Eq. 5. The divergence factor of the divergent supply chain is $divF$ and $\sigma_{d,i}$ is the risk period demand of stage $i = 1$. The lead time of echelon i is L_i.

$$Safety\ stock = Z \times \sqrt[4]{divF} \times \sigma_{d,i} \tag{4}$$

$$\sigma_{d,i} = \sigma\sqrt{L_i}\sqrt{1 + \alpha L + \frac{\alpha^2 L_i^2}{3}} \tag{5}$$

Safety stock for the upstream member (other than stage $i = 1$) is shown in Eq. 6 and $\sigma_{u,i}$ is the standard deviation of demand during risk period (See Eq. 7 for the estimation of standard deviation) of upstream members in the divergent supply chain.

$$Safety\ stock = Z \times \sqrt[4]{divF^3} \times \sigma_{u,i} \tag{6}$$

$$\sigma_{u,i} = \sigma\sqrt{\sum\nolimits_{j=L_{i-1}}^{L_{i-1}+L_i-1} (1 + j\alpha)^2} \tag{7}$$

In the simulation model, safety stock is calculated using Eqs. (4) and (6).

6 Experimental Results and Validation

The entire simulation models are run under the assumptions given in the above sections. The desired service level is set as 95%. Three-echelon supply chains with different divergence factor are simulated. The divergent factor is varied by varying the number of members in echelon $i = 1$. The divergent supply chains are operated under the safety stock determined using Grave's [2] method and proposed method. In particular, service level, lost sales cost and total supply chain inventory cost obtained after executing the simulation models under both methods are given in Table 2. The performance of the supply chain is analysed against the divergence of the supply chains. A comparison of performance measures can be easily obtained from the graphs given in Figs. 2 and 3. These graphs show that divergence factor has significant influence on the performance of the supply chain. This analysis shows the effect of divergence on supply chain performance, especially, in lost sales and it clearly support the incorporation of divergence factor in the safety stock required for managing the divergent supply chains. In real life, the supply chains are divergent and the proposed method provides a good estimates for the safety stock.

The average fill rate obtained from the proposed method is better than the fill rate obtained from Grave's [2] method. From the result obtained after execution of simulation models; it is very clear that the service level has not reached the desired value with the safety stock proposed by Grave [2]. But, in the case of proposed safety stock calculation with the incorporation of the divergence factor, equal and more than the set service level (95%) is obtained. In specific, there is an increment of service level within the range of 6.16–9.41% compared to the method suggested by Grave [2]. Comparison of service level under both methods is available in Fig. 2. The lost sales comparison is shown in Fig. 3 and there is a decrement of 32.35 to 38.02% in lost sales cost. As the divergence factor increases, the supply chain structure goes away from the serial structure. Figures 2 and 3 show that as divergence factor increases, the need for accommodation of divergence factor for safety stock determination is clearly visible.

6.1 Conclusion

This paper focused on safety stock estimation of multi-echelon divergent supply chain with non-stationary demand. Different three-echelon divergent supply chains are considered and simulation models are developed for the analysis. The divergence of the supply chain is measured using divergence factor. The proposed safety stock determination method uses the divergence factor. The performance of the supply chain is evaluated using the performance measures such as fill rate, total supply chain

Table 2 Comparison of performance measures

Three-echelon divergent supply chain				Safety Stock (*as per Grave, 1999* [2])			Safety Stock (*as per the proposed model*)		
No. of retailers	No. of distributors	No. of factory	Divergence factor	Lost sales cost	Total supply chain cost	Fill rate	Lost sales cost	Total supply chain cost	Fill rate
3	1	1	1.15	71,840	247,838	0.9213	44,601	198,840	0.9781
4	1	1	1.73	127,731.6	304,755.8	0.9046	52,917	257,843	0.9723
5	1	1	2.31	179,065.8	395,727.6	0.8936	121,124	359,867	0.9678
6	1	1	2.89	206,803.6	477,850.8	0.8887	143,627	411,907	0.9637
7	1	1	3.46	242,006.4	554,959.6	0.8834	162,767	485,438	0.9614
8	1	1	4.04	320,838.8	628,915.8	0.8795	180,854	546,757	0.9543
9	1	1	4.62	361,652	715,619.6	0.8766	229,654	668,675	0.9523
10	1	1	5.20	409,325	833,891	0.8701	253,672	792,256	0.952

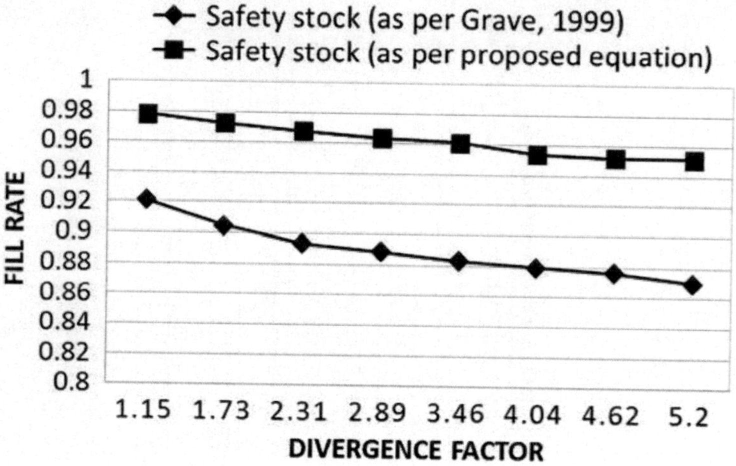

Fig. 2 Comparison of fill rate

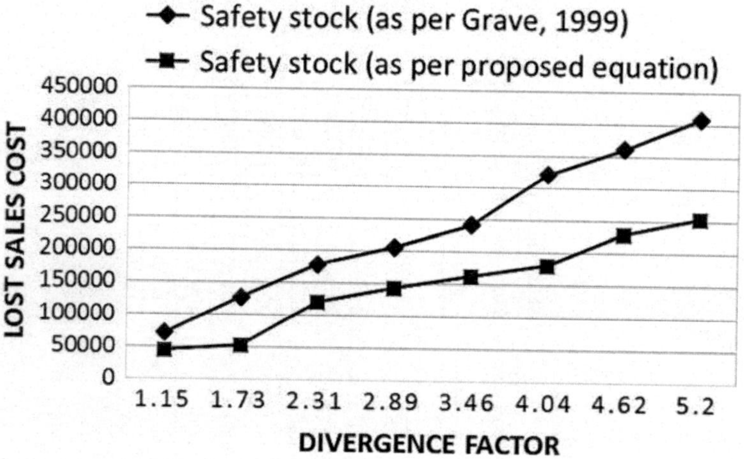

Fig. 3 Comparison of lost sales cost

inventory cost and lost sales cost. Safety stock determination method proposed in the literature for the serial supply chain with non-stationary demand is applied for the operation of divergent supply chains. But, it is found that the model proposed in this paper for determination of safety stock is providing better performance for the supply chains considered. The performance of the supply chain against the divergence is analysed and the analysis shows that estimation of safety stock based on divergence factor can give a good estimate of the safety stock for divergent supply chain. In particular, the proposed safety stock determination method helps the supply chains to meet demand on a timely manner. Further, similar analysis is required under different echelon levels and divergent factors to establish safety stock estimation for a general supply chain under non-stationary demand.

References

1. John JN, Sean PW (2009) Managing inventory in supply chains with non-stationary demand. Ins Operat Res Manag Sci 39(5):388–399
2. Stephen CG (1999) A single-item inventory model for a non-stationary demand process. Manufact Serv Operat Manag 1:50–61
3. Stephen CG, Sean PW (2008) Strategic inventory placement in supply chains: non-stationary demand. Manufact Serv Operat Manag 10(2):278–287
4. Morton TE, Pentico DW (1995) The finite horizon non-stationary stochastic inventory problem: Near-myopic bounds, heuristics, testing. Manage Sci 41:334–343
5. Ettl M, Feigin GE, Lin GY, Yao DD (2000) A supply network model with base-stock control and service requirements. Operat Res 48:216–232
6. Chen F, Song JS (2001) Optimal policies for multi-echelon inventory problems with Markov-modulated demand. Operat Res 49:226–234
7. Abhyankar HS, Stephen CG (2001) Creating an inventory hedge for Markov-modulated poisson demand: an application and a model. Manufact Serv Operat Manag 3:306–320
8. Treharne JT, Sox CR (2002) Adaptive inventory control for non-stationary demand and partial information. Manage Sci 48(5):196–202
9. Janssen L, Sauer J, Claus T, Nehls U (2018) Development and simulation analysis of a new perishable inventory model with a closing days constraint under non-stationary stochastic demand. Comput Ind Eng 118:9–22
10. Tunc H, Kilic OA, Tarim SA, Eksioglu B (2011) The cost of using stationary inventory policies when demand is non-stationary. Int J Manag Sci 39(4):410–415
11. Pauls-Worm KG, Hendrix EM, Alcoba AG, Haijema R (2016) Order quantities for perishable inventory control with non-stationary demand and a fill rate constraint. Int J Prod Econ 181:238–246
12. Srinivasan R (2001) Information sharing in a supply chain: a note on its value when demand is non-stationary. Manag Sci 47(4):605–610
13. Qiu R, Sun M, Lim YF (2017) Optimizing (s, S) policies for multi-period inventory models with demand distribution uncertainty: robust dynamic programing approaches. Eur J Oper Res 261(3):880–892
14. Dominguez R, Framinan JM, Cannella S (2013) Serial vs. divergent supply chain networks: a comparative analysis of the bullwhip effect. Int J Prod Res 52(7):2194–2210
15. Lovejoy WS (1990) Myopic policies for some inventory models with uncertain demand distributions. Manage Sci 36(6):724–738
16. Chopra S, Meindl P, Kalra DV (2010) Supply chain management: strategy, planning, and operation. Third edition, Pearson Education

Solving Unequal Area Facility Layout Problems with Flexible Bay Structure by Simulated Annealing Algorithm

Irappa Basappa Hunagund, V. Madhusudanan Pillai, and U. N. Kempaiah

Abstract This article proposes an application of Simulated Annealing algorithm for Flexible Bay Structure (SA-FBS) based Unequal Area Facility Layout Problems (UA-FLPs). Initially, MILP model of UA-FLPs with FBS is solved using LINGO software, due to complexity of the model, LINGO solver is not giving the optimal solution in a reasonable time. Then, the proposed SA-FBS is developed and applied to test the UA-FLPs reported in the literature. SA-FBS method has either given the same or better solution as compared to the solution reported in the literature. The proposed SA-FBS computational timings are competitive as compared to other meta-heuristics timings reported in the literature.

Keywords Unequal area facility layout problem · Flexible bay structure · Meta-heuristics · Simulated annealing

1 Introduction

The arrangement of facilities or resources on the shop floor for the manufacture of goods or to deliver the services is called the facility layout planning or design. The proper arrangement of facilities is a challenging task and this problem is usually formulated as Facility Layout Problems (FLPs). The various main objectives considered in the design of FLPs are minimizing the material handling cost (MHC) and production/service time, and maximizing the space utilization, and the safety of

I. B. Hunagund (✉)
Department of Mechanical Engineering, Government Engineering College Ramanagar, Ramanagar 562159, Karnataka, India
e-mail: iranna346@yahoo.com

V. M. Pillai
Department of Mechanical Engineering, National Institute of Technology Calicut, Calicut 673601, Kerala, India

U. N. Kempaiah
Department of Mechanical Engineering, University Visvesvarayya College of Engineering, Bangalore 560001, Karnataka, India

© Springer Nature Singapore Pte Ltd. 2021

P. M. Pandey et al. (eds.), *Advances in Production and Industrial Engineering*, Lecture Notes in Mechanical Engineering,
https://doi.org/10.1007/978-981-15-5519-0_7

workers and machines. In most of the FLPs, the main objective considered is the MHC. The material handling cost is a non-value adding cost. A proper arrangement of facilities on the shop floor is required to reduce the material handling cost and thus increases the productivity.

UA-FLPs represented on a continuous space are solved either with an assumption of larger floor area than the area required for the all facilities or with the floor area equal to the sum of area of all facilities. In the former case, facilities get clustered towards the centre of the plant. The clustering of facilities at the centre of plant area is overcome by considering the floor area equal to the sum of area of all facilities. In this case, facilities are arranged either as Slicing Tree Structure (STS) or in Flexible Bay Structure (FBS) or in Relaxed Flexible Bay Structure (RFBS). In STS, the plant area is partitioned both in vertical and horizontal direction, simultaneously. In FBS, plant area is partitioned either in vertical or horizontal direction, but not in both. In FBS, the width of vertical or horizontal bays is flexible depending on the sum of facilities' area within each bay, and also, the facilities are not allowed to span over multiple bays. The FBS formulation is made more relaxed by RFBS representation. In RFBS, extra area or extended space is considered for plant area to allow extra space within each bay.

Sample FBS, RFBS and STS layouts for ten unequal area facilities are shown in Fig. 1. Figure 1a represents FBS layout with four vertical bays in the layout. Figure 1b represents RFBS layout with four vertical bays and empty spaces in the layout. The empty spaces in the RFBS are shown by dark shadings. Figure 1c represents STS without formation of bays. The bay boundaries in FBS help to form aisle structure on the plant floor. Aisle structure is suitable when fork lift trucks or automated guided vehicles are used for material transportation from one facility to another. Hence, FBS is one of the important layout representations that researchers and practitioners continue to study [1]. Further, Simulated Annealing (SA) is simple meta-heuristic used to solve combinatorial problems. Unlike other population-based meta-heuristics, SA is simple to implement, and is a sequential search algorithm hence it is computationally more efficient one.

In the present work, initially, LINGO software package is used to code the MILP model of UA-FLP with FBS. LINGO solver is used to obtain optimum solution of small size problems and thus to check the effectiveness of the proposed Simulated Annealing algorithm for Flexible Bay Structure (SA-FBS). The rest of the paper is

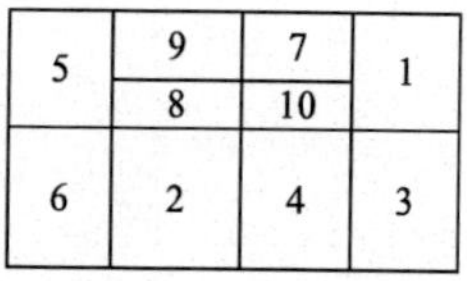

(a) Flexible bay layout

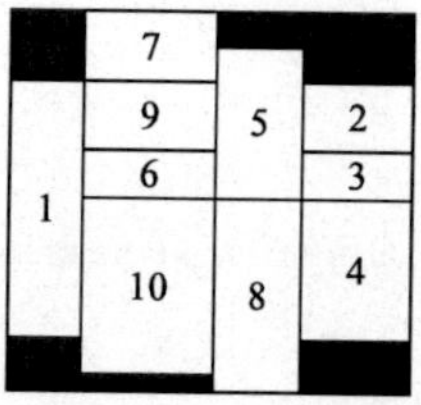

(b) Relaxed flexible bay layout

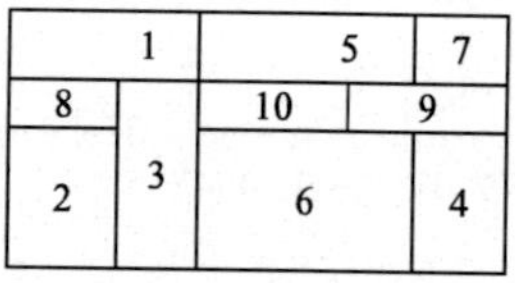

(c) Slicing tree layout

Fig. 1 A sample flexible bay, relaxed flexible bay and slicing tree structure layouts organized as follows: Section 2 gives the literature review; Section 3 discusses the problem environment and MILP model of UA-FLP with FBS. Section 4 discusses the simulated annealing heuristic and explains about solution encoding, neighbourhood search methods and SA-FBS parameter settings. Section 5 gives numerical experiments and computational results for the problems reported in the literature. Section 6 presents the conclusions and future research direction.

2 Literature Review

Much of earlier works in FLPs assume equal area for all facilities. Koopmans and Beckmann [2] are first to formulate the layout problem as a Quadratic Assignment Problem (QAP). QAP is NP-hard problem. Hence researchers use heuristics and meta-heuristics to solve QAP [3–6]. But, in most of practical situations each facility's area would be different. Hence, recent researches in layout design consider the Unequal Area Facilities Layout Problems (UA-FLPs). UA-FLPs are either represented in discrete or continuous space. Armour and Buffa [7] are first to deal with Unequal Area Facilities Layout Problems (UA-FLPs) by discrete space representation, and they applied the conditional pair-wise exchange method to solve UA-FLPs. Space Filling Curve (SFC) method to solve UA-FLPs by discrete space representation is reported in [8, 9]. Discrete space representation approaches generate irregular shape facilities as layout solution, which is difficult to implement practically. Hence, recent research works [10–25] on UA-FLPs consider continuous space representation to generate regular shape facilities in the layout.

UA-FLPs represented on continuous space are formulated as MIP model. MIP models give more realistic solution since they determine location coordinates of facilities and their sizes while designing the layout for required objective. Lacksonen [10], and Meller et al. [11] assume larger floor area than the area required for the all facilities to solve MIP model. The solution values by this approach are better but the facilities get clustered towards the centre of the plant and the solution contains empty spaces among the facilities leading to poor space utilization. Hence many researchers take the floor area equal to the sum of all facilities' area. Researchers [12–15] use various meta-heuristics to solve UA-FLPS with slicing tree structure (STS). Tate and Smith [16] have first time represented the UA-FLP by a flexible bay structure (FBS) and use Genetic Algorithm (GA) as the solution method. UA-FLP with FBS representation is more constrained than STS due to the creation of bays. Hence, optimal UA-FLP designs based on the FBS are expected to have a higher material handling cost than the corresponding optimal designs based on STS.

In 2006, Konak et al. [1] have formulated the UA-FLP with FBS as a Mixed Integer Linear Programming (MILP) model. In this formulation, the non-linear constraint of earlier MIP model is converted into linear constraint. This formulation is NP-complete and hence it has a limitation on the size of problems that can be solved optimally. Researchers use various meta-heuristics for solving this formulation. Researchers [17–20] used different meta-heuristics to solve formulation of

Konak et al. [1]. Researchers [18, 19, 21] used Relaxed Flexible Bay System (RFBS) by allowing dummy facilities with empty spaces in the bays. Konak et al. [1], and Gau and Meller [22] argues that the solution value of RFBS layout is expected to be lower than the solution value of FBS layout due to more flexibility in search space as RFBS layout provides extra or empty spaces in the bays. But, RFBS layout is expected to have more floor area requirement than UA-FLP with FBS. Kulturel-Konak and Konak [21] used Ant Colony Optimisation (ACO), Kulturel-Konak and Konak [18] used Particle Swarm Optimisation (PSO) and Ulutas and Kulturel-Konak [19] used Artificial Immune System (AIS) to solve the UA-FLPs by RFBS representation. Mazinani et al. [23] formulated the MILP model for dynamic environment UA-FLP with FBS and used GAMS software and GA to solve both static and dynamic UA-FLP with FBS. Meta-heuristics like GA, ACO, AIS and PSO used by above, various researchers are population-based parallel search algorithms hence they are computationally less efficient. Simulated Annealing (SA) is another simple meta-heuristic used to solve combinatorial problems. Application of SA meta-heuristic in equal area static and dynamic facility layout problems is made by [4–6]. Tam [24] and M-Y Ku et al., [9] used SA and SA based GA for general type UA-FLPs, but these SA applications are not for UA-FLPs based on flexible bay structure. Recently, Kulturel-Konak and Konak [25] developed a hybrid simulated algorithm to solve unequal area cyclic facility layout problems, and they combined the large-scale local search with simulated annealing. But, layout solution is not for the flexible bay structure (FBS). More recently, Hunagund et al. [20] used the SA to solve the dynamic environment UA-FLP with FBS but not for static environment problems. Hence, the extensive literature review reveals that flexible bay structure is one of the important layout representations as bay boundaries help to form aisle structure. The present study proposes a Simulated Annealing algorithm for Flexible Bay Structure (SA-FBS) to solve unequal area facility layout problems.

3 Problem Environment and MILP Model of UA-FLP with FBS

Research problem under consideration is to design and develop a layout plan for a manufacturing system operating in an environment where the annual demand is not fluctuating considerably. The problem considered here has unequal area facilities with rectangular shapes which are to be placed on a given plant floor area without overlapping of facilities. The minimum and maximum limits on the side lengths of the facilities are defined by aspect ratio. The design of block layout involves formation of flexible bays with objective of minimizing the material handling cost. The formation of flexible bays helps in better design of aisles structure in the layout. Good aisle structure helps the facilities to interact easily with material handling equipment. Hence, the developed layout plan can be directly used by an end user with small modification, if any required.

MILP model of UA-FLP with FBS: In this section, an MILP model of UA-FLP with FBS of Konak et al. [1] is presented.

Assumptions in the Model:

- Unequal area facilities with rectangular shapes are considered.
- Centre to centre distance between the facilities is measured considering the rectilinear distance.
- The minimum and maximum limits on the side lengths of the facilities are defined by aspect ratio.
- All the facilities allocated to a bay should fill the bay completely. The proposed model can solve problems with floor size more than or equal to the sum of area of all facilities. In case, floor size is more than the sum of facility areas, the excess space will be pushed to the east side of the plant floor.

Inputs to the Model:

- The material flow between facilities.
- The cost of moving unit material per unit distance between facilities
- The area of each facility
- The aspect ratio or shape constraint of each facility
- The maximum number of bays allowed in the layout and
- The size (width × height) of the plant floor

Notations:

Indexes;

$k = 1, 2, ..., K,$	where, K is the number of parts
$m, n = 1, 2, ..., N,$	where, N is the number of facilities
$r, s = 1, 2, ..., M,$	where, M is the maximum number of bays allowed

Input Parameters;

W	Horizontal length of floor in x–axis direction.
H	Vertical length of floor in y–axis direction.
A_m	Area of facility m.
α_m	Maximum aspect ratio of facility m.
$S_m^{\max} = \min\{H, \sqrt{A_m \alpha_m}\}$	Maximum allowable side length of facility m.
$S_m^{\min} = \sqrt{\frac{A_m}{\alpha_m}}$	Minimum allowable side length of facility m.
$f_{m \to n}$	Parts flow volume from facility m to facility n.
$f_{mn} = f_{m \to n} + f_{n \to m}$	Parts flow volume between facilities m and n.
$f_{mn} = \sum_{k=1}^{K} \zeta_{k,mn} \quad \forall m, n$	Part flow, if demand of product k and its route sheet are given.
$\zeta_{k,mn}$	Part flow between facilities m and n due to product k.

C_{mn}	Cost of transporting unit part per unit distance between facilities m and n.
MHC	The material handling cost.

Decision Variables;

b_r	Horizontal width of bay r.
l_{mr}	Height of facility m in bay r.
h_m^y	Vertical height of facility m in y-axis direction.
(x_m, y_m)	Centre coordinates of the facility m.
$D_{mn}^x = \lvert x_m - x_n \rvert$	Distance between the centre of facilities m and n in x-axis direction.
$D_{mn}^y = \lvert y_m - y_n \rvert$	Distance between the centre of facilities m and n in y-axis direction.
$I_{mr} = 1$	If facility m is allocated to bay r.
$= 0$	Otherwise.
$\varphi_r = 1$	If bay r is having facilities.
$= 0$	Otherwise.
$Y_{mn} = 1$	If facility m is above the facility n in the same bay.
$= 0$	Otherwise.

The MILP Model:

$$\text{Minimize } MHC = \sum_{m=1}^{N-1} \sum_{n=m+1}^{N} C_{mn} f_{mn} \left(D_{mn}^x + D_{mn}^y \right) \tag{1}$$

Subjected to:

$$D_{mn}^x \geq x_m - x_n \quad \forall n > m \tag{2}$$

$$D_{mn}^x \geq x_n - x_m \quad \forall n > m \tag{3}$$

$$D_{mn}^y \geq y_m - y_n \quad \forall n > m \tag{4}$$

$$D_{mn}^y \geq y_n - y_m \quad \forall n > m \tag{5}$$

$$\sum_{r=1}^{M} I_{mr} = 1 \quad \forall m \tag{6}$$

$$b_r = \frac{1}{H} \sum_{m=1}^{N} I_{mr} A_m \quad \forall r \tag{7}$$

$$S_m^{\min} \leq b_r \leq S_m^{\max} + W(1 - I_{mr}) \;\; \forall m, r \tag{8}$$

$$x_m \geq \sum_{s \leq r} b_s - 0.5b_r - (W - S_m^{\min})(1 - I_{mr}) \quad \forall m, r \tag{9}$$

$$x_m \leq \sum_{s \leq r} b_s - 0.5b_r + (W - S_m^{\min})(1 - I_{mr}) \quad \forall m, r \tag{10}$$

$$\frac{l_{mr}}{A_m} - \frac{l_{nr}}{A_n} - \max\left\{\frac{S_m^{\max}}{A_m}, \frac{S_n^{\max}}{A_n}\right\}(2 - I_{mr} - I_{nr}) \leq 0 \quad \forall r, m, n > m \tag{11}$$

$$\frac{l_{mr}}{A_m} - \frac{l_{nr}}{A_n} + \max\left\{\frac{S_m^{\max}}{A_m}, \frac{S_n^{\max}}{A_n}\right\}(2 - I_{mr} - I_{nr}) \geq 0 \quad \forall r, m, n > m \tag{12}$$

$$\sum_{m=1}^{N} l_{mr} = H\varphi_r \quad \forall r \tag{13}$$

$$S_m^{\min} I_{mr} \leq l_{mr} \leq S_m^{\max} I_{mr} + W(1 - I_{mr}) \quad \forall m, r \tag{14}$$

$$\sum_{r=1}^{M} l_{mr} = h_m^y \quad \forall m \tag{15}$$

$$y_m - 0.5h_m^y \geq y_n + 0.5h_n^y - H(1 - Y_{mn}) \quad \forall m, n \neq m \tag{16}$$

$$Y_{mn} + Y_{nm} \leq 1 \quad \forall m, n > m \tag{17}$$

$$Y_{mn} + Y_{nm} \geq I_{mr} + I_{nr} - 1 \quad \forall r, m, n > m \tag{18}$$

$$0.5h_m^y \leq y_m \leq H - 0.5h_m^y \quad \forall m \tag{19}$$

$$x_m,\ y_m,\ h_m^y,\ b_r,\ l_{mr},\ D_{mn}^x,\ D_{mn}^y\ \geq\ 0 \quad \forall m, n, r \tag{20}$$

$$I_{mr} \in \{0,\ 1\},\ \varphi_r \in \{0,\ 1\},\ Y_{mn} \in \{0,\ 1\} \ \forall m, n, r \tag{21}$$

The objective function is represented by Eq. (1) which minimizes the material handling cost of a UA-FLP with FBS. The objective function finds the facilities dimension and coordinates by considering the amount of material flow and cost of moving unit material per unit distance between facilities *m* and *n*. The absolute centre to centre rectilinear distance between facilities is ensured by constraints (2) to (5). The absolute value term $|x_m - x_n|$ is linearized by $D_{mn}^x \geq x_m - x_n$ and $D_{mn}^x \geq x_n - x_m$ where $D_{mn}^x \geq 0$. The assigning of each facility to only one bay is ensured by constraint (6). The width of each bay is computed using Eq. (7) based on the areas of the facilities assigned and the height of floor space. The upper and lower limits on the width of each bay is ensured by constraint (8) based on facilities assigned in each bay. The facilities centre coordinates along the plant floor's horizontal axis (x-axis coordinate) are determined by constraints (9) and (10). And also, the facilities are located within the limits of floor size and this aspect is ensured by constraints (7), (9) and (10). The facilities centre coordinate (y-axis coordinate) along the floor's vertical axis is determined by constraints (11) to (18), and also these constraints ensure that the facilities do not overlap in the vertical direction. The accommodation of facilities within the limit of floor space along the vertical axis is ensured by constraint (19). Constraint (20) ensures the non-negativity of continuous decision variables and constraint (21) puts the restrictions for binary decision variables.

Initially, the above MILP model is coded in LINGO 16.0 $\times$ 64 mathematical modelling software. LINGO solver is used to obtain optimum solution of small size problem by an exact method. In the second part of present work, a Simulated Annealing algorithm for Flexible Bay Structure (SA-FBS) is developed to solve the above UA-FLP with FBS model.

4 SA-FBS to Unequal Area Facility Layout Problems with Flexible Bay Structure

Simulated Annealing (SA) algorithm is a neighbourhood search meta-heuristic for global optimisation. The objective of global optimization is to find the globally best solution for models, in the presence of multiple local optima. There are many meta-heuristics like GA, Tabu Search (TS), ACO, PSO, AIS, etc. are reported in the literature for global optimization but all these are population-based parallel search algorithms. The SA algorithm is the one of meta-heuristic and it works based on probabilistic methods that avoid being stuck at local (non-global) minima. It is proven to be a simple sequential search algorithm but powerful method for large-scale combinatorial optimization.

In SA, the solution *A'* derived from a solution *A* is not only accepted if *A'* is better but it may also be get accepted if *A'* is worse than *A*. Worse solution is accepted with some acceptance probability. Boltzmann's law is used to determine this acceptance probability, It is given as $P(accept) = exp(-\Delta/(b \times T_i))$, where '*b*' is Boltzmann's constant and 'T_i' is the temperature at each iteration level according to cooling schedule. The value of 'T_i' is between $T_0 \leq T_i \leq T_F$. Where 'T_0' and 'T_F' are initial and final temperatures, respectively. $\Delta = z(A')\text{-}z(A)$, where Δ represents difference between the value of solutions *A* and *A'*. This is known as Metropolis acceptance rule which implies that the smaller the increase in the 'Δ' value, more likely the new solution is selected, and the lower the value of 'T_i', the less likely the new solution is selected.

1	2	3	4	5	6	7	8	9	10	11	12	13	14	15
0	0	5	6	0	0	0	7	8	2	0	3	4	1	0

(a) Sample encoding alternative-1.

1	2	3	4	5	6	7	8	9	10	11	12	13	14	15
5	6	0	7	8	2	0	0	3	4	1	0	0	0	0

(b) Sample encoding alternative -2.

Fig. 2 Sample encoding scheme for FBS layout

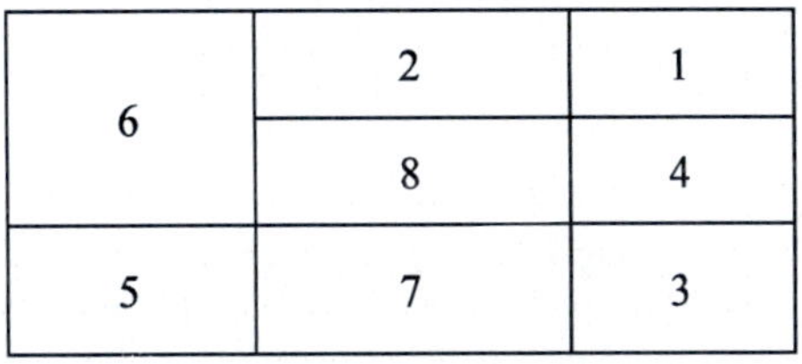

Fig. 3 FBS layout for encoded scheme of Fig. 2

4.1 *Solution Encoding*

The encoding scheme given in Ulutas and Kulturel-Konak [19] is used to encode the UA-FLP with FBS for the proposed SA-FBS solution method. This encoding scheme has a single segment containing the whole information regarding the name of facilities, facilities sequence and bay division points. When the numbers of bays (M) are not given, the length of the string is taken as (2 N-1), since the maximum number of bays in the solution can be equal to the number of facilities (N). The bay breakpoints are represented by '0's. Hence maximum bay breakpoints '0' can be (N-1). If the maximum number of bays (M) are given data, then the length of the string is ($N + M$-1).

The encoding scheme for 8-facilities with the consideration of maximum number of bays equal to the given number facilities (N) is shown in Fig. 2. This figure shows two sample encoded strings and both strings lead to the same layout configuration. The layout configuration for these strings is shown in Fig. 3. If $N = 8$, then length of string = 2 N-1 = 2 × 8–1 = 15.

4.2 *Configuration Change*

In the proposed work, to create neighbourhood solution, insert, swap and reversion operators as given in [26] are used. These three operators are used randomly with the equal probability of selection for each operator. That is, a random number r is generated between 0 and 1, if the r is between $0 < r < 0.33$, insert operation is carried out and if the r is between $0.33 \leq r < 0.67$, swap operation is carried

Before

1	2	3	4	**5**	6	7	**8**	9	10	11	12	13	14	15
0	0	5	6	0	0	0	7	8	2	0	3	4	1	0

After

1	2	3	4	5	6	7	8	9	10	11	12	13	14	15
0	0	5	6	7	0	0	0	8	2	0	3	4	1	0

(a) Insert operation for random numbers $i = 8$ and $j = 5$.

1	2	3	4	5	6	7	8	9	10	**11**	12	13	14	**15**
0	0	5	6	0	0	0	7	8	2	0	3	4	1	0

1	2	3	4	5	6	7	8	9	10	11	12	13	14	15
0	0	5	6	0	0	0	7	8	2	3	4	1	0	0

(b) Insert operation for random numbers $i = 11$ and $j = 15$.

1	2	3	**4**	5	6	**7**	8	9	10	11	12	13	14	15
0	0	5	6	0	0	0	7	8	2	0	3	4	1	0

1	2	3	4	5	6	7	8	9	10	11	12	13	14	15
0	0	5	0	6	0	0	7	8	2	0	3	4	1	0

(c) Insert operation for random numbers $i = 7$ and $j = 4$.

Fig. 4 Illustrations of insert operation for configuration change

out and if the r is between $0.67 \leq r < 1$, reversion operation is carried out for neighbourhood configuration creation. After generating the neighbourhood solution by using randomly selected operator, the algorithm checks the feasibility of solution; if the solution is infeasible then the randomly selected operation is repeated on the string until gets a feasible solution. The operation of these operators is explained below.

Insert Operation: In this operation, two random numbers i and j indicating the positions of element in a string, are generated between 1 and length of string. Figures 4a, b and c illustrate how insert operation changes the number of facilities within the bay, merges the bays, and creates the new bays for the encoded string-1 shown in Fig. 2a.

The insert operation removes the element in the position i of the string and then moves certain elements either leftward or rightward depending on values of i and j and then insert the removed element into the position j. If $i < j$, the element in the position i is removed and the elements from position $i + 1$ to j are moved one position leftward, then the removed element is inserted into position j. If $i > j$, the element in position i is removed and the elements from position j to (i-1) are moved one position rightward, then the removed element is inserted into position j.

Swap Operation: In this operation, two random numbers i and j indicating the positions of element in a string, are generated between 1 and length of string. The swap operation swaps the element in positions i and j of the string. Application of swap operation is shown in Fig. 5. Note that if the element of positions i and j are bay breaks '0' then the swap operation does not generate different neighbourhood solution from the current solution; in such a case new i and j are generated until the elements in i and j positions are not bay breaks '0'.

Before

1	**2**	3	4	5	6	7	**8**	9	10	11	12	13	14	15
0	0	5	6	0	0	0	7	8	2	0	3	4	1	0

After

1	2	3	4	5	6	7	8	9	10	11	12	13	14	15
0	7	5	6	0	0	0	0	8	2	0	3	4	1	0

Fig. 5 Illustrations of swap operation for random numbers $i = 8$ and $j = 2$

Before

1	2	3	**4**	5	6	7	8	9	10	11	**12**	13	14	15
0	0	5	6	0	0	0	7	8	2	0	3	4	1	0

After

1	2	3	4	5	6	7	8	9	10	11	12	13	14	15
0	0	5	3	0	2	8	7	0	0	0	6	4	1	0

Fig. 6 Illustrations of reversion operation for random numbers $i = 4$ and $j = 12$

Reversion Operation: In this operation, two random numbers i and j indicating the positions of element in a string, are generated between 1 and length of string. This operation reverses all the elements located from the position i to the position j. In this configuration change, the reverse operation sequentially removes the elements from positions i to j in current solution and places these elements sequentially into positions j to i in reverse manner to create neighbourhood solution. Application of reversion operation is shown in Fig. 6.

4.3 SA-FBS Parameters Settings and Flow Chart

The flow chart of the proposed SA-FBS heuristic method for UA-FLP with FBS is shown in Fig. 7 and the parameter settings for the proposed SA-FBS method are explained in the further section.

Perturbation methods: Three operators explained in Section 4.2 are used for generating neighbourhood solution from the current solution. The operators are used randomly with an equal probability of selection of an operator.

Starting Temperature (T_0): The starting temperature is set high such that virtually all configuration changes are accepted. In the proposed SA-FBS, the starting temperature is computed based on the initial configuration changes acceptance probability of 95%. Hence, the configuration change acceptance probability (P_c) is set to 0.95 in the present study. The initial temperature depends on problem size, flow volume between facilities, area of facilities and aspect ratio.

Annealing: Common functions used, in the literature for computing the temperature at state *i*, are: Arithmetic function, Geometric function, Logarithmic function and Inverse function. The proposed SA-FBS algorithm uses the geometric function which is given by: $T_{i+1} = \gamma \times T_i$, where $i = 0, 1, \ldots$, and γ = Constant < 1. To allow the system to cool slowly, the value of 'γ' (cooling rate) is taken as 0.98 in the present study.

Epoch length (L): The number of feasible configuration changes at each temperature level is done by the expression; $L = aN^2$, where, 'a' is a constant number and 'N' is the number of facilities. In the present SA-FBS, the value of 'a' is set to 1.

Final temperature (T_F): In the proposed SA-FBS heuristic, the termination temperature is decided by the probability of accepting worse configuration change (P_f). The probability of accepting a worse solution at the termination of SA should be very low. In the present study, it is taken as $P_f = 1 \times 10^{-15}$. When the values of P_c, γ and P_f are known then the total number of temperatures decreasing cycles (CT) can be found using the expression given in [4]. The expression is:

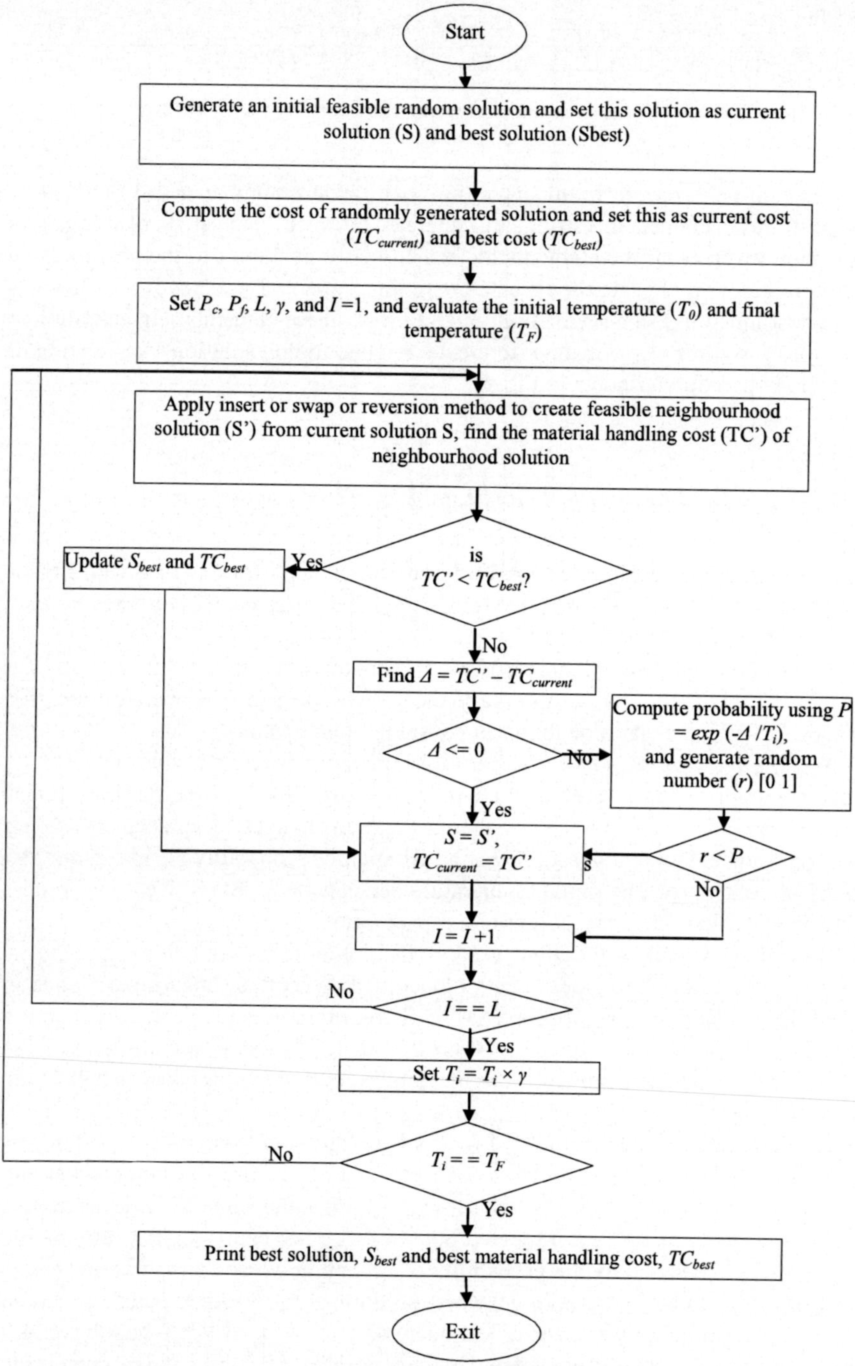

Fig. 7 Proposed SA-FBS flow chart for **UA-FLP** with **FBS**

$\gamma = (log\ (P_c)/log\ (P_f))^{1/(CT-1)}$; and the terminating temperature is computed using the expression, $T_F = T_0 \times (\gamma)^{CT}$.

Neighbourhood solution acceptance criteria: The neighbourhood solution generated is accepted using Metropolis criterion. This criterion has two ways of accepting the generated neighbourhood solution. These two ways are:

i. If the generated neighbourhood solution results in a net decrease in the objective function value, then it is accepted.

ii. If the generated neighbourhood results in a net increase in the objective function value, even then there is a chance of accepting the generated solution but this is with a probability of $exp\ (-\Delta/T_i)$, where 'Δ' is the difference in the objective function values, and T_i is the temperature at the corresponding state. In the proposed SA-FBS, the generated neighbourhood solution is accepted if a random number generated between 0 and 1 is less than $exp\ (-\Delta/T_i)$ otherwise the configuration change is rejected.

5 Numerical Experiments and Results Analysis

The performance of the LINGO solver-based method and the proposed Simulated Annealing for Flexible Bay Structure (SA-FBS) based method is tested using the problems reported in the literature. The MILP model of UA-FLP with FBS representation is coded in LINGO 16.0 × 64 software package. The SA-FBS algorithm is coded in MATLAB R2015b. For SA-FBS procedure, the encoded solution string length is, usually, (2 N-1), where N is the number of facilities. All the experiments are conducted on 64-bit PC with Windows 8.1Pro operating system having 3.2 GHz Intel(R) Core (TM) i5-4570 CPU and 8 GB RAM.

Initially, equal area static facility layout problem instants are tested to judge the performance of proposed methods. In the literature, equal area-FLPs are studied with discrete space QAP model. The UA-FLP with FBS is a special case of equal area discrete space QAP formulation when area and aspect ratio of all facilities are set to one. The data set of [3] is taken for equal area static facility layout problem. This data set consists of five period's data. These five-period data are taken as five independent equal area static layout problems and are named as Y9-1, Y9-2, Y9-3, Y9-4 and Y9-5 respectively for each period data. The input data consist of products demand in various periods and their operational sequence. These input data information are provided in Tables 1 and 2. In order to use the equal area data in continuous space MILP model of UA-FLP with FBS, we assumed unit area ($A = 1$) for each facility and minimum side length of each facility as one ($S^{\min} = 1$), and the plant floor size of 3 × 3 is considered (i.e., $W = 3$ and $H = 3$).

Further, the standard UA-FLPs available in the literature are used to test the performance of proposed SA-FBS. Table 3 gives the list of unequal area problems studied in literature either using MIP approach or slicing tree structure or flexible bay structure or relaxed flexible bay structure representation. The problem sizes

Table 1 Demand profiles in Yaman et al.'s [3] case

Part	Periods				
	(Y9-1)	(Y9-2)	(Y9-3)	(Y9-4)	(Y9-5)
1	10	35	90	40	55
2	30	50	25	65	20
3	45	15	40	70	15
4	70	80	55	90	85
5	85	60	70	20	30

Table 2 Operational sequence of products in Yaman et al.'s [3] case

Part	Machine operational sequence
1	01 → 03 → 05 → 07 → 02 → 07 → 09
2	01 → 04 → 02 → 05 → 06 → 08 → 09
3	01 → 05 → 07 → 08 → 05 → 06 → 02 → 09
4	01 → 02 → 04 → 06 → 07 → 08 → 02 → 03 → 09
5	01 → 07 → 06 → 04 → 02 → 08 → 03 → 05 → 06 → 09

vary from 7 to 35 facilities. Hence all these problems are taken as good benchmark problems for present study. The original input data of all these test problems can be found in the corresponding references. In the proposed study, all the problems are solved without any empty space inside the bays of layout and also no dummy departments are considered for FBS layout formation. In these problems, if the given plant floor area is more than the sum of areas of all facilities then the extra space is pushed towards the east side of the plant. Note that 'N' is the number of facilities, α is the maximum aspect ratio constraint, 'S^{min}' is the minimum side length constraint, 'W' is plant length in the x-direction and 'H' is plant length in the y-direction.

5.1 Results and Discussion of Equal Area Facility Layout Problems

In this section, the performance of proposed LINGO coded MILP model, and SA-FBS heuristic method of layout formation for UA-FLPs with FBS representation in continuous space are tested on equal area static FLP instants. In literature, these problems are studied with discrete space QAP model [6, 31, 32]. Table 4 shows the solution values, best solutions and CPU timings obtained by proposed LINGO and SA-FBS methodologies to UA-FLPs with FBS representation. LINGO coded method has given the global optimum solution values for all these five equal area problems. SA-FBS method is also given the optimal solutions in very less computational times compared to LINGO solver. In this table, best layout for each problem is given under

Table 3 Summary of unequal area facility layout problem data sets

Sl. No	Problem name	N	Shape constraint	Floor dimensions		Data references
				W	H	
1	O7	7	$\alpha = 4$	8.54	13	Meller et al. [11], and Komarudin and Wong [15]
2	O8	8	$\alpha = 4$	11.31	13	Meller et al. [11], and Komarudin and Wong [15]
3	O9	9	$\alpha = 4$	12	13	Meller et al. [11], and Komarudin and Wong [15]
4	VC10Rs	10	$S^{min} = 5$	51	25	Van Camp et al. [27]
5	VC10Ra	10	$\alpha = 5$	51	25	Van Camp et al. [27]
6	M11a	11	$\alpha = 5$	6	6	Meller [28], and Liu and Meller [13]
7	M11s	11	$S^{min} = 1$	6	6	Meller [28], and Liu and Meller [13]
8	M15a	15	$\alpha = 5$	15	15	Meller [28], and Liu and Meller [13]
9	M15s	15	$S^{min} = 1$	15	15	Meller [28], and Liu and Meller [13]
10	BA12	12	$S^{min} = 1$	10	6	Bazaraa [29]
11	BA14	14	$S^{min} = 1$	9	7	Bazaraa [29]
12	NUG12	12	$\alpha = 4$	4	3	Nugent et al. [30]
13	NUG15	15	$\alpha = 4$	5	3	Nugent et al. [30]
14	AB20	20	$\alpha = 4$	30	20	Armour and Buffa [7]
15	TAM20	20	$\alpha = 5$	40	35	Tam [24], and Gau and Meller [22]
16	TAM30	30	$\alpha = 5$	45	40	Tam [24], and Gau and Meller [22]
17	SC30	30	$\alpha = 5$	15	12	Liu and Meller [13]
18	SC35	30	$\alpha = 4$	16	15	Liu and Meller [13]

Table 4 LINGO and SA-FBS results for static equal area facility layout problems

Problem name	LINGO		SA-FBS		Best solution
	Solution value[a]	Run time (secs)	Solution value	CPU time (secs)	
Y9-1	2780	1067	2780	4.72	4-6-9|2-7-8|1-5-3
Y9-2	2640	2012	2640	4.87	9-8-3|6-7-5|4-2-1
Y9-3	2950	780	2950	4.87	3-5-1|8-7-2|9-6-4
Y9-4	3020	1423	3020	4.66	6-4-1|5-2-3|7-8-9
Y9-5	2200	2203	2200	4.90	3-8-9|5-7-6|1-2-4

[a]LINGO solver has given global optimum solutions to all above problems

Table 5 Comparison of proposed LINGO and SA-FBS results with the solution values reported in the literature for equal area facility layout problems solved using discrete space QAP

Problem name	Pillai et al. [6]	Chan et al. [31]	Tang and Abdel –Malek [32]	Yaman et al. [3] spiral-1	Yaman et al. [3] spiral-2	Best Known	Proposed SA-FBS/LINGO
Y9-1	2780	2850	2820	3630	3470	2780	2780
Y9-2	2640	2790	2980	3180	3350	2640	2640
Y9-3	2950	3160	3200	3690	3570	2950	2950
Y9-4	3020	3165	3100	3975	4065	3020	3020
Y9-5	2200	22,750	2355	3045	2975	2200	2200

best solution column. The best solution is encoded such that bays are separated by symbol (|) and the order of facilities within the bays are given from bottom to top.

Researchers [6, 31, 32] used the Yaman et al.'s [3] data for their approaches. All these authors used the discrete space QAP model and solved them with various exact and heuristic methods. Table 5 shows the comparison of material handling costs of proposed continuous space MILP model coded in LINGO and SA-FBS with the other approaches used in the literature to solve discrete space QAP model. Solution values obtained by the proposed methods are same as the best solution values published in equal area layout literature using discrete space.

5.2 *Results and Discussion of Unequal Area Facility Layout Problems Data Set*

The unequal area facility layout problems given in Table 3 are solved by LINGO solver. It has given solution values near to best-known solutions in literature for small size problems. In the case of larger size problems, solutions are very poor and the solver is interrupted after running a few minutes to hours for different problem sizes.

The poor results are due to a large number of binary integer decision variables in the model for large size problems and the solver uses the branch and bound exact method to solve the model. Hence, it is difficult to obtain better results for bigger size problems using LINGO solver. For SC35 problem, LINGO solver did not generate the feasible solution in a reasonable time, hence result is not reported for this problem. The solution values recorded for various problem sizes are given in Table 6.

To overcome the inferior solution by LINGO solver, SA meta-heuristic is used, and all the unequal area problems given in Table 3 are tested using proposed SA-FBS algorithm coded in MATLAB. The SA-FBS is initialized with the random feasible solution, and 10 runs are made for each problem. The number of iterations at each temperature level for 10-replications is set as N^2, where N is number of facilities. The results of proposed SA-FBS heuristic method after running 10-replications for various size problems are recorded in Tables 7 and 8.

Table 7 gives the best, worst, average solution values found and also gives the minimum, maximum and average CPU timings of the algorithm. Table 7 indicates that the variability of solution values is less and the percentage difference between the best and worst solutions is relatively very low and the solution procedure has very less computational timings. Therefore, the proposed SA-FBS heuristic method for UA-FLPs with flexible bay system layout formation is robust one. Table 8 gives the best solution in 10-run of SA-FBS in encoded form for different size problems.

6 Comparison of SA-FBS with the Other Approaches results to UA-FLP Based on FBS or RFBS

In this section, the performance of SA-FBS heuristic approach is compared with other approaches reported in the literature for solving UA-FLPs with FBS as well as with RFBS. Table 9 summarizes the comparison.

Comparisons are made with the GA of [16], MILP approach of Konak et al. [1], ASA of Wong and Komarudin [17], ACO of Kulturel-Konak and Konak [21], PSO of Kulturel-Konak and Konak [18] and AIS of Ulutas and Kulturel-Konak [19], and GA of Mazinani et al. [23]. To have consistency with these approaches, vertical bays are used in the present study. The proposed SA-FBS heuristic has given either better Material Handling Cost (MHC) than the one reported in the literature or the same as reported in the literature for UA-FLPs with FBS. The proposed SA-FBS algorithm has found the new best FBS solution for the problems O7, VC10Rs, VC10Ra, M11a, M15a, M15s and AB20. The SA-FBS best found solution values as compared with best-known solution values in the literature are shown by bold numbers in Table 9. The MILP approach of Konak et al. [1] has given optimum solution for problem O9 optimally and our SA-FBS algorithm also gives the same optimal solution. AB20 problem is one of the most studied test problems in the literature by many researchers. In the present study, a new best solution is found for this problem as compared to the best-known solution available in the literature. The RFBS based approaches are less

Table 6 LINGO solver results for UA-FLP with FBS

Problem name	O7	O8	O9	VC10Rs	VC10Ra	M11a	M11s	M15a	M15s
LINGO solution value	136.15	251.37	251.74	30,410.7	25,612.54	1518.67	1360	48,274.7	50,501
problem name	BA12	BA14	NUG12	NUG15	AB20 (4)	TAM20	TAM30	SC30	SC35
LINGO solution value	10,301.5	5561.58	292	677.458	5312.46	13,692.3	28,127.05	9520.51	–

Table 7 SA-FBS 10-run solution values and cpu timings for UA-FLP with FBS

Sl. No	Problem name	10-run solution value			10-runs CPU time (s)		
		Best	Worst	Mean	Minimum	Maximum	Average
1	O7	134.19	134.19	134.19	1.79	2.52	2.09
2	O8	245.51	247.85	245.62	3.24	3.92	3.58
3	O9	241.06	250.89	244.65	2.82	3.9	3.5
4	VC10Ra	21,456.83	21,662.27	21,465.11	4.1	5.11	4.5
5	VC10Rs	22,897.65	24,050.57	23,721.17	4.72	6.06	5.16
6	M11a	1197.76	1220.32	1205.14	6.86	7.94	7.35
7	M11s	1317.79	1409.83	1350.19	9.89	12.33	11.05
8	M15a	27,350.77	30,897.17	29,357.84	14.78	18.64	16.88
9	M15s	22,757.18	26,303.22	23,640.34	9.49	10.9	10.33
10	BA12	8786	9102.67	8801.83	9.43	11.85	10.59
11	BA14	5004.55	5042.29	5016.58	12.47	15.55	13.86
12	NUG12	262	276.5	267.85	7.28	8.82	8.15
13	NUG15	524.75	562.75	534.17	14.86	19.39	16.53
14	AB20	5178.2	5487.79	5293	41.44	73.71	52.34
15	TAM20	9003.82	9350.61	9144.95	44.88	84.9	59.58
16	TAM30	19,667.46	20,293.28	19,916.92	96.94	158.55	120.11
17	SC30	3731.92	4737.2	4040.04	99.56	187.6	125.08
18	SC35	4182.4	5025.78	4499.42	313.03	13,884.35	2983.88

constrained than FBS. Therefore, a solution found by RFBS approach is expected to have a better solution than the corresponding FBS solution. For VC10Ra, BA12, BA14, NUG12, TAM20, TAM30, SC30 and SC35 problems, the SA-FBS algorithm MHC is inferior to RFBS based MHC due to more flexibility in the availability of space for arranging the facilities on the plant floor. The percentage inferiority of MHC value for these problems are provided in the last but one column of Table 9. In fact, the proposed SA-FBS algorithm has found a new best solution for VC10Rs, M11a, M15a, M15s and AB20 problems as compared to the previously best-known solution with RFBS. In summary, the computational performance of proposed SA-FBS algorithm is excellent compared with other meta-heuristic approaches reported in the literature to solve UA-FLPs with FBS.

7 SA-FBS Results Analysis for Armour and Buffa's [7] AB20 Problem

Armour and Buffa [7] 20-facility AB20 problem is still considered as a fairly large problem in UA-FLPs. It is one of the most studied test problems in the literature, and

Table 8 Best solutions found by proposed SA-FBS

<table>
<tr><th>Sl. No</th><th>Problem name</th><th>Best solution of 10-run SA-FBS</th></tr>
<tr><td>1</td><td>O7</td><td>1-4-6-2|3-5-7</td></tr>
<tr><td>2</td><td>O8</td><td>2-1-4-7|5-8-6-3</td></tr>
<tr><td>3</td><td>O9</td><td>7-8|4-1-2|3-6-9-5</td></tr>
<tr><td>4</td><td>VC10Ra</td><td>3-5-8-10|9-4-7|2-6|1</td></tr>
<tr><td>5</td><td>VC10Rs</td><td>5-3|8-10-9|4-2|7-6|1</td></tr>
<tr><td>6</td><td>M11a</td><td>10-9-2|11-1-5-6-3|8-7-4</td></tr>
<tr><td>7</td><td>M11s</td><td>3-6-1-11|2-5-9-10|4-7-8</td></tr>
<tr><td>8</td><td>M15a</td><td>3-4-10-12-14-5|13-9-15|7-11-8-6-1-2</td></tr>
<tr><td>9</td><td>M15s</td><td>3-4-10-9-12-14-5|15|1|11-13-2|7|8|6</td></tr>
<tr><td>10</td><td>BA12</td><td>4|9-5-10|12-8-11|3|2|1|7-6</td></tr>
<tr><td>11</td><td>BA14</td><td>7-8-14|4|2|9-6-12-13|3-1|11-5-10</td></tr>
<tr><td>12</td><td>NUG12</td><td>8-4-7-11-9-12|1-5-6-10-2-3</td></tr>
<tr><td>13</td><td>NUG15</td><td>10-15-6-5-12|1-4-3-2-14-13-7-8-9-11</td></tr>
<tr><td>14</td><td>AB20</td><td>11-16|15-13-17|5-14-10-9-12|3-19|20-8-7-2-4-6|1-18</td></tr>
<tr><td>15</td><td>TAM20</td><td>19-9|2-14-10|15-12-18-3|4-8-11-7-16-13-6-1|20-17-5</td></tr>
<tr><td>16</td><td>TAM30</td><td>5-9|2-10|29-23|19-25-26-17|21-30-16-28-7-13-24-12-6|4-3-14-27-8-11-18-22-15|20-1</td></tr>
<tr><td>17</td><td>SC30</td><td>7-8-9-10-13-16|11-12-17|5-15-14|6-18-26|2-4-23-28|21-3-22|19-25-29-24|20-1-30-27</td></tr>
<tr><td>18</td><td>SC35</td><td>14-31-5-11-6|13-32-12-9-8-7-2|15-10-21-17-19|18-28-1-25-30-20|4-29|35-3-27-33-23|16-26-24-34-22</td></tr>
</table>

Table 9 Comparison of SA-FBS results with the results of other approaches reported in the literature for UA-FLP with FBS/RFBS

Problem name	Tate and Smith [16]	Konak et al. [1]	Wong and Komarudin [17]		Kulturel-Konak and Konak [21]		Kulturel-Konak and Konak [18]	Ulutas and Kulturel-Konak [19]	Mazinani et al. [23]	Best-RFBS known solution	Best-FBS known solution	Proposed SA-FBS best	Imp (%) over best-RFBS known	Imp (%) over best-FBS known
	FBS	FBS	FBS	RFBS	FBS	RFBS	RFBS	RFBS	FBS					
O7	–	134.63	–	–	–	–	–	–	–	–	**134.63**	**134.19**	–	**0.33**
O9	–	241.06	–	–	–	–	–	–	–	–	241.06	241.06	–	0
VC10Rs	23,671	22,899.7	22,899.7	–	22,899.7	–	22,899.7	22,899.7	22,899.7	22,899.65	**22,899.7**	**22,897.7**	0.01	**0.01**
VC10Ra	–	21,463.1	21,463.1	–	21,463.1	–	20,142.1	21,463.1	21,463.1	20,142.13	**21,463.1**	**21,456.8**	– 6.53	**0.03**
BA12	8861	8801.33	8786	8587.04	8801.33	8083	8587.04	8587.04	8801.33	8083	8786	8786	– 8.70	0
BA14	5080.1	5004.55	5004.55	4927.69	5004.55	4739.94	4843.37	4927.69	5004.55	4739.94	5004.55	5004.55	– 5.58	0
M11a	–	1225	1204.15	–	–	–	1201.12	–	–	1201.12	**1204.15**	**1197.76**	0.28	**0.53**
M11s	–	1317.79	1317.79	–	–	–	1317.79	–	–	1317.79	1317.79	1317.79	0	0
M15a	–	31,779.1	27,545.3	–	–	–	27,545.3	–	–	27,545.27	**27,545.3**	**27,350.8**	0.71	**0.71**
M15s	–	27,782	23,197.8	–	–	–	23,197.8	–	–	23,197.8	**23,197.8**	**22,757.2**	1.94	**1.94**
NUG12	–	265.5	262.003	–	262	–	257.5	–	262	257.5	262	262	– 1.75	0
NUG15	–	526.75	536.75	–	524.75	–	524.75	–	524.75	524.75	524.75	524.75	0	0
AB20	5743	–	–	–	5360.8	–	5336.36	5360.8	5183.52	5336.36	**5183.52**	**5178.2**	3.05	**0.1**
TAM20	–	–	–	–	–	–	8753.57	9003.82	–	8753.57	–	9003.82	– 2.86	–
TAM30	–	–	–	–	–	–	19,462.41	19,667.45	–	19,462.41	–	19,667.46	– 1.05	–
SC30	–	–	–	3679.85	–	3562.9	3443.34	–	–	3443.34	–	3731.92	– 8.38	–
SC35	–	–	–	3962.72	–	3750.74	3700.75	–	–	3700.75	–	4182.4	– 13.01	–

Imp (%) =100 × (the best-known MHC value - proposed SA-FBS MHC value)/minimum of (best known MHC value, proposed SA-FBS MHC value)

is a good benchmark problem to study the performance of the proposed SA-FBS. In this section, the AB20 problem is solved for different maximum aspect ratios (α^{max}) ranging from 3 to 1000 with 10-replications. Table 10 gives the best, worst, average solution values found and also gives the minimum, maximum and average CPU timings of the algorithm. Table 11 gives the SA-FBS 10-run best solution in encoded form for different maximum aspect ratios.

Table 12 gives the comparison of proposed SA-FBS best solution values with other researchers' results reported in literature for the AB20 problem with different aspect ratios. The comparisons are made with the GA of [19], the hybrid fuzzy model and genetic search of [33], the ACO of [21], the AIS of [19] and GA of [23]. These authors solved the AB20 problem for different maximum aspect ratios ranging from 1.70667 to 1000. Except for Ulutas and Kulturel-Konak [19] all other authors solved AB20 problem with the FBS representation, whereas Ulutas and Kulturel-Konak [19] solved with RFBS representation. For most cases of AB20, the proposed SA-FBS solutions are almost equal to earlier best-known solutions (i.e., only less by 0.04% to 0.06%). In fact, for the cases, $\alpha^{max} = 25$, $\alpha^{max} = 5$ and $\alpha^{max} = 4$, the new

Tabe 10 SA-FBS solution values and CPU timings for AB20 problem with different α^{max} values

α^{max}	10-run solution values			10-runs CPU timings		
	Best	Worst	Mean	Minimum	Maximum	Average
1000	1588.49	3287.33	2051.11	16.27	18.60	17.43
50	2382.74	3650.57	2775.5	16.10	19.11	17.82
25	3324.49	3552.92	3432.52	17.54	19.98	18.68
15	4045.58	4404.34	4101.19	21.26	24.46	23.09
10	4367.57	4833.81	4510.07	25.26	27.99	26.37
7	4720.36	5017.39	4860.29	28.96	33.39	30.79
5	5132.34	5564.80	5295.28	37.18	44.34	40.14
4	5178.2	5487.79	5293	41.44	73.7	52.34
3	5372.6	5663.25	5465.63	61.44	720	282.24

Table 11 Best solutions found for AB20 problem with different α^{max} values

<table>
<tr><th>α^{max}</th><th>Best solution of 20-runs SA-FBS</th></tr>
<tr><td>1000</td><td>1-18-5-20-8-7-6-2-4-19-3-10-14-9-15-12-17-13-16-11</td></tr>
<tr><td>50</td><td>18|1|5|20|8|7|6|2|4|19|3|10|14|9|15|12|17|13|16|11</td></tr>
<tr><td>25</td><td>11|16|13|17|12|15|10|14|3|19|4|2|1-5-9-7-8-20|6|18</td></tr>
<tr><td>15</td><td>11|16|17|12|15|13-3-14-10-9-19-4-2-7-8-20|1-6-5|18</td></tr>
<tr><td>10</td><td>11|16|17|12|15-13-14-10-9-19-3-1|5-20-8-7-2-4-6-18</td></tr>
<tr><td>7</td><td>11|16|17|13-12|15-14-10-9-3-1|5-19|20-8-7-6-2-4-18</td></tr>
<tr><td>5</td><td>11-13-16|15-14-9-17|5-10-12|3-19|20-8-7-2-4-6|1-18</td></tr>
<tr><td>4</td><td>11-16|15-13-17|5-14-10-9-12|3-19|20-8-7-2-4-6|1-18</td></tr>
<tr><td>3</td><td>20-18|6-8-7-4-2-1|5-19-3|12-9-10-14|17-13-15|16-11</td></tr>
</table>

Table 12 Comparison of SA-FBS results with the results of other approaches reported in the literature for AB20 problem

α^{max}	Tate and Smith [16]	Enea et al. [33]	Kulturel-Konak and Konak [21]	Ulutas and Kulturel-Konak [19]	Mazinani et al. [23]	Best-Known	Best known reference	Proposed SA-FBS best	% diff
1000	1638.5	1638.5	1638.5	1638.5	1587.91	1587.91	Mazinani et al. [23]	1588.49	−0.04
50	3009.5	3009.5	2706.5	2457.79	2381.86	2381.86		2382.74	−0.04
25	3535.1	3535.1	3526.5	3399.16	3391.95	**3391.95**		**3324.49**	**1.99**
15	4296.1	4140.5	4119.8	4119.8	4043.91	4043.91		4045.58	−0.04
10	4633.3	4440.7	4440.7	4440.7	4364.74	4364.74		4367.57	−0.06
7	5255	4793.5	4793.5	4793.5	4717.53	4717.53		4720.36	−0.06
5	5524.7	5397.6	5297.6	5297.6	5183.52	**5183.52**		**5132.34**	**0.99**
4	5743.1	5370.9	5360.8	5360.8	5183.52	**5183.52**		**5178.2**	**0.1**
3	5832.6	5594.3	5594.3	5594.3	5369.3	5369.3		5372.6	−0.06

best solutions are found and these solutions improved the earlier best-known values by 1.99%, 0.99% and 0.1% respectively. The SA-FBS best found solution values are shown by bold numbers in Table 12. The problem AB20 with $\alpha^{max} = 4$ is one of the most studied problem in the literature by many researchers. In the present study, a new best solution is found for AB20 with $\alpha^{max} = 4$ problem as compared to best-known solution reported in the literature.

7.1 SA-FBS Computational Effort Discussions

In this section, computational efficiency of proposed SA-FBS is compared with various meta-heuristic algorithms used by earlier researchers to solve the UA-FLPs with FBS. The CPU timings are not directly comparable due to different hardware used by various researchers for testing their algorithms. But the comparisons are made here with the description of hardware facilities used by various researchers.

The hardware used for the proposed SA-FBS algorithm is Windows PC with Intel(R) core(TM) i5-4570, 3.2 GHz, 8 GB RAM. The ASA of [17], ACO of [21] and PSO of [18] are tested using special high-performance computing hardware workstation. Authors [18, 21] in their published paper acknowledged that, All the experiments were conducted in High-Performance Computing Group facility available at the Pennsylvania State University. AIS of [19] is tested using Windows PC with 2.6 GHz Intel Xeon Processor and 2 GB RAM. In Table 13, it can be noticed that in spite of, basic level hardware used for testing the proposed SA-FBS algorithm, the CPU timings are very much promising in most of the test problem cases. Particularly, proposed SA-FBS CPU timings are better as compasred to Kulturel-Konak and Konak [21], and Kulturel-Konak and Konak [18] who used the professional high-performance Intel Xeon processor with 32 GB RAM workstation for all their experiments. Ulutas and Kulturel-Konak [19] also used the professional high-performance Intel Xeon processor with a RAM of 2 GB.

8 Conclusions

This paper has proposed a SA-FBS algorithm to solve unequal area facility layout problems with flexible bay structure. Also, an MILP model of UA-FLP with FBS is coded in LINGO mathematical modelling software and solved by LINGO solver. The performance of SA-FBS algorithm and LINGO coding are tested with the problems reported in the literature. For small size equal area facility layout problems, both LINGO and SA-FBS methods have given the same solutions, but for UA-FLPs with FBS, SA-FBS has given better solution than LINGO solver. The performance of SA-FBS algorithm is robust and could produce best FBS solutions. Proposed SA-FBS method has either given the same solution or even better solution for the test problems as compared to the best-known solution reported in the literature based on

Table 13 Comparison of SA-FBS CPU timings with the other approaches' CPU timings reported in the literature for UA-FLP with FBS/RFBS cases

Problem name	Average CPU times by other meta-heuristics in (s)				Best known time in (s)	Proposed SA-FBS average CPU time in (s)	Difference from best-known in (s)
	Wong and Komarudin [17]	Kulturel-Konak and Konak [21]	Kulturel-Konak and Konak [18]	Ulutas and Kulturel-Konak [19]			
O9	46.5				46.5	3.5	43
VC10Rs	61.1	43	3	2.1	2.1	4.5	−2.4
VC10Ra	60.6	34	2	1.5	1.5	5.16	−3.66
BA12	179.3	67	10	72.93	10	10.59	−0.59
BA14	357.9	108	19	56.56	19	13.86	−5.14
M11a	86.8	80	4		4	7.35	−3.35
M11s	79.3		3		3	11.05	−8.05
M15a	282	113	18		18	16.88	1.12
M15s	309.2		17		17	10.33	6.67
NUG12	126.7	36	5		5	8.15	−3.15
NUG15	352.4	61	14		14	16.53	−2.53
AB20(4)	1555		85		85	52.34	32.66
TAM20		226	104	200.18	104	59.58	44.42
TAM30		623	924	764.52	623	120.11	502.89
SC30	15,785.5	902	873		873	125.08	747.92
SC35	29,165.2	1185	1842		1185	2983.88	−1798.88

FBS. Regarding computational effort, though the proposed SA-FBS used the basic hardware facility for testing the algorithm, it has given the superior CPU timings than the other meta-heuristics tested on professional very high-performance hardware workstations. Finally, the research can be extended in future to consider dynamic environment UA-FLPs with FBS.

Acknowledgements Authors would like to thank Dr. Suresh Kumar M R, Assistant Engineer Public Works Department, Kerala State Government, India, for his guidance and help for coding the Mixed Integer Linear Programming model of the UA-FLP with FBS in LINGO software.

References

1. Konak A, Kulturel-Konak S, Norman BA, Smith AE (2006) A new mixed integer programming formulation for facility layout design using flexible bays. Opers Res Letrsss 34:660–672
2. Koopmans TC, Beckmann M (1957) Assignment problems and the location of economic activities. Econometrica 25(1):53–76
3. Yaman A, Gethin DT, Clarke MJ (1993) An effective sorting method for facility layout construction. Int J Prod Res 31(2):413–427
4. Baykasoglu A, Gindy N (2001) A simulated annealing algorithm for dynamic layout problem. Comp and Opers Res 28(14):1403–1426
5. McKendall AR Jr, Shang J, Kuppusamy S (2006) Simulated annealing heuristics for the dynamic facility layout problem. Comp and Opers Res 33:2431–2444
6. Pillai VM, Hunagund IB, Krishnan KK (2011) Design of robust layout for Dynamic Plant Layout Problems. Comp & Ind Engg 61:813–823
7. Armour GC, Buffa ES (1963) A heuristic algorithm and simulation approach to relative location of facilities. Mgm Sci 9:294–309
8. Islier AA (1998) A genetic algorithm approach for multiple criteria facility layout design. Int J Prod Res 36(6):1549–1569
9. Meei-Yuh K, Michael HH, Ming-Jaan W (2011) Simulated annealing based parallel genetic algorithm for facility layout problem. Int J Prod Res 49(6):1801–1812
10. Lacksonen TA (1997) Pre-processing for static and dynamic facility layout problems. Int J Prod Res 35(4):1095–1106
11. Meller RD, Narayanan V, Vance PH (1998) Optimal facility layout design. Opers Res Letrs 23(3–5):117–127
12. Castillo I, Westerlund J, Emet S, Westerlund T (2005) Optimisation of block layout problems with unequal areas: A comparison of MILP and MINLP optimization methods. Comp and Chem Engg 30:54–69
13. Liu Q, Meller RD (2007) A sequence-pair representation and MIP-model-based heuristic for the facility layout problem with rectangular departments. IIE Trans 39:377–394
14. Scholz D, Petrick A, Domschke W (2009) STaTS: a slicing tree and tabu search based heuristic for the unequal area facility layout problem. Euro J Oper Res 197:166–178
15. Wong KKY (2010) Applying ant system for solving unequal area facility layout problems. Euro J Oper Res 202:730–746
16. Tate DM, Smith AE (1995) Unequal-area facility layout by genetic search. lIE Trans 27:465–472
17. Wong KKY (2010) Solving facility layout problems with flexible bay representation and ant system algorithm. Exp Sys Appls 38:5523–5527
18. Kulturel-Konak S, Konak A (2011) A new relaxed flexible bay structure representation and particle swarm optimization for the unequal area facility layout problem. Engg Opt 43(12):1263–1287

19. Ulutas H, Kulturel-Konak S (2012) An artificial immune system based algorithm to solve unequal area facility layout problem. Exp Sys Appls 39:5384–5395
20. Hunagund IB, Pillai VM, Kempaiah UN (2018) A simulated annealing algorithm for unequal area dynamic facility layout problems with flexible bay structure. Int J Ind Engg Comput 9:307–330
21. Kulturel-Konak S, Konak A (2010) Unequal area flexible bay facility layout using ant colony optimisation. Int J Prod Res 39(7):1877–1902
22. Gau KY, Meller RD (1999) An iterative facility layout algorithm. Int J Prod Res 37(16):3739–3758
23. Mazinani M, Abedzadeh M (2012) Dynamic facility layout problem based on flexible bay structure and solving by genetic algorithm. Int J Adv Manufr Tech. https://doi.org/10.1007/s00170-012-4229-6
24. Tam KY (1992) A simulated annealing algorithm for allocating space to manufacturing cells. Int J Prod Res 30(1):63–87
25. Kulturel-Konak S, Konak A (2015) A large-scale hybrid simulated annealing algorithm for cyclic facility layout problem. Engg Opt 47(7):963–978
26. Ripon KSN, Glette K, Khan KN, Hovin M, Torresen J (2013) Adaptive variable neighbourhood search for solving multi-objective facility layout problems with unequal area facilities. Swarm and Evolu Compu 8:1–12
27. Van Camp DJ, Carter MW, Vanelli A (1992) A nonlinear optimization approach for solving facility layout problems. Euro J Oper Res 57(2):174–189
28. Meller RD (1992) Layout algorithms for single and multiple floor facilities. *Dissert. (PhD).* University of Michigan
29. Bazaraa MS (1975) Computerized layout design: a branch and bound approach. AIIE Trans 7(4):549–560
30. Nugent CE, Vollman TE, Ruml J (1968) An experimental comparison of techniques for the assignment of facilities to locations. Oper Res 16(1):150–173
31. Chan WM, Chan CY, Ip WH (2002) A heuristic algorithm for machine assignment in cellular layout. Comp & Ind Engg 44:49–73
32. Tang C, Abdel-Malek LL (1996) A framework for hierarchical interactive generation of cellular layout. Int J Prod Res 34(8):2133–2162
33. Enea M, Galante G, Panascia E (2005) The facility layout problem approached using a fuzzy model and genetic search. J Intell Manuf 16(2):303–316

Mechanical Characterization of Polycarbonate-Graphene Oxide (PCG) Nanocomposite

Jaskaran Singh, Suneev Anil Bansal, and Amrinder Pal Singh

Abstract Polymer matrix nanocomposite are of great interest due to its high specific strength, low cost, and ease of processing & synthesis. Various attempts have been made to improve properties of polymer matrix by introducing nano reinforcement. In present report, effect of low cost nanosheet reinforcement, Graphene Oxide (GO), in Polycarbonate (PC) matrix was studied for mechanical properties of PC-GO (PCG) nanocomposite. PC is used in various mechanical parts and structural applications. Low cost GO was synthesized by chemical oxidation route using low cost graphite flakes. To get better dispersion of GO in PC, solution mixing method was used. First, the thin film of PCG nanocomposite was prepared by mixing sonicated GO in Tetrahydrofuran (THF) and beads of PC. After that these sheets were extruded using an injection molding machine to synthesize dog-bone sample of PCG nanocomposites. Morphological studies of samples were performed using FE-SEM machine. Dog-bone samples were characterized using micro mechanical testing machine. PCG composite was prepared for 0.05, 0.1, and 0.2 wt.% of GO reinforcement in PC matrix. As the percentage of GO reinforcement increased, both tensile strength and elastic modulus of PCG nanocomposite increased. At 0.2 wt.% of GO tensile strength and elastic modulus was increased by 57 and 13%, respectively. GO reinforcement in PC showed better mechanical performance over pure PC.

J. Singh · S. A. Bansal · A. P. Singh (✉)
Department of Mechanical Engineering, UIET Panjab University, Chandigarh 160014, India
e-mail: apsinghpu@gmail.com

J. Singh
e-mail: apsinghpu@gmail.com

S. A. Bansal
e-mail: suneev@gmail.com

S. A. Bansal
Department of Mechanical Engineering, MAIT, Maharaja Agrasen University, Baddi, HP 174103, India

© Springer Nature Singapore Pte Ltd. 2021

P. M. Pandey et al. (eds.), *Advances in Production and Industrial Engineering*, Lecture Notes in Mechanical Engineering,
https://doi.org/10.1007/978-981-15-5519-0_8

1 Introduction

Two or more materials, having different types of properties, are combined in different ways to form composites that possesses superior properties than pristine matrix materials. Matrix phase has less strength whereas the reinforcement has more strength and stiffness than matrix, giving good mechanical and physical properties to composites. Thus composite materials possess better mechanical properties along with light weight, a combination difficult to find in conventional materials like metals. Weight of the finished part produced by composite has high strength to weight ratio because of the low density of composite as compared to bulk material. Uniquely, polymer composites are widely used for mechanical applications in aerospace, automobile, marine industry, etc., due to their light weight, better strength, and ease of processing [1, 2].

Polycarbonate (PC) finds variety of application like structural parts, body, automotive parts, helmets, hydrophobic surfaces, packaging, etc. [3, 4]. Various filler like carbon fibers, carbon nanotubes, glass fibers, etc., were reported to enhance properties of PC by synthesizing PC composite materials [5–10]. Still area of carbon nanoparticles reinforcement in PC is still in development. As Graphene/Graphene oxide reinforcement reported to produce excellent results with other matrix materials, GO in PC can produce exciting results [11, 12].

Graphene, sp^2 hybridized two-dimensional (2D) carbon nanostructure, has as an atomically thin sheet structure with carbon to carbon 0.142 nm. A 2D graphene sheet weighing only 0.77 mg/m^2 is 100 times stronger than the steel sheet of same size and thickness. The single-layered graphene sheet is one of the strongest materials; its breaking strength is 42 Nm^{-1} and 1.0 TPa Young's modulus. Graphene exhibits excellent thermal conductivity (5000 Wm^{-1} K^{-1}), extremely high transmittance at 97.7% over wide range of wavelengths and very large surface area of 2630 m^2/g [13]. Graphene, is extensively researched for application in various fields like battery storage, sensing, photo activity, composites, due to fascinating mechanical properties, electrical properties, and thermal properties [14–18].

Our group recently studied effect of graphene/CNT reinforcement in polystyrene and found carbon nanomaterial can enhance mechanical properties of polymer [19]. Effect of Graphene oxide (GO) reinforcement of epoxy resin was also studied and found that GO enhances modulus and hardness of epoxy-GO composites [20]. Composite synthesis requires a high quantity of GO at reasonable quality that can be suitably produced by wet chemical oxidation method. Although graphene was first synthesized from graphite precursor by simple mechanical cleavage method using an adhesive tape [21, 22].

The present work reports mechanical characterization of polycarbonate-Graphene Oxide (PCG) nanocomposite with the percentage weight variation of GO. GO was synthesized by chemical oxidation method. To achieve better dispersion solution mixing method was used. Dog-bone samples were synthesized and tested on micromechanical testing machine. Small percentage reinforcement of GO in PC tremendously improved mechanical properties of PCG nanocomposite.

2 Experimental

2.1 Materials

Commercial grade polycarbonate (PC), in form of beads, was procured from market. Received materials were put in vacuum over at 60 °C for 8 h to remove moisture and volatile substances. Graphite flakes were procedure from Sigma Aldrich as used as received. All reagents used were of AR grade and procured from either, Thermo Fisher Scientific, India or Sigma Aldrich, India. All reagents were used as received without any purification.

3 Synthesis of GO

Graphene oxide (GO) was prepared using wet chemical oxidation method reported earlier [23, 24]. In short, 1 g of graphite flakes were stirred in H_2SO_4 and H_3PO_4 acid mixture in 9:1 ratio (by volume) on magnetic stirrer. An oxidizing agent, $KMnO_4$ (six times by weight of graphite flakes), was slowly added and mixture was left on stirring condition till color changes to dark brown. A small amount of H_2O_2 was added to stop oxidation reaction. Solution was washed several times with mild HCl solution and then with De-ionized water to neutralize pH. Solution was dried in vacuum oven to get sheets of GO.

4 Synthesis of PC-GO Nanocomposite

Fabrication of Dog-bone shaped specimen was carried out in two steps. In first step, PC was mixed with GO using solution mixing method. GO was sonicated in Tetrahydrofuran (THF) using ultra-sonicator for 15 min. After complete dispersion of GO in THF, beads of PC were added in solution under constant stirring condition. Different composition samples were obtained by varying the weight percentage of GO in PC from 0–0.2 wt.%. The nomenclature of resulting composite was designated as PCG000 where last three numeric characters represent weight percentage of GO in PC multiplied by 100, e.g., PCG010 represents 0.10 wt.% of GO reinforcement in PC (Table 1). The solution was then dried on glass perti-dish at 70 °C for couple of hours to completely evaporate solvent.

In second step, Composite films were molded using injection molding machine shown in Fig. 1 [25–27]. Temperature of injection moulding machine was set at 290^0C and pressure settings were 6–10 psi. Composite films, obtained from solution blended method, were poured into the hopper of Injection moulding machine. After the conversion of solid-state film to molten state, molten material was pushed into the die with the pressure ram. Figure 2 shows actual digital image of Dog-bone samples.

Table 1 Table shows the results of different % of Graphene in compositions

S. no	Specimen name	GO wt.%age	Tensile strength (MPa)	Elastic modulus (GPa)	Tensile strength Ratio	Elastic modulus
1	PCG000	Pure	29.141	1.280	1.00	1.00
2	PCG005	0.05	26.762	1.072	0.92	0.84
3	PCG010	0.1	31.743	1.295	1.09	1.02
4	PCG020	0.2	45.694	1.440	1.57	1.13

Fig. 1 Digital photograph of Injection Moulding Machine

Figure 2a and b shows digital images of pure and composite samples, respectively. Figure 2c shows digital image of broken dog-bone sample after testing. All the procedure was repeated to prepare pristine PC sample using same procedure without mixing GO in it.

5 Results and Discussion

Solution mixing method for polymer composite synthesis is widely used due to improved dispersion of reinforcement in matrix [28, 29]. Solvent should be compatible for matrix and reinforcement [30]. THF was used as a solvent in present work as GO can be easily dispersed and PC can be dissolved in it. Regular color of synthesized

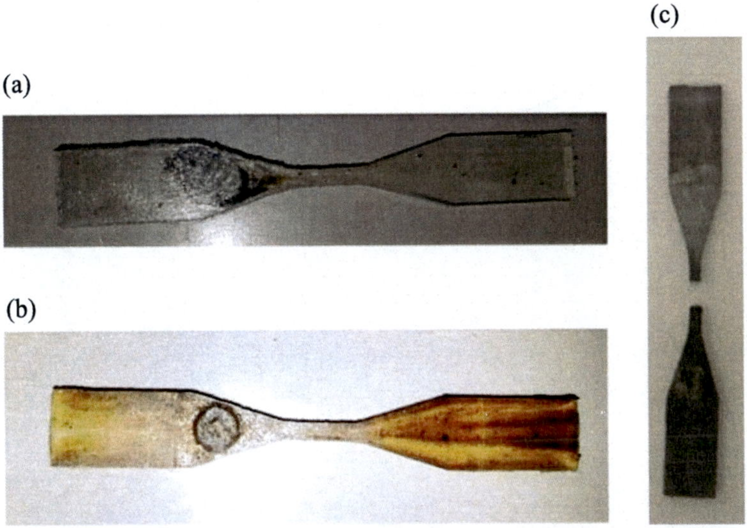

Fig. 2 Digital camera Images of Dog-bone shaped sample **a** Pure PC sample, **b** PC/Graphene (PCG) nanocomposites and **c** broken dog-bone sample after testing

specimen also points toward better dispersion of GO in PC matrix. Better dispersion enhances interaction of reinforcement and matrix to achieve higher mechanical properties.

Morphological studies of PCG nanocomposite samples were performed to understand reinforcement of GO in nanocomposite. Figure 3 PCG nanocomposite FE-SEM micrographs and presence of GO is marked by arrows. Shining spot in FE-SEM can be attributed to thin sheets of GO successfully embedded in PCG nanocomposite. Distribution of GO sheets has been fairly even too.

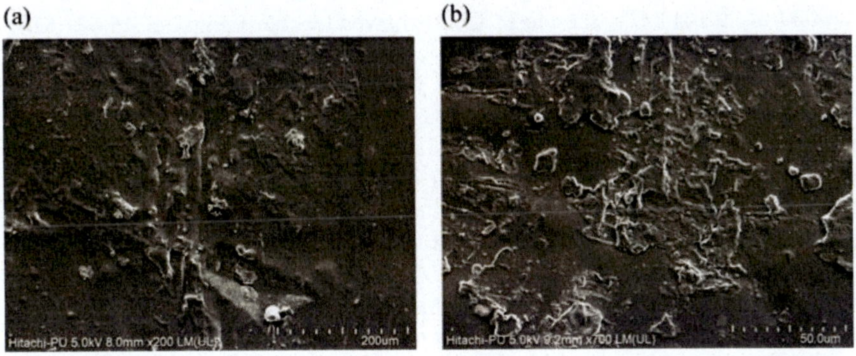

Fig. 3 FE-SEM micrograph of PC/G nanocomposite **a** Surface micrograph of PCG nanocomposite and **b** cross-sectional micrograph of PCG nanocomposite

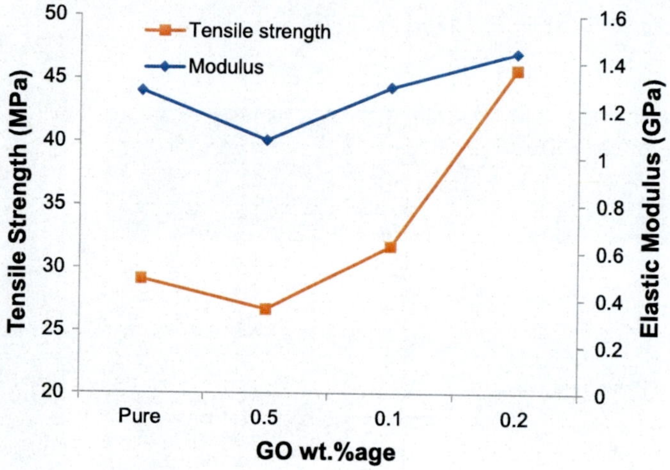

Fig. 4 Tensile strength and elastic modulus of different variations of graphene % in compositions

5.1 Tensile Strength

To characterize tensile behavior of PCG nanocomposite, a dog-bone sample was tested on micromechanical testing machine. Sample was prepared according to ASTM 638. Before testing samples were cleaned from burrs and extra materials. Surfaces of all samples were scratched with sand papers to achieve smooth finishing. Results of tensile testing are presented in Fig. 4 and Table 1.

Tensile strength of PCG upon addition of 0.05wt.% initially decreased. Decrease in tensile strength may be attributed to developments of few defects during synthesis as new materials phase was introduced. As GO concentration was further increased the tensile strength started to increase. Maximum tensile strength was achieved at 0.02 wt.% of GO in PCG020 sample. Comparing with tensile strength of pure sample PCG000 of 29.141 MPa, sample PCG020 showed tensile strength of 45.694 MPa that was 57% improvement from pure polycarbonate sample. A small reinforcement of 0.2 wt.% showed improved mechanical performance of PCG nanocomposite.

Elastic modulus characterization of PCG nanocomposites were also performed. A higher modulus represents lower strain while applying same load, i.e., better dimensional stability upon applying load. After an initial dip in modulus at 0.05 wt.% reinforcement of GO that may be attributed to small defects due to introduction of second phase modulus was increased in by further reinforcement. Modulus was increased from 1.280 GPa at pure PC to 1.440 GPa at PCG020 (GO 0.20 wt.%). It was a 13% improvement from pure PC. Once high strength GO reinforced at interface, GO gives strength to nanocomposite by increasing attachment matrix with high strength GO.

6 Conclusions

Following conclusions are drawn from the presented work:

- Solution blending was successfully used to synthesize Polycarbonate-GO (PCG) composite solution at room temperature for achieving superior mechanical properties.
- Casted thin sheets of PCG nano-composites were successful extruded Dog-bone shaped samples using Injection Molding Machine.
- Tensile strength and modulus of PCG nanocomposite at 0 wt%(pure), 0.5 wt%, 0.1 wt%, 0.2 wt% GO reinforcement were tested. Increase in tensile strength and elastic modulus of composite was observed with increase in concentration of GO in PC/Graphene composite.
- At 0.2 wt.% reinforcement of GO a rise in tensile strength and elastic modulus of 57% and 13% reported, respectively.

References

1. Gill AS, Kumar S (2016) Surface roughness and microhardness evaluation for EDM with Cu–Mn powder metallurgy tool. Mater Manuf Process 31:514–521
2. Saba N, Paridah MT, Jawaid M (2015) Mechanical properties of kenaf fibre reinforced polymer composite: a review. Constr Build Mater 76:87–96. https://doi.org/10.1016/j.conbuildmat.2014.11.043
3. Wang J, Li C, Zhang X, Xia L, Zhang X, Wu H, Guo S (2017) Polycarbonate toughening with reduced graphene oxide: toward high toughness, strength and notch resistance. Chem Eng J 325:474–484
4. Hirschbiel AF, Geyer S, Yameen B, Welle A, Nikolov P, Giselbrecht S, Scholpp S, Delaittre G, Barner-Kowollik C (2015) Photolithographic patterning of 3D-formed polycarbonate films for targeted cell guiding. Adv Mater 27:2621–2626
5. Kumar S, Lively B, Sun LL, Li B, Zhong WH (2010) Highly dispersed and electrically conductive polycarbonate/oxidized carbon nanofiber composites for electrostatic dissipation applications. Carbon N Y 48:3846–3857
6. Bagotia N, Choudhary V, Sharma DK (2017) Studies on toughened polycarbonate/multiwalled carbon nanotubes nanocomposites. Compos Part B Eng 124:101–110
7. Biswas KK, Ikueda M, Somiya S (2001) Study on creep behavior of glass fiber reinforced polycarbonate. Adv Compos Mater 10:265–273
8. Jang MG, Ryu SC, Juhn KJ, Kim SK, Kim WN (2018) Effects of carbon fiber modification with multiwall CNT on the electrical conductivity and EMI shielding effectiveness of polycarbonate/carbon fiber/CNT composites. J Appl Polym Sci, 47302. https://doi.org/10.1002/app.47302
9. Zhou S, Hrymak AN, Kamal MR (2018) Microinjection molding of multiwalled carbon nanotubes (CNT)-filled polycarbonate nanocomposites and comparison with electrical and morphological properties of various other CNT-filled thermoplastic micromoldings. Polym Adv Technol 29:1753–1764. https://doi.org/10.1002/pat.4282
10. Lee HY, Cruz H, Son Y (2019) Effects of incorporation of polyester on the electrical resistivity of polycarbonate/multi-walled carbon nanotube nanocomposite. J Compos Mater 53:1291–1298. https://doi.org/10.1177/0021998318801932

11. Fu R, Zhu M (2016) Synthesis and characterization of structure of Fe3O4@Graphene oxide nanocomposites. Adv Compos Lett 25:096369351602500. https://doi.org/10.1177/096369351602500604
12. Dalmis R, Cuvalci H, Canakci A, Guler O (2016) Investigation of graphite nano particle addition on the physical and mechanical properties of ZA27 composites. Adv Compos Lett 25:096369351602500. https://doi.org/10.1177/096369351602500202
13. Shah R, Kausar A, Muhammad B, Shah S (2015) Progression from graphene and graphene oxide to high performance polymer-based nanocomposite: a review. Polym Plast Technol Eng 54:173–183. https://doi.org/10.1080/03602559.2014.955202
14. Su BQ, Pang S, Alijani V, Li C, Feng X, Mu K (2009) Composites of graphene with large aromatic molecules, 3191–3195. https://doi.org/10.1002/adma.200803808
15. Bansal SA, Singh AP, Kumar S (2016) 2D materials: graphene and others. AIP Conf Proc 1728:020459. https://doi.org/10.1063/1.4946510
16. Yang M-Q, Xu Y-J (2016) Photocatalytic conversion of CO_2 over graphene-based composites: current status and future perspective. Nanoscale Horizons. 1:185–200. https://doi.org/10.1039/C5NH00113G
17. Cao L, Zhang D (2018) Preparation of graphene/polymer composite sponge for pressure-sensing application. Adv Compos Lett 27:096369351802700. https://doi.org/10.1177/096369351802700403
18. Zhu M, Fu R (2017) Synthesis and analysis of Mn 3 O 4 @Graphene nanocomposites for supercapacitors. Adv Compos Lett 26:096369351702600. https://doi.org/10.1177/096369351702600101
19. Bansal SA, Singh AP, Kumar S, Kumar A, Kumar S, Kumar N, Goswamy JK, Di Prima MA, Gall K, McDowell DL, Guldberg R, Lin A, Sanderson T, Campbell D, Arzberger SC (2018) Synergistic effect of graphene and carbon nanotubes on mechanical and thermal performance of polystyrene. Mater Res Express 5:075602. https://doi.org/10.1088/2053-1591/aacfc0
20. Bansal SA, Singh AP, Kumar A, Kumar S, Kumar N, Goswamy JK (2018) Improved mechanical performance of bisphenol-A graphene-oxide nano-composites. J Compos Mater 52:2179–2188. https://doi.org/10.1177/0021998317741952
21. Novoselov KS, Geim AK, Morozov SV, Jiang D, Zhang Y, Dubonos SV, Grigorieva IV, Firsov AA (2004) Electric field effect in atomically thin carbon films. Science (80-.) 306:666–669. https://doi.org/10.1126/science.1102896
22. Geim AK, Novoselov KS (2007) The rise of graphene. Nat Mater 6:183–191. https://doi.org/10.1038/nmat1849
23. Bansal SA, Singh AP, Kumar S (2019) Reinforcing graphene oxide nanoparticles to enhance viscoelastic performance of epoxy nanocomposites. J Nanosci Nanotechnol 19:4000–4006. https://doi.org/10.1166/jnn.2019.16336
24. Bansal SA, Singh AP, Kumar S (2018) High strain rate behavior of epoxy graphene oxide nanocomposites. Int J Appl Mech 10:1850072. https://doi.org/10.1142/S1758825118500722
25. Sulaiman S, Ismail N, Hamouda AMS (2004) Design and simulation of plastic injection moulding process. Anveshana's Int J Res Eng Appl Sci 12:100–110
26. Ecker JV, Haider A, Burzic I, Huber A, Eder G, Hild S (2019) Mechanical properties and water absorption behaviour of PLA and PLA/wood composites prepared by 3D printing and injection moulding. Rapid Prototyp J
27. Feldmann M (2016) The effects of the injection moulding temperature on the mechanical properties and morphology of polypropylene man-made cellulose fibre composites. Compos Part A Appl Sci Manuf 87:146–152
28. Gong L, Yin B, Li L, Yang M (2015) Nylon-6/Graphene composites modified through polymeric modification of graphene. Compos Part B Eng 73:49–56
29. Xu Z, Gao C (2010) In situ polymerization approach to graphene-reinforced nylon-6 composites. Macromolecules 43:6716–6723
30. Lago E, Toth PS, Pugliese G, Pellegrini V, Bonaccorso F (2016) Solution blending preparation of polycarbonate/graphene composite: boosting the mechanical and electrical properties. RSC Adv 6:97931–97940. https://doi.org/10.1039/C6RA21962D

Comparative Experimental Analysis of Machining Parameters for Inconel 825 on Cryogenic Treatment

Shivaji Vithal Bhivsane and Arvind L. Chel

Abstract This paper investigates machining of Inconel 825 with and without cryogenic treatment. The micromachining of Inconel 825 was done using CNC lathe and PVD coated (Multilayer TiAlN/TiN) carbide inserts as a cutting tool. Improvement in output parameters was observed when the workpiece material was treated cryogenically. The experiments were designed using the Taguchi method. A mathematical model has been established using the regression analysis of the experimental results. The independent variables in the developed mathematical model are cutting speed (Vc), feed (f), and depth of cut (d). The surface roughness and temperature are the dependent variables. The effect of independent variables on dependent variables with and without cryogenic treatment of the workpiece (i.e., Inconel 825) was analyzed. Nine experiments were conducted on a CNC lathe machine. Surface roughness of the workpieces was measured after machining the workpieces and the temperature was measured while machining of the workpieces. A mathematical model was formulated by using regression analysis. Based on the developed mathematical model surface finish was calculated.

Keywords Surface roughness · PVD coated carbide inserts · Cutting temperature · Tool wear

1 Introduction

The common properties in combination with low thermal conductivity, high rate of strain hardening, presence of hard and abrasive particles in the microstructure, and high chemical affinity make such alloys difficult to machine. Therefore, attempts

S. V. Bhivsane (✉)
BAM University, Aurangabad, Maharashtra, India
e-mail: bhivsane_s_v@yahoo.com

A. L. Chel
Mechanical Engineering Department, JNEC, Aurangabad, Maharashtra, India

© Springer Nature Singapore Pte Ltd. 2021

P. M. Pandey et al. (eds.), *Advances in Production and Industrial Engineering*, Lecture Notes in Mechanical Engineering,
https://doi.org/10.1007/978-981-15-5519-0_9

should be made to improve the machinability characteristics of nickel-based superalloys. White layer formation affects the surface integrity and decreases the machinability. White layer increases crack propagation decreases the surface quality and fatigue strength. As micromachining includes machining of components in microns to reduce temperature generation and minimum white layer formation. It is found that for high cutting speed and hard workpiece material, the machining results in the formation of the white layer. This results in hardness of workpiece material and affects for tool wear as well as tool life.

2 Literature Review

The generation of the white layer during machining has an adverse influence on surface integrity of the machined component [1, 2, 3]. The white layer is formed due to microstructural changes when recrystallization and strain hardening takes place due to high thermomechanical load [4]. It is known as the white layer as it resists standard etchants and appears white when seen under an optical microscope [5, 6]. The hard phase white layer leads the surface to become brittle hence causing initiation of crack and component failure [7, 8]. Formation of the white layer was noticed having a thickness of 2–4 μm which got reduced as cutting speed increased during turning [9]. High-speed machining of Inconel 825 at speed of 300 mm/rev is not possible as high tool wear is occurring and tool failure is rapid [10].

The lateral flow of material increased with increasing cutting speed. Tool wear during dry machining of Inconel 825 was characterized by adhesion, diffusion, plastic deformation, and catastrophic failure. The average flank wear increased with both cutting speed as well as with machining duration. However, the use of CVD multilayer coated improved the resistance to flank wear [11]. This effect was more pronounced at high cutting speed [12]. During machining it is observed that with an increase in cutting speed as well as with the progression of machining duration, the tool wear increases for both uncoated and coated inserts. However, the superiority of multilayer coated tool can be explained due to high wear resistance of coating as compared to its counterpart uncoated carbide inserts [13]. As the cutting speed increased, plastic deformation of chips increased due to rise in temperature, resulting in the material flow toward the side of the chips. Coated insert resulted in less plastic deformation and serrated chips than its uncoated counterpart. This is due to the generation of less cutting temperature owing to the presence of antifriction coatings like TiN, TiCN, and ZrCN in the coated insert [14, 15].

The rate of the tool wear in high-speed machining depends heavily on the chip–tool contact conditions. Tool wear is mainly by abrasion and some plastic deformation [16]. The difference in tool life between coated and uncoated CBN tool is neglected when Vc > 300 m/min [17, 18]. The surface roughness produced by coated CBN tool tends to be higher than the uncoated CBN tool [19]. Austenitization and Quenching, this stage is adopted for materials with a considerable amount of Carbon composition in it. But, as we can see Inconel 825 is a superalloy which does not contain Carbon composition to perform the austenitization operation, therefore this stage is omitted in our experiment [20]. Formation of this thermal barrier ceramic white layer on the surface of the machined part negatively affects its surface integrity and cutting tool life [21]. The deformed layer thickness obtained with higher cutting velocity (124 m/min) was found to be comparable for both uncoated and coated inserts [22].

From literature review it is understood that due to high depth of cut and feed which are usually in mm temperature generated is high and this results in white layer formation which decreases the fatigue life of material and hard surface formation which results in high tool wear and early failure of cutting tools and surface finish deteriorate [21, 22, 23, 24, 25]

To overcome this problem, it is beneficial to reduce the parameters into microns and perform these operations so that there will be minimum white layer formation and there will be minimum tool wear with longer tool life and it also increases surface finish of the workpiece.

3 Composition of Workpiece Material

Nickel-based superalloys are widely used in engineering applications especially in aerospace sectors. These alloys have good strength and corrosion resistance at high-temperature applications. The machinability of superalloys Inconel 825 as workpiece material is discussed in this paper. The composition of this material is as given in Table 1.

3.1 Advantages of Inconel 825

- High mechanical strength
- Resistance to oxidation and corrosion at elevated temperature,
- Good precipitation hardening ability and remarkable capability to withstand chemical
- Attack specially under sulphuric acid, phosphoric acid, and marine (chloride) environments.

Table 1 Material composition (Element % by mass)

Ni	Cr	Fe	Mo	Cu	Ti
46	19.94	25.62	3.26	1.53	**2.60**

3.2 *Applications of Inconel Alloys*

- Chemical and petrochemical processing
- Components where exposure to high mechanical stress and seawater are required
- Flare stacks on offshore oil platforms
- Gas turbines, rocket motors/engines, and spacecrafts
- Pollution control equipment
- Nuclear reactors
- Pickling equipment
- Food processing equipment

4 Cryogenic Treatment of Workpiece (Inconel 825)

Low-temperature treatment is one of the most promising methods to enhance the performance of the materials. The concept of metalworking at less than zero degrees Celsius is not new, but one that, if using correct procedures, can bring substantial economic benefits [1, 2, 3]. There is always a difference between working at low temperature and working at cryogenic conditions, keeping in view the strict controls of temperature in the treatment cycle. Depending upon the application of the temperature it may be classified as [17, 18]:

1. Cold treatment, which is also recognized as Shallow Cryogenic Treatment (SCT), is to gradually cool the workpiece in the range −84 C (189 K);
2. CT or Cryo-treatment may also be explored as Deep Cryogenic Treatment (DCT) is carried out at about −195 C (78 K).

Cryogenic Treatment (CT) and refrigerant-based cold treatments are essentially an extension of standard heat-treatment processes, rather, a complementary process to heat treatment that optimizes the material characteristics. CT is an inexpensive one-time treatment that influences the core properties of the component, unlike purely surface treatments [6, 4, 15]. CT process is needed to be identical in implementation to the existing conventional heat-treatment process, so that it can be applied to the materials to enhance mechanical properties as per requirement. This would require a lot of research work to optimize the process for various materials and to summarize the same on a platform [5].

Performing Cryogenic Heat Treatment on the specimens:

After preparing the samples, the cryogenic heat-treatment process was performed on all the nine specimens.

- The parameters for this heat treatment were as follows shown in Table 2

  Figure 1 illustrates all the stages performed in the cryogenic heat treatment:

Table 2 Stages of heat treatments and temperature ranges

Sr. no	Stages	Range of temperature	
		From	To
1	Descending stage	Room temp	– 193 °C
2	Soak period (for 24 h)	– 193 °C	– 193 °C
3	Ascending stage	– 193 °C	Room temp
4	Tempering process	150 °C	150 °C

Fig. 1 Prepared samples prior to cryogenic heat treatment

A typical cryogenic heat-treatment cycle is completed in four stages as shown in Fig. 2. A slow decent lowering down of temperature of the specimen from room temperature to –193 °C called as Descend. At temperature –193 °C the specimen is held to be cryogenically treated by using liquid nitrogen for the period of 24 h called as soaking period. After that the slow rate of change of temperature is recommended to return to room temperature from –193 °C, to avoid any kind of stresses or microcracking in the specimen called Ascend. In the tempering process, the specimen is tempered at 150 °C for 2 h to avoid brittleness, with a conventional heat treatment. It is the last step and takes 2 h.

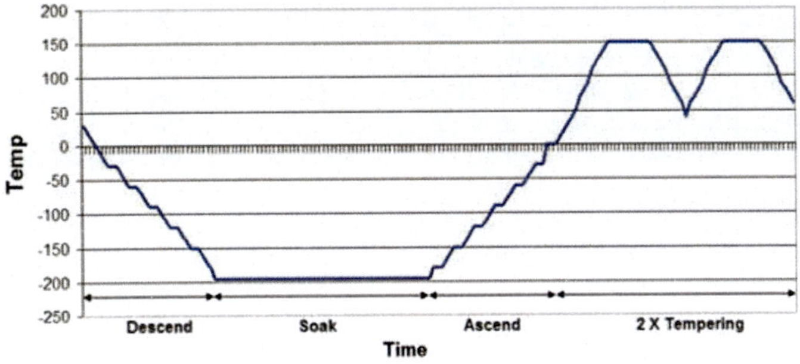

Fig. 2 Temperature versus time graph

5 Experimental Setup

The experimental work has been carried out at Anoop industry, M.I.D.C. Waluj, Aurangabad. This experiment is performed on a CNC lathe machine. CNC lathe machine is selected as it is required that material is to be removed in microns and precise feed is required. So as per requirement CNC lathe is selected.

After conducting the second experiment the final parameters have been decided. These parameters then entered into Minitab software (Minitab 71.ink) which provided nine different combinations on which actual experiment can be conducted. These parameters are shown in the following table.

In Minitab software (Minitab 71.ink) for deciding these parameters Taguchi methods have been used which provide us with the design of experiment and number of experiments to be conducted for satisfying our purpose. Taguchi methods are statistical methods, or sometimes called robust design methods, developed by Genichi Taguchi to improve the quality of manufactured goods, and more recently also applied to engineering, biotechnology, marketing and advertising.

The experiment is carried out as shown in Fig. 3, relative parameters on each blank of the test rod are given in Table 3. Initially a rough cut is taken, and then an actual cut is taken. This is done so that surface roughness can be calculated for parameters. Along with machining the temperature generated for each cut is also measured before and after cryogenic treatment by using a temperature gun. After the test has been carried out, surface roughness value (Ra) is measured by a surface roughness tester.

Fig. 3 Experimental setup

Table 3 Final parameters obtained by using Minitab software

Sr. no	Cutting speed (V_c) (m/min)	Feed (f) (mm/rev)	Depth of cut (d) (μm)
1	125	0.1	15
2	125	0.15	20
3	125	0.2	25
4	150	0.15	15
5	150	0.2	20
6	150	0.1	25
7	175	0.2	15
8	175	0.1	20
9	175	0.15	25

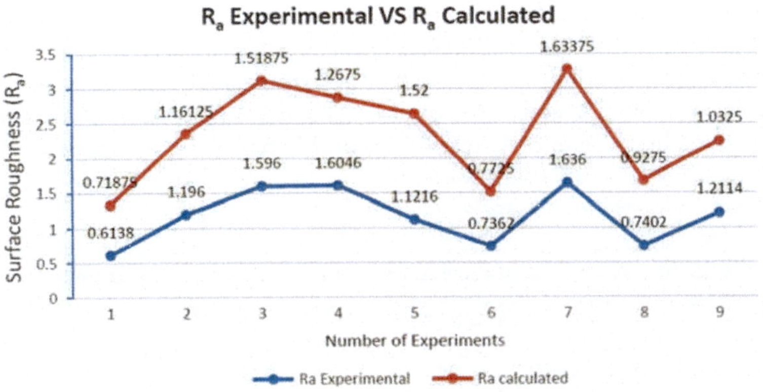

Fig. 4 Calculated and experimental surface roughness (Ra)

5.1 Cutting Tool

After studying different research papers and talking with experts it is recommended that PVD coated (Multilayer TiAlN/TiN) carbide inserts are preferred for this operation. ISO designation of TNMG 160,408 and this PVD coated carbide tool are based on carbide grades, ceramic thin-film technology and are effective for milling, threading, and grooving. The tough basis material of carbide and the innovative coating technology ensures excellent wear resistance and strong coating adhesion for long tool life and stable machining [10]. The PVD is, namely, Physical vapor deposition (PVD) describes a variety of vacuum deposition method which can be used to produce thin films and coatings [7, 8]. PVD is characterized by a process in which the material goes from a condensed phase to a vapor phase and then back to a thin film condensed phase. The most common PVD processes are sputtering and evaporation. PVD is used in the manufacture of items which require thin films for mechanical, optical, chemical, or electronic functions [9, 11].

Properties of cemented carbide are Density, g/cm3- 14.45, Vickers hardness-1850, Rockwell hardness-80, Young modulus (GPA)-630, Poisson's ratio- 0.31, Bulk modulus (GPA)- 646, Shear modulus (GPA)- 243.

Dimension for TNMG 160,408—Cutting edge length (mm)-16.0, Inscribed circle or height (mm)- 8.525, Thickness (mm)- 4.76, Hole diameter (mm)-3.81, Corner radius (mm)-0.8, Side clearance- 0o.

5.2 Preparation of the Specimens (workpiece)

- The Inconel 825 rod was measured 25 mm diameter and 460 mm length, it is hard to machine a specimen of such length in a CNC Lathe Machine, so Wire Cut operation was performed on the specimen.
- The specimens were cut at 90 mm length with 0.6 mm wire diameter, thus giving us five specimens of 1 mm allowance on either side of each specimen.
- It was decided that the work area will be at 25 mm length from either end of the specimen and the clamping area was given 40 mm in the middle. The reason for this was to maintain even surface for clamping area or else the specimen would have shown vibrations while machining.
- After this the specimen is ready for a rough cut to reduce the diameter from 25 to 23.3 mm.
- After this the specimen to subjected to Cryogenic treatment as follows.

5.3 Design of Experiment

After conducting the second experiment the final parameters have been decided. These parameters then entered into Minitab software which provided nine different combinations on which actual experiment can be conducted (Table 4).

Table 4 Process parameters and their levels

Level	Cutting speed (Vc) m/min	Feed rate (f) mm/rev	Depth of cut (d) (μm)
Low	125	0.1	15
Moderate	150	0.15	20
High	175	0.2	25

Table 5 Temperature generated, and surface roughness obtained from actual experiment

Sr. no	Cutting speed (Vc) (m/min)	Feed (f) (mm/rev)	Depth of cut (d) (μm)	Temp. in degree Celsius	Ra value (μm)	Temp. In degree Celsius	Ra value (μm)
				before cryogenic treatment		After cryogenic treatment	
1	125	0.1	15	34.4	1.15	31.3	0.6138
2	125	0.15	20	34.8	1.381	31.7	1.196
3	125	0.2	25	35.1	1.658	32.01	1.596
4	150	0.15	15	35.6	1.781	32.05	1.6046
5	150	0.2	20	35.1	1.419	32.4	1.1216
6	150	0.1	25	35	0.989	32.03	0.7362
7	175	0.2	15	35.7	1.758	31.9	1.636
8	175	0.1	20	34.4	0.860	32	0.7402
9	175	0.15	25	35	1.901	32.01	1.2114

6 Experimental Results and Mathematical Modelling

6.1 Experimental Result

After obtaining the required parameters test has been conducted on a CNC operated lathe machine. The relative parameter on each blank of the test rod is given in the table. Initially a rough cut is taken, and then an actual cut is taken. This is done so that surface roughness can be calculated for both before machining and after machining parameters. Along with machining the temperature generated for each cut is also measured by using a temperature gun. After the test has been carried out surface roughness value (Ra) is measured by a surface roughness tester. The following table shows the temperature generated for each speed and Ra value accordingly. During the test for each cut a separate insert is used so that the behavior of the tool for each parameter can be studied.

From Table 5 it is observed that the cutting temperature is reduced and surface finish is increased after cryogenic treatment of the workpieces.

7 Mathematical Modelling

The correlation between a criterion variable and a combination of the predictor variable, the statistical multiple regression method is applied. Since the independent variables of this study were cutting speed (Vc), feed (f), depth of cut (d), the dependent variable is surface roughness (Ra).

The theoretical surface roughness is calculated by using Minitab software 16 in which regression analysis is done and the equation for theoretical Ra value is obtained. By using this equation all the theoretical Ra values for each process parameters are calculated.

A mathematical model is developed using Regression analysis using MINITAB 16. Cutting speed, feed and doc interactions are considered while formulating the mathematical model to incorporate the combined effect of design variables of surface finish. Corresponding data generated for the model formulation is recorded.

The mathematical model formulated by regression analysis is:

$$Ra = -3.06 + 0.0204\,Vc + 0.074d + 21.3f - 0.073\,Vc \times f - 0.00045Vc \times d - 0.17f \times d \quad (1)$$

where Vc is cutting speed in m/min, d is depth of cut in mm, f is feed in mm/rev, Vc *d is product of cutting speed and depth of cut, Vc *f is product of cutting speed and feed, and d*f is product of depth of cut and feed.

8 Result and Discussion

The values obtained from the mathematical model formulated by regression analysis (Equation number 1) are given in Table 6.

Standardized regression coefficients (sometimes referred to as beta coefficients). They allow to directly compare the relative influence of the explanatory variables on the dependent variable, and their significance (Fig. 4 and Table 7).

Table 6 Calculation of theoretical and predicted surface roughness Ra

Sr. No	Cutting speed (V_c) in mm/rev	Depth of cut (d) in mm	Feed (f) in mm/rev)	V_c *f	V_c *d	f*d	Ra experimental (μm)	Ra predicted (μm)
1	125	15	0.10	12.50	1875	1.50	0.6138	0.71875
2	125	20	0.15	18.75	2500	3.00	1.1960	1.16125
3	125	25	0.20	25.00	3125	5.00	1.5960	1.51875
4	150	15	0.15	22.50	2250	2.25	1.6046	1.2675
5	150	20	0.20	30.00	3000	4.00	1.1216	1.52
6	150	25	0.10	15.00	3750	2.50	0.7362	0.7725
7	175	15	0.20	35.00	2625	3.00	1.6360	1.63375
8	175	20	0.10	17.50	3500	2.00	0.7402	0.9275
9	175	25	0.15	26.25	4375	3.75	1.2114	1.0325

Table 7 Analysis of variance

Source	DF	Adj SS	Adj MS	F-Value	P-Value
Cutting speed (Vc)	2	0.00577	0.002885	0.03	0.967
Feed (f) (mm/rev)	2	0.99255	0.496277	5.9	0.145
Depth of cut (d) (Micron)	2	0.10746	0.053732	0.64	0.61
Error	2	0.16811	0.084056		
Total	8	1.2739			

Table 8 Optimum values for input parameters

Optimum (Vc) (m/min)	Optimum (d) (μm)	Optimum (f) (mm/rev)	Ra predicted (μm)
125	15	0.1	0.71875

From the graph it can be seen that in majority of cases both surface roughness (Ra) values are similar but in some cases they are different.

The table given below gives the information about optimum values of cutting speed (Vc) is 125 m/min, feed (f) 01 mm/rev, and depth of cut (d) 15 mm. This means that by using these parameters the best possible result will be obtained (Table 8).

9 Conclusion

The paper narrates the improvement in surface finish and reduction in cutting temperature while machining cryogenically heat-treated Inconel 825 workpiece material using PVD coated carbide insert cutting tool. The result was obtained with and without cryogenic heat treatment and compared in Table 4.

The regression analysis was carried out for the obtained experimental result and the following key conclusions were drawn.

- The surface roughness gets affected by different values of cutting speed, optimum cutting speed obtained from Analysis of Variance was 125 m/min for Ra value 0.719 μm.
- The mathematical model obtained from regression analysis shows that the depth of cut is an insignificant parameter.
- Increase in feed rate deteriorates the surface roughness, hence optimum value for the feed rate is obtained as 0.1 mm/rev.
- Cutting temperature is reduced from 35.7 to 31.9 °C due to cryogenic heat treatment of the workpiece.
- PVD coated carbide insert as a cutting tool should be used for improved surface finish, with cryogenic heat treatment.

References

1. Alava LA (2008) Multistage: cryogenic treatment of materials: process fundamentals and examples of application, 10th Cryogenic. IIR International Conference, Praha, Czech Republic, April 21–25
2. Kollmer KP (1999) Applications & developments in the cryogenic processing of materials, the technology interface. Elec J Eng Technol 3(1). Available at: https://et.nmsu.edu/~etti/. Last accessed (Oct. 1, 2009)
3. Preciado M, Bravo PM, Alegre JM (2006) Effect of low temperature tempering prior cryogenic treatment on carburized steels. J Mater Process Technol 176:41–44
4. Gogte CL, Iyer KM, Paretka RK, Peshwe DR, Deep Subzero (2009) Processing of metals and alloys: evolution of microstructure of AISI T42 tool steel. Mater Manufac Proces 24(7&8), 718–722. 43
5. Darwin JD, MohanLal D, Nagarajan G (2008) Optimization of cryogenic treatment to maximize the wear resistance of 18% Cr martensitic stainless steel by Taguchi method. J Mater Process Technol 195:241–247
6. Bensely A, Venkatesh S, MohanLal D, Nagarajan G, Rajadurai A, Junik K (2008) Effect of cryogenic treatment on distribution of residual stress in case carburized En 353 steel. Mater Sci Eng, A 479:229–235
7. Gangopadhyay S (2014) Effect of cutting speed and tool coating on machined surface integrity of Ni-based super alloys. Procedia CIRP, pp 541–545
8. A. Thakur, S.Gangopadhayay, K.P.Maity : Surface Integrity Of Inconel 825 When Turning With Uncoated And CVD Coated Carbide Inserts, Precision, Meso, Microand Nano Engineering,(December 2013).
9. Mohanty A (2015, January) Investigation on some machinability aspects of inconel 825 during dry turning, Mechanical Engineering Department, National Institute of Technology Rourkela, pp 1–28
10. Ezugwu EO, Wang ZM, Machado AR (1999) The machinability of Nickel based Alloys. J Mater Proces Technol, 1–16
11. PVD coated carbide tool material, https://www.kyocera-unimerco.com [1] Arvind Andre (editor). Handbook of Plasma Immersion Ion Implantation and Deposition. New York: Wiley-Interscience
12. Thakur A (2014) Effect of cutting speed and CVD multilayer coating on machinability of inconel 825. Published by Maney on behalf of the Institute 30:516–523
13. Maity KP (2014) Tool wear and chip characteristics during dry turning of inconel 825. Procedia Materials Science, pp 2169–2177
14. Thakur A (2014) Effect of cutting speed and tool coating on machined surface integrity of Nickel base super Alloy. Procedia CIRP 14, pp 541–545
15. Akhbarizadeh A, Shafyei A, Golozar MA (2009) Effects of cryogenic treatment on wear behavior of D6 tool steel. Mater Des 30:3259–3264
16. Thakur DG (2009) A study on the parameters in high-speed turning of super alloy inconel 718. Mater Manufact Proces 24:497–503
17. Radebaugh R (2002) Cryogenics in the MacMillan encyclopedia of chemistry. MacMillan, New York
18. Liquid Nitrogen, Air Products & Chemicals Inc. (June 27 2009). https://www.airproducts.com/nr/rdonlyres/89de2726-10bb-4e89-94e962f11a23cfbc/0/sfgrm07.pdf, Allentown, USA, 2006, Last accessed
19. Zhou, J, Bushlya V, Avdovic P (2012) Study of surface quality in high speed turning of Inconel 718 with uncoated and coated CBN tools. Int J Adv Manuf Technol, 141–151
20. Thakur A, Gangopadhayay S, Maity KP, Precision, Microand M (2013, December) Surface integrity of inconel 825 when turning with uncoated and CVD coated carbide inserts, Nano Engineering
21. Zhirafar S, Rezaeian A, Pugh M (2007) Effect of cryogenic treatment on the mechanical properties of 4340 steel. J Mater Process Technol 186:298–303

22. Bensely A, Prabhakaran A, MohanLal D, Nagarajan G (2006) Enhancing the wear resistance of case carburized steel (En 353) by cryogenic treatment. Cryogenics 45:747–754. 42
23. Huang JY, Zhu YT, Liao XZ, Beyerlein IJ, Bourke MA, Mitchell TE (2003) Microstructure of cryogenic treated M2 tool steel. Mater Sci Eng A339:41–244
24. Moore K, Collins DN. Cryogenic treatment of three heat treated steels. Key Eng Mater 993, 86–87, 47–54
25. Totten GE (2006) Steel heat treatment handbook (Metallurgy and Techniques); Taylor & Francis

Optimizing Gas Injection Stir Casting Process Parameters for Improving the Ultimate Tensile Strength of Hybrid $Mg/(SiC_p + Al_2O_{3p} + Gr_p)$ Through Taguchi Technique

Jaspreet Hira, Alakesh Manna, Pushpinder Kumar, and Rohit Singla

Abstract In this experimental research work Taguchi technique has been utilized for finding out the optimal parameter combinations that result for better ultimate tensile strength of hybrid $Mg/(SiC_p + Al_2O_{3p} + Gr_p)$ MMC. The process parameters that are used for the analysis are the furnace temperature (°C), stirring speed (rpm) and stirring time (min). Also considering significant process parameters mathematical models related to the ultimate tensile strength (MPa) has been established. An $L_{27}(3^{13})$ orthogonal array has been utilized for finding out the optimum conditions. The optimal parametric combination where maximum ultimate tensile strength has been achieved are $A_3B_2C_3$, respectively. The results of this investigation result in selecting suitable casting process parameters in fabricating hybrid Mg-MMC of desired strength and longer durability.

Keyword Magnesium · Metal matrix composite · Taguchi · Orthogonal array · Mathematical model

---

J. Hira (✉) · R. Singla
Mechanical Engineering Department, Amity University, Uttar Pradesh, Noida, India
e-mail: jas.iitdelhi@gmail.com

R. Singla
e-mail: rsingla1@amity.edu

A. Manna
Mechanical Engineering Department, PEC University of Technology Chandigarh, Chandigarh, India
e-mail: alakeshmanna@pec.ac.in

P. Kumar
Mechanical Engineering Department, Thapar Institute of Engineering and Technology, Patiala, India
e-mail: pushpindergera@gmail.com

© Springer Nature Singapore Pte Ltd. 2021

P. M. Pandey et al. (eds.), *Advances in Production and Industrial Engineering*, Lecture Notes in Mechanical Engineering,
https://doi.org/10.1007/978-981-15-5519-0_10

1 Introduction

Metal matrix composites developed as an engineering material system that supports a wonderful combination of mechanical and chemical properties for immediate structural applications. Their enhanced mechanical properties including strength, stiffness, wear and damping capacities. Aluminium, magnesium, titanium and its alloys are the primary matrix materials used nowadays in modern industrial applications. The main matrix materials are aluminium, magnesium, titanium and its alloys. In addition to these hybrid matrix materials are the hard reinforcement particulates such as silicon carbide, aluminium trioxide, graphite, boron carbide, fly ash, red mud and zircon sand. The casting of hybrid metal matrix materials can be processed using various developed technique, i.e. stir casting, compo casting, liquid infiltration, vortex flow, surface deposition. The developed hybrid MMC's are further tested for its strength, durability, toughness and other material properties. Taguchi technique is a powerful tool to obtain information in a controlled manner and to analyze the effect of different process variable on response variable. Taguchi GRA Entropy technique has been utilized for achieving optimal level during fabrication of Al2024/red mud composite using two steps stir casting [1] Taguchi method was being utilized for predicting the influence of process parameters on erosion wear performance of Al7034 – T_6 composite reinforced with silicon carbide and alumina particles. Taguchi L_{16} orthogonal array was being utilized for numerous erosion trials and results concluded that the filler content silicon carbide and alumina particles and impact velocity are the most contributing factors in terms of erosion [2]. Utilizing Taguchi and regression analysis degree of sphericity of the primary phase during cooling slope casting of A356 Al alloy has been identified. The optimum degree of sphericity has been achieved with 60° angle of slope, 650 °C pouring furnace temperature, 60 °C wall temperature and 500 mm length of travel of the melted region [3]. Accumulative roll bonding has been utilized for fabricating alumina aluminium hybrid composite and the results found out that the uniformity of particle distribution has been improved. The tensile strength and percentage elongation showed increasing trend through utilizing this accumulative roll bonding process [4]. Taguchi technique has been utilized during thixoforming of A356 alloys, and pouring temperature 640 °C, slope angle 60 °C and cooling length 300 mm are the optimal values for higher hardness. The most influential parameter was the pouring temperature [5]. Homogeneous dispersal, smooth interfacial composition and grain enhancement result in elevated strength of AM60 based metal matrix nanocomposite strengthened with Al_2O_3 fibre and nano-sized Al_2O_3 particles [6]. Coating calcium on magnesium and its alloy during casting lowers the susceptibility to inflammation and makes the casting possible in the air forming protective film [7]. The strength and ductility of the Mg Matrix were enhanced by 22% when subjected to homogenization treatment at a temperature of about 900 °C for two hours duration, with the addition of Ti and B_2C powders as reinforcements [8]. The addition of SiCp/m + s + n was supportive to upsurge the ratio of DRXed nucleation and decrease the average grain size of the matrix thereby enhancing the tensile strength [9]. Based on the literature review above, it was obviously stated that the Taguchi method provides us the highest ideal parameters with a minimum number of experimental tests. In the present work, the Taguchi method applied for finding out the optimum settings for each parameter and its combinations for achieving ultimate tensile strength (MPa) of prepared hybrid Magnesium-MMC.

2 Experimental Details

2.1 *Material Preparation*

A stir-casting set-up, which consists of argon gas injection, has been designed and fabricated for prepare hybrid Mg-MMC. The magnesium was first preheated at a temperature of about 450 °C for maximum two hours before melting. Simultaneously the hard reinforcement particulates such as alumina (Al_2O_3), silicon carbide (SiC) and graphite (Gr) were also preheated at a temperature of about 1100 °C for a maximum of 2 h for removing absorbed hydroxides and other affluent gases. The preheating effect improves the wettability of the cast Mg-MMC. The induction furnace is raised above the liquid temperature, i.e. 710 °C, and retained for 30 min when the Mg-MMC has fully melted. The preheated particles added and mechanically mixed with the stirrer with average speed up to 300 rpm. The graphite addition improves the magnesium wettability. The 3.15 mm thick steel sheet has been used to fabricate the metal mould with size dimension 50 mm diameter and 150 mm long length with 8wt%SiC, 2wt% Al_2O_3 and 1wt% Gr in the metal matrix composite.

2.2 *Ultimate Tensile Strength (MPa)*

Figure 1 shows hybrid Mg-MMC's ultimate tensile strength specimen measured on an universal testing machine with the maximum testing capacity of 100 kN, the measurement of the load is displayed on five digit digital display with ± 1% of the indicated value. The maximum stress that hybrid Mg-MMC can withstand under uniaxial loading has been measured.

$$\sigma_{ut} = \frac{P_{max}}{A_o} \tag{1}$$

σ_{ut} is the maximum stress that hybrid Mg-MMC can withstand under uniaxial loading.

P_{max} is the maximum load per unit area A_o.

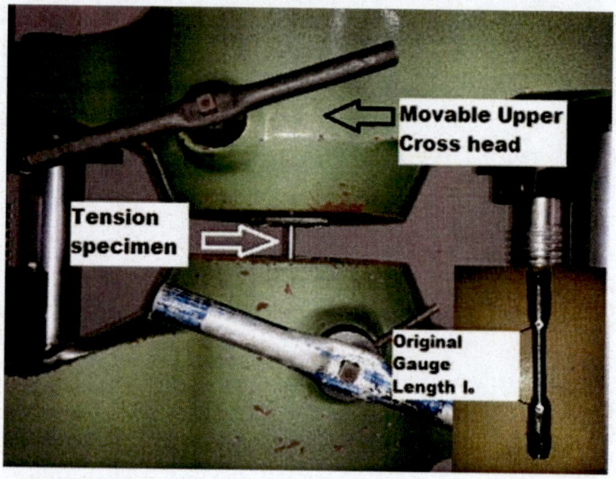

Fig. 1 Tensile strength test specimen

3 Optimization of Stir Casting Parameters and Development of Mathematical Model

The detail experimental investigations have been carried out in a pre-planned way to explore the significant parameters and optimize the parametric conditions during casting of Mg/(8wt%SiCp + 2wt%Al2O3p + 1wt%Grp)-MMC with nitrogen gas-assisted and argon gas injection liquid stir casting technique. An orthogonal L_{27} (3^{13}) array is employed for 3^3 factorial design and the analysis of variance is the workout to investigate the influence of furnace temperature, stirring speed and stirring time on fabricated Mg metal matrix composite performance mechanical properties of the ultimate tensile strength (MPa). Taking into account significant casting parameters and using multiple linear regression, mathematical models related to the ultimate tensile strength are established to explore the influence of different casting parameters for casting Magnesium/(8wt%SiC + 2wt%Al_2O_3 + 1wt%Gr) MMC.

3.1 *Planning for Experimentation*

The various sets of tests were carried out for fabrication of hybrid Mg-MMC on the developed set-up. For this purpose, it utilized three furnaces simultaneously. The first furnace is used for melting of Mg; the second furnace is used for heating and agitating of reinforced particles and the third furnace is used for backing of thin coated fire clay metal moulds. The workpiece samples are prepared by shaping, sawing; and turning, parting-off from cast Mg-MMC ingots for tensile testing, respectively. The ultimate tensile strength of the prepared sample has been measured utilizing Universal testing machine HT2107 A and average test results are taken for analysis. Robust based on the Taguchi technique employs an orthogonal array of L_{27} (3^{13}) for testing.

Table 1 Gas injection liquid stir casting parameters and their levels

Sr. no	Casting parameters	Levels		
		1	2	3
1	A: Furnace Temperature, °C	600	675	750
2	B: Stirring speed, rpm	160	230	350
3	C: Stirring Time, min	4	6	8

3.2 *Design of Experimental Plan Based on Standard Orthogonal Array*

Experimental plan development is a methodology for effectively studying the impact of several control variablesconcurrently through an experimental matrix plan called an orthogonal array. Based on facts of the number of control factors, their stages and the desire to study specific interaction, an orthogonal array can be built for a robust experimentation design. Table 1 shows the casting parameters and their level which has been used in the experimentation.

Table 2 represents the L_{27} (3^{13}) orthogonal as per planned and experimental design considered for experimental investigation. Experimentally acquired results based on this planned of experiments are utilized for optimization of gas injection liquid stir casting parameters to achieve better mechanical properties in terms of maximum ultimate tensile strength during casting of Mg/(8wt%SiC_p + 2wt%Al_2O_{3p} + 1wt%Gr_p) MMC. Initially twenty-seven experiments have been performed according to the $L_{27}(3^{13})$ orthogonal array and each set of test trials was repeated three times, i.e. total 81 experiments have been performed for the present set of investigations.

3.3 *$L_{27}(3^{13})$ Orthogonal Array*

S/N ratio (dB) for fabricated Mg/(8wt%SiC_p + 2wt%Al_2O_{3p} + 1wt%Gr_p)-MMC for S/N ratio, ANOVA and Mathematical model MATLAB version 7.10 (R2010 a) was employed and acquired results were used to develop the mathematical model.

3.3.1 Anova

Table:4 represents ANOVA and ‘F-test’ value for Ultimate tensile strength (Mpa) of Mg/(8wt%SiC_p + 2wt%Al_2O_{3p} + 1wt%Gr_p)-MMC. From Table 7, it is clear that the ‘F-test’ value against the parameter, furnace temperature (°C) is 31.2105, stirring speed (rpm) is 85.1841 and stirring time (min) is 26.3968 and all have a great influence on ultimate tensile strength (Mpa). Here, stirring speed is the most

Table 2 L_{27} (3^{13}) orthogonal array and actual parametric setting value

Exp. No	Coded level			Actual setting value		
	1	2	3	1	2	3
1	1	1	1	600	160	4
2	1	1	2	600	160	6
3	1	1	3	600	160	8
4	2	2	1	675	230	4
5	2	2	2	675	230	6
6	2	2	3	675	230	8
7	3	3	1	750	350	4
8	3	3	2	750	350	6
9	3	3	3	750	350	8
10	2	3	1	675	350	4
11	2	3	2	675	350	6
12	2	3	3	675	350	8
13	3	1	1	750	160	4
14	3	1	2	750	160	6
15	3	1	3	750	160	8
16	1	2	1	600	230	4
17	1	2	2	600	230	6
18	1	2	3	600	230	8
19	3	2	1	750	230	4
20	3	2	2	750	230	6
21	3	2	3	750	230	8
22	1	3	1	600	350	4
23	1	3	2	600	350	6
24	1	3	3	600	350	8
25	2	1	1	675	160	4
26	2	1	2	675	160	6
27	2	1	3	675	160	8

significant parameter as 'F-test' value is 85.1841. It is also clear that the maximum contribution, i.e. 27.365 by the interaction of the parameters furnace temperature (°C) and stirring speed (rpm).

Figure 2 shows the S/N ratio graph for different factor levels for ultimate tensile strength (MPa) of fabricated Mg/(8wt%SiC_p + 2wt%Al_2O_{3p} + 1wt%Gr_p)-MMC. From Fig. 2, it is clear that the optimal parametric combination for ultimate tensile strength (MPa) is $A_3B_2C_3$, i.e. 750 °C furnace temperature, 230 rpm stirring speed and 8 min stirring time.

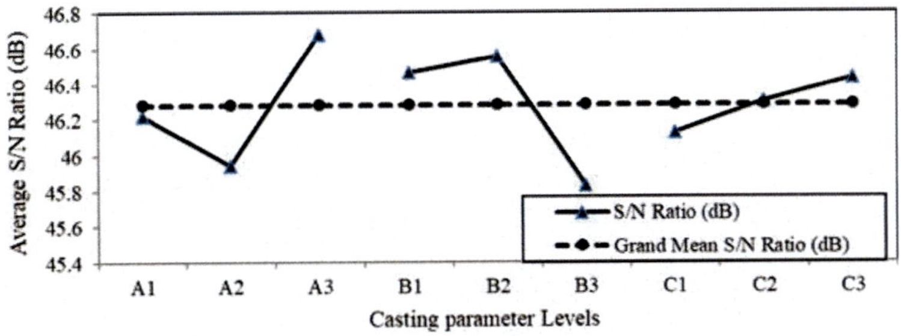

Fig. 2 S/N ratio (dB) for ultimate tensile strength (MPa)

3.3.2 Mathematical Model

Considering three parameters, namely furnace temperature (°C) as X_1, stirring speed (rpm) as X_2 and stirring time (min) as X_3 and using the Gauss elimination method, the mathematical model has been developed. The mathematical model for ultimate tensile strength of the fabricated Mg/(8wt%SiC + 2wt%Al_2O_3 + 1wt%Gr) –MMC is as follows

$$\begin{aligned} Y_{tensile\,strgth} = &-57.496682 + 0.8442709\,X_1 - 0.438245\,X_2 + 9.865764\,X_3 \\ &+ 0.00099947\,X_1X_2 - 0.016646\,X_1X_3 + 0.030814X_2X_3 \\ &- 0.000703\,X_1^2 - 0.000943\,X_2^2 - 0.389259\,X_3^2 \end{aligned} \quad (2)$$

$R^2 = 0.92$.

Where, R^2 stands for thhhe coefficient of multiple determinations. It processes the proportion of total variability of Y (i.e. response) obtained by using the regression variables X_1, X_2, X_3 in the model. Here, X_1 = Furnace Temperature, °C, X_2 = Stirring Speed, rpm; X_3 = Stirring Time, min.

From the above confirmation test results (Table 5), the findings achieved using the mathematical model are found to be in excellent agreement with the test outcome as the error proportion is always below 4.5%. Hence, the developed mathematical model can be successfully utilized for predicting the parameter value in advance for the fabrication of Mg/(8wt%SiC + 2wt%Al_2O_3 + 1wt%Gr)-MMC with desired tensile strength.

Table 5 Confirmation test results for ultimate tensile strength (MPa)

Confirmation test No	Experimental value	Calculated value Eq. (2)	% Error
1	205	210.807	2.75
2	207	216.553	4.41
3	210	214.462	2.08

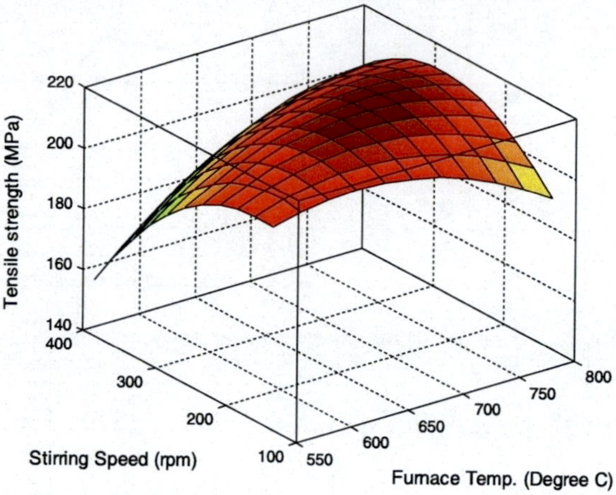

Fig. 3 Effect of furnace temperature and stirring speed on Ultimate Tensile strength (MPa) of prepared hybrid Mg-MMC

4 Results

4.1 Interaction Effect of Stirring Speed and Furnace Temperature

The variation of ultimate tensile strength with furnace temperature (°C) and stirring speed (rpm) for fabricated Mg/(8wt%SiC_p + 2wt%Al_2O_{3p} + 1wt%Gr_p)-MMC on developed gas injection liquid stir casting set-up is shown in Fig. 3. The ultimate tensile strength is high at moderate furnace temperature and moderate stirring speed (rpm).

4.2 Interaction Effect of Stirring Time and Furnace Temperature

The variation of ultimate tensile strength with furnace temperature (°C) and stirring time (min) for fabricated Mg/(8wt%SiC_p + 2wt%Al_2O_{3p} + 1wt%Gr_p)-MMC is shown in Fig. 4. It is clear that the ultimate tensile strength is high at high furnace temperature (i.e. 700–750 °C) and high stirring time (min).

4.3 Interaction Effect of Stirring Time and Stirring Speed

The variation of ultimate tensile strength with stirring speed (rpm) and stirring time Mg/(8wt%SiC_p + 2wt%Al_2O_{3p} + 1wt%Gr_p)–MMC is shown in Fig. 5. The ultimate tensile strength is high at moderate stirring speed (i.e. 200 rpm) and high stirring

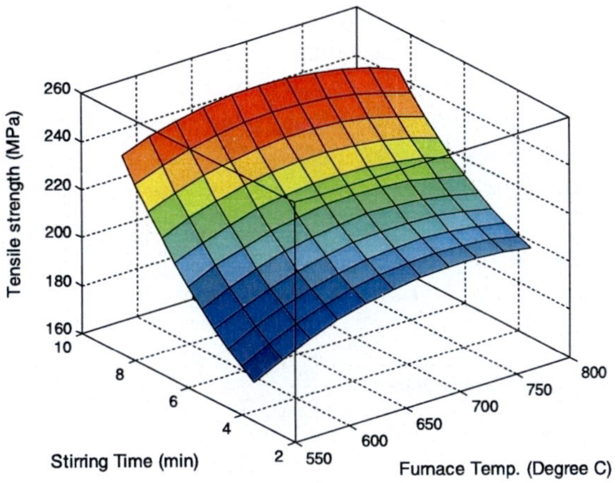

Fig. 4 Effect of furnace temperature and stirring time on Ultimate Tensile strength (MPa) of prepared hybrid Mg-MMC

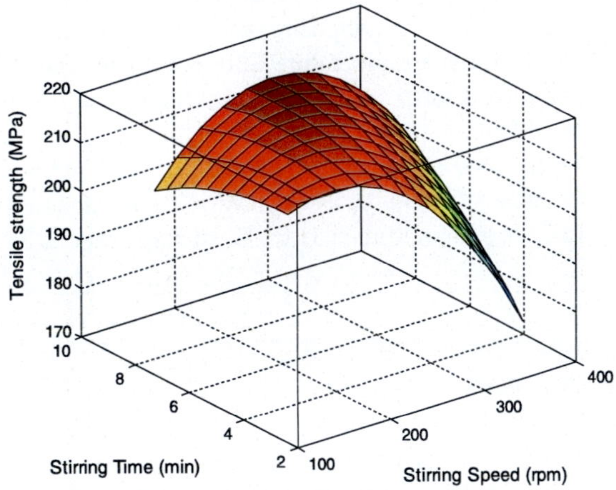

Fig. 5 Effect of stirring speed and stirring time on Ultimate Tensile strength (MPa) of prepared hybrid Mg-MMC

time (i.e. 8 min) and this interaction has the greater effect on ultimate tensile strength of fabricated Mg/(8wt%SiC + 2wt%Al_2O_3 + 1wt%Gr)-MMC.

4.4 Intensive Discussion

The fabrication of hybrid magnesium metal matrix composite with hard reinforcements was a tedious process, which involves a number of test trails for the achieving desired property. Taguchi proven to be an effective technique in achieving the desired results. An L_{27} (3^{13}) orthogonal array was employed, varied the parameters such as furnace temperature, stirring speed and stirring time for achieving the desired ultimate tensile strength property. The furnace temperature has found to be the most significant single parameter with 18.29% contribution while the maximum contribution of 29.31% found for the interaction of furnace temperature and stirring time parameters. These parameters found to have a great influence on ultimate tensile strength (MPa) of the fabricated MMC. The stirring speed with a contribution of 20.68% found to be the most significant single parameter, while the maximum contribution of 27.365% was found for the interaction of the parameter furnace temperature and stirring speed.

5 Conclusions

- The Taguchi optimization method has proven successful in carrying out the experiments to achieve greater ultimate tensile strength values of hybrid Mg-MMC.
- The optimal parametric combination for ultimate tensile strength (MPa) of hybrid Mg metal matrix composite with hard reinforcement particulates are $A_3B_2C_3$, i.e. 750 °C furnace temperature, 230 rpm stirring speed and stirring time 8 min. These parameters have a significant contribution in the ultimate tensile strength of the Mg-MMC
- From ANOVA for ultimate tensile strength (MPa) of fabricated Mg/(8wt%SiC_p + 2wt%Al_2O_{3p} + 1wt%Gr_p)-MMC, it is found that the F-test value for parameter furnace temperature is 31.2105, stirring speed is 85.18 and stirring time is 26.39.

References

1. Sharma A, Belokar RM, Kumar S (2018) Multi-response optimization of Al2024/red mud MMC using hybrid taguchi-GRA-entropy optimization technique. Mater Today: Proce 5(2):4748–4760
2. Pasha Mudasar BA, Kaleemulla M (2017) Taguchi approach to influence of processing parameters onerosive wear behaviour of Al7034-T6 composites. Trans Nonferrous Metals Soc China 27.10:2163–2171
3. Das P, Samanta SK, Das R, Dutta P (2014) Optimization of degree of sphericity of primary phase during cooling slope casting of A356 Al alloy: taguchi method and regression analysis. Measurement 55:605–615
4. Toroghinejad MR, Jamaati R, Nooryan A, Edris H (2014) The effect of alumina content on the mechanical properties of hybrid composites fabricated by ARB process. Ceram Int 40(7):10489–10498
5. Kumar SD, Vundavilli PR, Mantry S, Mandal A, Chakraborty M (2014) A Taguchi optimization of cooling slope casting process parameters for production of semi-solid A356 alloy and A356–5TiB2 in-situ composite feedstock. Procedia Mater Sci 5:232–241
6. Zhou J, Ren L, Geng X, Fang L, Hu H (2019) As-cast magnesium AM60-based hybrid nanocomposite containing alumina fibres and nanoparticles: Microstructure and tensile behavior. Mater Sci Eng A 740:305–314
7. Koltygin AV, Bazlova TA, Plisetskaya IV (2013) Effect of calcium on the process of production and structure of magnesium melted by flux-free method. Met Sci Heat Treat 54(9–10):540–544

8. Sahoo BN, Panigrahi SK (2016) Synthesis, characterization and mechanical properties of in-situ (TiC-TiB2) reinforced magnesium matrix composite. Mater Des 109:300–313
9. Shen MJ, Wang XJ, Ying T, Wu K, Song WJ (2016) Characteristics and mechanical properties of magnesium matrix composites reinforced with micron/submicron/nano SiC particles. J Alloy Compd 686:831–840

Microstructural Characterization of Aluminium Alloy 6061 Powder Deposit Made by Friction Stir Based Additive Manufacturing

Akash Mukhopadhyay and Probir Saha

Abstract Consolidation of powder inside Additive Friction Stir (AFS) tool poses a problem of hindering its application. Additive Friction Stir Processing (AFSP) is a simple modification of this technique which can negate this. A slightly modified version of this AFSP is used in the present work to successfully make a 3D deposit structure of height 5 mm and width 64 mm. Microstructural study was done both in longitudinal and transverse directions. Induced isotropy was predicted from similar microstructures in these directions. No layer effect was seen in a particular deposit layer, unlike layered Friction Stir Welding. Also, the microstructure was similar in the deposit along the build-in direction. Reduction in the number of voids that was observed in the deposit might help to increase its hardness and other mechanical properties.

Keywords Additive friction stir · Additive friction stir processing · 3D structure · Aluminium 6061 alloy powder · Microstructure · Isotropy

1 Introduction

Puleo et al. [1] stated that metal 3D parts fabricated by typical subtractive manufacturing process can lead to 90% scrap from the starting material. The optimal additive manufacturing process which can replace this subtractive manufacturing should require less material and should have the capacity to improve the strength-to-weight ratio. Selective laser sintering, laser beam melting, electron beam welding, layered friction stir welding are some of the additive manufacturing processes which can be used for metal-based 3D printing. Sing et al. [2] stated that Selective Laser Melting is known to produce parts with some porosity. High laser powers used in laser beam

A. Mukhopadhyay · P. Saha (✉)
Department of Mechanical Engineering, Indian Institute of Technology Patna, Bihta, Patna 801106, Bihar, India
e-mail: psaha@iitp.ac.in

A. Mukhopadhyay
e-mail: mjakash.ju@gmail.com

© Springer Nature Singapore Pte Ltd. 2021

P. M. Pandey et al. (eds.), *Advances in Production and Industrial Engineering*, Lecture Notes in Mechanical Engineering,
https://doi.org/10.1007/978-981-15-5519-0_11

melting may result in increased loss of metal through evaporation, excessive spatter formation, and lower surface quality due to balling [3]. In contrast, high temperature and gradient is generated in electron beam welding. It leads to rapid grain growth, anisotropic microstructure, stress concentrations in the parts produced which can also have typical cast defects [4]. Liquid–solid phase transformation associated with this process is the reason behind these problems [5]. Friction based solid-state additive manufacturing (SSAM) like layered friction stir welding eliminates all the aforesaid limitations [6]. However, to make the process more successful, it requires machining between each consecutive layer for good contact surfaces and as a result, layered microstructures are formed [4].

Additive Friction Stir (AFS) is a new-age solid-state manufacturing technology used for additive manufacturing. It is believed to avert most common problems faced during solid-state additive manufacturing technologies like Friction Stir Additive Manufacturing (FSAM), layered Friction Stir Welding, Friction Surfacing, etc. In AFS, solid rod or powder is used as filler material which is pushed through hollow tool onto the substrate. A solid deposit is fabricated from this filler material with the help of the friction stirring action of the tool against the substrate. Kandasamy et al. [7] performed AFS using AA6061-T6 aluminium alloy and AZ31B and AZ31-H24 magnesium alloy. While examining tensile strengths of deposited AA6061-T6, it has found to be similar to AA6061-O after solutionizing. According to them, the powder deposition rate is the controlling process parameter of AFS and it is governed by tool geometry and tool traverse rate. WE43 alloy as solid filler rod and powder form is used in AFS by Calvert et al. [4] Study of the microstructure revealed good intermixing between additive layers. Isotropic tensile properties in case of the deposited materials were also observed. AFS is performed by Rivera et al. [8] who have used solid filler rod of IN625 with the substrate of HY80. Electron Backscatter Diffraction (EBSD) results show fine and equiaxed grains in the microstructure as a result of grain refinement which was due to plastic deformation associated with this process. Daye et al. [9] added transition metals like Ti, Cr, Mn, Mo, Fe, Ni, si, etc., to make Powder Metallurgy aluminium alloys during AFS operation. The microstructure of the deposit showed similarity to that of the hot extruded powder. Bonding of the powder with the substrate was deemed good as fully dense microstructure was detected.

Majority of the works on AFS was found using solid rods as the filler material. In the meantime, works on powder-based AFS was done using only aluminium with transition metals and magnesium alloys. Literature works prove that AFS can be of great use to produce functionally graded structures. But, these powder-based AFS encounters a problem like consolidation of powders inside the hollow tool. This consolidated powder chokes the exit of the tool and hinders the process from reaching its full potential. The productivity also gets minimized because, for the successful implementation of the process, it requires to remove this consolidation from inside the hole so that the powder starts flowing again. Additive Friction Stir Processing (AFSP) is a slight modification of the process which was previously developed by the authors to avoid the issues faced. In AFSP technique, the powder is provided from the side of the tool instead of feeding through it. Authors were inspired by

Friction Stir Processing (FSP) to develop the process. But, the difference being the objective of the process is to form a 3D structure instead of surface modification of the substrate. The AFSP process is further altered slightly in the present work to make it more simple and convenient. The feasibility of this new AFSP process is checked by using Al6061 alloy powder.

The objectives of the present work were (i) to implement AFSP successfully in order to make deposit layers of Al6061 from Al6061 alloy powder onto a substrate of Al6061-T6 plate, (ii) microstructural characterization of the additive layers by investigating the optical microscope and scanning electron microscope images to find out the grain size, bonding between successive deposit layers and deposit and substrate, to check any induced superplasticity.

2 Experimental Details

2.1 Performing AFSP

The proposed AFSP process could successfully avert the problems faced during AFS. Sintering of powder inside the AFS tool happened due to the transfer of heat from the work-tool interface toward the inside of the tool.

The filler material used in the present work was aluminium alloy 6061 powder of particle size 50 μm. The chemical composition of aluminium alloy 6061 powder is shown in Table 1.

A vertical milling machine (Make: Tengzhou Wellon Machinery Co. Model: XL5036B) is used for the implementation of AFSP. Figure 1 shows the schematic and actual photograph of the process along with the final product.

The substrate used here is a 10 mm × 200 mm × 200 mm Al6061-T6 plate. The additive layers are formed onto this. Four tracks were deposited side by side each having width of 16 mm and height of 5 mm. The experimental setup which is shown in Fig. 1 consists of (i) a pipe which can deliver powder from the container to the substrate (ii) an AFSP tool which rotates on the substrate to perform the friction stirring. The probe-less cylindrical tool which consists of a concave bottom and a chamfered corner is used in this experiment. The schematic of the AFSP tool along with dimensions is shown in Fig. 2.

The container used in this experiment is filled with aluminium alloy Al6061 powder. The pipe which was fitted to it was used to make the powder flow toward the working zone. This flowing action took place due to the action of gravitational force. The Al6061-T6 substrate surface was pre-heated by the rotation of the tool for

Table 1 Chemical composition of the filler material (weight %)

Element	Al	Cr	Cu	Fe	Mg	Mn	Si	Sn	Zn
Al6061 alloy powder	95.86	0.07	0.34	1.83	0.80	0.14	0.70	0.10	0.16

Fig. 1 **a** The actual photograph of the AFSP experimental setup, **b** close-up view of the working zone and, **c** deposit on the substrate

Fig. 2 Schematic of the AFSP tool (All dimensions are in mm)

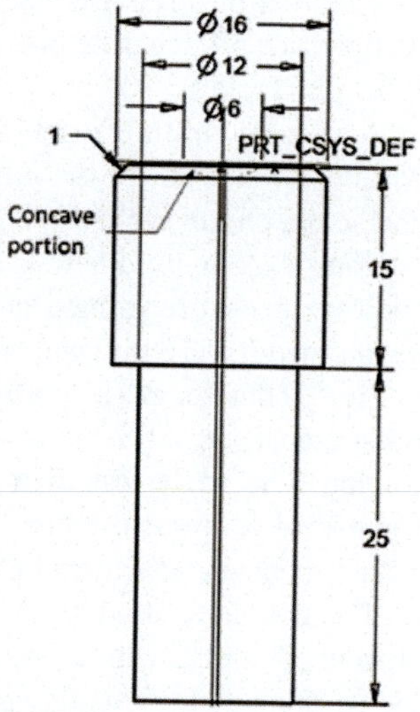

some time before the actual operation. This powder was plasticized by the friction stirring operation. Pre-heating of the substrate surface made the process easier. The dispersion of the powder particles from the working zone is also minimized by this pre-heating. The early phase of the operation shows a high tendency of the dispersion of the powder which is due to the centrifugal force generated by the rotation of the tool. This is believed to be due to the smoothness of the substrate surface. Pre-heating also helps in minimizing this dispersion during the early phase. The concave bottom surface of the tool also helped in this regard. But, it was seen that this dispersion got reduced as the deposit surface which was formed thereafter was textured. The powder was held more firmly to this textured surface. Filler powder easily passed to the working zone through the chamfered geometry at the corner of the tool.

The demonstration for the AFSP process is shown in detail in Fig. 3. The gap between the tool and the substrate surface or the tool and the deposited layer controlled the thickness of the layer. During the current experiment, this gap was set at 0.2 mm. A deposit layer was formed by the plasticization due to the friction stirring action of the AFSP tool. The friction stirring took place due to the friction between the rotating tool and the substrate or the rotating tool and the deposit layer. After the formation of one layer, the tool was taken to the side of this track and another layer was formed parallel to this. In this way, four parallel tracks were formed side by side. After all these steps, the tool was taken above the first layer at a height of 0.2 mm from the top surface of this layer. Then, the AFSP process helped to build the second layer above it. The powder left after the formation of one layer was used in the following layer. This cycle was repeated 25 times and a 3D structure with height 5 mm and width 16 mm is formed which is the same as the diameter of the tool. This structure is described here as a single track. Four such tracks were formed sidewise. Some trial experiments were done to finalize the process parameters and they were tool rotational speed of 1700 rpm and traverse speed of 22 mm/min.

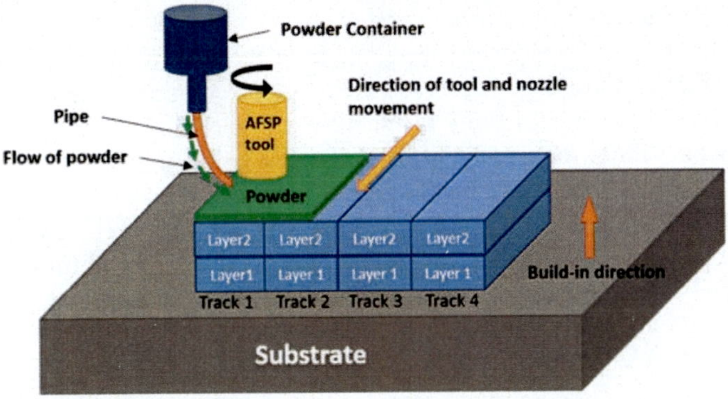

Fig. 3 Demonstration of AFSP

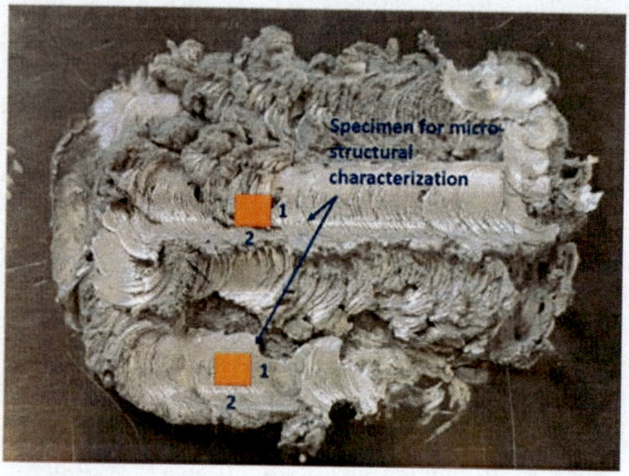

Fig. 4 Demonstration of different microstructural characterization samples cut from the workpiece

2.2 *Details of Microstructural Characterization and Micro-hardness Testing*

Samples needed for microstructural characterization were cut from the deposit and substrate. These samples had the cuboid geometry having a square face with the dimension of 5 mm and length equal to the total height of the substrate and the deposit. These specimens are cut from the workpiece as shown in Fig. 4 where the top view is given. The microstructures were checked in both longitudinal (1) and transverse (2) directions (as shown in Fig. 4).

The microstructural specimens were cut from the workpiece with the help of a wire-cut Electro Discharge Machine (EDM) of Electronica make and model of SPRINTCUT. The microstructures were obtained with the help of an optical microscope (model: STM6; make: OLYMPUS). The SEM images were taken with the help of a Field Emission Scanning Electron Microscope (make: HITACHI; model: S4800). Modified Poulton reagent (6 g CrO_3 in 20 ml H_2O; 15 ml HCl; 20 ml HNO_3; 1.25 ml HF; 1.25 ml H_2O) was used as the etchant and was applied for 20.

The same samples were subjected to micro-hardness testing both in longitudinal and transverse directions in a Vicker's micro-hardness testing machine (model: ZHμ/ST2000; make: Zwick/Roell) with 500 gf vertical load.

3 Results and Discussion

3.1 *Microstructural Characterization*

The aluminium alloy Al6061 deposit was made successfully over a substrate of Al6061-T6 plate having a thickness of 10 mm, length and width of 200 mm. Aluminium alloy Al6061 powder was deposited onto a substrate of Al6061-T6 plate.

The deposit and the substrate were subjected to microstructural characterization after the completion of the AFSP process.

Figure 5(I)a–d show the different deposit layer interfaces in longitudinal direction and Fig. 5(II)a–d show different deposit layer interfaces in transverse direction. Also, the micrograph of the substrate in longitudinal and transverse direction could be seen from Fig. 5(I)e and (II)e, respectively. Twenty-five layers of Al6061 powder were deposited on the Al6061-T6 substrate surface. The layers and their interfaces show similar microstructural properties along the deposit direction. For reference purpose, we have taken the adjacent interfaces of 2nd, 3rd, 4th; 9th, 10th, 11th; 16th, 17th, 18th; 23rd, 24th, 25th deposit layers. No visible difference in the microstructures could be seen from the figures showing the interfaces. At the same time, the deposit interfaces could not be differentiated from the optical microscope images. It indicated that the different layers were intermixed properly. Severe plastic deformation happened at the stirring zone and there was intimate contact of the tool with the powder particles and with the substrate surface during the formation of initial layers. Also, the similarity was also there between the images of Al6061 powder deposit and Al6061-T6 substrate. Hence, it could be predicted that the properties of the deposited layers could be comparable to those of the substrate. Furthermore, similarity could be seen in the micrographs in longitudinal and transverse directions. So, it can be said that the deposited material exhibit isotropic properties. Isotropic properties in the deposit enhance the flexibility in structural design. According to Yuqing et al. [10], layered microstructures were present in layered friction stir welding due to the presence of nugget zone, thermomechanically affected zone and heat-affected zone. But, there was no prevalence of layered structure in the deposit fabricated by AFSP. This is due to the presence of the whole width of the deposit layer just under the surface of the tool and similar thermomechanical history experienced by them due to friction stirring action. It was also observed, that the presence of voids got minimized from the Al6061-T6 substrate to the Al6061 powder deposit. This could again pertain to the thermomechanical effect of the tool on the powder deposit. As a consequence, the mechanical strength of the deposit got increased.

3.2 *Micro-hardness Testing*

Vicker's micro-hardness were measured in four spots in the deposit region both in longitudinal and transverse directions and the average of these values were calculated and compared to Al6061 alloy in the solutionized state. This comparison is shown in Table 2.

The similar values of micro-hardness in longitudinal and transverse direction prove the presence of isotropy which was predicted from the micrographs. Also, the micro-hardness values were comparable to Al6061-O. So, the mechanical strength was maintained.

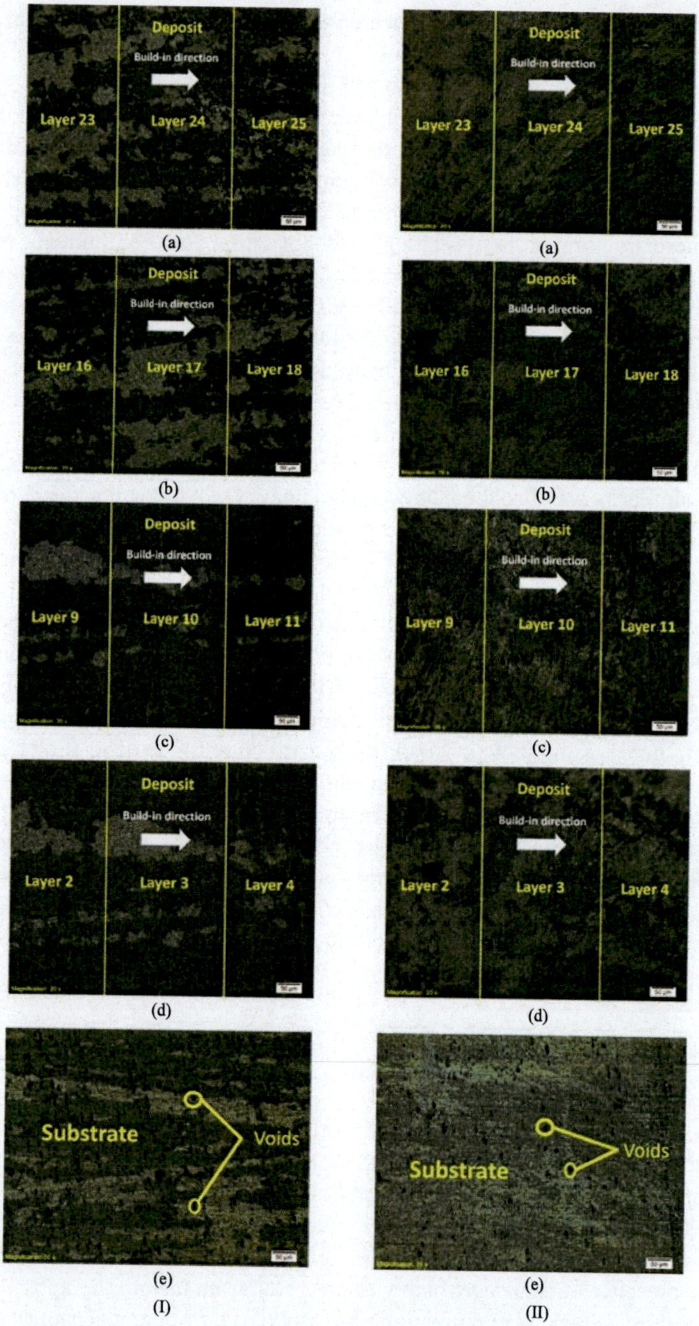

Fig. 5 Micrographs of **a–d** deposit and **e** substrate in (I) longitudinal and (II) transverse directions

Table 2 Comparison of micro-hardness of Al6061-O as per ASTM [11] and the deposit

Average micro-hardness (HV)		
Al6061-O	Deposit (Longitudinal)	Deposit (Transverse)
64	60.25	70.25

4 Conclusions

1. During AFS, choking of the mouth of the tool happens due to the consolidation of powder. Thus, the full potential applicability of the AFS process is hindered but this was successfully negated by AFSP. A slightly modified AFSP is used to fabricate 5 mm thick deposit consisting of four parallel tracks each having width of 16 mm from Al6061 alloy powder. A concave geometry at the bottom surface was given to the tool and the substrate surface was pre-heated in order to reduce the dispersion of powder from the working zone.
2. The concave geometry provided at the bottom of the tool proved to be helpful in reducing the dispersion of powder from the working zone. Also, in order to aid this purpose, the substrate was pre-heated.
3. No visible interfaces in the deposit were seen which proved good intermixing and bonding between layers.
4. Uniform microstructures in longitudinal and transverse directions prove the presence of isotropic properties in the deposit which is supported by the similar values of micro-hardness in longitudinal and transverse directions.
5. Presence of uniform microstructure in a layer proved the absence of layer-effects like in layered friction stir welding.
6. Similar microstructure was noticed in substrate and deposit layers. Moreover, the number of voids got reduced in the deposit from the substrate. This indicates similar or slightly enhanced mechanical properties in the Al6061 alloy deposit. This claim is again proved by results of micro-hardness testing.

Acknowledgements Authors would like to express their thanks to the SERB, CRG, Govt. of India, for the financial assistance provided for the project titled "Mechanical and micro-structural characterization of additive friction stirred (AFS) 3D structures made of Al6061 T6 aluminium powder." (Project File No. CRG/2018/001338)

References

1. Puleo SM (2016) Additive friction stir manufacturing of 7055 aluminum alloy. Honors thesis, University of New Orleans
2. Sing SL, Yeong WY, Wiria FE, Tay BY, Zhao Z, Zhao L, Tian Z, Yang S (2017) Direct selective laser sintering and melting of ceramics: a review. Rapid Prototyp J 23(3):611–623. https://doi.org/10.1108/RPJ-11-2015-0178

3. Mertens AI, Delahaye J, Lecomte-Beckers J (2017) Fusion-based additive manufacturing for processing aluminum alloys: state-of-the-art and challenges. Adv Eng Mater 19(8):1700003(1–13). https://doi.org/10.1002/adem.201700003
4. Calvert JR (2015) Microstructure and mechanical properties of WE43 alloy produced via additive friction stir technology. Master thesis, Virginia Tech
5. Palanivel S, Sidhar H, Mishra RS (2015) Friction stir additive manufacturing: route to high structural performance. JOM 67(3):616–621. https://doi.org/10.1007/s11837-014-1271-x
6. Palanivel S, Nelaturu P, Glass B, Mishra RS (2015) Friction stir additive manufacturing for high structural performance through microstructural control in an Mg based WE43 alloy. Mater Des 65:934–952. https://doi.org/10.1016/j.matdes.2014.09.082
7. Kandasamy K, Renaghan LE, Calvert JR, Creehan KD, Schultz JP (2013) Solid-state additive manufacturing of aluminium and magnesium alloys. In: Proceedings of materials science and technology, montreal, Quebec, Canada, pp 59–69
8. Rivera OG, Allison PG, Jordon JB, Rodriguez OL, Brewer LN, McClelland Z, Whittington WR, Francis D, Sue J, Martens RL, Hardwicke N (2017) Microstructures and mechanical behavior of Inconel 625 fabricated by solid-state additive manufacturing. Mater Sci Eng A 694:1–9. https://doi.org/10.1016/j.msea.2017.03.105
9. Daye WK, Pelletiers I, Thomas W (2018) Property development of new generation pm aluminum materials via innovative processing. Int J Powder Metall 54(2):35–51
10. Yuqing M, Liming K, Chunping H, Fencheng L, Qiang L (2016) Formation characteristic, microstructure, and mechanical performances of aluminum-based components by friction stir additive manufacturing. Int J Adv Manuf Technol 83:1637–1647. https://doi.org/10.1007/s00170-015-7695-9
11. ASM International Handbook Committee (Ed.) (1990) Properties and selection: nonferrous alloys and special-purpose materials. Asm Intl 2 ISBN 0–87170–378–5

Service Quality—A Case Study on Selected Hotels in Goa

Vallabh S. Prabhu Gaunker and Rajesh S. Prabhu Gaonkar

Abstract Goa is an international tourist destination where tourism is considered to be one of the main sources of revenue generation for the state. Hotel is considered to be a part of the hospitality sector in a service industry. Service and customer satisfaction are the prime objectives of any service-related organization. The service offered by every hotel is not uniform across all the hotels. Therefore, in order to standardize the service quality and reduce the gap, it is essential to fix service quality attributes which determine the customer satisfaction. A case study has been carried out on four hotels of five-star category across the state of Goa. The paper aims at measuring service quality and examining service quality attributes significant to the customer satisfaction. A SERVQUAL model was chosen for the conceptual frame work of the data. A feedback data of 200 respondents across hotels and travel websites were analyzed by gap, factor and the multiple regression analysis. The analysis results revealed a gap in the service quality perceived and expected by the customers. Exploratory factor analysis extract three factors of service quality, namely service reliability, staff assurance and physical facilities which significantly impact customer's satisfaction measuring 51.061% of total variance. The relationship between three extracted factors of service quality and customer satisfaction is expressed in the form of multiple regression equation. This study will help the managers as a guide to improve service quality and enhance the customer satisfaction across the state of Goa.

Keywords Service quality · Customer satisfaction · SERVQUAL · Expectations · Perception

V. S. P. Gaunker (✉)
Mechanical Engineering Department, Goa College of Engineering, Goa, India
e-mail: vallabh21@yahoo.co.in

R. S. P. Gaonkar
School of Mechanical Sciences, Indian Institute of Technology, Goa, India

© Springer Nature Singapore Pte Ltd. 2021

P. M. Pandey et al. (eds.), *Advances in Production and Industrial Engineering*, Lecture Notes in Mechanical Engineering,
https://doi.org/10.1007/978-981-15-5519-0_12

1 Introduction

Goa is an international tourist destination where tourism is one of the fastest growing sectors. It not only boosts state economy but also contribute to the growth of other sectors. It is a big employment generator and has a multiple impact on the state economy. During the year 2017, the state recorded a total of 0.85 million foreign tourists across 141 countries as per the data available from the Department of Tourism, Goa. Tourism sector alone plays a vital role in the economic growth of the state with a net GDP of 8.78% from the year 2011 to 2017. In the poll conducted by Smart Asia travel.com, Goa is ranked as the sixth most popular holiday destination in Asian region among the foreign tourist travelling to India.

This paper aims at measuring the service quality and examining the impact of service quality dimensions on customer satisfaction. A case study has been carried out on four hotels of five-star categories located along the coastline of south. The names of the hotels and customers are kept anonymous as per company privacy policies. The research was conducted for three months to collect the data.

As the research has no inclination towards lifesaving neither it is related to medical field, the ethical point of view was not taken into consideration. The Institute has no Ethical committee set up. The data collected for the research is validated and approved by guide Dr. Rajesh Prabhugaonkar Professor, Goa College of Engineering and Dr. Vinay Shirodkar, HOD Mechanical Engineering Department, Goa College of Engineering and the data is anonymized. The dissertation based on the research paper has undergone plagiarism check of Goa University.

A modified SERVQUAL instrument has been used to frame the questionnaire in order to measure and examine the service quality [1, 2]. Customer feedback data on perceived and expected service quality were recorded using a Likert scale on 1–5 score. The qualitative and quantitative analysis is performed using gap analysis, exploratory factor analysis and multiple regression analysis for quality measurement and hypothesis testing. The last section concludes recognizing the weaknesses and suggesting improvements necessary so that service providers undertake quality improvement measures for customer satisfaction.

2 Brief Literature Review

2.1 Service Quality

Service quality plays an important role for business performance and has been widely recognized in various literatures. It is a complex and subjective concept and has some distinct characteristics unlike product such as intangibility, heterogeneity and inseparability. It is defined as an assessment of how well delivered service confirm to customer expectations [1]. The gap in the service quality is the difference between perceived service (P) and the expected service quality (E). The gap measures how well the delivered service confirms to the customer expectations. The expectations act as a benchmark to which successive experiences are compared which results in the evaluation of satisfaction. The satisfaction depends on many factors like tangible things, facilities offered, responsiveness, staff empathy, service reliability, information accuracy, location, value for money, etc.

2.2 *Customer Satisfaction*

Customer satisfaction is considered as significant for any of service-related organization. Companies strive hard to improve their service level. A number of customer satisfaction indicators have been developed by different countries. These indicators are considered as a benchmark for measuring the performance. Therefore, measurement of service quality and its impact on customer satisfaction has been considered as crucial in the service-related industries including tourism industry.

It is concluded that providing good service quality helps to retain the customers, improve corporate image, firm survival and profitability [3]. Managers would focus on empathy, reliability, responsiveness and assurance to achieve customer satisfaction [4]. The study on Thai hotel industry revealed moderately low service quality than expectation [5]. A SERVQUAL model is implemented in four service organizations and it was observed that service quality has relationship to cost, profitability, retention and positive word of mouth [6]. SERVQUAL model used in the study came out with the findings that competence, courtesy along with assurance are the important attributes of perceived service quality [7]. Customer satisfaction level is examined in five-star hotel in Sri Lanka. Customers were satisfied on the factors like tangible, responsiveness and assurance of the service quality [8]. It is revealed that room cleanliness and comfort, convenience of location, prompt service, safety and employee friendliness are the important factors for customer satisfaction [9]. It was noted that staff quality, room quality and value determine the customer satisfaction [10]. Customer satisfaction is determined by cleanliness, security, value for money and courtesy determine customer satisfaction [11]. The perceived service quality of the customers is best explained by service quality dimensions like responsiveness, empathy, assurance and reliability [12]. This shows that the number of factors depends on measurement constructs. The analysis result revealed conformity of expected service with experienced service has significance influence on customer loyalty [13]. The result illustrates modified SERVQUAL dimensions extracted four subscales in the new model instead of eight in the initial model [14].

The SERVQUAL instrument has been used in various service industries to measure service quality. It is based on expectation perception standards [2]. It was originally based on ten service quality dimensions which later improved to five dimensions namely Tangible, Reliability, Responsiveness, Assurance and Empathy. The other scales which are developed based on SERVQUAL are SERVPERF [15], Technical and Functional Quality Models [16], HOLSERV [17].

3 Research Method

The research objective is to measure gaps in the service quality and to examine the factors that significantly impact customer's satisfaction from the selected four hotels of the five-star categories from Goa. The data from 200 customers is gathered through personal interviews and partially from the reviews obtained from travel sites such as Trip advisor, make my trip, go-ibibo using SERVQUAL questionnaire on a five-point Likert scale. A statistical software tool of Microsoft Excel and SPSS 24 is used to analyze the data for measuring service quality and hypothesis testing. Initially the data is tested for sampling adequacy and scale reliability. The quantitative analysis like gap analysis, factor analysis and multiple regression analysis is performed. The gap analysis measure average service quality based on five SERVQUAL dimensions. Exploratory analysis extracted three factors of service quality having significant relationship with customer satisfaction which is expressed in the form of equation.

3.1 Gap Analysis

The concept to measure service quality is based on obtaining average gap score of SERVQUAL dimensions for each of the item variable [2]. When perceived service quality is high, the gap score is less and customer satisfaction level is higher.

Figure 1 depicts Gaps in measuring service quality. The service quality is measured on five dimensions of SERVQUAL namely tangible, reliability, responsiveness, assurance and empathy. The perceived values based on these dimensions are 4.58, 4.56, 4.54, 4.51 and 4.49, respectively against an expectation score of five in each case. The average service quality perceived by the customer's is 4.54 as compared to their expectation level of five thus indicating less satisfaction. A negative gap of

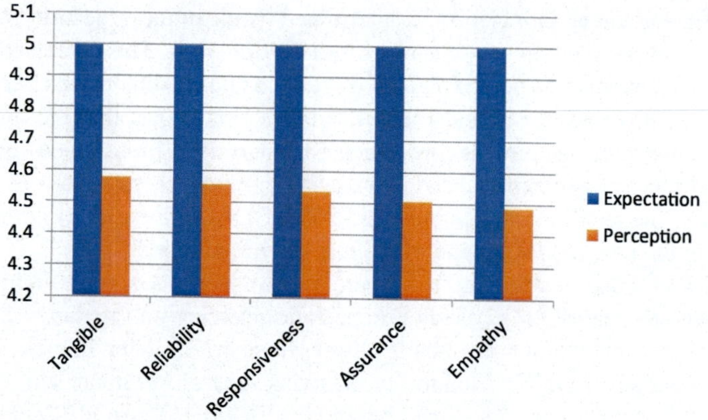

Fig. 1 Measuring gap in service quality

−0.46 reflect that the service quality perceived by the customer is lesser than their expectations.

3.2 *Reliability Analysis*

Reliability scale analysis test the reliability and validity of SERVQUAL model by computing the value of Cronbach's alpha for each of the item of service quality as depicted in Table 1.

Cronbach's alpha value for the overall scale of 19 items is 0.899 which is more than acceptable value of 0.7. It is observed that the item mentioned in Sr. No 11 shows a Cronbach's value of 0.901 if item is deleted. As a result it is not considered for further analysis. The content validity ensures that the measure includes an adequate and representative set of items that tap the domain of the concept.

3.3 *Factor Analysis*

Factor analysis is a data reduction technique used to identify correlations or patterns among the variables in a complex set of data. An exploratory factor analysis (EFA) was undertaken explore factor structure using Varimax rotation (orthogonal).

Table 2 depicts KMO (Kaiser–Meyer–Olkin) and Bartlett's test of Sphericity. KMO measures sampling adequacy and Bartlett's test gives significance of the study. KMO of 0.897 is determined which is more than acceptable value of 0.6 according to KMO measures. A significance p-value of 0.000 indicates that data is adequate to perform factor analysis and dimensions of the service quality is likely going to factor well.

After conducting EFA by principal component analysis (PCA) method, three components are with Eigen values greater than one are retained. The eigenvalues for the extracted factors are 6.775, 1.341 and 1.074. These three factors explain percentage cumulative variance of 51.061%. Factor one explains relatively large amount of total variance of 37.64% followed by other two factors as 7.452% and 5.969%, respectively as depicted in Table 3.

In multivariate statistics, screen plot is the line plot of the eigenvalues of factors. The elbow of the graph is to be found and factors to the left of these points are to be retained as significant factors for the analysis. Figure 2 depicts three components of service quality that are to be retained based on the eigenvalues greater than one. Eigen value signifies that a reasonable proportion of variance is explained by the three extracted factors.

Table 1 Item scale reliability

Sr. No	Item statements	Cronbach's alpha for 19 items	Cronbach's alpha if item is deleted
1	Physical facilities of the Hotel were visually appealing	0.899	0.897
2	Employees appear neat and well dressed		0.894
3	Physical environment of the hotel was clean		0.893
4	Hotel room were with comfort and ambience		0.890
5	Hotel food was good and variety in beverages		0.895
6	My reservation at the front desk handled properly		0.894
7	When employee promised to do something, they did it		0.893
8	Hotels performed the service right the first time		0.892
9	Hotels kept their records and billing accurately		0.892
10	Hotel employees gave prompt service		0.889
11	Employees were always willing to help		0.901
12	Employees make information easily available		0.893
13	Employees were never too busy to respond to request		0.892
14	The behaviour of employees-built confidence in customers		0.892
15	Feel safe in transactions with the employees		0.893
16	Employees were courteous and knowledgeable to answer inquiries		0.896
17	Employees understand customer's specific needs		0.897
18	Hotels gave customers individual attention		0.892
19	Employees gave personal service with customer's interest at heart		0.893

Table 2 KMO and Bartlett's test of sphericity

Kaiser-Meyer-Oklin measure of sampling adequacy		0.897
Bartlett's test of sphericity	Approximate Chi-Square	1322.382
	Degree of freedom (df)	153
	Significance (p-value)	0.000

3.3.1 Component Matrix with Varimax Rotation

Rotated component matrix is a significant part of factor analysis. Table 4 depicts rotated component matrix with factor loading coefficients. These coefficients give the strength of the relationship between variable and the factor. The value greater than 0.5 reflects more measure of service quality. The variables are grouped under three factors. Service quality factor one is titled as SERVICE RELIABILITY, having six variables with factor loading ranging from 0.561 to 0.637. Factor two is titled as STAFF ASSURANCE with seven variables and factor three is titled as PHYSICAL FACILITIES as deciding factors of service quality measure.

3.4 Multiple Regression Analysis

One of the targets of conducting this study is to investigate the relationship between service quality and customer satisfaction. Therefore, the hypothesis is proposed for the research and multiple regression analysis is performed. The analysis predicts relationship between the set of dependent variables with the independent variables. Customer satisfaction is taken as dependent and service quality attributes as independent variables. The data is analyzed by calculating mean of each variable.

The hypothesis proposed for the analysis is:

Null Hypothesis (H_0): The service quality dimensions have no significance on customer satisfaction.

Alternate Hypothesis (H_1): There is significant impact of service quality dimensions on customer satisfaction.

Model summary of the regression analysis determines how well the data fits in the regression model by computing values of multiple correlation coefficients (R), coefficient of determination (R^2) and standard error of the estimate (SEE). As depicted in Table 5, the correlation coefficient (R) is 0.641. The regression results showed that the service quality is a good significant predictor of customer satisfaction. The three predictor variables of service quality explain 41.10% of its variance in customer satisfaction. The adjusted R^2 which is goodness of fit of regression model is 0.402. Adjusted R^2 is intended to control overestimates of population which occurs either because of a small sample, small variables or high colinearity. The standard error of estimate is 0.514 which is actually the standard deviation of the residual.

Table 3 Total variance explained

Initial eigen values				Extraction sums of squared loadings			Rotation sums of squared loadings		
Component	Total	% of variance	Cumulative %	Total	% of variance	Cumulative	Total	% of variance	Cumulative%
1	6.775	37.640	37.640	6.775	37.640	37.640	3.107	17.262	17.262
2	1.341	7.452	45.092	1.341	7.452	45.092	3.048	16.934	34.196
3	1.074	5.969	51.061	1.074	5.969	51.061	3.036	16.865	51.061
4	0.941	5.228	56.288	–	–	–	–	–	–

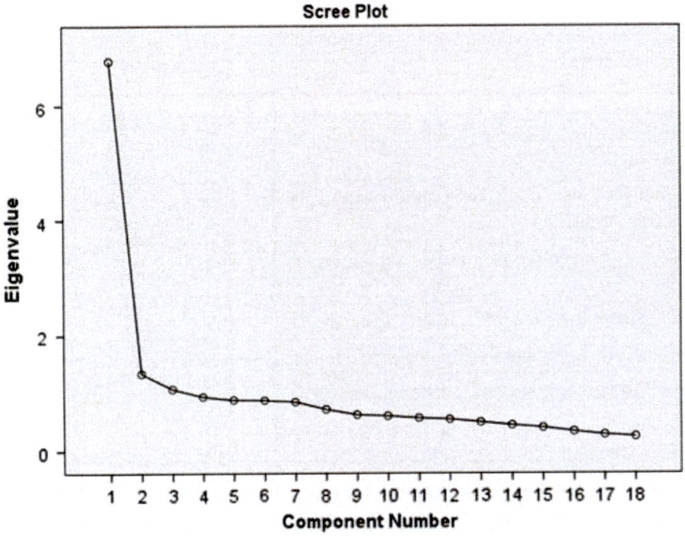

Fig. 2 Scree plot

As R^2 increases the SEE will decrease, meaning better fit of the model with less estimation error.

The ANOVA test is a way to find out if analysis results are significant and whether to reject the null hypothesis by computing p-values. ANOVA compares variances between populations. Any *p*-value less than alpha value of 0.05 means the null Hypothesis is rejected as there exists a significant relationship between dependent and independent variable. ANOVA shows that the independent measures of service quality extracted are statistically significant with customer satisfaction. Table 6 summarizes the multiple regression model with three predictors with $R^2 = 0.411$ at $F(3, 196) = 45.551, p < 0.05$. This indicates that regression model is a good fit of the data.

Table 7 illustrates the predictors of service reliability, staff assurance and physical facilities shows statistically significant relationship with customer satisfaction. The beta value implies that each added point to the independent variable increases customer satisfaction by beta value. The unstandardized beta coefficient value for service reliability is 0.254 for staff assurance 0.439 and for physical facilities as 0.344, respectively. Multiple regression model rejects the null hypothesis (H_0) with conclusion that there exists a significant relationship between service quality and customer satisfaction. The model is expressed in the Eq. (1).

$$\begin{aligned}\text{Customer Satisfaction} = & -0.194 + (0.254 \times \text{Service Reliability}) \\ & + (0.439 \times \text{Service Assurance}) + (0.344 \times \text{Physical Facilities})\end{aligned} \tag{1}$$

Table 4 Rotated component matrix

Sr. No	Item statements	Components		
		Factor 1	Factor 2	Factor 3
1	The behaviour of employees-built confidence in customers	0.637	–	–
2	Employees were courteous and knowledgeable to answer inquiries	0.627	–	–
3	When employee promised to do something, they did it	0.618	–	–
4	Hotels kept their records and billing accurately	0.606	–	–
5	Employees understand customer's specific needs	0.564	–	–
6	Hotel food was good and variety in beverages	0.561	–	–
7	Employees gave personal service with customer's interest at heart	–	0.712	–
8	Hotels gave customers individual attention	–	0.581	–
9	Physical facilities of the Hotel were visually appealing	–	0.556	–
10	Employees were never too busy to respond to request	0.460	0.555	–
11	Feel safe in transactions with the employees	–	0.534	0.464
12	My reservation at the front desk handled properly	–	0.533	–
13	Employees make information easily available	0.446	0.487	–
14	Hotels performed the service right the first time	–	–	0.767
15	Physical environment of the hotel was clean	–	–	0.716
16	Employees appear neat and well dressed	–	–	0.654
17	Hotel room were with comfort and ambience	–	0.471	0.653
18	Hotel employees gave prompt service	–	0.480	0.621

Table 5 Model summary

Predictor variables	Dependent variable	Correlation regression coefficient (R)	Coefficient of determination (R^2)	Adjusted (R^2)	Standard error of estimate (SEE)
Service reliability staff assurance physical facilities	Customer satisfaction	0.641	0.411	0.402	0.514

Table 6 ANOVA (Dependent variable: customer satisfaction)

Model	Sum of squares	Degree of freedom (d_f)	Mean square	F-statistics	Significance
Regression	36.117	3	12.039	45.551	0.000[a]
Residual	51.803	196	0.264	–	–
Total	87.920	199	–	–	–

[a]Predictors: constant, service reliability, staff assurance, physical facilities

Table 7 Regression coefficient (Dependent variable: customer satisfaction)

Model	Unstandardized coefficients		Standardized coefficients	t Statistics	Significance p-value
	Beta (β)	Standard error	Beta		
Constant	−0.194	0.411	–	−0.473	0.637
Service reliability	0.254	0.077	0.349	6.104	0.000
Staff assurance	0.439	0.072	0.349	6.104	0.000
Physical facilities	0.344	0.066	0.321	5.231	0.000

4 Results and Discussion

The study was conducted to determine gap in service quality and to examine service quality attributes having significant relationship to customer satisfaction. The customer expectation and perception of the service quality was examined. The study was conducted on four hotels of five-star category taking feedback of 200 respondents. The data is collected using SERVQUAL structured questionnaire. Analysis like gap analysis, exploratory factor analysis and multiple regression analysis were performed on the data received. The gap scores analysis results in low service quality perceived by the customers on all dimensions of service quality. The average service quality perceived by customer is measured 4.54 against an expectation score of 5 showing a gap of 0.46. This shows that the customers are less satisfied with the hotel service offered.

Preliminary analysis of the data is performed to test reliability and validity of the SERVQUAL for performing factor analysis. The model gave a satisfactory overall reliability of 0.901. Exploratory factor analysis after performing varimax rotation extracted three factors of service quality namely service reliability, service assurance and physical facilities. These three dimensions of service quality were explaining 40.2% of variance in the data.

Hypothesis is proposed and multiple regression analysis is performed to express significance relationship of extracted factors with the customer satisfaction. Multiple regression analysis explored relationship between service quality attributes to customer satisfaction. Analysis showed that these three factors are having statistically significant relationship with the customer satisfaction at 5% significance alpha level. The three independent variables of service quality significantly predict the dependent variable of customer satisfaction at, $F(3, 196) = 45.551, p < 0.05$ as shown in regression model. Therefore, the study suggests that manager of the service organization need to focus more on service reliability, service assurance and physical facilities for improving customer satisfaction.

5 Conclusions

Hotel managers should have knowledge of the market within which they operate to identify the needs of their target customer's. The hotel managers need to improve performance on all the service quality attributes so that the gaps can be minimized to increase customer satisfaction. Based on factor analysis, three important factors have been extracted which are to be given prime importance. They are Service reliability, staff assurance and physical facilities. Identifying these factors will help the managers to focus on key areas of service quality. The regression analysis helps to relate and predict the relationship between the three extracted factors called independent variables and dependent variable is called customer satisfaction. This will help managers to predict and alter the parameters as per the requirements. For further research a study can be conducted with larger sample data and more hotels to assess the service quality during peak seasons and rush hour. Also, a similar study can be done for budgetary hotels prefered by customers of lower income group.

References

1. Parasuraman A, Berry LL, Zeithaml VA (1985) A conceptual model of service quality and its implications for future research. J Market 49(4):41–50
2. Parasuraman A, Berry LL, Zeithaml AV (1988) SERVQUAL: a multiple item scale for measuring consumer perception. J Retail 64(1):12–40
3. Ladhari R (2009) A review of twenty years of SERVQUAL research. Int J Qual Serv Sci 1(2):172–198
4. Minh NH et al (2015) Service quality and customer satisfaction: a case study of hotel industry in Vietnam. Asian Soc Sci 11(10):1911–2025
5. Boon-itt S, Rompho N (2012) Measuring service quality dimensions: an empirical analysis of Thai hotel industry. Int J Bus Admin 3(5):52–63
6. Buttle AF (1996) SERVQUAL: review, critique, research agenda. Eur J Mark 30(1):8–32
7. Akan P (1995) Dimensions of service quality: a study in Istanbul. Managing service quality 5(6):39–43
8. Karunaratne WMKK, Jayawardena LNC (2010) Assessment of customer satisfaction in a five-star hotel-a case study. Trop Agric Res 21(3):258–265
9. Knutson B, Stevens P, Wullaert C, Patton M, Yokohama F (1990) Lodgeserve: a service quality index for the lodging industry. Hosp Res J 14(2):227–284

10. Choi TY, Cho R (2001) Determination of hotel guest's satisfaction and repeat patronage in the Hong Kong hotel industry. Int J Hosp Manage 20(3):277–297
11. Atkinson A (1988) Answering the eternal question: what does the customer want. Cornell Hot Restaur Admin Q 29(2):12–14
12. Tuan PN, Linh NTH (2014) Impact of service quality performance on customer satisfaction: a case study of Vietnam's five-star hotel. ABAC J 34(3):53–70
13. Jasinskas et al (2016) Impact of hotel service quality on the loyalty of the customers. Econ Res-Ekonomska istrazivanjz 29(1):559–572
14. Pakurar M et al (2019) The service quality dimension that affect customer satisfaction in Jordanian banking sector. Sustainability 11:1113
15. Cronin JJ, Taylor SA (1992) Measuring service quality: a re-examination and extension. J Market 56:55–68
16. Gronroos C (1984) A service quality model and its marketing implications. Eur J Mark 18(4):36–44
17. Mei WOA, Dean AM, White CJ (1999) Analysing service quality in hospitality industry. Manag Serv Qual 9(2):136–143

Influence of Nozzle Distance on Tool–Chip Interface Temperature Using Minimum Quantity Lubrication

Gurpreet Singh, Vivek Aggarwal, Sehijpal Singh, and Ajay Kumar

Abstract In Manufacturing Science machining play a very important role in making the finished product. There are several metal cutting operations like, turning, milling, drilling, shaping, grinding, etc., by which the product can be machined. During machining operation heat is produced due to rubbing of sharp edge tool and work material. As the cutting speed increases, the amount of heat at different rubbing sections increases. The foremost objective of coolant is to reduce the emission of heat using different cooling methods. Traditionally, cutting fluids and lubricants were applied in abundance to reduce heat with the aid of flood lubrication system. Undoubtfully, this approach of lubrication reduces the temperature while machining, contrary to this, it has a negative impact on the worker, environment and most importantly costlier in the present era of machining. Therefore it was essential to find an alternative to stated muddle that should be safer for the worker, environmental, and economical too. So, as to keep these facts in mind the minimum quantity lubrication was applied in turning operation, which uses a very smaller extent of cutting fluid along with air to contribute cooling and lubrication action. In the present investigation, minimum quantity lubrication containing very low quantity of vegetable-based lubricant and air pressurized stream was utilized to cater heat generation in orthogonal machining. The outcomes of thermal reading at insert chip articulation were compared in dry and least coolant conditions. From experimental observations, it was revealed that MQL had significantly reduced the heat generation by 10–30% contrary to the dry turning of EN-31 steel. The maximum cutting temperature was recorded during dry machining when the nozzle was positioned at 55 mm gap from the cutting zone. It was evaluated that the nozzle distance of 30 mm gap was proved significant as compared to other distance. Furthermore, it was reported that the process capability of turning

G. Singh (✉) · V. Aggarwal
Department of Mechanical Engineering, IKGPTU, Main Campus, Kapurthala, Punjab, India
e-mail: gssingh410@gmail.com

S. Singh
Department of Mechanical Engineering, G.N.D.E.C, Ludhiana, Punjab, India
e-mail: gssingh410@gmail.com

G. Singh · A. Kumar
Department of Mechanical Engineering, Chandigarh University, Gharuan, Punjab, India

© Springer Nature Singapore Pte Ltd. 2021

P. M. Pandey et al. (eds.), *Advances in Production and Industrial Engineering*, Lecture Notes in Mechanical Engineering,
https://doi.org/10.1007/978-981-15-5519-0_13

operation was enhanced using minimum quantity lubrication due to its economical behavior, better cooling and lubrication action and environmental friendly nature.

Keywords Nozzle distance · Heat generation · Tool–chip interface temperature · K-type thermocouple · Dry machining · Minimum quantity lubrication · EN-31 steel

1 Introduction

Machining is the way toward utilizing slicing instruments to evacuate some measure of a bit of material (metal, wood, plastics, fired, and so forth.) to correctly shape it for a planned utilize. This utilization of the physical activity of cutting apparatuses is otherwise called subtractive assembling. The essential machines utilized in machining are the motor machine (metal machine), processing machine (both level and, most regularly, vertical), penetrate press, and grating processors. These and different machines can be either manual or mechanized. Most robotized machines have CNC (PC numerical control) and are equipped for creating exceptionally perplexing, exact, and complex parts with a high level of repeatable exactness for any number of uses.

During machining operation heat is produced as an effect of rubbing across sharp edge cutter and solid work material. As the cutting speed increases, the amount of heat at different contact points increases. The sole objective of cutting fluid is to reduce the amount of heat generation which can be achieved by different methods. Traditionally cutting fluid and lubricant were applied in bulk to reduce this unwanted heat using conventional flood lubrication. This flood lubrication reduces the temperature during machining but on the other hand it also has a negative impact on the worker, environment, and costlier in the present era of machining. Therefore, the other alternative to reduce the cutting tool temperature in different machining processes is to be investigated, which should be safe for the worker, environmental friendly as well as economical [1–3]. So as to keep these facts in mind the MQL was applied in turning operation. A Report on the fabrication of automotive parts in Europe concluded that that the cost related to cutting fluids and lubricant was stated approximately twenty percent of the total manufacturing cost as equivalent to cutting tools (7.5%) cost. It indicated that the lubricant cost was predominately larger than tooling cost [4, 5]. As indicated in the Table 1 there is a huge difference in the toxicity ranges of permissible exposure level and actual measured level of cutting fluid discharged by industries. It was further felt a need to reduce cutting fluids consumption, so as to make the process cost-effective, environmental friendly, and safe for the worker. Also, it was suggested to improve the conventional flood lubrication methodology possibly by the use of technology that consumes fewer lubricants. U.S firm dealing with citizen healthcare (NIOSH) evaluated that more than 10 lakhs peoples in the U.S.A are exposed to cutting fluids. Different categories of engineers, hardware mechanics, metalworkers, and other machine administrators and setters have the best contact with these liquids. Specialists might be uncovered by skin contact, breathing in (taking in), or ingesting

Table 1 Highlights of PEL level and actual level of cutting fluid [4]

<table>
<tr><th>S. no</th><th>Agency</th><th>Threshold level of toxicity in cutting fluids (mg/m^3)</th><th>Actual toxicity level observed in industries (mg/m^3)</th></tr>
<tr><td>1</td><td>Occupational safety and health administration (OSHA)</td><td>5</td><td rowspan="2">20–90</td></tr>
<tr><td>2</td><td>National institute for occupational safety and health (NIOSH)</td><td>0.5</td></tr>
</table>

(gulping) particles, fogs, and mist concentrates. MWF fogs may likewise cause a few respiratory sicknesses, including asthma, bronchitis, and excessive touchiness pneumonitis. Introduction to some MWFs and additionally their added substances may cause malignancy [4]. The well-being perils of MWFs rely on the kind of liquid utilized, and in addition the added substances and contaminants that might be available in the liquid. The extravagant use MWF should be taken into consideration besides the traditional ways of cooling and numerous other coolant strategies were developed in later years to eliminate heat from the cutting zone [6].

The traditional way of using of coolants during machining promotes frequent technological and environmental complications and avoidance of MWF can be a better alternative. On the other side, higher depth of cut and harder material machining cannot be performed under dry conditions because of the fact that it shortens tool life [7].

During the situations, where complete dry machining fails to be meet production targets at higher cutting speed and was not economical than during such conditions the minimal cutting fluid cooling technique may be adopted. It supplies the sprayed pattern of cutting fluid jet over the tool–workpiece interface that minimizes the heat generation and consequently the cost of machining as well [8]. It was concluded that proper utilization and controlling MQL parameters ensured safer machining environment as well as improved process capability particularly in machining heat generation, tool wear, and product quality [9]. Experimental investigation on the existence of tool sharpness and surface quality employing various cutting inserts during machining of Inconel-718 concluded that the high-speed MQL turning aided with coated tool had enhanced the surface quality and life of tool as compared to wet and dry machining [10]. While machining AISI 9310 alloy steel with uncoated carbide insert it was stated that MQL assisted system reduced the approximate chip–tool interaction temperature by 10% than wet machining. In addition to this quality of work surface was also boosted as a result of less wear and vandalism to tooltip contact [11].

The lowest range of machining variables such as cutting speed, feed rate, depth of cut, and insert roundness evolved minimum expenditure of specific power during machining of EN-31 steel having 50 mm diameter and 500 mm length utilizing diamond shaped carbide insert [12]. While Turning the material AISI 304L aided with six different categories of cutting fluids the results concluded that 8% canola based cutting fluid at extreme pressure overcome the machining performance of other fluids [13]. Investigations on EN-31 turning using soybean oil and mineral oil MQL expressed that vegetable oil performed equivalent to mineral oil during aspects of surface finish. Also vegetable oil MQL machining had minimized the severity of environmental degradation on the consequences of fewer fumes, mists, and oxides as compared to mineral oil [14].

Experiment results during machining of Inconel 718 assisted with synthetic ester cutting fluid MQL examined that the MQL machining was able to reduce the accumulation of heat generation, force during cutting. In addition this length of tool–chip interaction and quality of chip characteristics like thickness and breakage improved compared to dry machining environment [15]. Experimental investigation on machining hardened stainless steel using coated carbide inserts confirmed that the vegetable oil can be used as cutting fluid because of its superior lubrication and high-pressure performance capability. It was also revealed that MQL can be a good technique for cooling coated carbide cutting tool at various machining parameters [16]. It was analyzed that while machining of TC11 titanium alloy with uncoated carbide tool that the machining forces within MQL environment differ by a minute margin of ten percentage as compared to dry cutting [17]. Research results during processing of AISI 1045 steel conducted with nanofluid MQL mixed with vegetable-based oil and ester oil as base fluids revealed that the application of graphite oil-based nanofluid MQL lowered the cutting force and temperature significantly [18]. Study on FEMA which analyzed the impact of MQL during turning of JIS S45C steel using TiCN-coated cermet tool exhibited that model applied in the experimentation evaluated the cutting forces and temperature effectively were influenced by utilizing MQL [19]. Experimental observations on turning Ti6Al4V machined with tungsten coated carbide tool revealed that lowest nose wear and Surface roughness was observed during MQL + PTFE and MQC + 15% Graphite combinations. However there was no significant difference in heat carrying capacity among MQL, MQC + 5% graphite, MQC + 10%, and MQC + 15% graphite cooling strategies [20].

Experimental observation analyzed that on increasing doc the temperature increases. The least temperature was recorded in cryogenic followed by MQL, wet and dry environment. For selected material and machining conditions the Cryogenic cooling was superior to dry, wet, and MQL in terms of tool wear, surface finish, and chip formation [21]. Results indicated that mixing MOS_2 to vegetable oil enhanced the tool life, surface quality, and cutting forces. It was observed that adhesion, abrasion, and diffusion were the significant wear mechanisms during machining of Inconel 718 under different lubrication cooling conditions. Worst performance was examined during dry cutting of Inconel 718. The various performance characteristics like chippings of the cutting edge and fractures were observed during experimentation [22]. The investigation results of machining medium carbon steel using carbide tool concluded that the most significant factor for an optimum response was nozzle diameter followed by air pressure, secondary nozzle angle, oil flow rate, and primary nozzle angle. The reported optimum values of mentioned parameters were listed as

1 mm, 19 bar, 15°, 80 ml/h and 20° respectively. Nozzle diameter was predominant on the surface roughness and main cutting force during machining operation [23].

1.1 Fundamentals of MQL

The foundation of Minimum quantity lubrication (MQL) is a jet of air mixed with lubricated oil targeted toward the cutting zone in a controlled manner. The principle involves that the liquid droplet is fragmented and spread in an airflow streaks through a properly designed nozzle and guided toward the machining zone of higher cutting temperature. The compressed jet of air assisted bio-based lubricant absorbs heat as well as provides cooling action at tool–chip interface. The cooling action is promoted by high-velocity air jet and lubrication action is supposed to be achieved due to viscosity of bio-based cutting oil. Further the exact mechanism of cooling and lubrication action of MQL is yet a challenging issue for the researcher. It is of utmost importance that the quantity of lubricating oil should be minute consisting of bio-based oil with compressed air to protect the environment, operator, and machined surface from negative effects of flood lubrication. The consumptions oil using MQL in industrial applications was reported in the range of approximately 10–100 ml per hour [6]. Contrary to this, a huge quantity of metal cutting fluid at a larger flow rate is utilized for machining during traditional flood lubrication system. But in MQL system pressurized air and atomized oil droplets impinged straightly into tool–chip junction.

Basically, there are two different types MQL supply system like external and internal MQL System [8]. In an external system which is same as the conventional flood lubrication system, where cutting fluid is supplied through a nozzle at a defined pressure and flow rate at a suitable distance from tool–work interface. However in external MQL systems pressurized air with cutting fluid is supplied through specially designed external nozzle system positioned at an appropriate gap from the machining zone having very less quantity of cutting oil. But on the other hand, in internal MQL system the cutting fluid and pressurized air is supplied through the tool itself and there is no need of specially designed external nozzle as compared to the earlier system. Both of the systems have their own advantages and disadvantages depending upon machining applications. The external system is utilized in orthogonal cutting like turning, milling, drilling, sawing and grinding, etc. Similarly internal system is applied in drilling but these days it is also practiced during turning operation too. The major advantage of internal system is that there is no need of specially designed nozzle, however, this system has more complexity than the external system. Leakage, rusting, accessories, fitting and design of nozzle are key functionaries of the external system (Table 2).

Table 2 Comparison of traditional cooling and MQL

Sr No	Cooling environment	Flow rate	Pump requirement	Re-filtration	Chips types	Severity of corrosion	Environmental hazards	Cooling cost	Fumes
1	Wet cooling	8-10L/min	High-pressure pump	Yes	Wet	High	High	higher	High
2	MQL	50–300 ml/h	Low-medium	No	Near dry	Low	Less	Low	Less

1.2 Heat Generation and Measurement

Evolution of heat while machining is phenomenal due to mechanical rubbing, shearing, and deformation of work material caused by sharp edge cutting tool of harder material. There is the formation of various heat zones within close proximity of cutting tool and workpiece. The temperature during machining may be low as 100 °C and even higher up to 1000 °C. The temperature may vary in different cutting zones like primary, secondary, and tertiary as well. The amount of heat generated is distributed among four major components like tool, chip, workpiece, and the fourth environment too. The mechanism of heat transfer is amalgamation on the basis of heat transfer phenomenon consisting of conduction, convection, and radiation. The majority of heat is passed to chips and some part of the heat is conducted to workpiece, further it is radiated to the environment too. The highest temperature is recorded at rubbing of tool–chip contact and it is the need of the hour to measure this elevated temperature. In this connection there are several techniques by which temperature can be measured depending upon the machining operation. The few of the methods are listed as below.

- Infrared thermometer
- Thermocouple
- Pyrometer
- Temperature sensitive paints
- Infrared camera.

But in present investigation standard K-type chrom-alumul thermocouple was utilized for measuring the temperature at tool–chip during various parametric levels and nozzle distance. The hole was drilled on cutting insert with the help of EDM drilling at a distance of 2 mm from nose radius, so as to record the cutting temperature during the machining operation. The schematic diagram for temperature measurement is shown in Fig. 1.

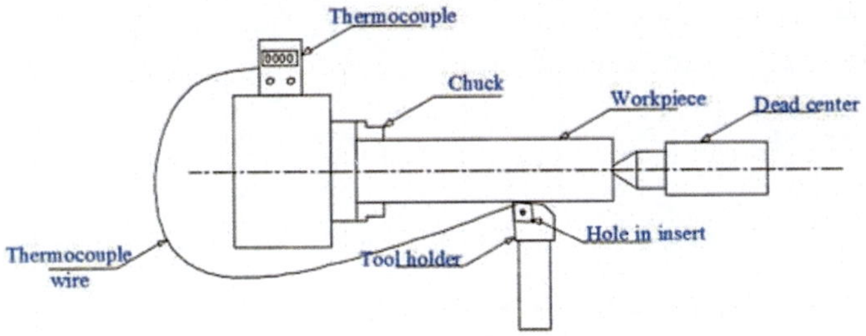

Fig. 1 Schematic diagram of temperature measurement

2 Methodology of Work

According to foundation ponder, it was seen that MQL explores had been experienced by changing the cutting parameters, materials, and the oils having distinctive stream rates. The majority of the temperature-related investigations were finished by utilizing thermo-photographic strategy, inserted thermocouple and infrared thermometer. According to writing study it was seen that less research accounted for on the impact of fluctuating the spout separate on cutting temperature has been performed. In this way in present the examinations the impact of changing the spout separate on temperature amid transforming task has been mulled over. The major finding in this research was to investigate the following task.

- Investigation into the effect of process variables on tool–chip heat evolution
- Significance of changing nozzle stand-off distance on heat generation
- Comparative analysis of temperature in Dry and MQL machining.

The trial work was completed by plain turning of EN-31 Steel rod of eighty mm diameter and one meter length on powerful machine of 10 H.P. Instrument chip interface temperature was recorded in dry, flood, and MQL conditions utilizing standard K-type thermocouple. The process variables of turning operation like speed, feed rate, and profundity of slice were differentiated to confirm its influence on tool–chip coherence heat generation. The nozzle distance was additionally been fluctuated to check its effect on instrument chip interface temperature.

2.1 *Description of Experimental Set up*

The experimental setup contains the following apparatus.

(a) Lathe machine (10 H.P)
(b) Compressor
(c) Thermocouple (K-type)
(d) EN-31 Steel as work material
(e) Cutting tool insert
(f) MQL Nozzle
(g) Mixing chamber
(h) Flow regulating valve
(i) Pressure gauge
(j) Soybean oil as a Lubricant.

Table 3 indicates the chemical composition of work material EN-31 round bar used in an investigation. It contains alloying elements like chromium and manganese that enhances its hardenability and corrosion resistance and further extends its application in the fabrication of die components. The EN-31 steel has wide applications like slip gauges, tapes, ejector pins, and roller bearings.

Table 3 Anatomy of work material

Contents	C	Si	Mn	Cr	Co	S	P
Wt%	0.9–1	0.10	0.30	1.2	0.025	0.040	0.04

3 Experimental Investigation

The full factorial design was used for evaluating the performance at different process parameters. There are two feed, seven speeds, two depth of cut, and four cutting environments. In addition to machining conditions such as dry machining and MQL machining the nozzle distance varied at a distance of 15, 30, and 55 mm. The experimentation was performed by machining of work material EN-31 steel rod having 80 mm in diameter and 1000 mm length. Prior to the final observations, the trial tests were practiced to ensure the accuracy of the work. During experimentation the cutting parameters like speed, feed were varied and accordingly the output parameter like temperature and surface roughness are recorded.

The cutting temperature was measured by chrom-alumul thermocouple at different cutting parameters and nozzle distance. The hole was made drilled on cutting insert with the help of EDM drilling at a distance of 2 mm from nose radius, so as to record the cutting temperature during the machining operation. Initially it was confirmed by the infrared thermometer to find the spot where maximum temperature exists. Further the literature was studied in this connection to find out the exact position of highest temperature while turning. It was found the maximum temperature incurred at tool–chip interface due to mechanical friction. Also the temperature profiles were studied from the past research prior drilling hole on cutting insert. The diameter of the drilled blind hole was 1 mm and the thermocouple wires were connected to chip welded in the blind hole to avoid burning. The Fig. 2 illustrates the photographic view of an experimental set for research including MQL accessories and Fig. 3 shows the cutting insert with the blind hole on the rake face. Further the workpiece was uniformly distributed among all experiment during measurement of temperature which is clearly visible in Fig. 4. The temperature reading was recorded with the help of digital meter as shown in Figs. 4.

4 Results and Discussions

Experimental observation obtained during turning of work material EN-31 is discussed as per an experimental plan mentioned in Table 4. The machining temperature was recorded at dissimilar extents of machining criterion and nozzle distance. The main concern of present research was to investigate the optimum machining parameter levels and nozzle distance from the cutting zone. The graphs were plotted as per the observation obtained from the experimentation conducted in different machining conditions. Furthermore the other aim of this experimentation was to

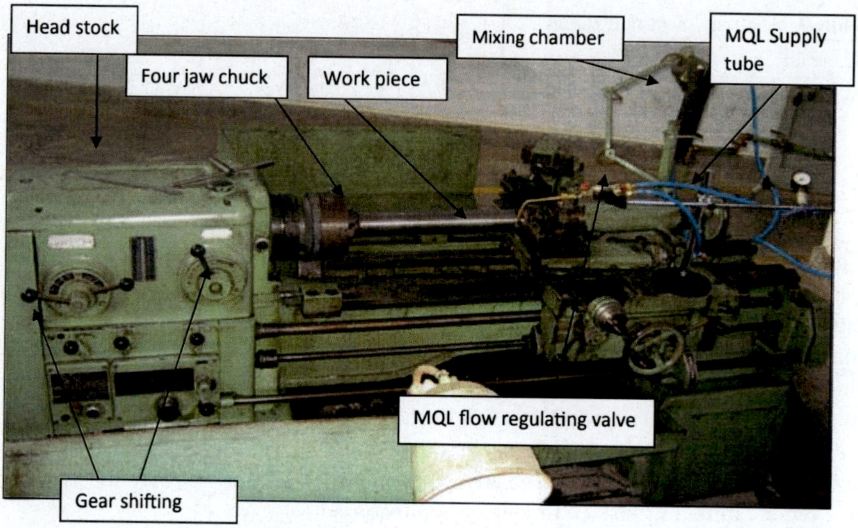

Fig. 2 Image of experimental setup

Fig. 3 Photographic view of cutting tool

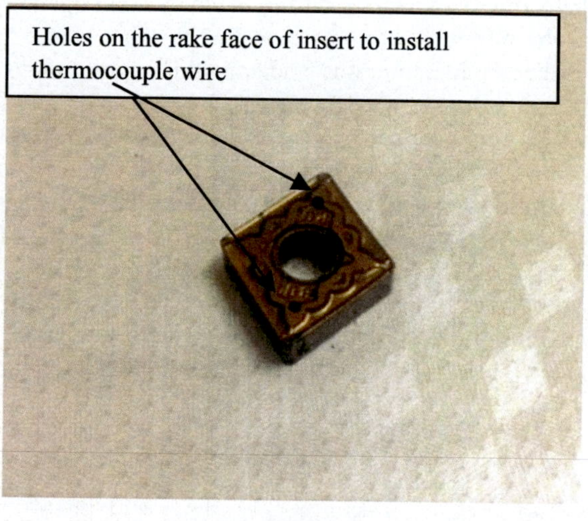

examine the role of changing the nozzle gap from the tool–chip interface zone. It was of prime importance to investigate the nozzle distance where the cutting temperature was least. In this section the following results have been discussed and plotted in the form of graphs.

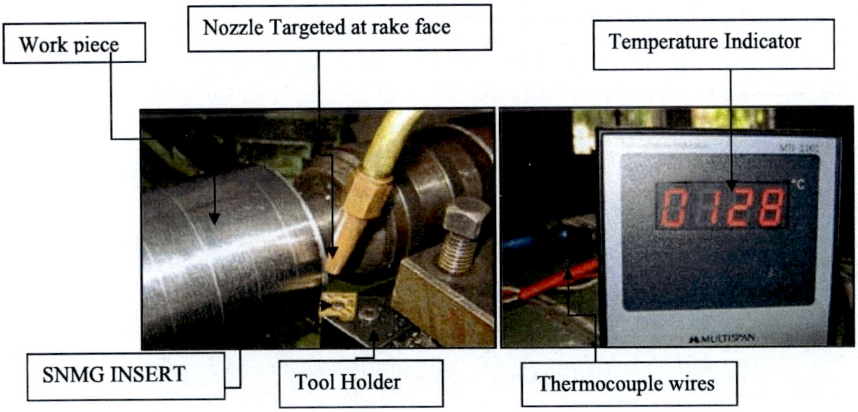

Fig. 4 Setup for temperature measurement

Table 4 Detail of process parameters

Sr. no	Velocity (rpm)	Feed rate (mm/rev)	First nozzle stand-off (mm)	Second nozzle stand-off (mm)	Third nozzle stand-off (mm)
1	142	0.063	15	30	55
2	200	0.063	15	30	55
3	320	0.063	15	30	55
4	500	0.063	15	30	55
5	640	0.063	15	30	55
6	800	0.063	15	30	55
7	1000	0.063	15	30	55
8	142	0.088	15	30	55
9	200	0.088	15	30	55
10	320	0.088	15	30	55
11	500	0.088	15	30	55
12	640	0.088	15	30	55
13	800	0.088	15	30	55
14	1000	0.088	15	30	55

- Impact of process parameters on thermal characteristics of machining processes.
- Comparison of machining temperature in different machining environment.
- Impact of nozzle stand-off on heat generation at distinct levels of process parameters during MQL.

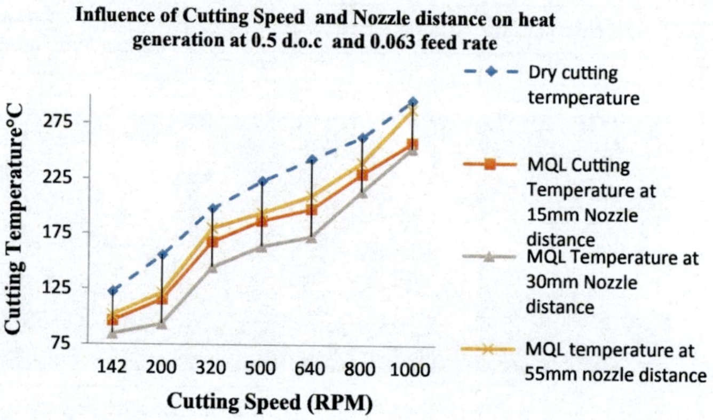

Fig. 5 Impact of cutting speed and nozzle distance on temperature

Figure 5 indicated that as the speed increases the cutting temperature increases rapidly due to more friction at tool–work interaction. Results revealed that maximum temperature was recorded during dry machining environment and minimum temperature was measured in MQL condition at nozzle distance of 30 mm. It can be also concluded that cutting speed, nozzle distance, and lubrication environment influenced the temperature generation. Further it was examined that nozzle distance of 30 mm gap was proved significant as compared to other distance. The maximum and minimum temperature recorded during dry conditions at 1000 rpm was approximately 288 and 126 °C at minimum rpm but on the other hand extreme and least temperature during MQL at 30 mm nozzle distance was recorded as 230 °C at 1000 rpm and 78 °C at 142 rpm, further it was nearly equal to 276 °C during 1000 rpm and 102 °C at 142 rpm while placing nozzle at the gap of 55 mm distance. Hence there was a temperature difference of 46 °C among variation of nozzle distance and 58 °C between dry cutting and MQL machining. In addition to this, while machining at least rpm the temperature difference among dry and MQL at 30 mm nozzle distance was examined as 44 and 24 °C amid dry vs. 55 mm nozzle distance.

Finally it was also observed that nozzle positioning very near and too-far from the machining zone would result in higher cutting temperature. Hence variation of nozzle distance was a significant factor affecting the heat generation during the machining operation. As shown in Figs. 5 and 6, it was evident that the cutting temperature increases as cutting speed increases. This was due to the fact that an increase in cutting speed would lead to more friction or rubbing at tool–work interface consequently generated more heat. The graph also indicated the effect of feed rate and depth of cut on cutting temperature that on increasing the feed rate the temperature raises.

It is clear from Fig. 7 that the cutting temperature increases rapidly, when the depth of cut was varied as compared to feed rate and speed. The maximum temperature was approximately 400 °C during dry cutting at 0.088 mm/rev feed rate and 1 mm depth of cut. Similarly it was recorded nearly 300 °C at 0.063 mm/rev feed and 0.5 mm depth of cut. So it was concluded that there was a difference of almost 100 °C during dry

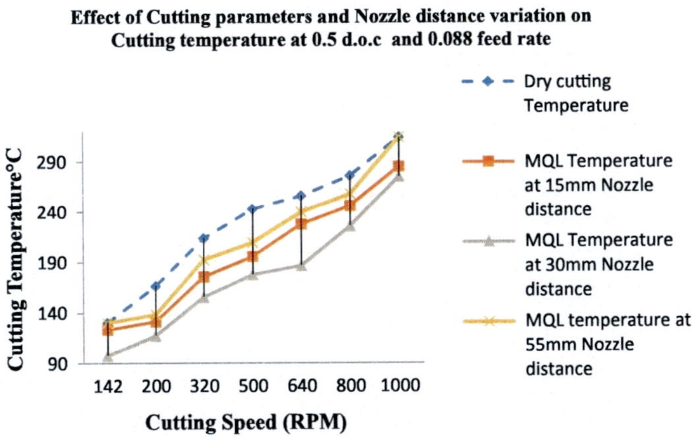

Fig. 6 Effect of cutting speed on the temperature at 0.5 mm doc and 0.088 feed rate

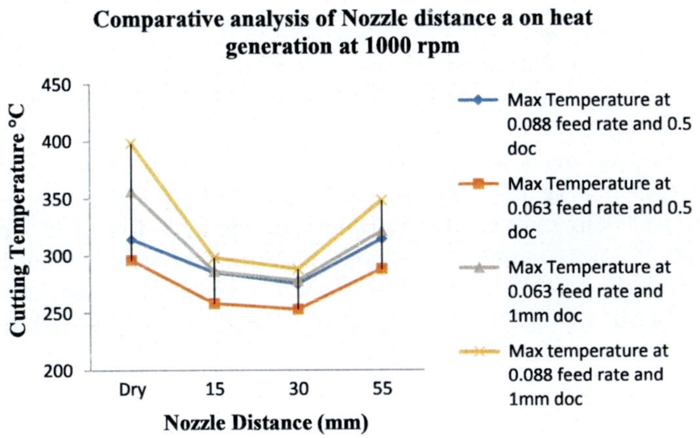

Fig. 7 Impact of nozzle distance on Temperature

machining environment. On the other hand maximum temperature measured during 30 mm nozzle distance was 298 °C at 0.088 mm/rev feed rate and 1 mm depth of cut and minimum temperature recorded was 247 °C while machining at 0.063 mm/rev feed and 0.5 mm depth of cut. Hence it can be stated that temperature difference during dry machining was larger than MQL. Further the impact of depth of cut was more as compared to feed rate during these machining conditions.

From experimental observations, it was also revealed that MQL significantly reduced the heat generation by 10–30% contrary to dry turning of EN-31 steel. The increase in cutting temperature while varying depth of cut may be due to increases in the shear area ahead of cutting tool as a result more penetration of tool into workpiece raises the heat generation. From Fig. 7, It was evaluated that temperature during dry machining was higher than other machining conditions using MQL. When placing nozzle distance at 15 mm the recorded temperature was more than 30 mm nozzle distance. Similarly positioning of the nozzle at 55 mm resulted in temperature almost identical to dry machining; which meant that placing nozzle at 55 mm distance was not significant for minimizing the heat generation amid other nozzle distance. This phenomenon may be due to splashing of MQL fluid jet targeted from far distance toward cutting zone.

5 Conclusions

As per the experimental results the following conclusions were drawn.

(a) The maximum cutting temperature was recorded during dry machining condition but on the other hand during MQL environment at 55 mm nozzle distance from cutting zone.
(b) While machining at least rpm the temperature difference among dry and MQL at 30 mm nozzle distance was examined as 44 °C and it was 24 °C amid dry versus MQL at 55 mm nozzle distance that indicates nozzle distance play important role in reducing cutting temperature.
(c) There was favorable significance of varying nozzle distance on machining temperature during MQL conditions irrespective of dry machining.
(d) MQL machining reduced the cutting temperature by 8–10% at 15 mm nozzle distance, 10–18% during 30 mm and 6% while the nozzle was positioned at 55 mm as compared to dry machining at 0.5 doc parameter level.
(e) MQL machining enabled a reduction in the cutting temperature by 16–18% during 15 mm nozzle distance, 23–28% during 30 mm, and 10–12% while positioning nozzle at 55 mm away from cutting zones contrast to dry machining executed at 1 mm depth of cut process parameter level.
(f) The experimental results exihibted the influence of depth of cut, feed rate, and speed on cutting temperature in all machining environment.

6 Future Scope

- Experimentation can be extended further using ionic fluid (gaseous based) MQL instead of mineral oil or vegetable-based lubricants.
- Investigation on the influence of mixing ratio of nanofluids to MWF and vegetable-based bio-oil can be further studied as novelty in this direction.
- Effect of mixing ratio of different cutting fluids and on surface quality, tool wear, and environmental friendliness can be examined in other machining processes like milling, drilling, and grinding in future.

References

1. Khan MMA, Dhar NR (2006) Performance evaluation of minimum quantity lubrication by vegetable oil in terms of cutting force, cutting zone temperature, tool wear, job dimension and surface finish in turning AISI-1060 steel. J Zhejiang Univ Sci 1790–1799 ISSN: 1009-3095
2. Attanasio A, Gelfi M, Giardini C, Remino C (2006) Minimal quantity lubrication in turning effect on tool wear. Wear 260:333–338
3. Autret R, Liang Y (2003) Minimum quantity lubrication in finish hard turning. G.I.T, Atlanta, Georgia, USA
4. U.S. Department of Health and Human Services; Occupational exposure to metalworking fluids (1998) NIOSH publication no. 98–102
5. Klocke F, Eisenblatter G (1997) Dry cutting. Ann CIRP 46(2):519–526
6. Bennett EO, Bennett DL (1998) Occupational Airway diseases in the metalworking industry. Tribol Int 18(3):169–176
7. Sreejith PS, Ngoi BKA (2000) Dry machining: machining of the future. J Mater Process Technol 101:287–291
8. Braga DU, Diniz AE, Miranda GWA, Coppini NL (2002) Using a minimum quantity of lubricant and a diamond coated tool in the drilling of aluminium silicon alloys. J Mater Process Technol 122:127–138
9. Itoigawa F, Childs THC, Nakamura T, Belluco W (2006) Effects and mechanisms in minimal quantity lubrication machining of an aluminum alloy. Wear 260:339–344
10. Kamata Y, Obikawa T (2007) High speed MQL finish-turning of Inconel 718 with different coated tools. J Mater Process Technol 192:281–286
11. Khan MMA, Mithu MAH, Dhar NR (2009) Effects of minimum quantity lubrication on turning AISI 9310 alloy steel using vegetable oil based cutting fluid. J Mater Process Technol 209:5573–5583
12. Abhang LB, Hameedullah M (2010) Chip-tool interface temperature prediction model for turning process. Int JEST 2(4):382–393
13. Ozcelik B, Kuram E, Cetin MH, Demirbas E (2011) Experimental investigations of vegetable based cutting fluids with extreme pressure during turning of AISI 304L. Tribol Int 44:1864–1871
14. Singh G, Singh S, Singh M, Kumar A (2013) Experimental investigations of vegetable & mineral oil performance during machining of EN-31 steel with minimum quantity lubrication. IJRET 02(06):1030–1037
15. Rahim EA, Sasahara H (2011) An analysis of surface integrity when drilling Inconel 718 using palm oil and synthetic ester under MQL condition. Mach Sci Technol Int J 15:76–90
16. Elmunafi M, Kurniawan D, Noordin MY (2015) Use of castor oil as cutting fluid in machining of hardened stainless steel with minimum quantity of lubricant. Procedia CIRP 26:408–411
17. Qin S, Li Z, Guo G, An Q, Chen M, Min W (2016) Analysis of minimum quantity lubrication (MQL) for different coating tools during turning of TC11 titanium alloy. Materials 9:804
18. Su Y, Gong L, Chen D (2015) An investigation on tribological properties and lubrication mechanism of graphite nano-particles as vegetable based oil additive. J Nano-mater
19. Jamaludin AS, Hosokawa A, Furumoto T, Koyano T, Hashimoto Y (2017) Evaluation of the minimum quantity lubrication in orthogonal cutting with the application of finite element method. Int J Mechan Mechatron Eng IJMME-IJENS 17(01):104–109

20. Sartori S, Ghiotti A, Bruschi S (2018) Solid lubricant-assisted minimum quantity lubrication and cooling strategies to improve Ti6Al4V machinability in finishing turning. Tribol Int 118:287–294
21. Sivaiah P, Chakradhar D (2018) Effect of cryogenic coolant on turning performance characteristics during machining of 17–4 PH stainless steel: a comparison with MQL, wet and dry machining. CIRP J Manuf Sci Technol 21:86–96
22. Marques A, Suarez MP, Sales WS, Machado AR (2019) Turning of Inconel 718 with whisker-reinforced ceramic tools applying vegetable-based cutting fluid mixed with solid lubricants by MQL. J Mater Process Tech 266:530–543
23. Zaman PB, Dhar NR (2019) Design and evaluation of an embedded double jet nozzle for MQL delivery intending machinability improvement in turning operation. J Manufact Process 44(2019):179–196

Investigation of Kerf Characteristics Using Abrasive Water Jet Cutting of Floor Tile: A Preliminary Study

Ramesh Chand, Vishal Gupta, N. K. Batra, and M. P. Garg

Abstract Abrasive water jet (AWJ) is a technology concerned with cutting of difficult materials and performs cutting without changing the mechanical and physical properties of material. Floor tile is widely used in various commercial and industrial filed. One of the major applications of floor tile is its use in decoration of floor. Cutting of floor tile is very difficult by conventional machining methods. In the present work cutting of floor tile was done by the use of advanced abrasive water jet technology. The effects of input process parameters (water pressure (WP), nozzle transverse speed (NTS), and abrasive flow rate (AFR)) on the output response (kerf characteristics and surface roughness) were investigated while performing cutting of the floor tile. It was found that top kerf width is reduced with increase in the WP, NTS, and AFR. Further, it was observed that with the decrease in WP and NTS kerf, taper angle is reduced significantly.

Keywords Water jet · Abrasive · Nozzle transverse speed · Water pressure · Floor tile

1 Introduction

Abrasive water jet machining (AWJ) is the technology concerned with cutting of materials difficult to be machined and cutting without reducing the properties of material [1]. The AWJM is typically used in industry to cut ceramics and hard materials [2]. AWJM is used to cut any material whether it is conductive or non-conductive.

R. Chand · N. K. Batra
Mechanical Engineering Department, Maharishi Markandeshwar Deemed To Be University, Mullana Ambala, Haryana, India

V. Gupta (✉)
Mechanical Engineering Department, Thapar Institute of Engineering and Technology, Patiala 147004, India
e-mail: vishal.gupta@thapar.edu

M. P. Garg
Mechanical Engineering Department, DAV University, Jalandhar, India

© Springer Nature Singapore Pte Ltd. 2021

P. M. Pandey et al. (eds.), *Advances in Production and Industrial Engineering*, Lecture Notes in Mechanical Engineering,
https://doi.org/10.1007/978-981-15-5519-0_14

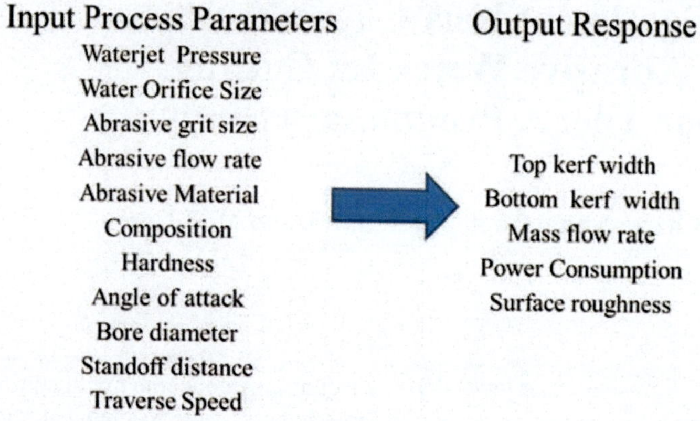

Fig. 1 Effect of process parameters on output response

It does not require any lubrication as in the case of other machines, for example, in lathe or milling, for cooling, etc. It uses a high pressure water jet to cut the material by means of erosion. In AWJM, a single tool is used for various processes like drilling, cutting, contouring, etc.

Various researchers [3–5]. Across the globe have investigated the effect of AWJM parameters on the kerf characteristics MRR and surface roughness using a brittle and ductile material as shown in Fig. 1. AWJM is widely used for brittle materials or for materials difficult to cut by conventional machining process.

In AWJM of microgroove and polishing of glass using stagnation under the nozzle, and stagnation area (machining area) is controlled by the tapper angle of v-shaped mask on the work surface [6]. If the abrasive particle impact angle is 90° irrespective of the velocity, the Sphericity at the top surface is equal and the depth of crater and erosion rate are maximum. If impact angle decreases from 90°, the sphericity, depth of crater, and erosion rate decreases in machining of titanium alloys with AWJM [7].

Water pressure and nozzle transverse speed were the only input parameters that were significantly affecting the surface roughness in AWJ machining of Makrana white marble [8]. As impact angle decreases, the MRR decreases and the curvature of width of cut increases. While increase in water flow rate did not have any significant effect on metal removal rate. Generally the depth of cut increases with the multi-passes cutting and increase in water pressure [9]. On increasing the abrasive flow rate and water pressure, the depth of cut increases because kinetic energy increases, while increasing the transverse speed and standoff distance there is a little variation of depth of cut of borosilicate glass [10].

Water flow rate and SOD decreases the kerf width and surface roughness, and also there is increase in transverse speed and material thickness, also the kerf width and surface roughness decrease with AWJ machining of TRIP sheet steels [11]. During drilling of a hole in glass plate MRR increases and taper angle decreases with increase in SOD and chemical composition of polymer with slurry of AWJ. The polymer in the slurry creates bonding strength of water molecules with abrasive particle due to which scattering of the jet is reduced [12]. With the increase in the transverse speed surface roughness is increased but kerf width decreases. While increase in abrasive flow rate, water pressure, and abrasive velocity, the material removal rate increases machining of ceramics plates with AWJM [13].

Increase in transverse speed the MRR increases while increase in percentage of tungsten carbide tends to decrease the MRR because at higher percentage of WC there is increase in hardness of composites, Abrasive Water Jet Machining of aluminum/tungsten carbide composites [14].

It was found that the water pressure and traverse speed had a more profound effect on the depth of cut than the standoff distance and abrasive flow rate in abrasive water jet during cutting of ceramic materials [15]. At high water pressure, low traverse speed, and low standoff distance, kerf taper is minimum of graphite/epoxy and glass epoxy [16]. Increase in transverse speed, standoff distance, and water pressure, the kerf angle of granite increases. However, increase in abrasive(coarse) flow rate has no significant effect on kerf angle, but with increase in fine abrasive flow rate the kerf angle also increases [17].

Ceramic tiles are hard material, which is difficult to machine by conventional machining process. Nowadays instead of marbles, ceramic tiles are commonly used for flooring or decorative purposes because of their various advantages like less weight, finishing is not required, easy to maintain, and require less time to arrange on the floor. Floor tiles are machined by the conventional machining technique like cutter, drilling, power hacksaw [18, 19]. These processes have various disadvantages like accuracy, jamming of tool, dust, etc. To overcome the previously mentioned problems, in this work, attempt has been made to introduce the AWJM on floor tile. It has been found that AWJM produces a very good surface during cutting of floor tile and a surface free from dust.

2 Materials and Methods

In this study, ceramic floor tile with dimensions 305 mm × 305 mm was used as workpiece material. The important mechanical property and chemical composition of the ceramic tile used in this work are shown in Table 1. The experiments were conducted on the OMAX 5555 machine as shown in Fig. 2. Various process parameters that were involved in the AWJM produced the effect on the cutting results like kerf characteristics, surface roughness, etc. [20]. In these experiments, nozzle transverse speed, abrasive flow rate, and water pressure were taken as the input process parameters. Table 2 shows the input process variable along with their levels.

Table 1 Important mechanical properties and chemical composition of ceramic floor tile

Elements	Lime	Silica	Alumina	Alkali	Hardness	Young modulus
Weight (%)	1	65	31	3	6–8 Mohs hardness	384 ± 2 MPa

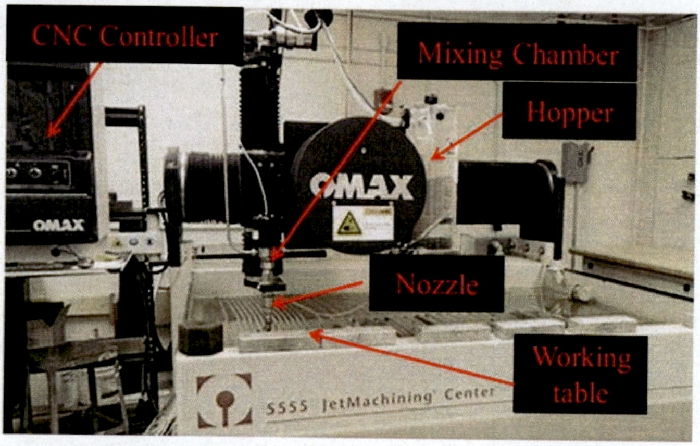

Fig. 2 Pictorial view of the AWJM setup

Table 2 Process parameters and their range

Process parameters	Level 1	Level 2	Level 3	Level 4	Level 5
NTS (mm/min)	100	150	200	250	300
WP (MPa)	150	200	250	300	350
AFR (g/min)	80	120	160	200	240

Experiments were performed according to the run order as shown in Table 3. Table 3 also represents the output response values w.r.t. the run order. Kerf width was measured by profile projector which is available in our laboratory and Kerf angle was measured by the Eq. 1

$$\theta = tan^{(-1)}(\frac{W_t - W_b}{2t}) \tag{1}$$

where W_t is the top kerf width, W_b is the bottom kerf width and "t" is the thickness of the work piece. The machined surface produced by AWJM is shown in Fig. 3.

3 Results and Discussions

In AWJM the material is removed by the stream of abrasive which is coming from abrasive jet nozzle. In mixing chamber, the momentum of water jet is transferred to the abrasive which leads to an increase in the velocity of abrasive particles. So finally, high stream of abrasive particles mixed with water strike on the workpiece and material removal occurs.

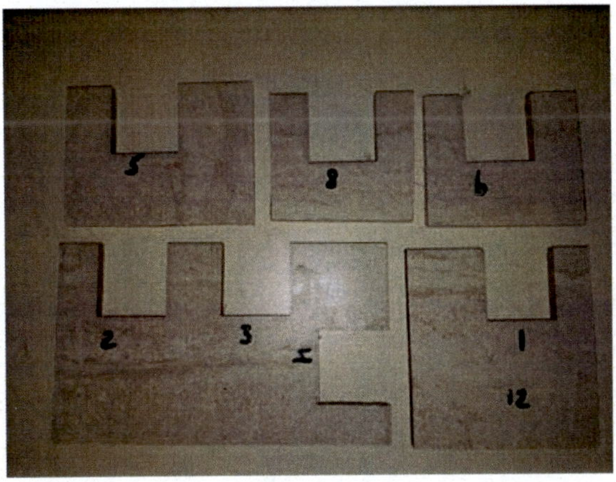

Fig. 3 Machined surface produced by AWJM

Table 3 Run order to perform the experiments and kerf characteristics value

Run order	WP	AFR	NTS	Kerf Width (mm)		Kerf angle
	(MPa)	(g/min)	(mm/min)	Top	Bottom	(Degree)
1	150	160	200	40.42897	40.6188	0.679
2	200	160	200	40.42302	40.6176	0.696
3	250	160	200	40.41707	40.6164	0.713
4	300	160	200	40.41112	40.6152	0.73
5	350	160	200	40.40517	40.614	0.747
6	250	80	200	40.41971	40.63	0.752
7	250	120	200	40.41839	40.6232	0.733
8	250	160	200	40.41707	40.6164	0.713
9	250	200	200	40.41575	40.6096	0.694
10	250	240	200	40.41443	40.6028	0.674
11	250	160	100	40.57707	40.7431	0.594
12	250	160	150	40.49707	40.67975	0.654
13	250	160	200	40.41707	40.6164	0.713
14	250	160	250	40.33707	40.55305	0.773
15	250	160	300	40.25707	40.4897	0.832

3.1 *Effect of Process Parameters on Top Kerf Width*

Effect of water pressure on top kerf is shown in Fig. 4. Also, Fig. 4 shows that with increase in the water pressure top kerf width reduced significantly. It is because of the higher water pressure, the jet kinetic energy increases which resulted in a decrease in the top kerf width [18–20]. Figures 5 and 6 also show that kerf width is decreased with increase in the nozzle transverse speed and abrasive flow rate. The negative

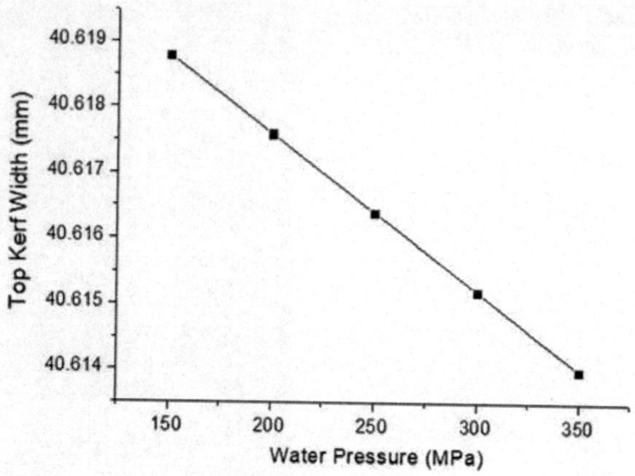

Fig. 4 Effect of WP on top kerf width

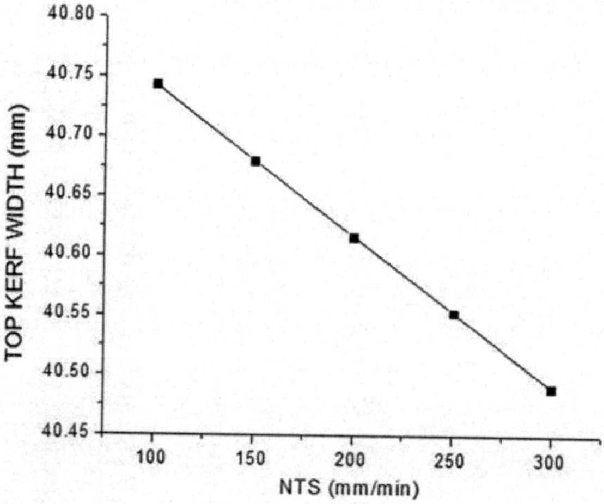

Fig. 5 Effect of NTS on top kerf width

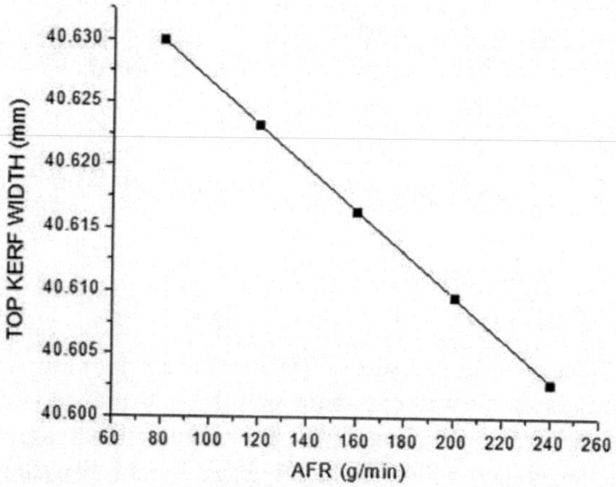

Fig. 6 Effect of AFR on top kerf width

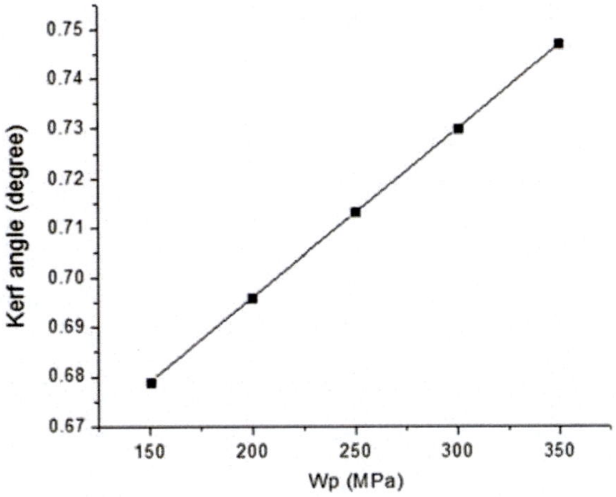

Fig. 7 Effect of WP on kerf angle

effect of traverse speed on the kerf width is due to the fact that a faster passing of abrasive water jet allows fewer particles to strike on the target material and hence generates a narrower slot [18–20].

3.2 *Effect of Process Parameters on Top Kerf Taper Angle*

Effect of water pressure on kerf taper angle is shown in Fig. 7. With increase in the water pressure kerf angle is increased significantly. It may be due to the fact that the outer rim of the diverged jet still has sufficient energy to cut material which may lead to larger kerf angles obtained at higher water pressure [18–20].

It was found that with decrease in the nozzle transverse speed, kerf taper angle is reduced as shown in Fig. 8 which is favorable in AWJM. It may be due to the fact that at lower nozzle transverse speed the widening of jet decreases as a result lower kerf is produced at low speed. The effect of abrasive flow on the kerf angle is shown in Fig. 9. It has been found that with increase in the abrasive flow rate kerf taper angle is decreased significantly.

4 Conclusion

The following conclusions have been made on the basis of the present research work:

1. Ceramic floor tiles can be easily machined by AWJM.
2. Top kerf width can be reduced with increase in the speed of the nozzle jet, abrasive flow rate, and water pressure.

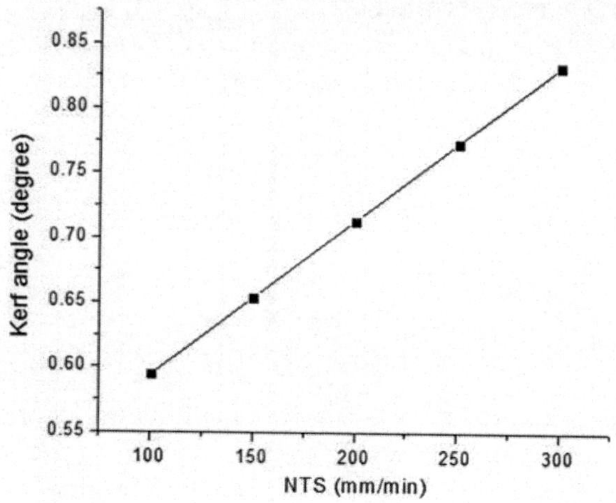

Fig. 8 Effect of NTS on kerf angle

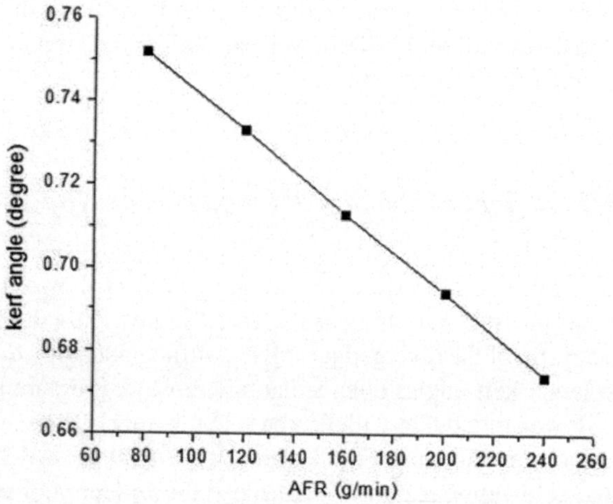

Fig. 9 Effect of AFR on kerf angle

3. Kerf angle reduces with the decrease in the nozzle transverse speed, water pressure, and increase in the abrasive flow rate.
4. Out of all the selected parameters, only water pressure and nozzle transverse speed significantly affect the surface roughness in AWJ machining of Makrana white marble.

References

1. Wang J, Guo DM (2003) The cutting performance in multipass abrasive waterjet machining of industrial ceramics. J Mater Process Technol 133(3):371–377
2. El-Hofy HA-G (2005) Advanced machining processes: nontraditional and hybrid machining processes. McGraw Hill Professional
3. Babu MK, Krishnaiah Chetty OV (2006) A study on the use of single mesh size abrasives in abrasive waterjet machining. Int J Adv Manuf Technol 29(5–6):532
4. Jegaraj JJR, Babu NR (2005) A strategy for efficient and quality cutting of materials with abrasive waterjets considering the variation in orifice and focusing nozzle diameter. Int J Mach Tools Manuf 45(12–13):1443–1450
5. Pi V (2007) An introduction to Abrasive water jet (AWJ) machining
6. Matsumura T, Muramatsu T, Fueki S (2011) Abrasive water jet machining of glass with stagnation effect. CIRP Ann 60:355–358
7. Kumar N, Shukla M (2012) Finite element analysis of multi-particle impact on erosion in abrasive water jet machining of titanium alloy. J Comput Appl Math 236:4600–4610
8. Gupta V et al (2014) Investigation on Makrana white marble surfaces machined by abrasive water jet machining. Int Conf Newest Drifts Mechan Eng
9. Hashish M (1984) A modeling study of metal cutting with abrasive water jets. Trans ASME 88:106
10. Aich U, Banerjee S, Bandyopadhyay A, Das PK (2014) Abrasive water jet cutting of borosilicate glass. Proc Mater Sci 6:775–785
11. Kechagias J, Petropoulos G, Vaxevanidis N (2012) Application of Taguchi design for quality characterization of abrasive water jet machining of TRIP sheet steels. Int J Adv Manuf Technol 62:635–643
12. Palleda M (2007) Study of tapper angles and material removal rates of drilled hole in the abrasive water jet machining process. J Mater Process Technol 189:292–295
13. Hcheng H, Chang KR (1994) Material removal in abrasive jet cutting of ceramics plates. J Mater Process Technol 40:287–304
14. Kumar KR et al (2018) Characterization and optimization of Abrasive Water Jet Machining parameters of aluminum/tungsten carbide composites. Measurement 117:57–66
15. Chen L (1996) Kerf characteristics in abrasive waterjet cutting of ceramic materials. Int J Mach Tools Manuf 36:1201–1206
16. Shanmugam DK, Masood SH (2009) An investigation on kerf characteristics in abrasive waterjet cutting of layered composites. J Mater Process Technol 209:3887–3893
17. Karakurt I et al (2011) Analysis of the kerf angle of the granite machined by the Abrasive water jet (AWJ). Ind J Eng Mater Sci 18:435-442
18. Gupta V et al (2013) Analysis of kerf taper angle in abrasive water jet cutting of Makrana white marble. Asian J Eng Appl Technolo 35–39
19. Wang J, Kuriyagawa T, Huang CZ (2014) An experimental study to enhance the cutting performance in abrasive waterjet machining. Mach Sci Technol 7(2):191–207
20. Gupta V et al (2014) Minimization of kerf taper angle and kerf width using Taguchi's method in abrasive water jet machining of marble . Proc Mater Sci 6:140–149

Effects of Process Parameters on Surface Roughness, Dimensional Accuracy and Printing Time in 3D Printing

Rajat Jain, Shivansh Nauriyal, Vishal Gupta, and Kanwaljit Singh Khas

Abstract In this study, the interaction of different process parameters on printing time, surface roughness and dimensional accuracy of a 3D printed object are investigated. The impact of different parameters on the performance of the 3D printer is also investigated and their optimal values have been identified. From the study, it was found that layer height, infilled density and printing speed are major influential factors affecting the surface roughness, printing time and dimensional accuracy of the 3D printed piece.

Keywords 3D printer · Infill density · Printing speed · Layer height · Printing time · Surface roughness

1 Introduction

Rapid Prototyping is an additive manufacturing technology which was commercialized during the early 1980s. This technology is used to build a physical part directly from a 3D CAD (Computer-Aided Design) file. The parts are built by using the layer by layer manufacturing technology [1–3]. Application of Rapid Prototyping technology has greatly reduced product development costs. Designers use Rapid Prototyping technique to determine product characteristics like shape, manufacturability and surface finish.

In order to manufacture the precise parts, it is necessary to have a good surface finish. A good surface finish on the part helps in eliminating the post-processing costs [4]. There are many commercial Rapid Prototyping systems available in the market such as Stereolithography (SLA), Selective Laser Sintering (SLS), Laminated Object Manufacturing (LOM), Fused Deposition Modelling (FDM), Solid Ground Curing

R. Jain · S. Nauriyal · V. Gupta (✉)
Mechanical Engineering Department, Thapar Institute of Engineering and Technology, Patiala 147004, India
e-mail: vishal.gupta@thapar.edu

K. S. Khas
Product and Industrial Design Department, Lovely Professional University, Phagwara, India

© Springer Nature Singapore Pte Ltd. 2021

P. M. Pandey et al. (eds.), *Advances in Production and Industrial Engineering*,
Lecture Notes in Mechanical Engineering,
https://doi.org/10.1007/978-981-15-5519-0_15

(SGC) and Three-Dimensional Printing (3DP), etc. Out of these, FDM accounts for almost half of the Rapid Prototyping systems built in the market [5].

FDM is a plastic extrusion process in which the plastic filament is heated to its melting point and it is deposited in the form of very thin layers onto a platform via a heated nozzle. FDM process is simple, which makes it suitable for being chosen by users. It has many advantages such as low equipment cost, inexpensive tooling, a wide range of material availability, acceptable dimensional accuracy. However, it has also got several disadvantages such as low mechanical strength, difficulty in obtaining thin walls and poor surface finish. A significant amount of research is required on surface finish and process optimization which improves the performance of manufactured parts [6].

In the FDM process, there are many factors which influence the surface finish. These consists of material properties (melting temperature of the material, viscosity, density, type of material used, mechanical properties), chamber properties (pressure and temperature inside the chamber, vibrations, position of platform, position of extruder, system coordinates, heat evacuation), extruder properties (speed of printing, diameter of nozzle) and deposition characteristics (building direction, wall thickness, layer height, orientation, external geometry) [7].

Due to the high number of parameters available, factors which were more convenient and economical were chosen. Earlier research was made that layer height and part orientation were the most significant factors that affect the surface quality of the parts [6].

In this experiment, our main emphasis is laid on parameters like extrusion temperature, the diameter of the nozzle, infill density, speed of printing and melting temperature which affects the surface finish of the parts build through the FDM process.

In the FDM process, generally, nozzle diameters of size 0.25, 0.4, 0.6 and 0.8 mm are available as standards. In the FDM 3D printers, the material is in the form of wire spools which is available in different diameters (0.75–3 mm). Generally, the nozzle temperature which is used in melting the material ranges from 210 to 260 °C. Further, the layers are laid down according to different infill patterns, as per the CAD design and the dimensions of the part (Fig. 1).

The basic steps in the fused deposition modelling process are given below.

2 Literature Review

Pérez et al. [6] used ANOVA (Analysis of Variance) and Graphical methods to determine the critical parameters that influence surface roughness in the FDM process. In the above research, PLA samples were used and critical parameters were determined and validated using DOE and ANOVA techniques. From the above research, it was also pointed out that parameters like printing speed, printing path and oven temperature do not have a higher effect on the surface roughness. The conclusion made through this research is that layer thickness is the most important factor influencing the surface roughness.

Hasan et al. [8] have found that the poor quality of surface finish in the FDM process depends upon various parametric factors such as layer ovality, improper Z-resolution (layer height), layer filling. Optimized parameters were obtained using the Design of Experiment (DOE) approach. Shrinkage was also determined as the major factor in determining the surface finish as there was a temperature difference of about 217 °C between the bed and molten material while using the PLA material.

Jayanth et al. [9] have found that layer height was the most impactful parameter which causes a poor surface finish on the 3D printed part. The part which was made of the ABS material was immersed in a solution of Acetone (99%) and Dichloroethane (98%). The solutions improve the surface finish by dissolving the asperities and reducing their height hence improving the surface finish.

Huang et al. [10] studied different parameters like layer thickness, processing quality, filling mode and processing. Optimal factors which influence the surface roughness were proposed using the Taguchi's design of experiments techniques. The factors which influence the side roughness and the top roughness were obtained during this research. Chari et al. [11] used the L9 orthogonal array to carry out the experiments.

Nancharaiah et al. [12] investigated the other affecting parameters like raster angle, road width and air gap on road width on surface roughness and dimensional accuracy of the 3D printed piece. He used the ANOVA and DOE technique to determine the factors and their interactions with one another. Through his experimentation, he concluded that layer height and road width have a higher effect on surface roughness, while the air gap has a larger effect on dimensional accuracy in comparison to the surface finish.

Alafaghani et al. [13] have printed a total of 18 samples in order to determine the effect of individual process parameter on the quality of the printed piece. It was found that the dimensional accuracy has a higher dependence on the direction, extrusion temperature and layer height rather than the infill pattern, printing speed and infill percentage. In order to improve the dimensional accuracy of the critical parts, the dimension should be parallel to the layer orientation instead of the direction.

3 Material and Methods

The material used in this FDM process is PLA (Polylactic acid). It is insoluble in water and is biodegradable. It is also a bioactive thermoplastic. The properties of the material are shown in Table 1 [14].

For the preparation of 3D CAD model, a well-known software SolidWorks 2013 is used and also for converting the 3D CAD into STL format. Cura 2.6.2 which is inbuilt software of "Ultimaker" 3D Printer is used for slicing the STL file into the G code file.

Table 1 Properties of the material

Property	Value
Technical name	Polylactic acid (PLA)
Chemical formula	$(C_3H_4O_2)_n$
Melt temperature	158–170 °C
Typical injection moulding temperature	178–240 °C
Heat deflection temperature (HDT)	49–52 °C
Tensile strength	60–65 MPa
Flexural strength	49–111 MPa
Specific gravity	1.3
Shrink rate	0.36–0.40%

The STL file is stored in a card reader. It is inserted in the reader slot of the 3D printer. The printer has been specified with layer resolution upto 20 μ (0.001 inch) and the maximum operating temperature at the nozzle is 280 °C. The printer is backed up by an open filament system capable of printing with any material of 2.85 mm thickness.

The three input parameters studied are:

- Infill Density: percentage of the object's interior volume that is filled with material.
- Layer Height: the thickness of each layer constituting the object.
- Printing Speed: The speed at which printing happens.

In order to determine the best parameters for the specimen, we printed a cube of 20 mm^3 × 20 mm^3 × 20 mm^3 for different infill densities, layer height and printing speed with the help of "Cura" software. The input parameters used in this work and their levels are shown in Table 2.

Due to this vast range in parameters, our experiment had been long and time-consuming. Due to an increase in the experimental trials, the experiment was not even cost-effective. So to overcome this problem we have used a statistical software named as "Minitab 18" which basically provides us with a matrix of our parameters.

The matrix has been created by assuming the three values of parameters in between our range and it has been created using Taguchi design of experiments (DOE).

Since there are three process parameters each with three levels in this printing, if a full factor experimental design had been used, there would be a total of 81 runs and it would have been too expensive [6].

Table 2 Parameters ranges

Parameter	Level 1	Level 2	Level 3
Layer height (mm)	0.06	0.1	0.2
Infilled density (%)	20	50	100
Printing speed (mm/s)	50	100	150

Table 3 L_9 orthogonal arrays for the three process parameters used in this experiment

Run order	Layer height (mm)	Infilled density (%)	Printing speed (mm/s)
1	0.06	20	50
2	0.06	50	100
3	0.06	100	150
4	0.1	20	100
5	0.1	50	150
6	0.1	100	50
7	0.2	20	150
8	0.2	50	50
9	0.2	100	100

To overcome this problem, one of the strategies is Taguchi's orthogonal scheme [15]. An L_9 orthogonal was selected for the experimentation as shown in Table 3. The data summary for the output responses is given in Table 4. Parts fabricated by the 3D printers are shown in Fig. 2.

After getting the adequate parameter values the following properties were optimized one by one.

(a) Printing Time—It was determined by the help of Ultimaker Cura software. After designing the 3D model when the STL file was imported, the software sliced the design and provided the actual time taken by the Ultimaker to print that model.
(b) Surface Roughness—The study of surface roughness (Ra) is measured by Mitutoyo SJ-301 roughness tester having least count 0.01 μm is used.
(c) Dimensional Accuracy—The dimension of our model was determined by the help of Mitutoyo PJ-A3000 Profile Projector.

Table 4 Data summary for output parameters

Run order	Printing time (min)	Surface roughness Ra (μm)	
		Top	Side
1	101	1.31	3.695
2	71	1.67	5.28
3	141	1.84	5.545
4	36	1.93	6.9
5	35	3.78	6.765
6	172	7.55	14.94
7	16	5.4	13.635
8	38	2.42	13.255
9	52	1.93	13.1

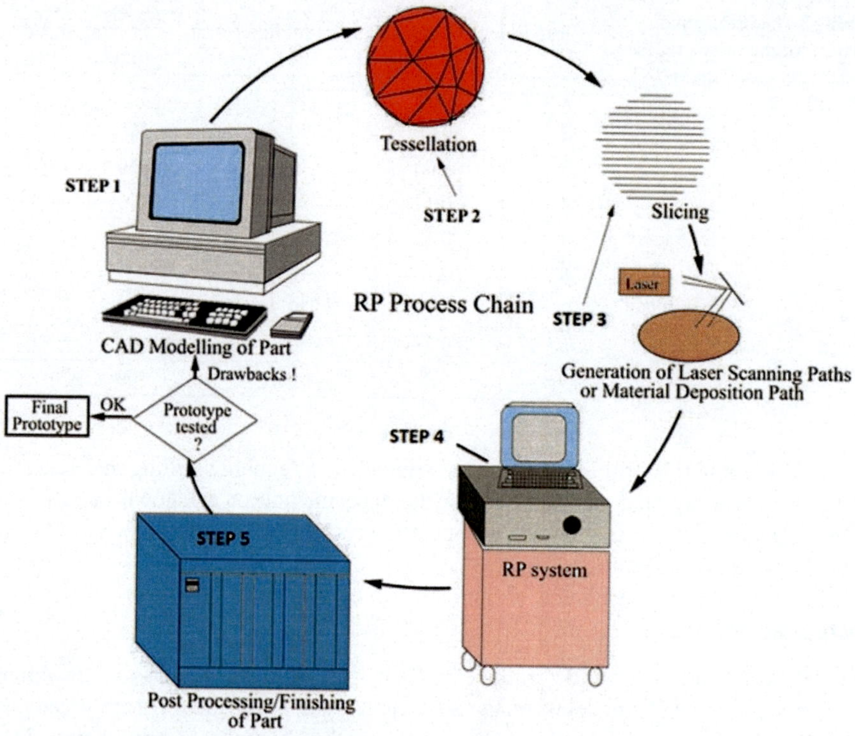

Fig. 1 Basic steps in the FDM process [7]

4 Results and Discussions

After conducting the experiments with different settings of input parameters i.e. layer height, infilled density and printing speed, the values of output parameter i.e. printing time, surface roughness and dimensional accuracy were recorded and plotted as per Taguchi design of experiments methodology.

4.1 Effect of Layer Height and Infilled Density on Surface Roughness

Figure 3 shows that with the increase in the infilled density, the surface roughness is also increased. For 0.06 and 0.10 mm layer height, the surface roughness increases, but for 0.20 mm layer height the surface roughness decreases as there is an increase in layer height and according to the requirement, there is a need for minimum surface roughness and smooth surfaces.

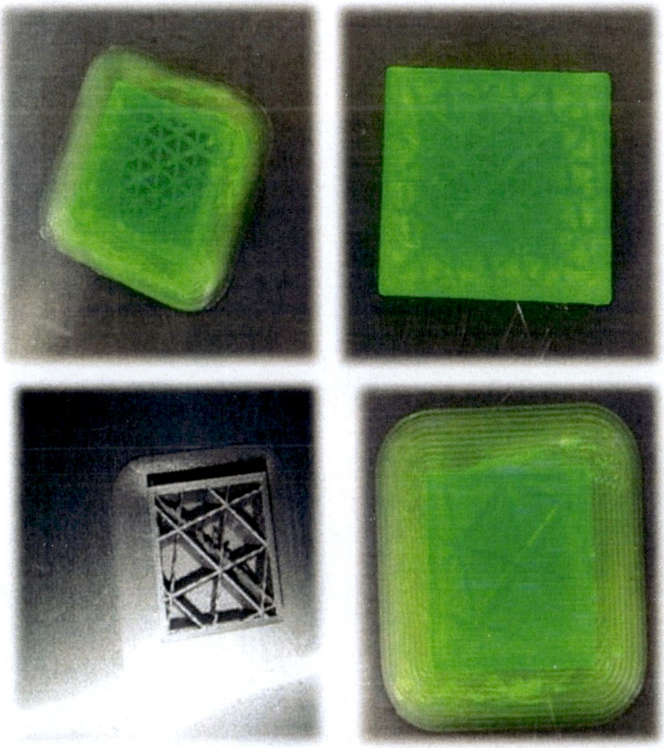

Fig. 2 Specimen samples after fabrication

4.2 *Effect of Layer Height and Printing Speed on Surface Roughness*

Figure 4 shows the effect of two parameters, printing speed and layer height on surface roughness. It has been found that for 0.06 mm of layer height, surface roughness increases with an increase in printing speed. Further, for the layer height i.e. 0.1 and 0.2 mm, the surface roughness is optimized.

Figure 4 also shows that the comparison plot, the pattern of the graph is not uniform. If we compare both the plots, then the average printing speed i.e.100 mm/s would be the best option in both the cases.

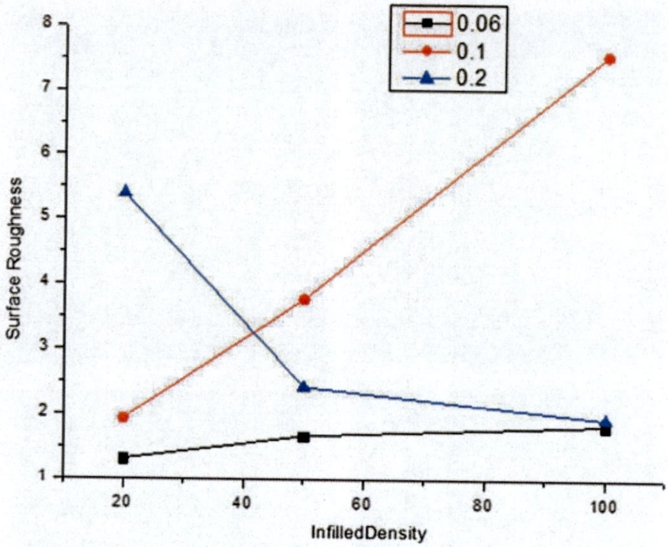

Fig. 3 Effect of layer height and infilled density on surface roughness

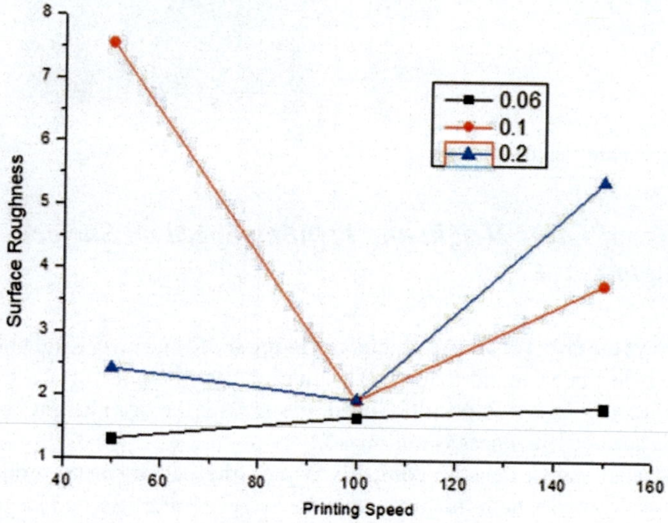

Fig. 4 Effect of layer height and printing speed on surface roughness

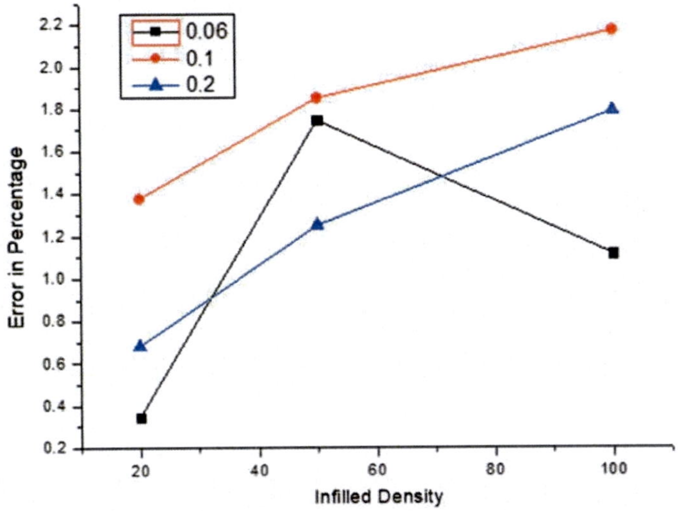

Fig. 5 Effect of layer height and infilled density on error in percentage

4.3 *Effect of Layer Height and Infilled Density on Dimensional Error (in Percentage)*

Figure 5 represents the effect of two parameters, layer height and infilled density on dimensional error (in percentage). For 0.10 and 0.20 mm layer height, the graph shows a clear view that with the increase of infilled density the percentage of dimensional error also increases and in the case of 0.06 mm layer height, there is also an increase in the error percentage at some extent but after that, on increasing the layer height there will be a decrease in dimensional error.

4.4 *Effect of Layer Height and Printing Speed on Dimensional Error (in Percentage)*

Figure 6 shows the effect of two parameters, layer height and printing speed on dimensional error (in percentage). It has been found that there is a lot of variation for the 0.06 mm layer height, the dimensional error increases with an increase in printing speed. For the 0.10 mm layer height, the dimensional error decreases with an increase in speed but if speed is increased thereafter to some extent, it then starts to increase. For 0.20 mm layer height, the dimensional error increases with an increase in speed but after increasing the speed to some extent, the dimensional error starts to decrease. The need is to get less the printing time and the surface roughness and dimensional error should be minimum [16].

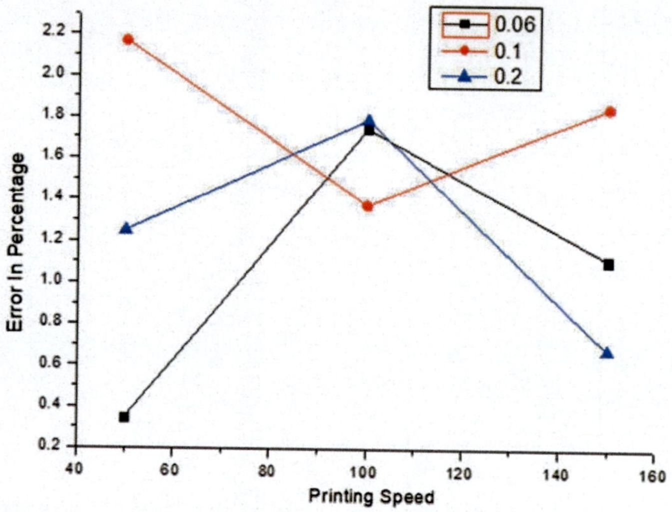

Fig. 6 Effect of layer height and printing speed on error in percentage

Table 5 Optimized value of the parameters

Parameter	Value	Parameter	Value	Parameter	Value
Layer Height	0.06 mm	Infilled Density	50%	Printing Speed	50 mm/s

4.5 *Optimization for Surface Roughness and Accuracy*

The optimized parameters of the 3D printer were determined by the help of graph and data analysis which is generated by the experimental investigation. For this Minitab 17 was used. The optimized values of the input process variables are given in Table 5.

5 Conclusion

The major objective of this work is to investigate the printing performance of the Ultimaker 3D printer. All the experiment trials and planning were executed using the Taguchi design of experiments (DOE) technique [17]. In total, 9 runs were undertaken in this experimental investigation. The following conclusions are drawn on the basis of the performance of machining characteristics studied in the present work namely, printing time, surface roughness and dimensional accuracy.

When the speed is increased up to the limit of 75 mm/s and layer height and infilled density remains unchanged, and it is directly proportional to the surface roughness of the model. Moreover, the rapid increment in the error progressed on increasing the layer height from 0.10 to 0.20 mm.

References

1. Mikołajewska E et al (2014) 3D printing technologies in rehabilitation engineering. (2014)
2. Rengier F et al (2010) 3D printing based on imaging data: a review of medical applications. Int J Comput Assist Radiol Surg 5(4):335–341
3. Lipson H, Kurman M (2013). Fabricated: the new world of 3D printing. Wiley, New York
4. Vasudevarao B, Natarajan DP, Henderson M (2000) Sensitivity Of Rp surface finish to process parameter variation.
5. Anitha R, Arunachalam S, Radhakrishnan P (2001) Critical parameters influencing the quality of prototypes in fused deposition modelling. J Mater Process Technol 118(1–3)
6. Pérez M, Medina-Sánchez G, García-Collado A, Gupta M, Carou D (2018) Surface quality enhancement of fused deposition modeling (FDM) printed samples based on the selection of critical printing parameters. Materials 11:1382
7. Chua CK, Leong KF, Lim CS (2003) Rapid prototyping: principles and applications. World Sci 1
8. Hasan MH, Ahmed S, Voldman R, Mehany M (2018) Parametric effect on surface finish of three-dimensional printed object. Int J Eng Mater Manuf 3(2):98–104
9. Jayanth N, Senthil P, Prakash C (2018) Effect of chemical treatment on tensile strength and surface roughness of 3D-printed ABS using the FDM process. Virtual Phys Prototyp 13(3):155–163
10. Huang X, Shen AZ, Yang AS (2018) Effect of fabrication parameters and material features on surface roughness of fdm build parts. In: 3rd joint international information technology mechanical and electronic engineering conference (JIMEC 2018), Atlantis Press
11. Chari VS, Venkatesh PR, Krupashankar, Dinesh V (2018) Effect of processing parameters on FDM process. AIP Conf Proc 1943:1
12. Nancharaiah T, Raju DR, Raju VR (2010) An experimental investigation on surface quality and dimensional accuracy of FDM components. Int J Emerg Technol 1(2):106–111
13. Alafaghani A, Qattawi A, Alrawi B, Guzman A (2017) Experimental optimization of fused deposition modelling processing parameters: a design-for-manufacturing approach. Proc Manuf 10:791–803, ISSN 2351-9789
14. Lanzotti A et al (2015) The impact of process parameters on mechanical properties of parts fabricated in PLA with an open-source 3-D printer. Rapid Prototyp J 21(5):604–617
15. Mohamed OA, Masood SH, Bhowmik JL (2017) Experimental investigation of creep deformation of part processed by fused deposition modeling using definitive screening design. Addit Manuf 18:164–170
16. Pandey PM, Reddy NV, Dhande SG (2003) Improvement of surface finish by staircase machining in fused deposition modelling. J Mater Process Technol 132(1–3):323–331, ISSN 0924-0136
17. Gupta V et al (2014) Minimization of kerf taper angle and kerf width using Taguchi's method in abrasive water jet machining of marble. Proc Mater Sci 6:140–149

Optimization of Process Parameters on MRR During Face Milling of Rolled Steel (AISI1040) Using Taguchi Method

Kulwinder Singh, Anoop Kumar Singh, and K. D. Chattopadhyay

Abstract Mass production at low cost in industry can be obtained by reducing the manufacturing time. In this study, a systematic approach based on Taguchi's philosophy is proposed to optimize the metal removal rate (MRR) during face milling of rolled steel (AISI1040). Experiments are performed on vertical milling center (HURCO-VM10) using face milling cutter of 80 mm diameter. Cutting velocity, feed per teeth, and depth of cut are considered as quantitative parameters. Whereas rolling direction, cutter offset, and soaking time are considered as qualitative parameters under cutting strategy. Experiments are conducted on the basis of Taguchi's L_{27} orthogonal array and MRR is calculated. Signal to noise (*S*/*N*) ratio is calculated to evaluate the optimum levels of process parameters. Analysis of variance (ANOVA) is conducted to identify the significant parameters, optimum settings, and percent contribution of each process parameter. The outcomes of ANOVA reveal that feed per teeth and cutting velocity are the main process parameters contributing to MRR. A confirmation test is performed to compare the predicted and experimental results of MRR. Results show that there is a significant improvement in the MRR by adopting the proposed optimization technique.

Keywords Taguchi · Analysis of variance (ANOVA) · Signal to noise ratio (s/n ratio) · MRR · Face milling

1 Introduction

In this modern era, customer demands high quality and low cost product along with good service. So, it is a big challenge in manufacturing sector to fulfill all the requirements of customer. Among all the factors, cost plays a vital role in sale and purchase of the component. Direct and indirect costs are the two main segments of total manufacturing cost. Under direct cost, manufacturing time contributes a lot because of direct involvement of machines and man power. Thus, to reduce the cost

K. Singh (✉) · A. K. Singh · K. D. Chattopadhyay
Chitkara University Institute of Engineering and Technology, Chitkara University, Punjab, India
e-mail: kulwinder.singh@chitkara.edu.in

© Springer Nature Singapore Pte Ltd. 2021

P. M. Pandey et al. (eds.), *Advances in Production and Industrial Engineering*, Lecture Notes in Mechanical Engineering,
https://doi.org/10.1007/978-981-15-5519-0_16

of product, high production rate without disturbing other performance characteristics is targeted in the industry.

Further, under various manufacturing processes, machining process has its own significance to generate finished product with tight tolerance. Machining is a metal removal process, used to shape the final product. Cutting tool is used to remove the material in the forms of chips. During cutting, rotary and linear motions are given to cutting tool and workpiece, respectively depending upon machining operation. In vertical milling machine, rotary cutting tool fixed on vertical spindle fed against the workpiece to remove the material. Parts of different size and shapes can be manufactured on milling machine. Many operations can be performed on milling machine such as face milling, end milling, slot milling, drilling and boring, etc. [1]. Face milling operation is commonly used operation in various industrial sectors such as automobile, aerospace, press tools, and shipbuilding, etc. To obtain high production rate in face milling, rate of material removal should be high. The amount of material removed by cutter per unit time is known as material removal rate (MRR). High MRR depends upon the selection of proper input parameters along with their settings levels. On the basis of experience, it is a difficult task to set the machine at optimum settings without disturbing other quality characteristics.

Currently, many types of optimization techniques (mathematical or statistical) are used to optimize the process parameters. In this study, Taguchi method is used to optimize the MRR during face milling. As per Taguchi philosophy, it is a loss if outcomes are deviating from the target. Loss is measured in terms of signal to noise ratio (*S/N* Ratio) [2–4]. Taguchi method is used worldwide for the optimization of process parameters. First of all, input parameters along with their levels are selected on the basis of experience and literature review. On the basis of the number of parameters and their levels, orthogonal array is selected in design of experiments (DOE). Taguchi proposed a partial factorial design technique to reduce the number of experiments. Also, repetition of experiments is recommended (minimum three times) to minimize the variability in outcomes. On the basis of outcome values, mean of each level for a particular parameter is calculated. Further optimum level of each parameter is selected as an optimum setting. Taguchi focuses on variation in results. Thus, analysis of variance (ANOVA) is performed on data values. ANOVA gives the information of significant parameters along with its contribution to variation [5–9].

2 Experimental Method

2.1 Experimental Setup

In this study, rolled steel (AISI1040) is chosen as work material because of its vast applications in the manufacturing industry. It is a medium carbon steel with the chemical composition shown in Table 1. Specimens of size 48 × 48 × 12 mm are

Table 1 Chemical composition

Material	Chemical Composition (wt. %)				
	C	Mn	Si	S	P
AISI 1040	0.42	0.9	0.31	0.05	0.05

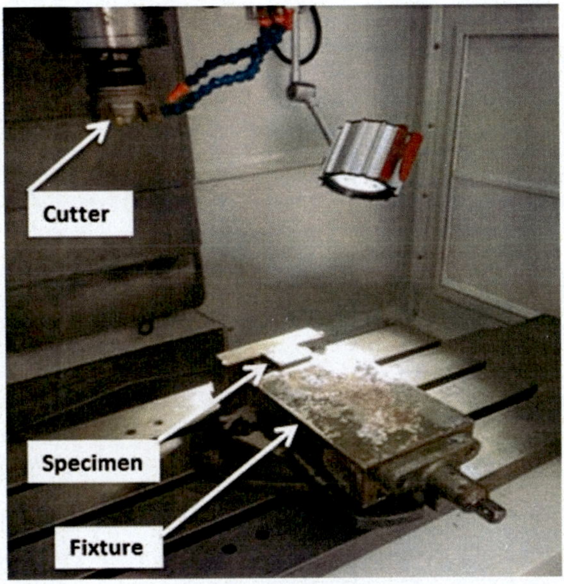

Fig. 1 Experimental setup

prepared as per L_{27} orthogonal array. Face milling operation is performed on vertical milling machine (HURCO-VM10) with the help of carbide insert shown in Fig. 1. The cutter of diameter 80 mm with equally spaced seven numbers of cutting inserts is used for face milling. Initial and final weight of specimen is required to calculate the MRR along with machining time. Weight is measured on digital weighing scale shown in Fig. 2.

2.2 *Selection of Process Parameters*

Process parameters are selected on the basis of literature review and machining conditions. Machining parameters such as cutting velocity, feed per teeth, and depth of cut are the main input parameters in metal removal process.

Apart from machining parameters other three parameters (rolling direction, cutter offset, and soaking time in recovery) are also selected as process parameters. These parameters have a significant role on various quality characteristics such as MRR,

Fig. 2 Weight measurement

surface finish, and burr growth, etc. [10]. Rolled steel behaves like anisotropic material. Strength of rolled sheets varied with respect to rolling direction (along and across). Moreover, stresses generate in the material due to rolling process. Thus, to relieve the stresses, recovery process (low temperature annealing) at 170 °C is performed before machining by varying the soaking time. Soaking time depends upon the thickness of workpiece. Generally, one hour per inch of thickness is taken into consideration [11]. Range of soaking time is fixed based on the literature review. Further effect of cutter path on the machinability is also studied. Cutter offset, i.e., distance of cutter axis w.r.t. left edge of workpiece is varied [12]. Detail of process parameters along with levels values is shown in Table 2.

3 Experimental Results and Discussion

3.1 Evaluation of S/N Ratio

Experimentation is performed as per L27 orthogonal array shown in Table 3. Each experiment is replicated three times. After machining, final weight of each specimen is measured and MRR is calculated using Eq. (1).

Table 2 Detail of face milling process parameters

Symbol	Process parameter	Unit	Level 1	Level 2	Level 3
A	Cutting velocity	m/min	100	150	200
B	Feed per teeth	mm/teeth	0.05	0.1	0.15
C	Depth of cut[a]	mm	0.4	0.8	1.2
D	Rolling direction	degree	0	45	90
E	Cutter offset	mm	12	24	36
F	Recovery-soaking time	min	30	45	60

[a]Depth of cut is completed in multi-passes with each cut of 0.2 mm

$$\text{MRR} = (\text{Initial weight}-\text{Final weight})/\text{Machining Time} \tag{1}$$

Response is measured in the form of *S*/*N* Ratio because average value of response does not measure the variability. Taguchi method focuses on the variability to control the process. It considered the variability (dispersion in data) as quality loss and proposed the *S*/*N* Ratio to measure it. Taguchi says that if the output value is deviating from target, it will be considered a quality loss (*QL*). So variability is measured as mean square deviation of the MRR (higher the better quality characteristic) values in terms of *QL* using Eq. (2), further converted into *S*/*N* Ratio using Eq. (3). Results of Average MRR and *S*/*N* Ratio are shown in Table 3.

$$QL = \frac{1}{n}\sum_{k=1}^{n} 1/y^2 \tag{2}$$

$$S/N\text{Ratio} = -10 * \log(QL) \tag{3}$$

where y represents the MRR value of kth replication and n represents the number of replications.

3.2 Determination of Optimal Level

Mean value of *S*/*N* Ratio for each level of distinct process parameter is calculated to study the effect of process parameters on MRR. Higher *S*/*N* Ratio among different level of each parameter is considered the optimum level. Results for optimum level are presented in Table 4. Effects of parameters on MRR are also shown in Fig. 3. Levels of process parameters for optimum MRR are cutting velocity at level 3, feed per teeth at level 3, depth of cut at level 2, rolling direction at level 1, cutter offset at level 1, and soaking time at level 1. In Table 4, range denotes the difference between maximum and minimum mean value for levels of a particular parameter. It exposes the contribution of parameter. Further, on the basis of range value, rank is designated

Table 3 L_{27} orthogonal array with results

Experiment no.	Process parameters						Response	
	A	B	C	D	E	F	Avg. MRR (gm/min.)	S/N ratio (dB)
1	1	1	1	1	1	1	3.76	11.49
2	1	1	1	2	2	2	2.97	9.46
3	1	1	1	3	3	3	3.06	9.71
4	1	2	2	1	2	3	7.81	17.85
5	1	2	2	2	3	1	7.07	16.99
6	1	2	2	3	1	2	6.94	16.82
7	1	3	3	1	3	2	10.49	20.42
8	1	3	3	2	1	3	11.54	21.24
9	1	3	3	3	2	1	11.17	20.96
10	2	1	3	3	3	2	5.52	14.84
11	2	1	3	1	1	3	5.66	15.04
12	2	1	3	2	2	1	5.73	15.16
13	2	2	1	3	1	1	11.52	21.22
14	2	2	1	1	2	2	10.24	20.20
15	2	2	1	2	3	3	9.52	19.57
16	2	3	2	3	2	3	17.04	24.63
17	2	3	2	1	3	1	16.69	24.45
18	2	3	2	2	1	2	17.19	24.70
19	3	1	2	2	2	3	9.19	19.26
20	3	1	2	3	3	1	10.23	20.19
21	3	1	2	1	1	2	9.23	19.30
22	3	2	3	2	3	2	13.14	22.37
23	3	2	3	3	1	3	15.27	23.68
24	3	2	3	1	2	1	15.07	23.56
25	3	3	1	2	1	1	21.58	26.67
26	3	3	1	3	2	2	22.04	26.86
27	3	3	1	1	3	3	22.11	26.89

to each parameter. So feed per teeth with range of 9.15 is designated with first rank followed by cutting velocity, depth of cut, soaking time, cutter offset, and rolling direction with range of 7.10, 1.35, 0.63, 0.53, and 0.42, respectively. The optimum parameter/level combination is identified as $A_3B_3C_2D_1E_1F_1$. During optimization of process parameters it is observed that to achieve maximization of MRR, higher feed rate (level 3) and higher cutting velocity (level 3) should be applied.

Table 4 Effects of process parameters on MRR

Main effects of process parameters on MRR						
Levels	A	B	C	D	E	F
1	16.10	14.94	19.12	19.91	20.02	20.08
2	19.98	20.25	20.47	19.49	19.77	19.44
3	23.20	24.09	19.70	19.88	19.49	19.76
Range	7.10	9.15	1.35	0.42	0.53	0.63
Rank	2	1	3	6	5	4
Optimum level	A3	B3	C2	D1	E1	F1

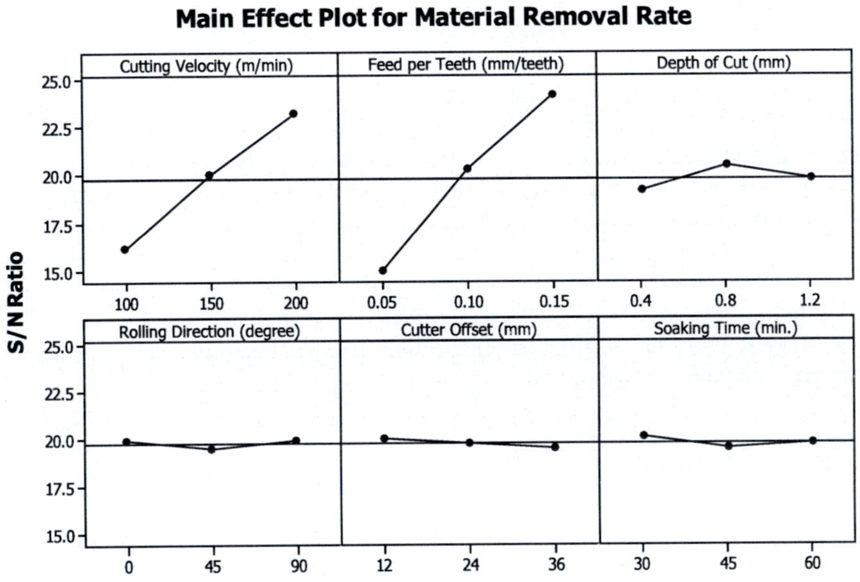

Fig. 3 Main effect plot for MRR

3.3 *Analysis of Variance (ANOVA)*

ANOVA is a statistical technique to identify the significant factor and their contribution on the basis of variation. Sum of squared deviation of individual parameter (SS_P) and total sum of squared deviation (SS_T) is evaluated on the basis of *S*/*N* ratio listed in Table 3. Then, SS_T, *F*-ratio, and contribution percentage (ρ) of each parameter on its output are calculated according to Eqs. 4, 5, and 6, respectively.

$$SS_T = \sum_{i=1}^{n} (n_i - n_n)^2 \tag{4}$$

where n represents the number of experiments in L_{27} matrix, n_i denotes mean value of response for ith experiment, n_n is the total mean of S/N ratio.

$$F\text{-ratio} = \frac{(\text{MSS})_P}{(\text{MSS})_e} \tag{5}$$

where $(\text{MSS})_P$ and $(\text{MSS})_e$ denotes the mean sum of square of individual parameter and error respectively.

$$\rho = \frac{SS_P}{SS_T} \tag{6}$$

SS_P represents the sum of square (variance) of individual factor whereas SS_T is the total sum of square.

Significance level of an individual parameter is determined on the basis of critical value of F-test at 5% significance level. Critical value of $F_{0.05,2,14}$ (before pooling) and $F_{0.05,2,20}$ (after pooling) is listed as 3.74 and 3.49, respectively in F-test table. Table 5 represents the ANOVA results along with contribution percentage of each parameter. All the non-significant parameters are pooled in error to increase the effectiveness of ANOVA. After pooling again new F-ratio, sum of square, and percent contribution are calculated. Results reveal that among significant parameters, feed per teeth is contributing highest with 60.53%, cutting velocity with 36.11% and depth of cut with 1.14%. Error has negligible contribution of 2.2% thus interaction effects of process parameters are neglected.

Table 5 Results of ANOVA

Results of ANOVA after pooling							
Parameter	Degree of freedom	Sum of square	Mean sum of square	F ratio (old)	F′ ratio (new)	New sum of square	Percent contribution
A	2	227.25	113.62	239.08	212.64	226.18	36.11
B	2	380.26	190.13	400.06	355.81	379.20	60.53
C	2	8.22	4.11	8.65	7.69	7.15	1.14
D	0	0.00	0.00	Pooled	Pooled	Pooled	Pooled
E	0	0.00	0.00	Pooled	Pooled	Pooled	Pooled
F	0	0.00	0.00	Pooled	Pooled	Pooled	Pooled
Error	20	10.69	0.53			13.89	2.22
Total	26	626.42	24.09			626.42	100.00

Table 6 Confirmation test results

	Initial process parameters	Optimal process parameters	
		Predicted	Experimented
Level	$A_1B_1C_1D_1E_1F_1$	$A_3B_3C_2D_1E_1F_1$	$A_3B_3C_2D_1E_1F_1$
Average MRR (gm/min.)	3.76	25.81	22.16
S/N ratio (dB)	11.49	28.23	26.89
Improvement in MRR (*S/N* ratio) = 15.40 dB			

3.4 *Confirmation Test*

Finally, in parametric design of Taguchi, confirmation test is performed on the basis of optimal settings to validate the predicted model, shown in Eq. (7).

$$\Psi_{\text{opt}} = \Psi_m + \sum_{k=1}^{n} \left(\Psi_{m_i} - \Psi_m\right) \tag{7}$$

where, Ψ_{opt} is the predicted optimum MRR, Ψ_m is the overall mean of MRR, Ψ_{m_i} is the mean of optimum level and n is the number of process parameter. Results of confirmation test are shown in Table 6. MRR is improved by 15.40 dB when the optimum process parameters ($A_3B_3C_2D_1E_1F_1$) are used instead of initial process parameters ($A_1B_1C_1D_1E_1F_1$). Thus confirmation test validates the success of Taguchi method in optimization of MRR.

4 Conclusion

This study presents a systematic methodology proposed by Taguchi, to optimize the face milling operation for metal removal rate (MRR). The results of ANOVA indicate that cutting velocity, feed per teeth, and depth of cut are significant parameters. Among all process parameters, cutting velocity and feed per teeth are found to be the main contributors in variation (36% and 60% respectively). Total depth of cut is completed with multi-passes of 0.2 mm, therefore showing negligible effect. Confirmation test based on optimal settings reveals the improvement of 15.40 dB as compared to initial settings. Thus, MRR of 22.16 gm/min. can be attained by setting the cutting velocity and feed per teeth at 200 m/min. and 0.15 mm/teeth, respectively. Improvement in MRR indicates that optimization of any process can be done easily by adopting Taguchi method.

References

1. Ghosh A, Malik AK (2009) Manufacturing science. East-West Press Private Limited, New Delhi
2. Taguchi G (1990) Introduction to quality engineering. Asian Productivity Organization, Tokyo
3. Ross PJ (1988) Taguchi techniques for quality engineering. McGraw-Hill, New York
4. Montgomery DC (1991) Design and analysis of experiments. Wiley, Singapore
5. Singh AK, Kumar S, Singh VP (2014) Optimization of parameters using conductive powder in dielectric for EDM of super Co 605 with multiple quality characteristics. Mater Manuf Process 29(3):267–273
6. Parashar V, Purohit R (2017) Investigation of the effects of the machining parameters on material removal rate using Taguchi method in end milling of steel grade EN19. Mater Today: Proc 4(2):336–341
7. Rao RV, Pawar PJ (2010) Parameter optimization of a multi-pass milling process using non-traditional optimization algorithms. Appl Soft Comput 10(2):445–456
8. Lin TR (2002) Optimisation technique for face milling stainless steel with multiple performance characteristics. The Int J Adv Manuf Technol 19(5):330–335
9. Fedai Y, Kahraman F, Kirli Akin H, Basar G (2018) Optimization of machining parameters in face milling using multi-objective Taguchi technique. Tehnički glasnik 12(2):104–108
10. Aoda YF, Jabur LS (2014) Effect of the rolling direction and draft on some of the mechanical properties for the medium carbon steel. Int J Sci Res 3(12):2425–2431
11. Upadhyay GS, Upadhyay A (2007) Material science and engineering. Viva Books, UK
12. Olvera O, Barrow G (1998) Influence of exit angle and tool nose geometry on burr formation in face milling operations. Proc Inst Mech Eng, Part B: J Eng Manuf 212(1):59–72

A New Permanent Magnet Type Magnetorheological Finishing Tool for External Cylindrical Surfaces Having Different Outer Diameter

Ajay Singh Rana, Talwinder Singh Bedi, and Vishwas Grover

Abstract An improved magnetorheological finishing process has been developed with three permanent magnets for nano finishing the external surface of cylindrical workpieces. The cylindrical permanent magnets used in the developed tool are placed at an angle of 90° from each other. The three cylindrical permanent magnets are positioned in such a way that all three maintain an equal working gap with the surface of cylindrical workpiece. Finite Element (FE) analysis of the entire setup has also been performed in the Maxwell Ansoft V13 software to observe the dispersal of magnetic field density in the working gap. In the current study, the preliminary experimentations have been carried out to evaluate the finishing capability of the present developed tool. Experiments have been conducted over the external cylindrical workpiece made of copper which can be used as an electron discharge machining (EDM) electrode. After the experimentations of 45 min over the entire cylindrical workpiece made of copper, the average surface roughness R_a gets reduced from 224 to 67 nm with negligible surface defects which confirm the finishing performance of the developed finishing tool.

Keywords Magnetorheological finishing · Roughness · Peaks · External diameters · Variable · Surface

A. S. Rana (✉)
RIMT University, Mandi Gobindgarh, Punjab 147301, India
e-mail: ajayrana@rimt.ac.in

T. S. Bedi
Indian Institute of Technology Ropar, Rupnagar, Punjab 140001, India

V. Grover
Ajay Kumar Garg Engineering College, Ghaziabad, Uttar Pradesh 201009, India

© Springer Nature Singapore Pte Ltd. 2021

P. M. Pandey et al. (eds.), *Advances in Production and Industrial Engineering*, Lecture Notes in Mechanical Engineering,
https://doi.org/10.1007/978-981-15-5519-0_17

1 Introduction

In today's world, industrial components require very precise surface finishing [1, 2]. To attain this specified finished surface, various traditional as well as advanced finishing processes have been developed in the last decade. Conventional finishing processes like grinding, honing, lapping, and ball burnishing, etc., have been developed to acquire surface finishing over various internal or external surfaces. But the traditional finishing processes have limitations that they do not have control over finishing forces [3]. Due to the uncontrollable finishing forces, various problems like surface or subsurface defects, heat generation, etc., persist through these processes [4–6]. To overcome these limitations and acquire good control over the finishing forces, various advanced finishing processes have been flourished in the last two decades [7]. These advanced finishing processes make use of magnetic field for finishing operation. In the presence of magnetic field, abrasives perform the finishing operation in these processes. By regulating the magnitude of magnetic field in these advanced finishing processes, finishing forces acting by the abrasives over the workpiece surface can easily be controlled. A lot of researchers had developed magnetorheological (MR)-based finishing processes for various applications [8–13]. These advanced finishing processes make use of smart fluid, i.e., magnetorheological (MR) polishing fluid to enhance for finishing the surfaces. Under the action of magnetic field, carbonyl iron particles [CIPs] forms the chain like pattern, grip the abrasives, and perform the finishing [14]. By regulating the magnitude of magnetic field, strengthening of the formed CIPs chains can be controlled and in this way, finishing forces can be controlled [13].

As the present scenario, there is a lack of MR finishing tool for finishing of different external diameter of cylindrical workpieces. For example, if the size of cylindrical workpiece varies from 30 to 50 mm or even more there is no permanent magnet tool available based upon MR polishing fluid for finishing such variable diameter surfaces. Thus to counteract these circumstances, an improved finishing tool has been made for finishing external cylindrical surfaces of different outer diameter.

2 Finite Element Magnetostatic Simulation

In order to clearly visualize the distribution of magnetic field between the working gap, finite element-based simulation has been done using Maxwell Ansoft V13 software as shown in Fig. 1.

In this simulation, relative permeability of permanent magnet of 1.009, copper workpiece of 0.999991, and MR polishing fluid of 5 were taken. The distribution of magnetic field density in the present developed setup, obtained through finite element (FE) analysis using Maxwell Ansoft V13 software as shown in Fig. 1. The simulation results determined that the magnitude of magnetic flux density was higher over the permanent magnet surface than on the cylindrical surface of copper workpiece. This

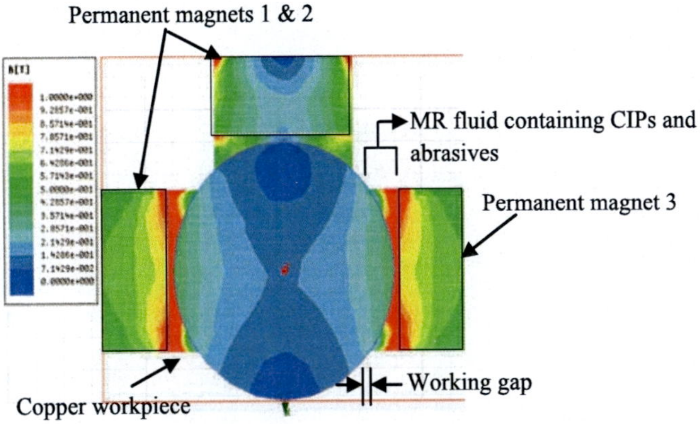

Fig. 1 Distribution of magnetic flux density in the present developed setup

means that MR fluid gets stuck properly on the surface of permanent magnet. Due to non-ferromagnetic nature of workpiece material, the CIPs of MR polishing fluid does not stick on it. This is a prerequisite condition for finishing of ferromagnetic external surfaces. In this tool design, the combined effect of three permanent magnets would be helpful for achieving a better and uniform surface quality on the workpiece surface.

3 Mechanism of Material Removal

One of the most important task in attaining high-quality surface finishing is to understand the process of removal of nanomaterial during polishing. The material gets removed due to mechanical abrasion phenomenon. The schematic of material removal from initial surface to final finished surface during finishing with the current developed tool is shown in Fig. 2.

Initially, the external cylindrical surface of workpiece was not having a high-quality surface finish. This is due to the uncontrollable finishing forces acted by the rigid abrasive stone as a result the surface roughness is not uniform. This nonuniformity in surface roughness was due to the formation of surface defects such as roughness profiles, cavities, etc.

The MR polishing fluid under magnetic field effect forms a semisolid type structure. It consists of nonmagnetic abrasive particles held tightly by the CIPs chains structure as shown in Fig. 2. When these abrasive particles continuously move over the workpiece and finally convert the major roughness peaks into a high-quality surface finishing is obtained as shown by different finishing stages in Fig. 2.

Fig. 2 Different stages of material removal mechanism during finishing of cylindrical copper workpiece

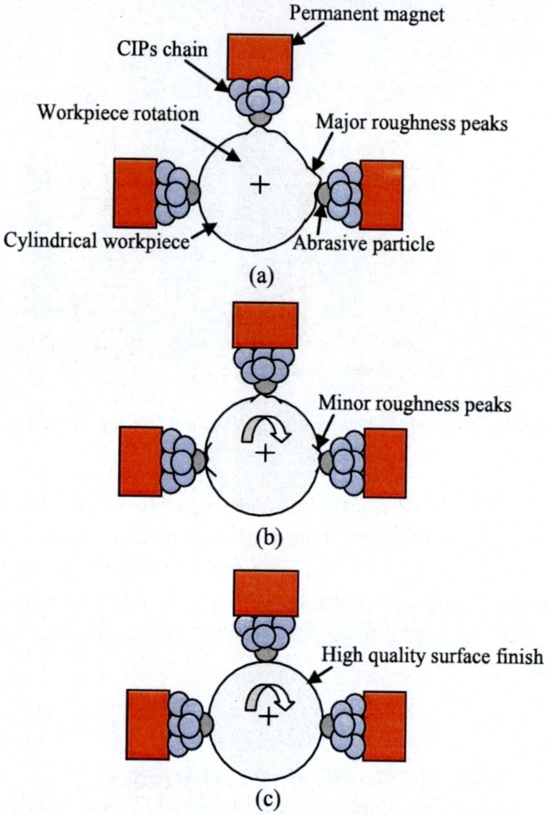

4 Schematic of Novel Tool Design Along with Its Mountings

This novel tool design has the advantage to be flexible according to the size of the cylindrical workpiece. In this design, the tool is capable to finish the workpiece having outer diameter ranging from 30 to 50 mm. The three permanent magnets (made of NdFe35) are placed at an angle of 90° to each other as shown in Fig. 3. The three permanent magnets of diameter as 25 mm were individually mounted on the separate cylindrical bush made of brass. The cylindrical bush was further attached to shaft by means of a rectangular bracket. With the help of rotating wheel, the magnets can be easily adjusted by means of metric threads. The workpiece was fixed in the rotating chuck of a lathe machine. The rectangular bracket along the assembled magnets was fixed on the tool post portion of lathe machine. The linear movement of magnets was provided through lathe machine, whereas the workpiece was rotated by the rotation of chuck of lathe machine.

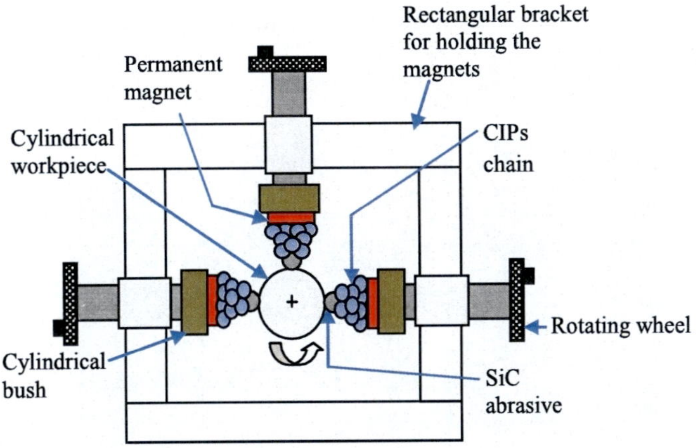

Fig. 3 Schematic of novel tool design along with its mountings

5 Preliminary Experimentation

To check the feasibility of the developed tool, it is important to prove it practically. Therefore, to observe this aspect, a preliminary experimentation has been performed with the novel developed tool. The developed tool was mounted on the tool post of lathe machine. The cylindrical workpiece used in this study was made of copper which can be used as an electron discharge machining (EDM) electrode for making a cylindrical cavity in the dies. A high-quality surface is needed on this cylindrical workpiece for getting a defect free cavity on the die material. During experimentation, the cylindrical workpiece was fixed with the help of rotating chuck of lathe machine. The actual photograph of an experimental setup during finishing of copper workpiece is shown in Fig. 4. The experimental conditions like tool linear speed of 50 mm/min and workpiece rotational speed of 500 rpm were used during finishing. The total length of workpiece to be finished was 50 mm.

The composition of MR fluid for finishing of copper workpiece was taken as by volume concentration, i.e., 23% of silicon carbide (SiC) abrasives with mesh size 800, 17% carbonyl iron particles (CIPs) with mesh size 400 and 60% base fluid (20% grease and 80% paraffin oil). This similar composition of MR polishing fluid was already taken [15] during finishing of flat copper workpiece. The initial surface roughness profiles before polishing and finished surface roughness profiles after polishing were measured by using Surftest SJ-210 (Mitutoyo). During measurement, the cutoff length as 0.25 mm and filter as Gaussian was taken. The surface structure of both initial and finished surface was taken by scanning electron microscopy (SEM) images. The actual photograph of an experimental setup during finishing of copper cylindrical workpiece is shown in Fig. 4.

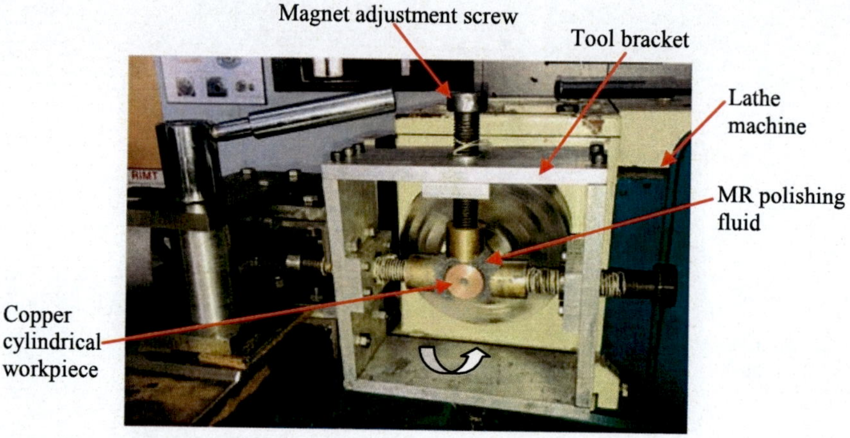

Fig. 4 Actual photograph of an experimental setup during finishing of copper cylindrical workpiece

6 Result and Discussion

After experimentation, the surface roughness value changes from initial surface (R_a = 224 nm) to final surface (R_a = 67 nm) after 45 min of finishing time with the current developed tool is shown in Fig. 5.

Correspondingly, the surface roughness profiles and SEM images of initial and final finished surface are shown in Fig. 6.

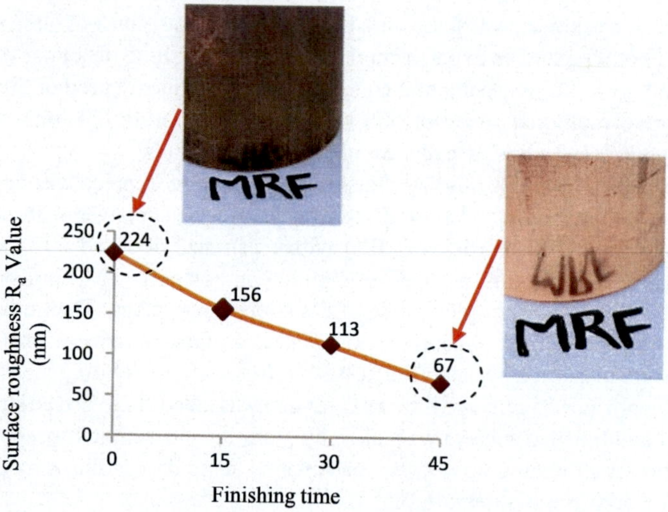

Fig. 5 Change in surface roughness (R_a value) after 45 min of finishing

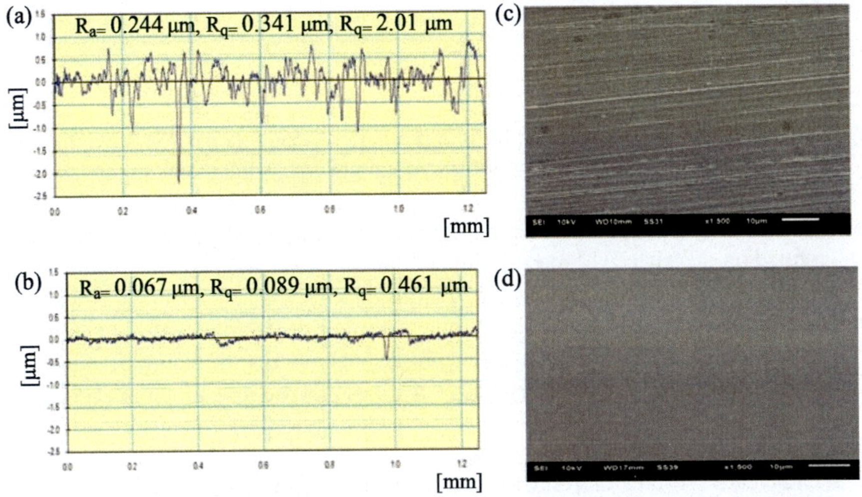

Fig. 6 Surface roughness profiles of **a** initial surface profile and **b** finished surface profile, SEM images of **c** initial and **d** finished surface after 45 min of finishing time

The copper cylindrical workpiece was initially finished by the traditional finishing operation such as grinding. The initial surface roughness (R_a) value was 224 nm. It consists of a major roughness peak on the circumference of copper cylindrical workpiece as shown in Fig. 2a. During 15 min of finishing operation takes place with the present developed MR tool, the SiC abrasives eroded the major roughness peaks and further reduced the surface roughness value, i.e., $R_a = 156$ nm (it is represented by the first finishing stage). After finishing, the roughness peaks get reduced and converted to minor roughness peaks (Fig. 2b).

The cutting of roughness peaks is because of the continuous movement of gripped SiC particle over the cylindrical surface of copper workpiece by mechanical abrasion phenomenon. Further the new MR polishing fluid was applied between the working gap for next 15 min of finishing, i.e., second finishing stage. The average surface roughness value R_a decreased to 113 nm from 156 nm. The change in R_a value in second finishing stage was less because the minor roughness peaks have the lesser shearing tendency as compared to major roughness peaks. In the third finishing stage, i.e., next 15 min of finishing, MR polishing fluid was again applied within the working gap. The average surface roughness value R_a reduced to 67 nm from 113 nm. This results in a better surface quality as compared to the initial ground surface (Fig. 2c). After further finishing, the surface roughness remained almost same. Hence, the surface roughness finishing was shown after 45 min of finishing.

7 Conclusions

The following conclusions are drawn:

- The present developed tool with three permanent magnets has the advantage to be flexible for the cylindrical workpieces having outer diameter ranges from 30 to 50 mm.
- From the magnetostatic simulation, it is observed that the MR polishing fluid gets retained on the surface of magnet rather than on cylindrical surface of workpiece which is useful for high-quality surface finish.
- Shear force exerted by the abrasives on the surface of workpiece is found mathematically higher than the resisting shear force generated by the workpiece material which confirms the finishing capability of the developed finishing tool for copper workpiece.
- The surface roughness changes from initial $R_a = 224$ nm to final $R_a = 67$ nm, as it can clearly visualized in surface roughness profiles and SEM images.
- The present developed tool (flexible type) can also be useful for finishing various cylindrical surfaces made of aluminum, stainless steel, etc.

References

1. Sankar MR, Jain VK, Ramkumar J (2009) Experimental investigations into rotating workpiece abrasive flow finishing. Wear 267:43–51
2. Balogun VA, Mativenga PT (2017) Specific energy based characterization of surface integrity in mechanical machining. Procedia Manuf 7:290–296
3. Benardos PG, Vosniakos GC (2003) Predicting surface roughness in machining: a review. Int J Mach Tools Manuf 43:833–844
4. Gupte PS, Wang Y, Miller W, Barber GC, Yao C, Zhou B, Zou Q (2008) A study of torn and folded metal (TFM) on honed cylinder bore surfaces. Tribol Trans 51:784–789
5. Salmon SC (1992) Modern grinding process technology. McGraw Hill, New York, USA
6. Hashimoto F, Yamaguchi H, Krajnik P, Wegener K, Chaudhari R, Hoffmeister H, Kuster F (2016) Abrasive fine-finishing technology. CIRP Ann-Manuf Technol 65:597–620
7. Bedi TS, Singh AK (2016) Magnetorheological methods for nanofinishing—a review. Part Sci Technol 34(4):412–422
8. Kordonski W, Shorey A (2007) Magnetorheological (MR) jet finishing technology. J Intell Mater Syst Struct 18:1127–1130
9. Jha S, Jain VK (2004) Design and development of the magnetorheological abrasive flow finishing (MRAFF) process. Int J Mach Tools Manuf 44:1019–1029
10. Sadiq A, Shunmugam MS (2009) Investigation into magnetorheological abrasive honing (MRAH). Int J Mach Tools Manuf 49:554–560
11. Singh AK, Jha S, Pandey PM (2011) Design and development of nanofinishing process for 3D surfaces using ball end MR finishing tool. Int J Mach Tools Manuf 51:142–151
12. Grover V, Singh AK (2018) Improved magnetorheological honing process for nanofinishing of variable cylindrical internal surfaces. Mater Manuf Process 33(11):1177–1187
13. Bedi TS, Singh AK (2018) Development of magnetorheological fluid based finishing process for finishing of ferromagnetic cylindrical workpiece. Mach Sci Technol 22(1):120–149

14. Singh AK, Jha S, Pandey PM (2015) Performance analysis of ball end magnetorheological finishing process with MR polishing fluid. Mater Manuf Process 30(12):1482–1489
15. Kansal H, Singh AK, Grover V (2018) Magnetorheological nano-finishing of diamagnetic material using permanent magnets tool. Precis Eng 51:30–39

Influence of Nanoparticle Addition (TiO_2) on Microstructural Evolution and Mechanical Properties of Friction Stir Welded AA6061-T6 Joints

Tanvir Singh, S. K. Tiwari, and D. K. Shukla

Abstract Attempts were made in the present study to evaluate the effect of titanium oxide nanoparticles addition on microstructural evolution and mechanical properties of friction stir welded 6061-T6 aluminum alloy joints. Optical microscopy and scanning electron microscopy was utilized to evaluate the microstructures of the produced joints (nanocomposites) and to ascertain the distribution of titanium oxide nano-range particles in the processed zone. Results reveal that the produced nanocomposites have a uniform distribution of titanium oxide nanoparticles across the perpendicular x-section of a welded processed zone via Zener-pinning effect occurred due to the presence of titanium oxide nanoparticles that helps to prevent the coarsening of grains accompanied by recrystallization throughout the friction stir welding process, resulted in significant grain size reduction. With the increase in volume percentage of titanium oxide nanoparticles, a remarkable increase in the microhardness was noticed. It was also noticed that the ultimate tensile strength and the resistance to wear of produced nanocomposite can be significantly increased with the addition of titanium oxide nanoparticles as compared to parent metal joints. The corresponding mechanical properties' results were correlated with microstructure and fractography.

Keywords Friction stir welding · Aluminum alloy · Nanocomposites · Microstructure · Mechanical characteristics

1 Introduction

Nowadays, weight reduction and less fuel consumption are the two most important demands in the marine, aerospace, and automobile industries. In order to fulfill such demands, soft and light metals as aluminum-alloys are the most suitable [1]. Heat-treatable aluminum-alloys especially 6061-T6 aluminum alloy is most commonly employed in marine frames, pipelines, aircraft, marine, and construction,

T. Singh (✉) · S. K. Tiwari · D. K. Shukla
Department of Mechanical Engineering, Dr. B.R Ambedkar National Institute of Technology, Jalandhar, India
e-mail: tanvirsingh3@gmail.com

© Springer Nature Singapore Pte Ltd. 2021

P. M. Pandey et al. (eds.), *Advances in Production and Industrial Engineering*, Lecture Notes in Mechanical Engineering,
https://doi.org/10.1007/978-981-15-5519-0_18

etc., because of high corrosive resistance, high material specific strength, and ability of weld [2]. Metal-matrix composites, particularly particulate metal-matrix composites that are fabricated by reinforcing reinforcement particles, either in fiber form (glass fiber, carbon fiber, etc.) or in particle form (titanium oxide, silicon carbide, etc.) into their material-matrix (aluminum, magnesium, copper alloys, etc.) are the new latest formed materials [1–3]. These metal-matrix composites have exceptionally higher conductivity to thermal resistance, high ratio of strength to weight, more corrosive resistance, more stiffness than the base alloy which makes them the center of attraction to many industries such as, automobile, aerospace, marine, nuclear, transportation, and so on [3, 4]. However, uniform distribution of reinforcement particles in the aluminum substrate and their control is complicated and difficult to achieve via conventional methods for surface modifications [5, 6]. Previously, thermal spraying and laser beam methods were adopted by numerous researchers to produce nanocomposites which lead to reducing mechanical characteristics properties via the occurrence of unenviable second-phase secondary particles [6, 7]. Because these methods were operated at very high temperatures, it is very difficult to stop the unwanted reactions generally (chemical) occurred in-between the reinforcement second-phase particles and base matrix that forms the detrimental secondary phases [7]. In order to avoid these limitations, a process is required for the production of nanocomposites which doesn't reach their melting point of the base material. In correspondence to the above problems, Friction Stir Welding is the most suitable process for the fabrication of nanocomposite on aluminum-matrix which is patented and proven via conducting experiments at The Welding Institute (TWI) in the UK in December 1991 [8, 9]. In this process, due to the application of mechanical force applied by the rotating tool, the tool is plunged at the interface between the joints. Because of which large amount of frictional heating was produced between the tool and workpiece, as a result the material gets soft. Therefore, when the tool moves forward along the joint line a large amount of plastically deformed material was produced which is recirculated and forged from front to the back of the rotating tool and in turn leads to plastic deformation-results in joining in solid state [9, 10]. In contrast to conventional techniques used for the fabrication of composites, friction Stir welding provides enormous advantages (a) The joining in solid state via friction Stir welding results in the absence of melting and chemical reactions occurred between reinforcement particles and the matrix, (b) the severe plastic deformation during friction Stir welding leads to vigorous mixing and refinement of microstructural characteristics in the parent alloy, (c) it leads to proper dispersion of nanoparticles in the aluminum alloy matrix due to adequate amount of localized heat produced via friction Stir welding tool stirring action. Attempts were made in the present study to produce titanium oxide nanoparticles based 6061-T6 aluminum alloy nanocomposites on-base aluminum-matrix via friction Stir welding. The effects of percentage in volume of titanium oxide nanoparticles on microstructure and characteristics (mechanical) of heat-treatable 6061-T6 aluminum-based alloy nanocomposite produced using friction stir welding was also studied.

2 Experimental Procedure

Thin sheets of 6061-T6 aluminum alloy of 2.5 mm thickness were employed to conduct friction Stir welding in this study. The reinforcing particles titanium oxide nanoparticles were used in different volume %, like 0.2, 0.3, and 0.4. Titanium oxide nanoparticles with 99.99% purity and >30 specific surface area were used in the study. The average grain size of as-received titanium oxide nanoparticles was 15 nm which was assessed by using transmission electron micrographs (Fig. 1). The rotational tool speed of 2000 rpm, tool traveling speed of 70 mm/min, and 0° tilt of tool were used in friction Stir welding. The Friction Stir welding tool with the cylindrical shoulder of 15 mm diameter and cylindrical pin (with a round bottom, to reduce wear and increase tool life) of length and diameter 2.30 and 5 mm was utilized. Friction Stir welding was conducted on the dedicated vertical milling machine. Samples for microstructural evolutions study were prepared by employing the standard procedure of metallography using etchant (Keller's reagent) and afterward examined using an optical and scanning electron microscope.

Testing corresponding to microhardness was carried out from the middle of the weld nugget area of produced nanocomposites (normal to the Friction Stir welding direction), by taking a distance of 0.5 mm down from the upper portion and applied the load of 200 g for a 15 s. Mechanical properties' regarding the produced nanocomposites were also calculated. Samples for the tensile test were cut according to ASTM-E8M-11 and tests were done by utilizing 2 mm/min of cross-head velocity.

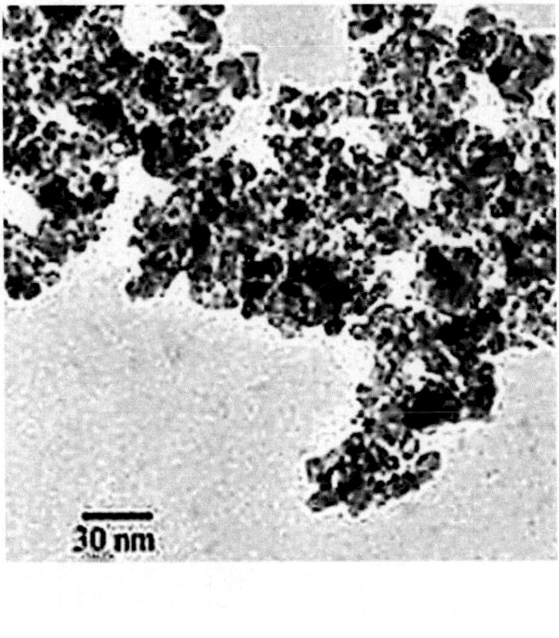

Fig. 1 Transmission electron micrograph of titanium oxide nanoparticle

3 Results and Discussion

3.1 Microstructural Characterization

It was observed that the size of the nugget zone is analogous to the Friction Stir welding cylindrical tool pin diameter and length of 5 mm and 2.30 mm, respectively. The cross-sectioned nugget zone of fabricated nanocomposites perpendicular to the Friction Stir welding direction is depicted in Fig. 2. From Fig. 2, it can be seen that the uniform distribution of titanium oxide nanoparticles was observed which is attributed to the vigorous stirring action provided by the tool, fragmentation process, and the generation of more nucleation sites due to nanoparticles addition. Scanning electron photomicrographs of produced aluminum-based titanium oxide nanocomposites and as-received base parent matrix as illustrated in Fig. 3. It can be noted that the dispersion of titanium oxide nanoparticles in the nugget zone is more evident and uniform during friction Stir welding. This can be attributed to the following reasons, (a) occurrence of severe plastic deformation and continuous dynamic recrystallization in the weld zone [8], (b) presence of nano subsized titanium oxide particles which significantly influenced the grains size in the nugget zone by providing hindrance to the grain boundaries motion via Zener-pinning effect [9, 11]. However, it is clearly inferred from Fig. 3 that titanium oxide nanoparticles are more homogeneously distributed in the nugget zone at 0.4% in comparison to 0.2 and 0.3% of aluminum-based titanium oxide nanocomposites fabricated via friction stir welding. Also, when the volume percentage of titanium oxide nanoparticles was increased up to 0.4%, there was a noticeable drop in the grain size which is generally due to large amount of heat generated and high strain rate because of which dynamic recrystallization granularly was

Fig. 2 Scanning electron photomicrograph depicts the dispersion of titanium oxide nanoparticles on aluminum-matrix

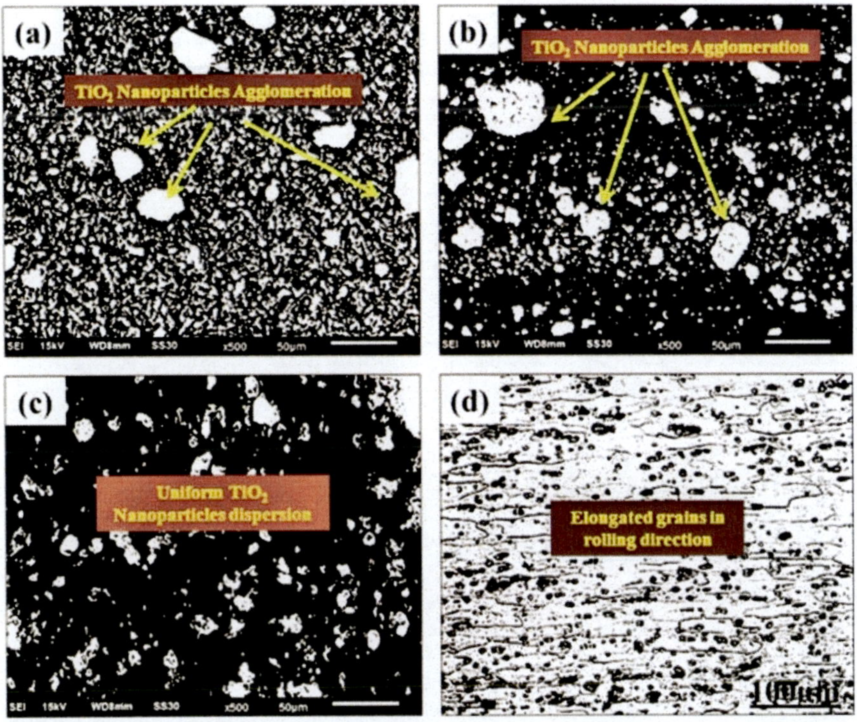

Fig. 3 Scanning electron micrographs of produced aluminum-based titanium oxide nanocomposites, **a** 0.2%, **b** 0.3%, **c** 0.4%, and **d** As-received 6061-T6 aluminum alloy

decreased with increment in strain rate and temperature during process and leads to uniform distribution in nugget zone [10, 12].

3.2 Microhardness Measurements

The profiles regarding the microhardness distribution were obtained from the processes area (nugget) of produced aluminum-based titanium oxide nanocomposites and base alloy matrix are shown in Fig. 4. It is well understood that the microhardness variation in the processed nugget zone merely depends upon the existence of titanium oxide nanoparticles and also their uniform dispersion [13, 14]. It is worth mentioning that, due to the large volume fraction of titanium oxide nanoparticles in the nugget zone their expansion to volume rate is quite high which leads to increase in microhardness to 125 HV which is higher as compared to the as-received base material (55 HV). This was attributed to high rotational speed (2000 rpm) of the Friction Stir welding tool due to which shoulder produced sufficient amount of localized heat and mechanical plunge force that was employed to cause reinforcement particles more

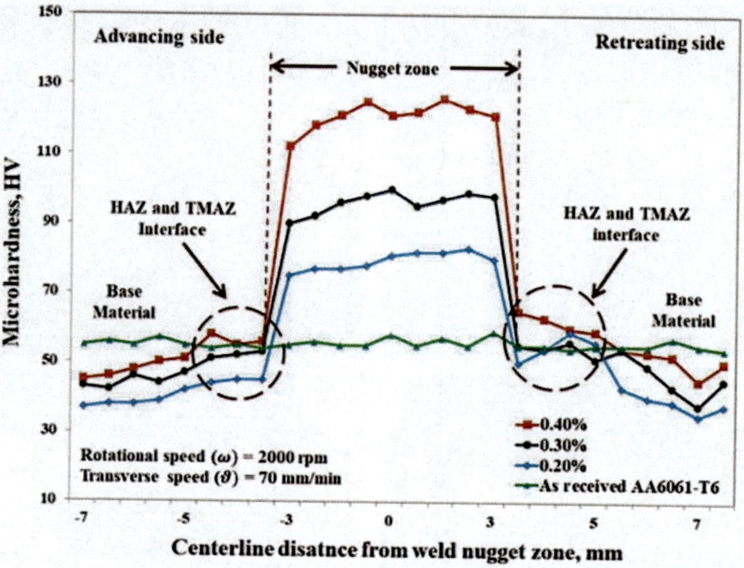

Fig. 4 Microhardness variation evaluated along the transverse section of nugget zone from centreline for aluminum-based titanium oxide nanocomposites and base material

easily encased by soft deformed plasticized material and circulates about the friction Stir welding tool. The increase in the volume fraction of titanium oxide nanoparticles causes more refinement of grains in the nugget zone and reduction in content to possess ductile matrix which deteriorates the ductility of produced nanocomposites [13–15]. Moreover, the microhardness values were significantly increased with the increment in a volume percentage of titanium oxide nanoparticles.

The addition of titanium oxide nanoparticles has a remarkable effect (pinning) on the grain boundaries movement which leads to grain refinement in the processed zone area (nugget). It has been noted that friction Stir welding technique is more alike to severe plastic deformation in which localized heat produced and strain rate had a considerable impact on refining the grains. Due to these phenomenons, the dynamic recrystallized granularly would decrease via a reduction in temperature or enhancement in strain rate during the process [2, 3].

4 Mechanical Properties Evaluations

The mechanical properties evaluation for produced aluminum-based titanium oxide nanocomposites along with that of base alloy matrix is graphically shown in Fig. 5. It is worth mentioning that, the properties of nanocomposites depend on the (a) grains size, (b) dislocation density, (c) % of agglomerated particles in the welded zone, and (d) bonding quality [14–16]. It can be observed that the tensile properties of

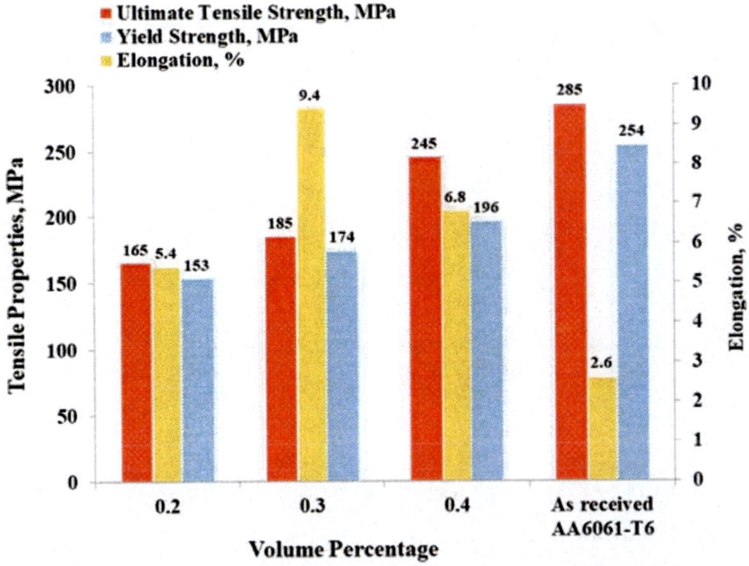

Fig. 5 Tensile properties of aluminum-based titanium oxide nanocomposites and base material

the produced aluminum-based nanocomposites were significantly increased due to increment in maximum volume % of titanium oxide nanoparticles. This is attributed to the presence of titanium oxide nanoparticles that pile-up the boundaries pertaining to grains and prevent the growth of the grains resulting in smaller grains that lead to strength and hardenability at maximum [15, 16].

It is worth mentioning that increment in volume percentage of titanium oxide nanoparticles and decrement in matrix volume results in (a) increment in interparticle spacing area between titanium oxide nanoparticles and the aluminum-matrix because of which large interfacial area occurred for dislocation punching to take place and leads to increment in tensile strength [17], (b) more load transferred from weak aluminum-matrix across the matrix/reinforcement particles interface to the higher stiffness hard ceramic-titanium oxide nanoparticles would occur that leads to increase in strength of the aluminum-matrix [16], (c) generation of thermal-induced dislocation density in the aluminum-matrix [18], (d) increment in work hardening rate which leads to increase in elastic modulus, macroscopic yielding, and tensile strength with lower ductility which is attributed to early onset of void nucleation with increase in volume fraction of titanium oxide nanoparticles [17, 19]. The scanning electron photomicrographs for fractured portions of produced aluminum-based titanium oxide nanocomposites and the aluminum base matrix is shown in Fig. 6.

Fracture appearance of the base matrix reveals the occurrence of deep small dimples which indicates the mode of ductile fracture and affirms the improvement in ductility [13, 20]. The decrease in tensile properties of 0.2 and 0.3% based titanium oxide nanocomposites is attributed to the presence of large shallow dimples with overlay and separation of reinforcement particles layer, whereas for 0.4% based

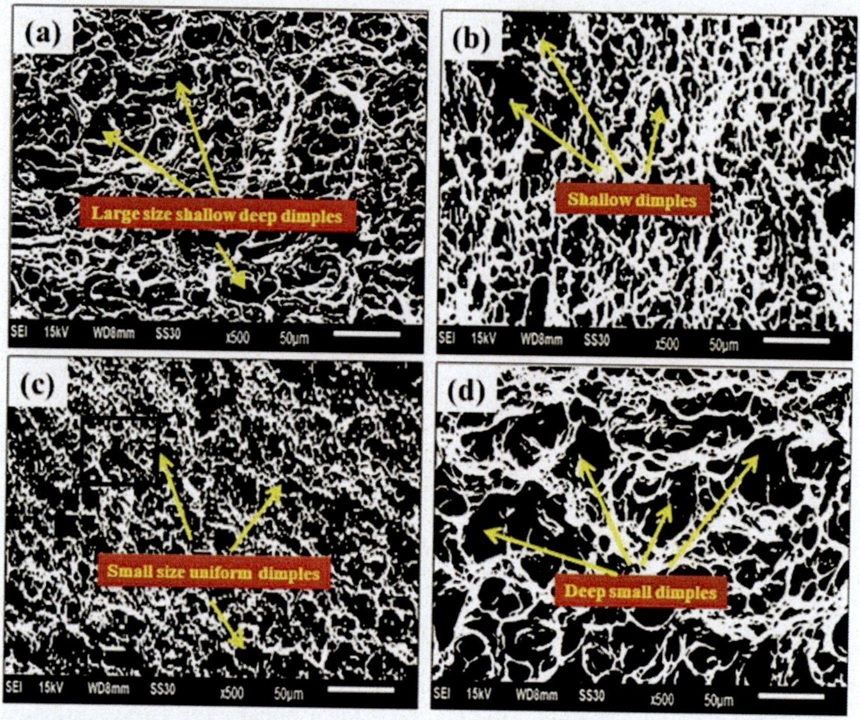

Fig. 6 Scanning electron photomicrographs of fractured surface for aluminum-based titanium oxide nanocomposites, **a** 0.2%, **b** 0.3%, **c** 0.4%, and **d** As-received 6061-T6 aluminum alloy

titanium oxide nanocomposite which consists of small size dimples and results in an increase of tensile strength. These outcomes justify the variations in tensile strength of 0.2, 0.3, 0.4% based titanium oxide nanocomposites.

5 Conclusions

With titanium oxide nanoparticles as reinforcement particles, 6061-T6 aluminum alloy based nanocomposite using friction Stir welding was successfully produced. The impact of titanium oxide nanoparticles on microstructural and mechanical characteristics of 6061-T6/titanium oxide nanocomposite produced with the help of friction stir welding process was investigated. The summary of the major outcomes for the current study is presented below:

1. 6061-T6 aluminum alloy based titanium oxide nanocomposites were successfully produced.
2. Titanium oxide nanoparticles were more homogeneously distributed in the nugget zone at 0.4% as compared to 0.2 and 0.3% of aluminum-based titanium oxide

nanocomposites fabricated via friction stir welding due to high amount of heat produced and strain rate because of which dynamic recrystallization granular would decrease via increment in strain rate and temperature during process and leads to uniform distribution in nugget zone.
3. Due to the large volume fraction of titanium oxide nanoparticles in the nugget zone, their expansion to volume rate is quite high which leads to increase in microhardness to 125 HV which is higher than as-received base material (55 HV).
4. Tensile properties of the produced 6061-T6 Al-titanium oxide aluminum-based nanocomposites were increased with the increase in the maximum 0.4% of titanium oxide nanoparticles.

References

1. Liu FC, Ma ZY (2008) Influence of tool dimension and welding parameters on microstructure and mechanical properties of friction-stir-welded 6061–T651 aluminium alloy. Metal Mater Trans A 39(10):2378–2388
2. He J, Ling Z, Li H (2016) Effect of tool rotational speed on residual stress, microstructure, and tensile properties of friction stir welded 6061–T6 aluminum alloy thick plate. Int J Adv Manuf Technol 84(9–12):1953–1961
3. Kaczmar J, Pietrzak K, Wlosinski W (2000) The production and application of metal matrix composite materials. J Mater Process Technol 106(1):58–67
4. Chawla KK (2012) Composite materials: science and engineering. Springer Science & Business Media, Heidelberg NY
5. Gupta M, Mohamed FA, Lavernia EJ (1990) Solidification behavior of Al-Li-SiCp MMCs processed using variable co-deposition of multi-phase materials. Mater Manuf Process 5(2):165–196
6. Mabhali, LAB, Pityana SL, Sacks N (2010) Laser surface alloying of aluminum (AA1200) with Ni and SiC powders. Mater Manuf Process 25(12):1397–403
7. Hu Q, Zhao H, Li F (2016) Effects of manufacturing processes on microstructure and properties of Al/A356–B_4C composites. Mater Manuf Process 311–292
8. Bodaghi M, Dehghani K (2017) Friction stir welding of AA5052: the effects of SiC nano-particles addition. Int J Adv Manuf Technol 88:2651–2660
9. Mishra RS, Mahoney MW (2007) Friction stir welding and processing. ASM Int
10. Karakizis PN, Pantelis DI, Fourlaris G, Tsakiridis P (2018) Effect of SiC and TiC nanoparticle reinforcement on the microstructure, micro-hardness, and tensile performance of AA6082-T6 friction stir welds. Int J Adv Manuf Technol 95:3823–3837
11. Devaraju A, Kumar A (2011) Dry sliding wear and static immersion corrosion resistance of aluminum alloy 6061–T6/SiCp metal matrix composite prepared via friction stir processing. Int J Adv Res Mech Eng 1(2):62–68
12. Mirjavadi SS, Alipour M, Emamian S, Kord S, Hamouda AMS, Koppad PG, Keshava-murthy R (2017) Influence of TiO_2 nanoparticles incorporation to friction stir welded 5083 aluminum alloy on the microstructure, mechanical properties and wear resistance. J Alloys Comp 712:795–803
13. Tjong SC (2007) Novel nanoparticle-reinforced metal matrix composites with enhanced mechanical properties. Adv Eng Mater 9(8):639–652
14. Nikoo MF, Azizi H, Parvin N, Naghibi HY (2016) The influence of heat treatment on microstructure and wear properties of friction stir welded AA6061-T6/Al_2O_3 nanocomposite joint at four different traveling speed. J Manuf Process 22:90–98

15. Jayabalakrishnan D, Balasubramanian M (2018) Eccentric-weave friction stir welding between Cu and AA6061-T6 with reinforced graphene nanoparticles. Mater Manuf Process 33(3):333–342
16. Baghchesara MA, Production AH (2014) Microstructural investigation of A356 aluminium alloy based magnesium oxide particles reinforced metal-matrix nanocomposites. J Ceramic Process Res 15:418–423
17. Cioffi F, Fernández R, Gesto D, Rey P, Verdera D, Doncel GG (2013) Friction stir welding of thick plates of aluminum alloy matrix composite with a high volume fraction of ceramic reinforcement. Compos A: Appl Sci Manuf 54:117–123
18. Prakash T, Sivasankaran S, Sasikumar P (2015) Mechanical and tribological behaviour of friction-stir-processed Al 6061 aluminium sheet metal reinforced with Al_2O_3/0.5Gr Al_2O_3/0.5Gr hybrid surface nanocomposite. Int J Adv Manuf Technol 80(9–12):1919–1926
19. Tanvir Singh, S.K. Tiwari, D.K. Shukla, (2019) Friction-stir welding of AA6061-T6: The effects of Al2O3 nano-particles addition. Results Mater 1:100005
20. Tanvir Singh, S.K. Tiwari, D.K. Shukla, (2020) Preparation of aluminium alloy-based nanocomposites via friction stir welding. Mater Today Proc 27:2562–2568

An Analysis on the Advanced Research in Additive Manufacturing

Gautam Chandra Karar, Ratnesh Kumar, and Somnath Chattopadhyaya

Abstract Additive manufacturing (AM) is a process in which materials are added to the other material in the form of layers through CAD/CAM instead of removing as conventional production process. This design driven technology is a new approach in the manufacturing. AM is advantageous in comparison with conventional production by reduction of material consumption and time, controlling and optimizing the production parameter for better performances of the products. It is also used for coating of different materials on elements or parts to prevent corrosion and wear. A composite layer can be formed on elements or parts. This paper represents the research, developments, and applications of additive manufacturing process. Efforts are made to get the knowledge about the materials selection, process parameters, and their optimization for friction surfacing through systematic review of research articles, technical notes, etc. The conventional techniques are less effective and large quantity of material converted to scrap material after comparatively long production process. AM can be used for highly complex structures as it has a high degree of design freedom. In contrast to pure additive processes, an additive-subtractive process called hybrid process is used commercially. The review of several research papers provides an idea of emerging solid-state surface coating process based on friction surfacing to improve properties and grain structure of the coating and substrate.

Keywords Additive manufacturing · Friction surfacing · Metal matrix composites · And optimisation

G. C. Karar (✉)
IIT(ISM) Dhanbad, Dhanbad, India
e-mail: karar410@gmail.com

R. Kumar
Mechanical Engineering Department, Birla Institute of Technology, Mesra, Ranchi, India
e-mail: Ranchi.ratan_876@yahoo.co.in

S. Chattopadhyaya
Indian Institute of Technology (Indian School of Mines, Dhanbad, Dhanbad, India
e-mail: somuismu@gmail.com

© Springer Nature Singapore Pte Ltd. 2021

P. M. Pandey et al. (eds.), *Advances in Production and Industrial Engineering*,
Lecture Notes in Mechanical Engineering,
https://doi.org/10.1007/978-981-15-5519-0_19

1 Introduction

Manufacturing is the base for economic development of the modern progressive world. Also it is progressed immensely with the advancement of technology. In usual production, material removal machining or cutting reduction manufacturing techniques are less effective and large quantity of material converted to scrap material after comparatively long production process. The drawbacks of conventional process are eliminated by a different approach to design and manufacturing called additive manufacturing. In the 1980's, models and prototypes are developed for 3D object by AM with the help of CAD [1]. Additive manufacturing first emerged in 1987. Initially AM is concentrated on models and prototypes. With the advancement of technology, the use of AM in production is increased. M. K. Thompson et al. explore the possibilities, compels, and financial thought associated with design for AM [2]. It is mainly design oriented optimized production process having opportunity for flexibility of design. Additive Manufacturing can be used for highly complex structures. Additive Manufacturing is the ideal technology for making prototypes during the development phases of a product and thereby reducing the time required for product development. In additive manufacturing, elements or parts are fabricated generally by layer wise material addition using CAD/CAM technology by applying allied energy sources such as power beam heating, mechanical friction heating, chemical, and electro-chemical, etc. AM is advantageous in comparison with usual production by reduction of material consumption and time, flexible controlling and optimizing the production parameter for better performances of the products. During the last twenty years, additive manufacturing has been progressed rapidly due to maximum flexible utilization of electron beam and laser beam coupled with CAD/CAM. Consequently, parts produced using these processes exhibit cast microstructures often with serious compositional inhomogeneities due to segregation. Also, with these processes, fabrication of parts in multi-materials and compositionally graded materials is restricted to few metallurgically compatible material combinations. Also, part fabrication with these processes typically takes a very long time as the deposition rates are very low [3]. But solid-state processes for material addition can be able to overcome many of the above problems [4]. Generally, additive manufacturing is of three types. 3D printing, a new approach of AM, is used for fabricating non-metallic part directly. The additive manufacturing is expected to change the conventional mode of production from mass to customized product at present. The Welding Institute, UK [5], have proposed that additive manufacturing consists of friction welding such as rotary and stir. Additive Manufacturing can use metal (ferrous and nonferrous) alloy, composite material, ceramic, metal powder, and polymer for the deposition of coating [6]. Hopkinson et al. [7] proposed friction surfacing is an advanced solid-state process of the various solid-state processes within additive manufacturing for developing deposits on parts having strong metallic bonding at the interface [4]. Friction surfacing has been successfully used by various researchers for developing corrosion and wear resistant layer in a wide range of materials [8–11]. The process can be utilized to recover defects due to use of metallic components [12]. Friction surfacing may be used for making multi-track and multilayer deposits [4]. In contrast to the pure additive processes, there are a few commercial processes where additive plus subtractive or hybrid processes is used. In this process, the dimensions of material layers so formed initially is slightly larger than the actual slice area and is then shaped to its precise slice area using CNC machining [4]. High-power lasers in conjunction with a powder-feeding technique have been applied to develop functionally graded materials (FGMs) by successive deposition of different clad layers. The large differences in thermal and physical properties between metals and ceramics cause serious problems in the fabrication of metal/ceramic FGMs by laser cladding [13]. Since the heating and the cooling rates are rapid during laser cladding, the cracks form more easily not only at the interfacial regions between different constituents but also within the ceramics. The metal matrix composites (MMCs) exhibit high wear resistance, fatigue resistance, good creep properties, high stiffness, high dimensional stability with high strength and high elastic modulus, high electrical, and thermal conductivity compared with other base material or other composites. But higher manufacturing cost and difficult to machining are the major limitations with MMCs [14, 15]. In the recent advancements in manufacturing industry, friction surfacing, a solid-state method is used to make MMCs more feasiblely to use in different field of action. Excellent bond strength between the coating and base metal, required microstructure of different materials can be obtained by optimizing of process parameters. There are different types of optimisation in AM such as design optimization and topology optimisation. The process planning optimization is applied for optimization of finishing state and geometrical accuracy. There is a logical tool which establishes a relation between the additive manufacturing input parameters to the output results. There is a relationship among market, making, materials, and metrology, i.e., 4Ms of AM [16]. The process of fabrication of layers to form 3-D object is different for different 3-D printers though all follows additive manufacturing processes.3D printer's robotic arms is shown in Fig. 2. There are two most common 3-D printing technologies such as (i) selective laser sintering (SLS) (ii) fused deposition modelling (FDM). In case of SLS, a high-powered laser is used to melt small plastic particles, metal, ceramic, or glass powders into a mass. But for FDM, thermoplastic materials are injected through indexing nozzles onto a platform. Stereo lithography (SLA) is the most common technology using UV laser or similar power source [17]. The outcome of this review paper is to explore the knowledge about AM and express its challenges in the domain of conventional production [18].

2 Additive Manufacturing

Initially a model is fabricated using 3D CAD process, but that can be obtained straight way in case of additive manufacturing. The complex 3D objects are produced easily using CAD data directly in AM technology. But other than AM process, a total study of the part geometry is carefully needed to produce object. In case of AM, components are developed by depositing materials coating having thin cross-section according to original CAD data. The Generic AM processes involve 8 stages such as CAD, STL convert, File transfer to machine, Machine setup, Build, Remove, Post-process, and Application [19].

2.1 Processes

ISO/ASTM52900:2015 standard classify AM processes into seven categories: (1) binder jetting (BJ); (2) directed energy deposition (DED); (3) material extrusion (ME); (4) material jetting (MJ); 5) powder bed fusion (PBF); (6) sheet lamination (SL); and (7) vat photo polymerization (VP) [16]. LS, LM, and LMD are very much flexible AM processes that can be used for a vast array of metals, alloys, ceramics, and metal matrix composites (MMCs) [20]. From a large number of literature study, it is concluded that additive manufacturing depends on material and process and controlled by both powder properties (i.e., composition, particle shape and size, packing density, flow ability, etc.) and manufacturing criterion (i.e., laser type, laser power, powder layer thickness, etc.) [21–24]. In case of product development, rapid prototyping (RP) was widely used. But RP is unable to explain modern technological applications efficiently. 3D printing has brought a new way to rapidly prototype objects. Rapid prototyping is vital in concept design in order to finalize the design of a new product. Due to this, the processes are considered as fast and effective product development method within industry. 3D printing is capable of mass production, since it fulfills most of its characteristics. 3D printing technology only successfully fulfilled five out of eight characteristics for mass production. Previous researchers have studied to establish accuracy throughout the process [25]. Buswell et al. [26] proposed, rapid production, rapid prototype fabrication, and solid-state process, AM is of the one processing group.AM is used in non-industrial sector such as furniture production. It is an innovative approach [6]. Micro-fabrication is a new technology which efficiently produces functional micro-components for microelectronics systems [27]. Garg et al. [28] proposed a relation between the AM processing parameters and the output product parameters by programming genetically. The porous structures on Titanium can be fabricated by advanced AM techniques such as EBM, SLM, and SLS [29]. Ramiz S.M. Samarjy et al. proposed a different type of additive manufacturing in which molten droplets of metal is transferred to a substrate. The molten droplets are produced due to the remote fusion cutting of a waste material sheet by laser for recycling. The deposition of droplets is controlled by diverging droplet jet through a small angle of 5°. It is the combining processes of laser remote fusion cutting and additive manufacturing. The process is shown in Fig. 1 [30]. Xuewei Fang et al. proposed a Fused-coating based additive manufacturing (FCAM) process in which 3D metal parts are formed by adding the material in layers by using the CAD model. It is an effective process of making metal parts using cheap and easily available material efficiently and having less value apparatus [31].

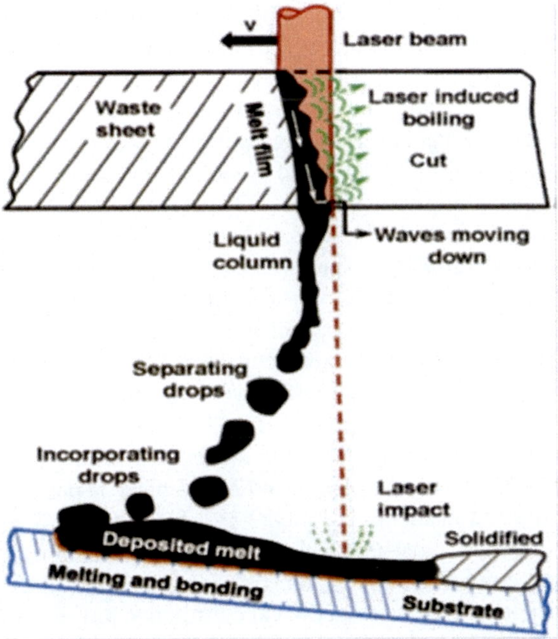

Fig. 1 The main mechanisms of the novel technique (side view)

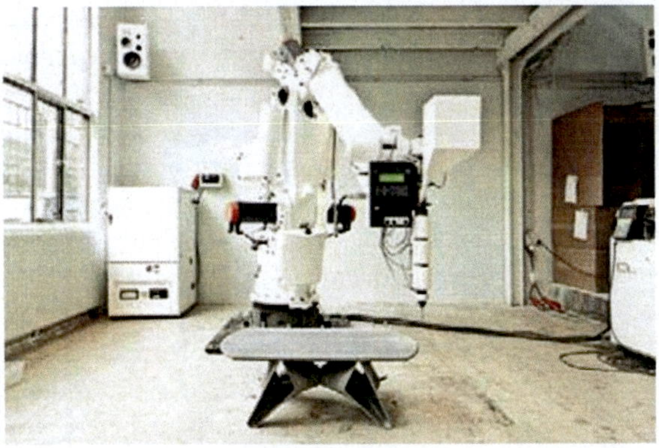

Fig. 2 Furoc 3D printer's robotic arms [6]. RFC of the waste sheet and then AM on the substrate by drop jet [30]

The traditional manufacturing of ceramic components is difficult due to high cost and high energy consumption. Ceramic On-Demand Extrusion (CODE) is a freeform AM process in which ceramic particles like zirconia powder in aqueous suspensions are deposited in layers and then radiated by infrared radiation after each layer to form parts having high density. Ghazanfari et al. [32] applied this process for the production of several sample parts [33]. Laser additive manufacturing (LAM) is an effective process for producing ceramic components with pure form and having good mechanical properties. SLM is used to produce ceramic components from ceramic powers(Al2O3). Compared with SLM process, laser engineered net shaping (LENS) manufactured ceramic components having cubic structures, Al_2O_3 efficiently [34]. In Friction stir processing (FSP), substrate is moving horizontally and tool is rotating vertically on the substrate. Kumar et al. [35] and Panaskar and Sharma [36] proposed FSP for the deposit of nano-composite. FSP is developed for the manufacture of surface composite. R.S. Mishra et al. successfully fabricated Al/SiC surface composites [37]. The FSP also used by several researchers for manufacturing of functionally graded composite material (FGCM) [38–40]. But FGCM manufacturing is limited to make functional grades and composition, micro structure, and material property are uncontrollable. Therefore controlling is needed for FGM to get useful components. The manufacturing of aluminum and TiC functionally graded composite is controlled by mathematical model. Bobbio et al. [41] proposed directed energy deposition additive manufacturing process for manufacturing functionally graded materials having changeable layerwise composition. He studied the result of experiment and analytical review of FGM to analyse material from Ti-6Al-4 V to Invar to improve design of FGM with nonlinear gradient. The microstructure, composition, phases, and microhardness of FMG depend on position in FMG. Sharmaa et al. [42] proposed mathematical model by which volume fraction of tailor-made FGCM is controlled under favorable process conditions in order to gradients in mechanical properties. Also cracks and voids formed initially are eliminated by multiple-passes. Leitão al. [43] measured local constitutive properties by using digital image correlation. They also proposed new AM method and AM integrated method in order to meet present challenges in the manufacturing domain by research society and industries [44–50]. Several researchers had developed composite materials and corresponding additive manufacturing processes to manufacture reinforced parts [51–54]. The composite AM are of two main categories (i) direct composite AM which can febrticate composite parts directly (ii) indirect composite AM which is applied to febrticate complicated composite components using core material which is soluble. There is a scope of producing composite parts by FDM, a direct composite AM. But there is a drawback in the production of composite parts due to the absence of reinforced polymer filament [54]. Currently short fibers are used for reinforcement to develop new composite materials. Hence the fabricated composite components differ with traditional composite parts by properties. Also modern FDM machines are generally not amicable for new materials. To overcome the problem, Y. Zhang et al. proposed that direct composite AM method is developed to new direct composite AM method in which lot of materials having reinforced continuous carbon fiber is deposited. Also present orientation optimization methods can not be directly applied to new FDM process due to its special characteristics [55].

Different additive manufacturing processes are shown in the Fig. 3. There are some typical AM technologies at present such that (i) Selective laser melting (SLM) (ii) Laser Cladding (LD) (iii) Electron beam melting (EBM) (iv) Wire and arc additive manufacturing (WAAM) for metal [56]. The major researches are targeted on powder bed and powder feed methods due to good geometric accuracy is to be well maintained by powder bed and feed method. The feed rate of powder for powder feed additive manufacturing method for manufacturing large metal parts is low as compared to wire-feed AM method due to the energy is directly concentrated at a particular

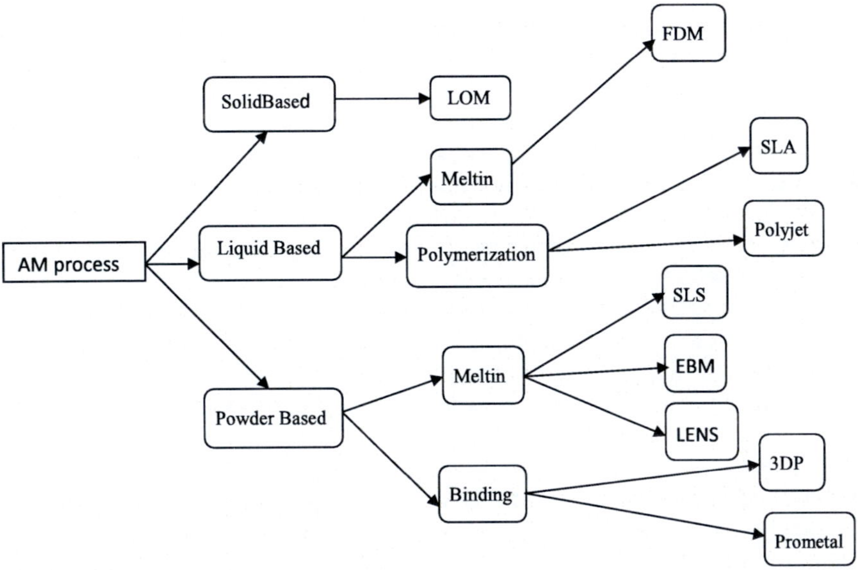

Fig. 3 Different AM processes

point [57]. A large costly complex metal part is manufactured by Wire-feed additive manufacturing method. Arc, laser, and electron beam are used as energy sources for metal deposition but wire is used as material in wire-feed AM. In case of wire and arc-welding AM, electric arc is used to make components. In case of WAAM research, surface roughness, internal stress, microstructure combining has been studied [58]. In case of Wire and laser AM, the energy source is laser while additive material is metal wires. But the laser power control wire-feed rate. Wire cannot be fully melted at higher wire-feed rate but will be melted partially at higher melt pool temperature. In case of wire and electron beam AM process, an electron beam creates molten pool in a high vacuum atmosphere [59]. In case of Wire-feed AM process, the main industrial challenges are (i) the residual stress and distortion due to excessive heat input, (ii) poor part accuracy, and (iii) poor surface finish [60].

There are two modes of approach of metal droplets deposition AM process such that (i) drop on demand (DOD) and (ii) continuous jetting. Tin, lead, zinc droplets having 0.17–0.6 mm diameter are generated in DOD approach [61, 62]. The drawbacks observed in droplets deposition method are micro-void and cold lap. Several researches have concentrated on reducing costs of manufacturing parts that can be made by metal AM processes. But continuous research of modern metal AM process is also needed for the development of few areas of production [63].

Xuewei Fang et al. proposed a novel fused-coating based AM (FCAM) in which time, cost, and energy are saved. The metal parts are made by the deposition of selective material by layers continuously in case of FCAM process. The effect of pressure, viscosity and surface tension control the flow of molten metal over the substrate [64].

FCAM has many advantages, (i) micro-void and cold lap are eliminated by making dense parts, (ii) Alloys such as Sn–Pb alloy, aluminum alloy, copper alloy, and even steel can be fabricated, (iii) The scarcity of the molten metal droplet deposition can be avoided by FCAM. In FCAM process, the molten mass flow rate is controlled by modern pressure control unit, source of energy is induction heating and bar or wire or block of material is used as the material. The residual stress and distortion can be relieved due to the excessive heat input to the material. FCAM method requires material and equipment having comparatively low cost than other methods. For FCAM process, the solidification and spreading time are very important parameter. The process of spreading and solidification in FCAM includes flow of fluid, change of phase, and transfer of heat. The further study should be taken on the development of the simulation analysis for the FCAM technology and to optimize the resolution of the building, part accuracy, and the processing parameters. Xuewei Fang et al. is focused on alloy having low melting point. Hence manufacturing equipment at high temperature and upgraded software required to fabricate 3D complex and metal parts having high melting point [31]. Fang et al. [65] also proposed that the width and thickness of the coating layer is dependent on processing parameters. The processing parameters are /. It is proved by preliminary experiment that the metal parts can be fabricated efficiently having good metallurgical bonding.

The natural resources of production, lead time during manufacturing and production costs for the improvement of the performance and quality of the endproduct is reduced by LAM process. The design of more complex products depends on advanced materials to fulfill the required need of product. But Laser additive manufacturing (LAM) process can use advanced materials and considerable energy to meet the industrial demands. Hence metal alloys are processed effectively by LAM processes [66]. LAM is modified to fabricate or renovate functional metal parts commercially but the developing LAM for metallic material is really a significant challenge. Also LAM is applied to develop the condition of outer layer of metallic component in order to develop wear and corrosion resistance, resistance due to oxidation and resistance due to fatigue upto a certain limit [67, 68]. Moreover, LAM can be applied to fabricate functionally graded materials (FGMs) due to its capability of handling different materials at a time [69, 70] However, the Laser additive manufacturing processes are applied with limitation mainly to repair a part. LAM can be enforced for manufacturing alloys due to its better control during solidification as compared to conventional casting [71, 72]. Fritz Klockea et al. studied the development of hot-work tool steels. The steels are developed thermally and mechanically in order to meet existing and new application. This development depends on steel manufacturing process and production route. Therefore researchers are motivated to evolve new LAM to upgrade the performance of hot-work tool steels. Powder and wire based DMD belongs to LAM. Also powder-based DMD is more energy efficient than wire-based DMD [73]. SLM can be effective for prototype but DMD is effective for the production of bigger dies due to high deposition rates. SLM improved the relationship between microstructure and mechanical properties of hot-work tool steels (H13) effectively as compared to DMD [71, 74]. The strength of parts produced by SLM is high in comparison with SLS which fabricate parts containing a number of voids [75]. The conventional manufacturing process of the ceramic parts takes place with high energy exhaustion, more tooling cost, and a long manufacturing time. The conventional price of ceramic part is two to five times more as compared to a steel part. Now laser engineered net shaping (LENS) being an AM, is more advantageous as compare to conventional manufacturing methods due to good cooling effect, more efficient and more scope of component rebuilt. Also it can fabricate highly pure ceramic parts having good mechanical performance with less time and using less energy [34]. Silva. R. J. et al. propose a specific new method for additive manufacturing by using plasma instead of the other processes used in the industries for metal parts fabrication. This additive manufacturing technology is focusing on perfection in the deposition of layers, adaptability in fabricating parts, and lesser cost and savings due to material waste abatement [76].

Cold spraying (CS) is an AM process in which solid-state particles are deposited. In this case the velocity of powder particles is increased to its high value by using a high-speed gas stream and forming a dense layer on a substrate with strong metallic bond during impact without solidification defects [77–79]. Xuemei Wanga et al. observed that bonding strength is maximum at 45° spray angle and it is increased as the spray angle is decreased from 90° but deposition efficiency decreases with decreasing spray angle from 90°. A relationships between parameters regarding bonding and operating parameters (spray angles, impact velocity, pre-heating temperature) are conformed. These relationships are helpful for the concept of mechanisms regarding bonding and strengths of deposits due to spray at different angles and an optimized spray angle can be describe [80]. Now-a-days CS is taken as an advanced process for AM due to low particle temperature, low oxidation, and residual stress [81].

Generally AM processes for metal consists of two types: (i) powder bed fusion method (ii) directed energy deposition (DED) method. In the area of precision measurement and surface nature, the parts fabricated by powder bed fusion are comparatively good than that of DED. But the production of large parts is not preferable by powder bed fusion process due to unsatisfactory feed rate and the build volume is not so enough [82, 83]. For comparatively higher dimensional correctness, powder and laser-based DED is applied in most cases [84]. The deposition rate of wire and arc-based DED is high as compared with the powder and laser-based DED. It is observed that the deposition efficiency of powder and laser-based DED is only 14% whereas deposition efficiency of wire and arc-based DED has a higher value up to 100% [85, 86]. It is also observed that wire and arc-based DED is more energy efficient than powder and laser-based DED and it is as high as 90% and that of powder and laser-based DED is 30–50% [85–86]. More over powder and laser-based DED process are more costly than that of wire and arc-based DED process. The scope of the DED system is to carry out dissimilar metal deposition. The change of material of a structure made of more than one kind of material during fabrication locally in this process is possible [88, 89]. But it is difficult to change the material locally in case of powder bed fusion process. Also the cladding process in which dissimilar metals are deposited by welding belongs to AM and different materials are deposited on the surface of the product because of the requirement of high heat resistance, corrosion,

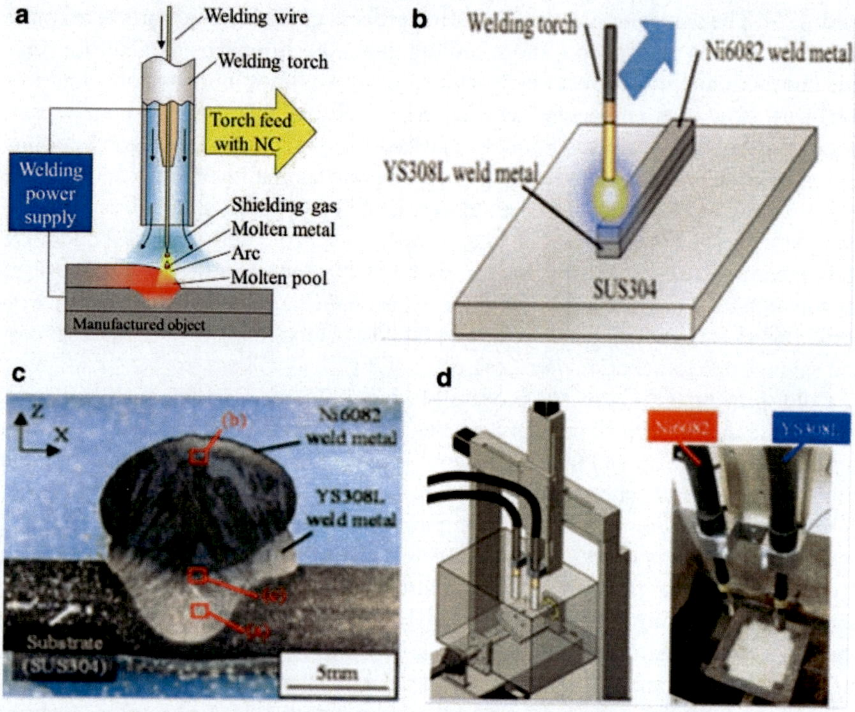

Fig. 4 **a** Process mechanism [90], **b** Deposition of dissimilar metal [90], **c** The built object cross-section [90], **d** AM system having double welding torches [90]

and abrasion resistance [90]. The process of deposition of dissimilar metals is shown in the Fig. 4a–d.

TakeyukiAbe and HiroyukiSasah had studied the deposition of Inconel 600 and SUS304L. The addition of Inconel 600 to SUS304L with cladding by welding is broadly used in the fabrication of parts in which high heat and corrosion resistance are required. Inconel 600 is superior to stainless steel, SUS304L with respect to high heat and corrosion resistance. Also Inconel 600 is more costly than stainless steel, SUS304. TakeyukiAbe and HiroyukiSasah were proposed the wire and arc-based AM system with torch path planning method to carry out deposition of dissimilar metal to manufacture functional structures. Several researchers have been attempted the deposition of dissimilar metals using powder and laser DED process [88, 89]. Hence, the DED process using wire material is more acceptable than using powder material for unlike metal impeachment. AM machine is modified for using two types of material by connecting two welding units and two welding torches to the AM machine. The surface is made by Ni6082 of the fabricated object and YS308L made the inner structure. Also this structure is free from heat and corrosion [90].

Titanium alloys are the most significant progressive materials having outstanding composition and excellent behavior regarding corrosion [91–96]. The use of titanium alloys is controlled due to high cost. The high cost of traditional titanium part encourages lots of study of various manufacturing processes having low cost including powder metallurgy and near net shape techniques [91–95, 96–100]. It is revealed that the AM processed product is economically superior compared to conventionally processed material [101]. Again Bo Yin et al. proposed the use of CaF_2 on the Ti-6Al-4 V parts fabricated by Wire and arc additive manufacturing (WAAM). The presence of CaF_2 to Ti-6Al-4 V improved the ultimate tensile strength (UTS) by 12% than the unmodified Ti-6Al-4Vat room temperature. The elongation (EL) in length and simultaneously decrease in area of the reformed components are lesser than those of the unreformed components because of pores and commencement of crack because of stress concentration. Therefore some operations are to be developed to eliminate pores and thereby upgrade the mechanical properties of parts during WAAM process. Hence CaF_2 offers the modification of microstructures and thereby mechanical properties of parts instead of pores during WAAM process. The fabrication of Ti components by FSW, an AM process, has showed remarkable outputs [102, 103].

Direct laser metal deposition (DLMD) is an additive manufacturing process controlled by laser power for manufacturing metallic and functional parts in which the metal powder is passing through a nozzle. DLMD process is applied directly for the fabrication of near net shape parts from costly metallic powders having high melting point [104]. DLMD is more advantageous than rapid alloy prototyping methods. It is observed that different types of metals having high melting points such as titanium alloys [105–108], steels [109, 110–124], nickel base super alloys [114, 116–119], cobalt base alloys [120, 121] can be fabricated by DLMD. In case of laser manufacturing, the simultaneous heating and rapid solidification process takes place and thereby a fine grain structure with direction-oriented material properties can be obtained [122–122]. From several observations on grain growth during AM processes, it is expressed that the powders having fixed composition were used to manufacture one kind of layer or parts. The microstructure and grain growth are developed gradually through layer by layer deposition by DLMD and thereby the functional properties of the graded steel changes. QiangWang et al. studied the development of microstructure, orientation of grain growth, and grain boundary misorder of graded steel by DLMD. It is possible to deposit three different alloy compositions in one large sample having different grades by DLMD and a kind of stainless steel is used as a substrate. They successfully manufactured a graded steel through 15 layers by direct laser metal deposition. Alloy powder 1 formed the first 5 layers (firstgrade), alloy powder 2 formed the next five layers, i.e., 6th to 10th (second grade), and alloy powder 3 formed the layer 11 to layer 15 (third grade). The newly formed graded steel were analysed by XRD, OM, and EBSD in order to observe the phase development, grain structure, and grain growth, respectively. It is observed from the outcome of the experiment that Austenite (FCC) in the first two grades improved to Ferrite (BCC) with a little Austenite phase. The columnar dendrites are present in the first and second grade having growth location horizontally toward the laser deposition. The finer microstructure with equiaxed grains is observed in the third grade. Also first two grades consist of preferred microstructure with small angle grain boundaries (about 2°). There was no preferred microstructure in the third grade with large-angle grain boundaries from 50 to 60°. Hence microstructure and the development of grain growth location can be obtained by DLMD in case of a graded steel [124].

Thompson et al. [126] proposed the use of x-ray computed tomography (XCT) accompanied with additive manufacturing in various fields of engineering. In case of advanced manufacturing and metrology, both technologies are used more and a lot of research has been taken place with increased use of both technologies for a vast domain of applications within areas of research. Mankovich et al. [127] fabricated a model skull by the first jointly use of XCT and AM in 1990 and XCT has improved far ahead from its roots as a reverse engineering equipment. Hirsch et al. [128] proposed the use of generic performance in case of inspection methods for additive manufacturing. Townsend et al. [129] proposed for an extreme appreciation of the additive manufacturing process and the improvement of surface texture, specifications, and standards of an AM parts. AM processes are challenging in case of the measurement and nature of surface texture. They observed that the surfaces of metal PBF parts are extremely irregular. Rivera et al. [130] proposed an innovative thermo-mechanically controlled Solid State Additive Manufacturing (SSAM) process that contributes a different way to fusion-based additive manufacturing processes for improving parts having purified equiaxed grain morphology. Brøtana et al. [131] proposed that AM has become an effective production process for manufacturing of molds due to flexibility in mold design. In order overcome fatigue due to heat a new invented paths are possible with AM because of the flexibility of design.

Data fusion is generally a computational process in which data from different sensors are to be combined efficiently with precise manner and it is used essentially for the measurement of data of multi-sensor.In there study, Rao et al. [132] assumed that data fusion indicates the completion of all the pre-processed work. Gaussian process (GP) residual approximation (RA) fusion (GPRA) has been explained the fusions of heterogeneous data. There are other two improved models such as b-spline and wavelet which give adequate fusion results. They also applied weighted fusion in most areas instate of RA fusion framework. It is demanded that weighted fusion can be applied effectively for homogeneous data. Also weighted fusion may consist of (i) parametric weighted fusion (ii) non-parametric weighted fusion.

2.2 Process Parameters on Coating

Coating is a layer of material on another material surface to preserve the surface from wear and corrosion or to repair the damaged surface owing to reduce the cost of replacement of the product. Different types of coating can be formed by AM process. Materials may be of two forms (i) powder (ii) wire and the heating source may be of three forms (i) laser, (ii) electron beam, (iii) arc discharge.

The hardness and wear resistance of the H13 tool steel surface is improved by Laser cladding of NiTi on H13 tool steel surface [133]. It is observed that the CoCrMo coated samples exhibit higher wear resistance than that of bare Al alloy substrates and Ti6Al4V coated samples. But the CoCrMo and Ti6Al4V coated samples were more corrosive than that of the uncoated Al alloys substrates. TheCoCrMo and Ti6Al4V coatings exhibit more anti-corrosive in comparison with the plain Al alloy substrates in a 3.5 wt. %NaCl solution due to the reduction of corrosion densities considerably for the CoCrMo and Ti6Al4V coated samples. Li et al. [134] proposed that the surface of 2024-T351 aluminum alloy joint is coated by an aluminum coating using cold spraying (CS) to protect the joint from corrosion as well as to improve the tensile strength of the joint. Also the surface of magnesium alloys [78] and aluminum alloys [79–136] are protected by using CS. The scope of hard, anti-wear coatings on light weight alloys containing aluminum is encouraged PEO to come to the industry [77–135]. Rech et al. [136] proposed that the plasma electrolytic oxidation (PEO) process decreases the weight loss in comparison to the pure substrate during the deposition of 100 μm thick mixture of α-Al_2O_3 and γ-Al_2O_3 and mullite phases on 6061 aluminum alloy. The coatings of the mixture of α-Al_2O_3, γ-Al_2O_3 and mullite phases is comparatively hard than the aluminum substrate, but there may be formation of crack. ŁukaszKaczmarek et al.controlled the chemical constituents of deposited low-friction anti-wear coatings in order to maximize the coefficient of friction by changing substrate polarization with fixed deposition time and temperature below 470 K. Hence it is established that the possibility of high adhesion of the depositing gradient coatings to the aluminum alloy substrate in one operation and thereby anti-wear properties is modified [136]. Vaezi, M. fabricated a Ni-Gr/Al composite coating on Al alloy and steel substrates by the mixed powder of Ni-Gr and Al [27]. The Ni-Gr/Al composite coating can be modified to multiple-layers having Al_3Ni_2 and Al_3Ni as inter metallic and becomes a heavy coating due to annealing at 450 °C for four hours [138]. Gorunov and Gilmutdinov [139] studied the stainless austenitic steel coatings developed by the supersonic laser deposition. Inaccurate laser power creates disturbance in the stability of deposition process and hence cracking of coating. The coating melts when the surface temperature is above 1300 °C due to the laser heating. Taylor [140] produced different Ni-based super alloys coatings by changing the operation parameters such as temperature, pressure, and spraying distance with ranges 700–850 °C, 3.5–4 MPa, and 20–60 mm, respectively, having 500 mm/s as traverse speed of gun. They studied the micro structural and hence mechanical properties of Ni-based super alloys coatings on carbon steel substrates by cold sprayed. The cold spray technology is applied for the repair of cracked steel sheets due to better level of adhesion achieved by optimal operation parameters. A wide range of thickness of the coating can be produced by CS, an AM process. As the operating temperature is comparatively low, the cooling shrinkage is very low in this process. Chen et al. [141] developed a coating thickness simulation model and coordinated in offline programming software. It is proposed that the average coating thickness obtained from numerical and experimental results changes and hence the change of deposition efficiency and material loss in the coating deposit. A numerical model based on standard experimental results using parameters such as spray angle, traverse speed of nozzle, and scanning step for single coating profile was developed. At present cold spraying (CS) is also widely used to prepare particles reinforced aluminum matrix composites (PRAMC) coatings due to its high deposition efficiency. Li et al. proposed the limitations of size and arrangement of reinforcements in coatings in cold-sprayed PRAMC. Hence the scope of development of the microstructure of PRAMC coatings is limited. Hodder et al. applied FSP to improve the as-sprayed Al/Al_2O_3 coating and observed the result of refinement and rearrangement of Al_2O_3 grains without studying the effect of reinforcements. Ashrafizadeh et al. proposed that the wear protection of WC-Ni MMC coating is high due to equally arranged WC particles. But the porosity of the deposit followed by FSP was developed because of inadequate flow of coating material. Morisada et al. developed cemented carbide (WC–CrC–Ni) coating by FSP. The error in the cemented carbide coating is eliminated but the hardness is improved to 2000 HV. Zahmatkesh et al. applied FSP on the Al/A12O3 coating formed by air plasma spray to acquire an uniform arrangement of Al_2O_3 particle. But the bonding between surface nano-composites and substrate are outstanding. Hence, Huang et al. [142] concentrate on the microstructure of the composite coating after FSP application. The tribological nature could be advanced due to reinforcements in cold-sprayed PRAMC. Triantou et al. proposed that COF increases when the ceramic particles Al_2O_3 present in the coating increases. Again Huang et al. [142] observed that the microhardness increases from 92.5 to 185.7 HV of as-cold-sprayed SiC_p/A15056 composite coating followed by FSP because homogenous rearrangement and refining of SiC in the FSPed coating and COF from 0.25 to 0.39 and thereby improving the mechanical properties [143]. There are different thermal spraying methods of coating but these coatings are competent to intensify the corrosion resistance upto certain limit due to the strong bonding between the coating and substrate. Micro-arc oxidation (MAO) process develops a good metallurgical bonding of the coating with the substrate [144]. From the literatures study more effecting MAO parameters were selected [145–147]. There are lots of independent primary and secondary MAO parameters which influences the nature of the coating of aluminium alloy. KamalJayaraj et al. [148] proposed that the MAO parameters are optimized in order to obtain coating having minimum porosity with maximum hardness by using response surface methodology (RSM). It is observed that Duplex stainless steels (DSS) have outstanding mechanical properties and corrosion resistant [149, 150]. Cronemberger et al. [151] studied the corrosion resistance of as-cast DSS under different rates of cooling. DSS can have excellent resistant to dense chloride environments but expensive as compared to normal austenitic grades. Therefore it limits the practical use of DSS. Hence upgraded corrosion resistant materials are used on carbon steel substrate by additive manufacturing process in order to increase its corrosion resistance and thereby saving costs. Sá Brito et al. [152] used electric arc spraying in order to deposit a lot of metallic coatings on the carbon steel substrates. But coatings made by thermal spray behaves like an outstanding corrosion protection because of the enhance microstructure having less porosity. It is observed that High velocity oxy-fuel (HVOF) sprayed coatings on Fe-based material can exibit an excellent corrosion protection in the chloride solution as compared to AISI 304L steel [153]. Liang et al. [154] explained that coatings having better chemical balance offer good corrosion resistance to the corroding surroundings. Zhang and Liu [155] improved anticorrosion claddings onto carbon steel substrates for protection by laser cladding process. Pajukoski et al. [156] introduced corrosion protection layer by laser cladding methods. Rezaeiet al. [157] proposed appropriate process parameters for the coating deposition in order to develop the spoil and fracture protection and thereby the life of the coating layers is increased. The effective parameters of the coatings formation are optimized by the improved numerical model. Holmberg et al. [158] studied the failure of titanium nitride (TiN) coating on steel surfaces. Adem Kurt et al. proposed that FSP decreased the grain size and thereby increased the hardness of the material under processed. The heat input increases due to increase of rotation speed of mechtrode and low traveling speeds of substrate and thereby affects the coating thickness, grain size, and arrangement of the reinforcing particles.In case of solid-state process, bonding between coating and base metal are excellent. The micro hardness of the aluminum coating increased remarkably with increasing travelling speeds [159]. Zhang et al. [160] explained the coatings of Ti-Si-Nitride on commercially pure titanium substrate with three different ratios of Ti-Si by Laser Engineered Net Shaping (LENS). It is observed that LENS employes resistance to the oxidation of the material and capable of processing of complex shapes.It is observed that hardness and wears resistance of the coating is directly affected by the microstructural variations and phase changes. Addition of Si increases solidification rate and affected the mechanical properties due to transformation of dendritic microstructure of the coatings. The hardness of coating having 10% Si was higher than that of coating having 0% Si. Also the addition of Si may take part in a tribological reaction and develop tribological film which behaves as lubricating layers to these coatings.It is observed that both samples having 10% Si and 25% Si bears a reduced wear rate whereas sample having 0% Si had comparatively high wear rate. Hence the tribological properties of the coatings are improved with the addition of Si. Chen et al. [161] developed 3D numerical model for the simulation of coating growth process and heat convey in cold spray process. The model so developed helps us to analyze the coating thickness as well as the distribution of temperature under the effect of several working parameters. Du et al. [162] suggested a new highly efficient metal deposition additive manufacturing (AM) process, called "metal fused-coating additive manufacturing (MFCAM)," for the fabrication of metal components. The MFCAM process is suitable to produce complex prototypes with savings of time and processing cost. But, this new metal AM process is complicated because of the combined effect of heat transfer, mass and momentum, and chemical reaction. The simultaneous melting and resolidification processes can affect remarkably on the distribution of thermal stress, dimension perfection, grain structure, and mechanical properties of the end product. Yao and Wang [163] proposed coating of sand product by AM to improve the samples surface. The surface roughness of sample is reduced by the using of both (i) refractory coating based on alcohol (ii) refractory coating based on the water. Arne et al. [164] suggested that SLM-densified 316 L austenitic steel has a finegrained microstructure with elongated grains in process direction having no preferred crystallographic orientation. Zhang et al. [165] studied the microstructure of laser additive manufactured (LAM) Ti–5Al–2Sn–2Zr–4Mo–4Cr alloy. The microstructure evolution is highly affected thermally by the LAM process.

2.3 Process Parameters and Control

The quality of product, time and cost of production, bond strength, etc., are to be controlled by some parameters during AM process. The error is to be reduced as far as possible in case of highly precision manufacturing processes. Hence different process can be modified by the varying the parameters through proper study.An innovative multisensory regulation system using numerous sensors has the ability to minimize errors remarkably. There are three control levels of AM process. For inside cascade1 control, the intensity sensor controls laser power. The intermediate control cascade 2 measures the order of exposure. The top level control cascade 3 studies the topography of the surface using the IDM sensor [166]. The three control levels of AM process is shown in Fig. 5. Z.Q. Wang et al. studied the outcome of operating parameters on microstructure and mechanical properties of 304 L coating made by AM [167]. In case of SLM technology, the properties of the parts rely on the operating parameters such as laser energy, coating thickness, and scan speed. Owing to many advantages, there are some drawbacks in the SLM technology though the process parameters are optimized. One of the main disadvantages is the process-oriented defects like pores and thereby the material performance is reduced under cyclic loading conditions [168]. Gong et al. [169] observed the significant outcome of process parameters of EBAM on the part quality of Ti-6Al-4V. Murr et al. [170] observed that changes of parameters affect the porosity developed due to EBAM process and also be used for the changes of microstructure in the end product. Puebla et al. [171] and Murr et al. [170] observed that the procesing parameters effect in such a way that the rise of the scanning speed indicates a reduction of relative density and simultaneously reducing micro indentation hardness. Wang et al. [172] observed that

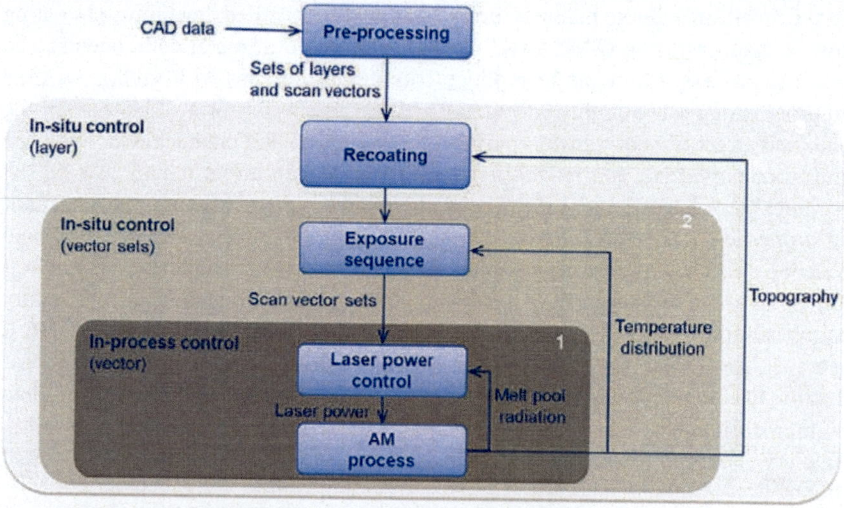

Fig. 5 Control cascades of the additive manufacturing process [178]

the rise of the scanning speed conforms the fine grain and simultaneously decreasing microindentation hardness and the scanning speed of actual beam is governed by the speed function (SF). John Romano et al. studied a model using laser melting technic and applied this thermal model to study the nature among commonly used powder materials including Ti6Al4V, Stainless Steel 316 L, and 7075 Aluminum powders. It was found that the requirement of beam power for steel powder beds is higher than that of titanium powder beds to form a melt pool. But the requirement of beam power for aluminum powder beds is higher than that of both titanium and steel powder beds [173]. Oyelola et al. [174] advised that the machining of Ti-6Al-4V like other titanium alloys, requires low cutting speed and high coolant pressure to acquire excellent results due to their nature. The quantification and nature of surface texture is demanding during AM processes. A vast range of scales are required for surface features and the selection of instrumentation and measurement scales is challenging [175]. Calignano et al. [176] exhibit the effects of process parameters on the surface finish of direct metal laser sintered (DMLS) AlSi10Mg surfaces. AlSi10Mg has better strength, good corrosion resistance, low density, and high thermal conductivity compared with other alloys. SEM micrograph of SLM A1Si10Mg part (as build) is shown in Fig. 6 and SEM micrographs of EBM Ti6AL4V part (as built) is shown in Fig. 7a–d. The relationship between AM process parameters and surface texture is complex [140, 81, 168, 177].

2.4 Hybrid Process

Hybrid Additive Manufacturing till date is known as the object-oriented combination of laser cladding and CNC operation within a single machine environment. Hybrid additive manufacturing has been attracting a more attention during the last half a decade or so but it is still in its infancy. It refers to a special machine that carries an additive metal deposition nozzle having capable of the latest three dimensional printing together with the axis control and metal-cutting capabilities of modern machining centers. The basis of the hybrid-additive manufacturing is the combination of additive and subtractive manufacturing [179, 178]. During last few years, the combination of LAM or electron beam and CNC milling have been proposed by hybrid manufacturing systems [75]. These hybrid-additive manufacturing systems are applied for the repair of tools and component part, hard facing, and other occasions in hot condition and thereby reduces cost of repair [180, 181]. Chunjie Huang et al. [182] proposed an advanced hybrid manufacturing in which Cold Spray (CS) and Friction Stir Processing (FSP) are combined to form Cu–Zn coating having high-strength grain. The CS coating consists of an enlarged grain with 78.42% of low-angle grain boundaries and FSP consists of grains with 90.47% of high-angle grain boundaries and thereby effective mechanical properties have been improved. Liu et al. [183] observed the deposition of Al-Si coatings on magnesium alloy AZ31B by LC, FSP, and hybrid technology (LC-FSP).It is observed that coatings of Al-Si to the AZ31B magnesium alloy substrate becomes compact, uniform, crack, and

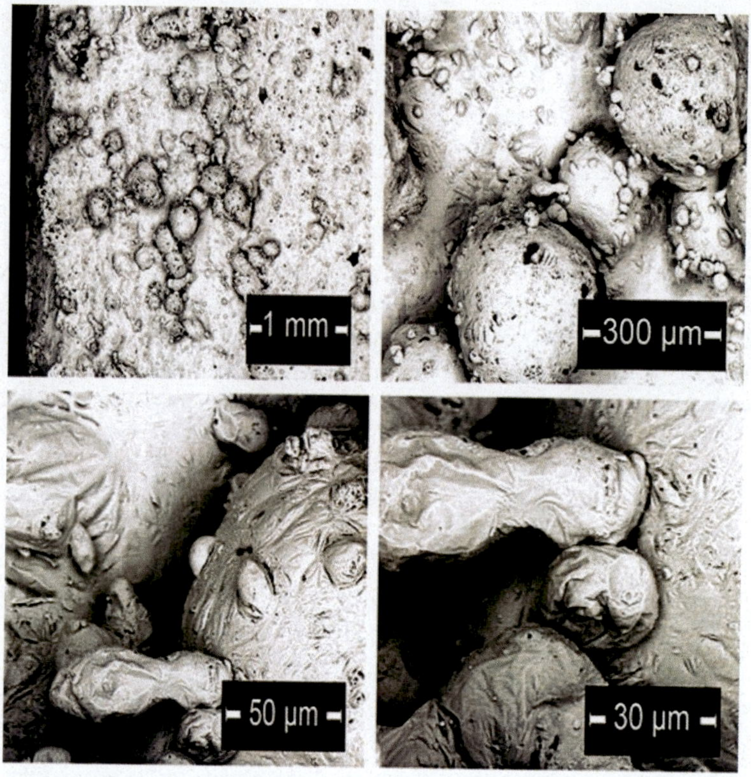

Fig. 6 SEM micrograph of SLM A1Si10Mg part (as build) SEM micrographs of EBM Ti6AL4V part (as built) [178]

porous free, strong metallurgical bond between the coating and substrate, excellent corrosion resistance were fabricated by hybrid technology (LC-FSP) in comparison with LC coatings. The pure AZ31B magnesium alloy substrate is more corrosive than coatings having multiplicative Mg intermetallic compounds. The corrosion potential ofAZ31B magnesium alloy increases by 30.6% due to coating of Al-Si by LC-FSP compared with bare AZ31B specimen whereas about 14.7% compared with LC specimen. The Al-Si coatings prepared by LC-FSP showed flat and smooth surface in comparison with LC coatings. Therefore, the coating formed by ultra-fine grain showed outstanding performance over the conventional coarse grain. Some researchers have studied coating by laser cladding for surface treatment. But there is a major problem of refinement the grain of laser clad coating from coarse grain to the fine-grain. These disadvantages of laser cladding are removed by applying friction stir processing (FSP) after laser cladding. At present, FSP targeted soft metals like Mg alloys and Al alloys for the microstructural refinement. Some researchers employed FSP on high-strength materials like stainless steel and tool steel. Hence FSP becomes an effective way for grain refinement and development of bonding

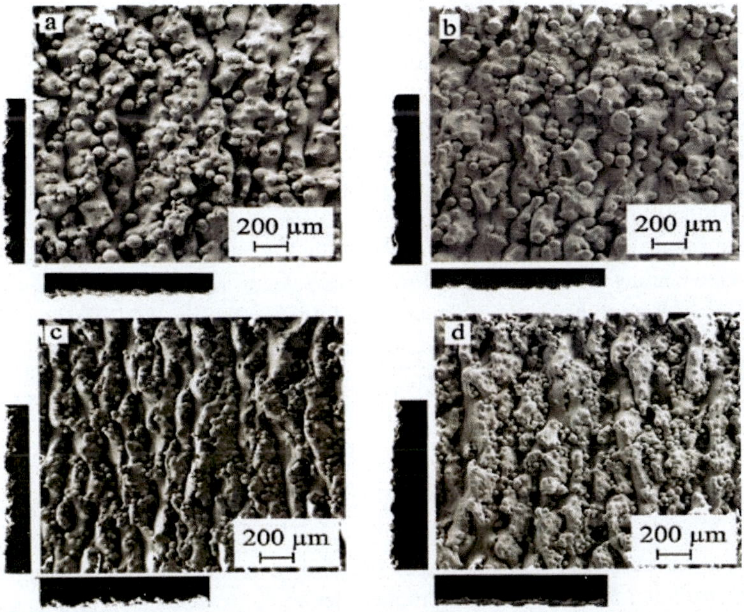

Fig. 7 **a** 45–100 μm powder and 70 μm thickness, **b** 45–100 μm powder and 50 μm thickness, **c** 25–45 μm powder and 70 μm thickness, **d** 25–45 μm powder and 50 μm thickness [178]

between substrate and coating. Xiong, et al. [184] developed a layer of an ultra-fine structured Ni–Al–WC by interlocking bonding with austenitic stainless steel using the combined application of laser clad (LC) and friction stir processing (FSP). At the beginning, laser produced coating on the austenitic stainless steel substrate and then FSP is applied on the coating to obtain an ultra-fine structured layer. Also they introduced some quantity of WC particles into the layer by a WC–Co FSP tool and hence wear resistance increases. Chew et al. [185] improved the fatigue strength of the laser cladded samples by dry cutting as end operation. Hence the hybrid additive-subtractive manufacturing technology would be a challenging technology for the forth coming industrial applications. Serres et al. [186] proposed that alternative hybrid machining process chain occurs with less atmospheric effect and reduce production time and hence to increase production efficiency. The several researchers agreed that the machining on layered material is restricted due to optimization of manufacturing. Salonitis et al. [187] proposed that machining of laser cladded parts causes reduction in the residual stress and distortion. Wang et al. [188, 189] studied the results of cutting parameters in thermal spraying coatings. The surface morphology of machine part determines the working nature of their surfaces. Nieslony et al. [190, 191] observed that the surface topography depends on the nature of cladded layers. Böß et al. [192] estimated cutting forces and the nature of surface of the finished end component during his study of re-contouring. Peirong Zhang et al. proposed the use of regular cutting of LC parts is relatively time consuming and suggested to develop the surface purity after machining [193]. Zahmatkesh et al. [194] applied FSP on Al/Al_2O_3 coating formed by air plasma sprayed, for uniform arrangement of Al2O3 particles in the coating and also the surface nano-composites so formed revealed outstanding bonding with the substrate. Triantou et al. [195] proposed that coating formed by as-sprayed having COF of 0.25 which is less than FSP coating having COF 0.39 because refined SiC particles are distributed homogeneously in the Al matrix due to FSP after as-sprayed coating. Huang et al. [142] suggested that the change of COF for the FSPed coating is lower than that of the as-sprayed coating for the refinement of SiC particles and the presence of wear debris after FSP. The microhardness increases rapidly from 92.5 to 185.7 HV due to FSP than as-sprayed coating because of homogeneous rearrangement and change of grain size of SiC particles in the FSPed coating.

2.5 *Solid-State Process*

There are various solid-state processes of additive manufacturing. Among them, friction surfacing is an encouraging modern technology for depositing material on another material and they are bonded metallurgically in solid-state. Advanced friction surfacing technology includes welding technology, monitoring technology, and testing technology.In this process, a movable rod is pushed on the substrate by an enforced vertical load. Due to frictional heat, a viscoplastic material film is developed at the face of the rod. Due to the applied axial force and temperature, a diffusion of viscoplastic material occurs internally resulting in a metallurgical bonding between the plasticized material and the substrate. Heat transfer takes place into the substrate through this plasticized film and thereby a viscoplastic interface is developed between the rotating consumable rod and the deposited film. The thickness of the layer increases due to continuous heat conduction. By the application of horizontal movement of the substrate, the consumable material is deposited on the substrate surface in the form of a layer. It is a thermo-mechanical process and the grain structure of consumable rod may be refined and thereby properties of the material may be changed. The friction surfacing was first patented in 1941, by Klopstock et al. [196] as a metal coating process. proposed friction surfacing (FS) being a solid-state method is used to fabricate layers of material. Mishra et al. [175] proposed solid-state technologies are at present in advanced stage and authentic substitute to regular processes. Friction surfacing permits deposition of various dissimilar materials as combinations.

2.5.1 Thermo-Mechanical Process

Friction surfacing integrated advanced science and current engineering technology and can be applied to new fields. The feed rate and nature of surfacing can be well controlled with this method and also reducing the equipment cost and increasing the commercial acceptance of friction surfacing process. A portable friction surfacing systems can be used for repairs of worn parts without changing position like railway track. The friction surfacing processes are pollution free as there are no harmful fumes, spatter, and radiation during operation. In this process energy is used efficiently because the heat is developed where it is actually required. Principles of friction surfacing are shown in the Fig. 8. Rafi et al. proposed that heat is passing through the consumable rod due to the friction and thereby a temperature gradient is developed which indicates the stage of deformation [197]. Bedford et al. [198] studied the auto hardening of high-speed steels where tempering continues after coating. Also steadiness of autohardening is the important properties of friction surfacing. Figure 9 below shows the separation of mechtrode material from mechtrode due to rubbing to fabricate coating or flash when substrate moves in contact with rotating mechtrode. The coating attains the temperature of the region just below the mechtrode before deposit to cold substrate.

Portugal scholar Gandra et al. [199] proposed that FS material requires pre-heating to soften the materials due to higher melting point or thermal conductivity and hence acquire the rate of plastic deformation. ûThe cooling technology increases the advantage of friction surfacing. The rapid cooling is significant for the micro structural transformations through dynamic recrystallization during regulation of precipitation hardening. American scholar Mishra and Ma [200] studied the deposition of materials in under water condition. Comparing deposition performance under atmospheric condition, underwater deposition was found to be more efficient. But underwater friction surfacing has limitations as it can't be applied horizontally. The combinations of dissimilar metal are generally difficult due to brittleness of intermetallic compound formation. Friction surfacing solved these problems because it is a solid-state process and giving freedom of choosing materials combination, i.e., a lots of metallic materials can be deposited on a lots of metallic substrates. Various researchers studied the use of FS for depositing a wide range of materials such as tool steel, inconel,

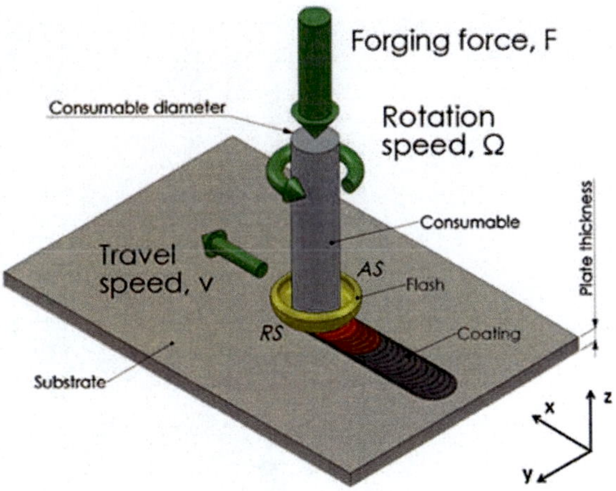

Fig. 8 Principles of friction surfacing [199]

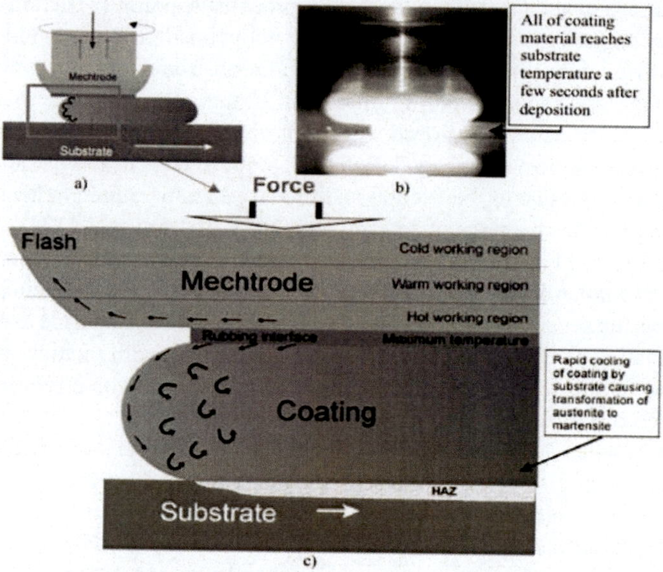

Fig. 9 Thermo-mechanical events, transfer of coating material to substrate during friction surfacing [198]

stainless steel, aluminum, etc., on mild steel to fabricate corrosion and wear resistant coatings [201–203]. Friction surfacing machine is shown in Fig. 10. The FS process can also be applied for renovate of defective or service-broken metallic parts instead of depositing coatings [202]. Reddy et al. [204, 205] proposed that friction surfacing is applied to produce aluminium metal matrix composites on aluminium and titanium substrates. Chandrasekaran et al. [206] studied FS of (a) tool steel,

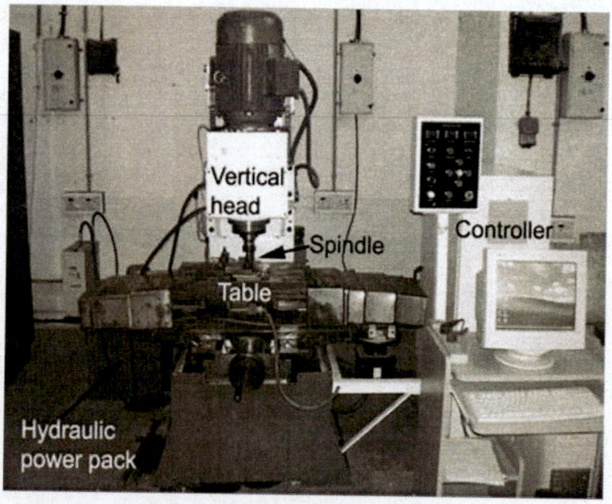

Fig. 10 Friction surfacing machine [222]

inconel, aluminum, and titanium mechtrode onto mild steel substrates and (b) stainless steel, mildsteel and inconel onto aluminum substrates. They observed that tool steel and inconel were formed a strongly bonded coating with steel while aluminum formed a coating at high contact pressures due to axial force. Titanium could not fabricate coating under the testing conditions. It is observed that Stainless steel, mild steel, and inconel developed required coatings onto aluminum substrates, but only SS could not formed intermetallic compound with Al. The frictional heat generated for melting of Al, guide the development of brittle, i.e., unexpected intermetallic compounds between the coating and substrate. The coating of aluminum onto MS substrate by friction surfacing required successful control of parameters to obtain an excellent coating and adhesion between the substrate and aluminum. Hiroshi Tokisue et al. [207] studied the monolayer and multilayer coating of dissimilar materials by friction surfacing. Here, 5052 aluminum alloy plate is used as a substrate and 2017 aluminum alloy bar as a consumable material. The deposit has inclined toward the retreating side because of a combined effect of the rotational direction of consumable rod and the direction of surfacing. In case of multilayer deposit, it is not possible to observe the difference between the first and second deposits, and also the second deposit do no effect the first deposit. Aluminium matrix composites were also effectively deposited on aluminium and titanium substrates during friction surfacing [208, 209]. The corrosion resistance of steel is not so very good and also the oxide layer so formed on steel is noncompact. To overcome this problem a thin film of aluminum is deposited on steel because it forms a compact oxide layer when oxidized and thereby save the substrate steel from corrosion and abrasion [210, 211]. An effective combinations of axial load and rotation speed can fabricate a good coating (other parameters were kept constant). Here small Fe particles are dispersed in the Al deposited on steel. Actually the deposit is a mechanical mixture of aluminum and iron [212]. Mishra R. S. et al. proposed that FSP is very effective solid-state method to form surface metal matrix composite with particles dispersed in the matrix and the bonding between metal-matrix composite and metal substrate is strong. At present, friction stir processing (FSP) has been acknowledged and used extensively to develop surface composite film on metallic substrates [213]. FSP is also suitable for different kinds of aluminum alloys and other metals due to absent of interfacial reaction and phase formation. Also grains are refined in surface coating by FSP. FSP is comparatively simple and easily controllable due to the absence of complicated high-energy laser device or high-voltage electron accelerator. Hence it is advantageous [214]. Friction assisted method such as friction surfacing and friction stir process are modern developed technique that can be applied to fabricate surface MMCs of different matrix materials and reinforcement in order to meet wide industrial application. But further development of the process is needed for the distribution of the secondary phase [215]. Saravanakumar et al. [216] tried to improve the wear resistance of copper surface by the synthesis of AlN reinforced copper matrix composites (CMCs) on pure copper substrates through friction stir processing (FSP). AlN particles were dispersed uniformly in the surface composite. The sliding wear characteristic of copper is improved by copper matrix composites (CMCs) reinforced by different ceramic particles [217, 218]. But the ceramic particles present in copper matrix composites (CMCs) decrease the ductility and toughness of the material. Therefore the surface of copper is reinforced by ceramic particles in order to transform it into a surface composite layer and thereby the surface of copper developed wear resistance and also the ductility of the inner copper matrix is retained [219, 220]. Copper has good thermal conductivity and workability but it has poor hardness and wears resistance. Also it has poor resistance to assertive atmosphere that can be effectively improved by the coating of ZnO. The coating of ZnO passivates the substrate copper from further oxidation [221]

S. Godwin Barnabas et al. proposed the fabrication of coating of stainless steel on ductile iron by FS for adverse uses. The process parameters are optimized in order to get excellent bonding strength and good coating integrity of deposit. This process extends their scope for other dissimilar metal combinations in order to protect the material against wear and corrosion [223, 224]. G. Barnabas studied the deposit of stainless steel on ductile iron by friction surfacing and it proves that the process is superior due to the presence of dense and fine grain structure of ferrite and pearlite on ductile iron. It is also proposed that the parameters are choosed in such a way so as to control heat addition. Purity of the coating is outstanding having strong interfacial bond [224]. Chandrasekaran et al. [225] proposed that stainless steel and tool steel are deposited on mild steel with suitable combinations of parameters. Scott F. Miller studied that the deposition of Monel 400 (Cu Ni–alloy) to the HY-80 steel surface by friction surfacing process for corrosion resistance is successful. But that deposition can be done by electroplating or by using brush, first layer is formed by copper then the final layer by nickel and thereby a poor corrosion resistance is developed. Therefore new friction stir process can be extended to coating and forming with a consumable tool [226]. D. K. Sahoo and B. S. Mohanty studied the coating of copper on mild steel formed by friction surfacing method. In this study, they observe mass deposition rates at intermediate layer and compare energy efficiencies of intermediate layers formed by different method.In this process coating hardness and overall coating efficiency increases [227]. J. J. S. Dilip proposed a multi-track multilayer deposit, as shown in Fig. 12a–c below, applying friction surfacing. From microstructural analysis of the multi-track multilayer deposit in mild steel; it is observed that inter-track and inter-layer bonding is outstanding. The deposit consists of a fine grained microstructure of ferrite and pearlite. The optimum mechtrode position for deposition may depend on track thickness, mechtrode diameter, and mechtrode material [4]. Figure 13a, b show the Optical and SEM microstructures of multi-track multilayered friction surfaced mild steel deposit, respectively. Figure 13c shows the lack of bonding at a track interface.

O. G. Riveraa et al. proposed a substitute way called Additive Friction Stir (AFS) manufacturing to fusion-based additive manufacturing for improving the microstructure of the material. In this case, grain refinement and densification during layer deposition are produced by AFS in IN625 and thereby improving mechanical properties (YS, UTS, etc.) [130]. M. V. Kumar and S. V. Satish studied the Friction Stir Surfacing on copper surface by the High Speed Steel (HSS) flat tool having 10% Cobalt. The equiaxed grains of copper solid solution changed to very elongated grains of copper solid solution due to heavy compressive load and hot forging action

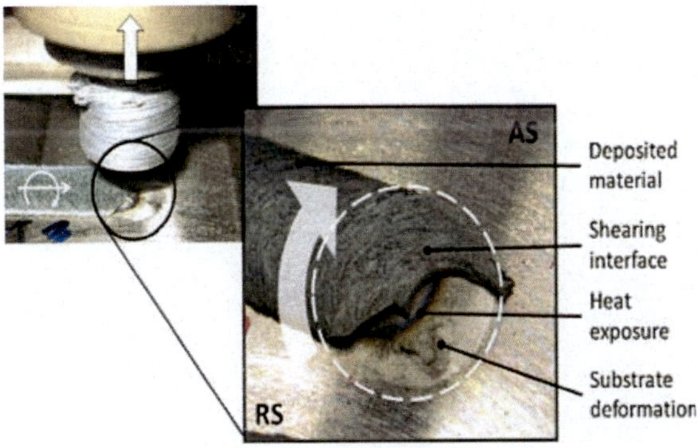

Fig. 11 Substrate is deformed plastically due to mechtrode materialis [228]

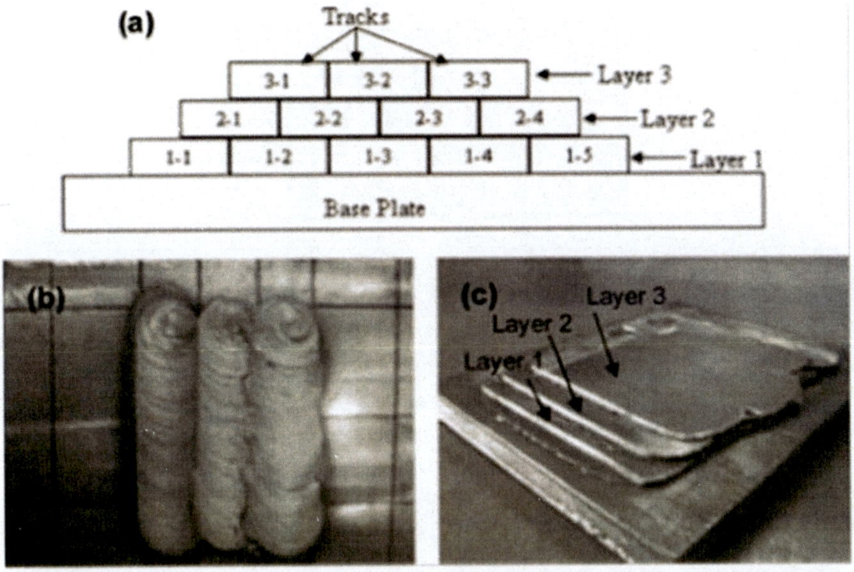

Fig. 12 **a** Material deposition for multi-track multilayer friction surfacing. **b** Multi-track friction surfaced deposit. **c** Multi-track multilayer friction surfaced deposit [4]

on the material. There were no distorted grains in the friction stir surfaced region of the material and also absent of cracks, voids, and scratches compared to parent metal surface [229]. Hiroshi Tokisueet al. studied coating of 2017 BE-T4 aluminum alloy on 5052 P-H34 aluminum alloy by friction surfacing. The friction surfacing was occurred by contact between coating rod and plate for few seconds and then plate moves. For multilayer friction surfacing, the surfaced material of the first layer was cooled to a room temperature after surfacing and after that, the second layer was formed. The surfacing efficiency of second layer of the 2017 alloy was magnificently

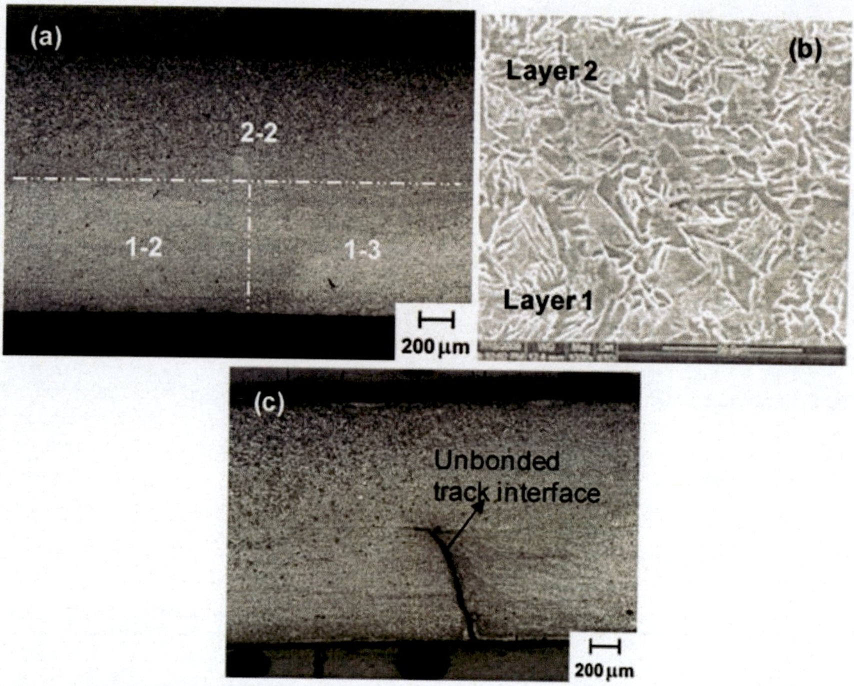

Fig. 13 **a** Optical and **b** SEM microstructures of multi-track multilayered friction surfaced mild steel deposit. **c** Lack of bonding at a track interface [4]

greater than that of the 5052 alloy and 2017 alloy monolayer surfaced material [207]. V. P. RAJU proposed that AISI M2 tool steel is deposited on low carbon steel which is low cost material by friction surfacing process in order to achieve hard surface and thereby reducing raw material cost and machining process time [230]. J. Gandraa et al. proposed that the rates of deposition and specific energy efficiency are to be considered for friction surfacing in order to compare coating processes.The substrate is deformed plastically due to mechtrode material is rubbed against substrate from the advancing to the retreating side and is shown in Fig. 11 [228]. J. Gandra et al. also observed that the mechanical properties, i.e., tensile strength, bending strength, and wear resistance of AA6082-T6 coatings over AA2024-T3 plates by friction surfacing with no porosity [211]. Muhammed Danish et al. proposed the deposition of aluminum alloy6061 on mild steel by FS method in order to protect steel from surface damage caused by corrosion, oxidation, and wear [231]. K. Prasad Rao et al. proposed the coatings of Stellite6 on steel substrate by FS. They also compared the friction surfaced coatings of Stellite6 with coatings formed by GTA and PTA processes. Also Friction surfaced layer exhibit more hardness than GTA and PTA coating and Stellite6 cast rod [132]. K. Prasad Rao et al. was analyzed the mechanism of bonding of coating by studying thermal profile data obtained at the coating/substrate interface by infrared thermography [214]. K. Prasad Rao et al. also

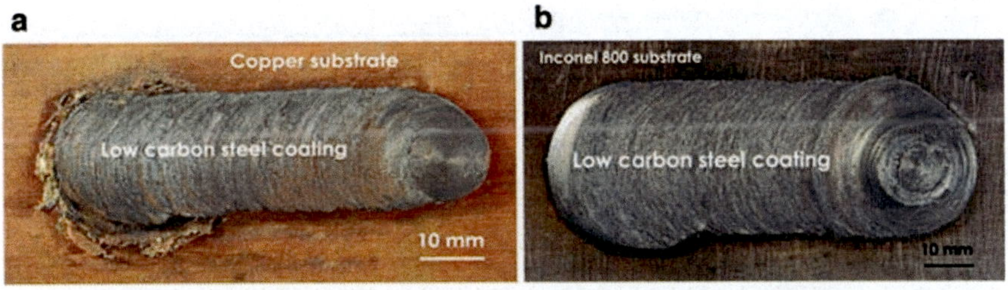

Fig. 14 **a** Coating of steel over copper [215]. **b** Coating of steel over inconel [215]

observed the coating of different nonferrous materials by friction surfacing such as aluminum, copper, magnesium (ZM21), Inconel 800, and titanium alloy (Ti-6Al-4V) as the substrates and low carbon steel, aluminum alloy (AA6063), pure copper and titanium as the consumable rods. Steel developed continuous coating over copper and Inconel 800 as shown in Fig. 14a, b but it is difficult for aluminum, magnesium (ZM21), and titanium alloy (Ti-6Al-4V) due to low strength at higher temperature. The coatings may be discontinuous in case of few materials. Steel may be deposited over aluminum substrate. Commercially pure (CP) copper and CP titanium are unable to deposit over any substrate material. AA6063 can be coated over CP aluminum substrate [215]. The microstructure of steel coating on copper substrate are shown in the Fig. 15a, b. The microstructure of steel coating on Inconel substrate are shown in the Fig. 16a, b.

H. Khalid Rafi et al. studied thermal profiles by IR thermography for the time-dependent heat generation in case of coating, consumable rod, substrate, and the interface during friction surfacing. The retreating side indicates higher temperature whereas advancing side indicates lower temperature. Figure 17a–c below indicate the thermal profiles at different stages [232].

Margam Chandrasekaran et al. observed that tool steel and inconel developed strong coating on mild steel surface by friction surfacing but aluminum can be coated at high contact pressures and titanium could not be coated. Stainless steel, mild steel, and inconel developed strong coating on aluminum but a brittle and undesirable compounds are formed at the interface between stainless steel coating and aluminum substrates [206]. Ramesh Puli et al. studied AISI 410 coating on mild steel substrates by friction surfacing. The coating is very hard (average hardness of 460 HV) due to fully martensitic microstructure. The bonding between AISI 410 coating and mild steel substrates is outstanding.In this way mild steel is protected by the coating from corrosion and wear [217]. Ramesh Puli and G. D. Janaki Ram compared the coatings made by AISI 316L on mild steel by FS and MMAW processes. Friction surfaced coatings showed good protection of corrosion in comparison with that of MMA coatings due to the absence of δ-ferrite. Also friction surfaced coatings of AISI 316L indicate better pitting corrosion resistance than that of AISI 316L bulk material [218]. Martensitic stainless steel, AISI 440C is an effective material for wear resistance coating. But coating is not possible with that material by conventional processes. Ramesh Puli and G. D. Janaki Ram proposed that the layers of AISI 440C

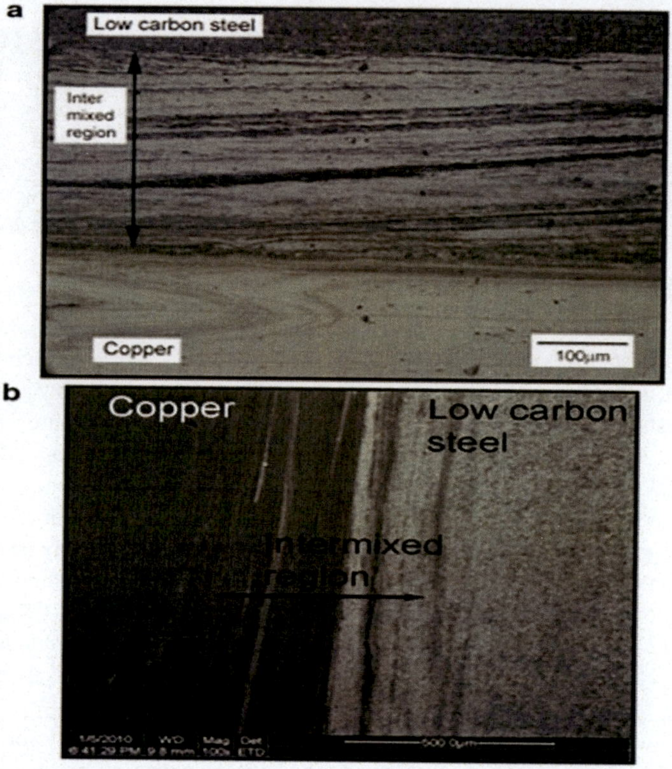

Fig. 15 **a** and **b** Intermixed region of steel coating on copper substrateare [215]

can be developed on low carbon steel substrates by friction surfacing. Also coatings of AISI 440C due to friction surfacing showed lower hardness and wear resistance but showed better corrosion resistance as compared to heat treated bulk material. The inter-track bonding of multiple overlapping tracks is outstanding [219]. Dai Nakama et al. observed that the hardness and wear resistance of the deposit of AZ91 was more than that of the AZ31 magnesium alloy base metal. Hence friction surfacing increases the wear resistance [220].

Compairing all the processes, solid-state is the effective process among them because there is no need of external heating source. Also the microstructure and hence mechanical properties of the coating has been modified by the frictional heat and subsequent cooling. Generally heat treatment is not required for the improvement of mechanical properties. Some materials can not be coated by other process except solid-state process, i.e., friction surfacing. It is material and time saving process.

Fig. 16 **a** and **b** Interface of steel coating on Inconel substrate [215]

2.5.2 Microstructural Evalusion of Coating

During friction surfacing consumable material is converted into a viscoplastic state and deposited into the substrate. After deposition, heat is convected and radiated to the surrounding environment during cooling and conducted into the substrate. During cooling, the grain structure of consumable material changes and forms homogeneous coating. Also chemical composition and phase transformation or modification of second phase particle of consumable material can take place.

Steel

H. Khalid Rafi et al. observed the microstructure of the deposit of austenitic stainless steel (AISI 310) and tool steel (H13) on mild steel by optical microscopy. At the deposit edges, the bonding between the coating and substrate is not perfect. In case of stainless steel coating, fine equiaxed grains are observed indicating recrystallization. In case of tool steel deposit, fine grained ferrite matrix having uniformly arranged carbide particles are observed. Figure 18a–c below are shown the microstructure of the deposits of austenitic stainless steel, tool steel, and unbonded region at the edge of the coating, respectively. The hardness varies across the interface for both the steel. The hardness is maximum near the HAZ but decreases with the distance from the HAZ. Same trends occur in case of mild steel substrate [233].

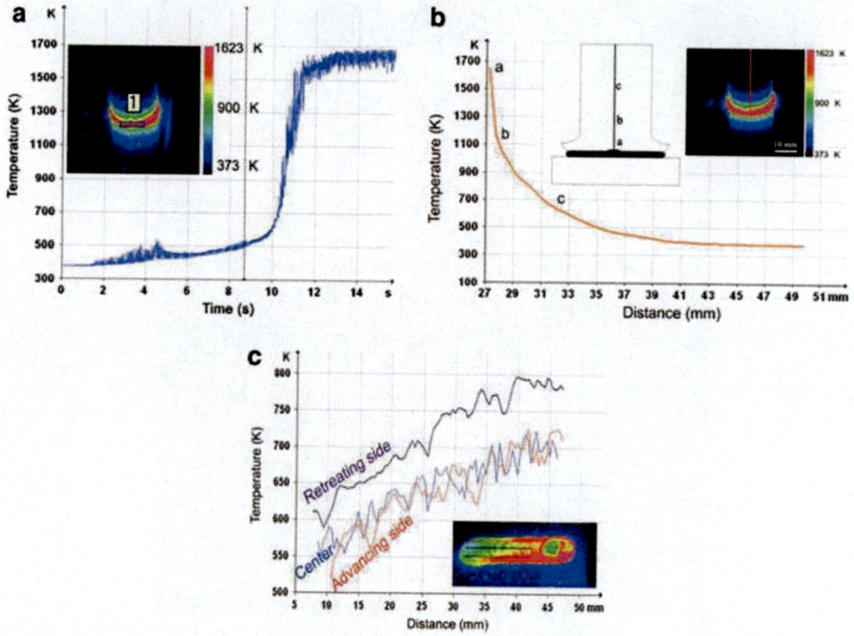

Fig. 17 **a** The consumable rod/substrate along the rotating consumable rod (AISI 304) [232], **b** surface temperature distribution, during the retreating, center, and advancing side [232]. **c** Thermal profile on interfacing the interface during the heating phase [232]

H. Khalid Rafi et al. also observed the microstructure of the deposit of tool steel (H13) on low carbon steel by OM. A large quantity of heat so produced during friction surfacing is dissipated so quickly through mechtrode and substrate that heat can not pass into the area outside the contact zone and hence enormous heating near the contact zone. There will be carburized layer or present of oxide in the interface region. The coating consists of fine equiaxed austenite grains having size between 2 and 10 μm obtained from annealed tool steel (H13) rod having size between 50 and 60 μm. Carbide particles $M_{23}C_6$ and M_7C_3 orginally present in the matrix of mechtrode are absent in the coating due to the higher temperature (>1100 °C). The cooling rates during FS are sufficient to prevent conversion from austenite to martensite. The coating has higher hardness (around 58 HRC) than tool steel rod having hardness of 20 HRC [213]. Stellite6 is very much useful cobalt-based alloy having unique wear resistance due to the presence of hard carbide phase in a CoCr alloy matrix. K. Prasad Rao et al. observed the coating of Stellite6 on steel (0.12%C) substrate by friction surfacing. Friction surfaced coating showed chemical compositional homogeneity, uniformly distributed more fine carbide particles and more Co throughout the matrix than gas tungsten arc (GTA) and PTA coatings and cast rod. The hardness of friction surfaced coating of Stellite6 is more than GTA and PTA coatings and cast rod. In FS coating, FCC Co phase is not transformed to HCP Co phase due to the process temperature 417 °C [282]. D2 steel, rich in carbon and chromium,

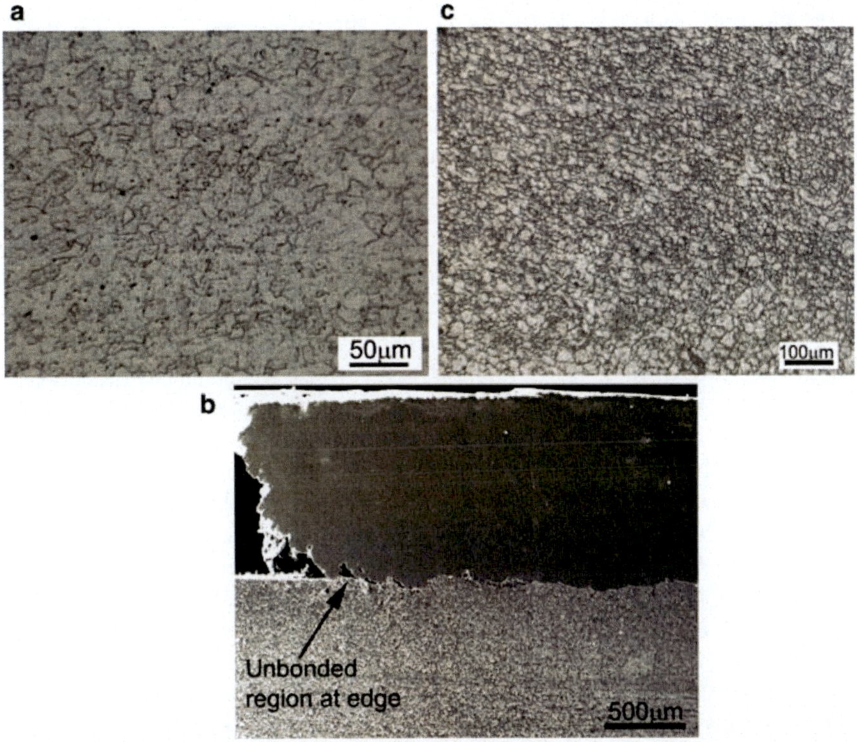

Fig. 18 **a** Microstructure of stainless steel deposit [222], **b** Unbonded region at the edge of deposit [222], **c** Microstructure of tool steel the coating [222]

is widely used to develop the properties and the formation of coating on low carbon steel can extend its life. R. Sekharbabu et al. observed that there is no any decarburized layer or oxide insertion at the interface between the FSed coating of D2 steel and low carbon steel and thereby avoiding chances of cracks formation and porosity development. FSed D2 steel coating consists of refined martensitic microstructure transformed from austenite of D2 steel during friction surfacing process. Also the larger carbides of D2 steel are fragmented to finer carbides having size from 0.1 to 1 μm, in the coating during friction surfacing process. Due to the presence of refined martensitic microstructure and carbides, hardness and wear resistance of coating increases. The coating showed micro-hardness (700 Hv) while that of D2 steel is less [234]. H. Khalid Rafi et al. observed that the coating/substrate interface between AISI 304 and low carbon steel substrate exhibit very good interfacial bonding and free from cracks and porosity. The coating due to friction surfacing showed refined equiaxed grains having an average size of 5 μm obtained from consumable AISI 304 having grain size of 40 μm due to discontinuous dynamic recrystallization as severe plastic deformation takes place during friction surfacing process. The material properties will remain the same as there are no carbides in the fine grained microstructure

[235]. Ramesh Puli et al. studied the coating of AISI 410 over mild steel substrates by friction surfacing in presence of effective inert gas atmosphere to prevent the coating surface from oxidation.It is revealed from microstructural study that there is excellent bonding at the interface through out width of the coating excluding at the boundary which is common in friction surfaced coatings. The coating consists of martensitic microstructure in as-deposited state having no $M_{23}C_6$ carbides which was present in consumable rod (AISI 410). From XRD study, it is observed that the carbide peak intensity in the coating is very very less as compared to consumable rod. The hardness of coating (460 HV) is very much higher than that of consumable rod (AISI 410) whose average hardness is 260 HV. Also shear strength of the friction surfaced coating is 380 MPa which is more than minimum specified strength (140 MPa) for clad plates in ASTM A263 [217]. Ramesh Puli et al. also compared coatings of AISI 410 over low carbon steelplates by FS and MMAW. The coatings of AISI 410 by FS is fully martensitic microstructure having some undissolved $M_{23}C_6$. The ferrite in the consumable rod is transformed to martensite in FS coating through austenite. The δ-ferrite is absent in FS coatings. The MMAW consists of untempered martensite microstructure and good quantity of δ-ferrite. Hence FS is better for coating [236]. Ramesh Puli et al. also studied austenitic stainless steel (AISI 316L) coatings over mild steel substrate. The consumable rods of AISI 316L were taken as hot rolled and annealed condition. Friction surfaced AISI 316L coating exhibits continuous and strongly bonded coating/substrate interface. The FS coatings exhibit fine wrought microstructure of average grain size $9 \pm 3\ \mu m$ which is less than consumable rod ($60 \pm 10\ \mu m$) due to friction surfacing and no elongated sulfides inclusions, no δ-ferrite but high dislocation density.In comparison, MMAW coatings showed austenite microstructure with skeletalδ-ferrite [218]. Ramesh Puli et al. observed the friction surfaced coating of martensitic stainless steel over a low carbon steel plate. The interface between coating and substrate was not uniform,a little wavy and physically continuous. The FS coating exhibits martensite, retained austenite and $M_{23}C_6$ phases which is obtained from consumable rods of AISI 440C consisting of ferrite and a more number of coarse and fine $M_{23}C_6$ carbides. The carbide particles are fewer and smaller in FS coating. The peak temperature attained at 1150 °C during friction surfacing. FS coating exhibits the uniform chromium levels are present over the coating thickness and thereby preventing coating material from mixing with the base material. The hardness of the AISI 440C coating (590 ± 10 HV) is more than that of AISI 410 (460 HV) and showed good corrosion resistance due to higher amount of retained austenite [219]. V. Pitchi Rajuet al. studied the FSed coating of tool steel M2 onto low carbon steel. In this coating, some quantity of tool steel M2 is mixed with low carbon steel, close to interface and thereby ensuring strong metallurgical bonding at the interface. Also there is unbounded region at the edges. The microstructure of coating showed fine carbides in tempered martensite and the microstructure of the layer below the interface of the low carbon steel showed fine pearlite and ferrite in the matrix. The coating is free from porosity, cracks, and voids. The microhardness of the coating showed about 750 HV which is little more than M2 tool steel material (748HV) due to martensite and carbides and also greater than the coating of martensitic stainless steel AISI 440C on a low carbon steel plate (590 ±

10 HV) [237]. V. Pitchi Raju et al. also observed the friction surfaced coating of tool steel M2 onto mild steel. The coating showed excellent hardness (821 HV in average) and 391 HV in average at the interface and hence heat treatment is not required after deposition to enhance hardness [238]. J. J. S. Dilip et al. studied the microstructure of multi-track multilayer deposit and multi-material deposit by friction surfacing. A multi-material deposition mild steel using single-track with alternating layers of austenitic stainless steel AISI 316 and AISI 410 by friction surfacing is analysed. The deposit exhibits a mixture of ferrite and pearlite due to dynamic recrystallization during friction surfacing. It is observed that outstanding bonding exists between individual layers. The microstructure of the top layer of a single-track multilayer deposit exibit fully martensitic with no distinguishable carbide particles. The microhardness gradually drops from the top-most layer (420 HV) to the bottom-most layer (260 HV) due to over-tempering of the martensite [4].

Aluminiumalloy, Copper, and Steel

J. Gandra et al. studied the microstructure of the coating of aluminium alloy on aluminium plates by FS. The coating exhibits equiaxial homogeneous microstructure with grain refinement ($4.2 \pm 1.8\ \mu m$) as compared to consumable rod microstructure ($6.7 \pm 3.3\ \mu m$). The interface is strongly bonded with free from porosity and intermetallic compounds. The edges of the deposits are not properly bonded. The hardness of the coating near the surface is maximum (91 HV0.2) [210]. Muhammed Danish et al. observed the coating of aluminium alloy 6061 on mild steel plates by friction surfacing. The hardness value of coating is small near to the interface and large far from the interface due to recrystallization of the grains [211]. Hidekazu Sakihama et al. studied the coating of 5052 BDS-Faluminum alloy on 5052 P-H34 aluminum alloy plate. The both edges of the deposits are incomplete. The grain of the coating was finer as compared to the consumable rod and substrate. The substrate near the deposit exhibits fine structure while slightly coarser structure in peripheral areas due to heat than substrate [239]. Hiroshi Tokisue et al. observed the layer of 017BE-T4 aluminum alloy over 5052 P-34 aluminum alloy plate. In this case, 2017 aluminum alloy is used instead of 5052 aluminum alloy as coating alloy due to higher strength at higher temperature. For both the monolayer and multilayer surfaced materials, a fine lamellar structure was observed in the deposit than both the coating material and substrate [207]. Tarunkumar jujare et al. observed the coating of copper on mild steel. The compounds $FeCu_4$ and Fe_{17} Cu_3 are present in the interface. The microstructures of the deposit are refined at interface from consumable rod which has equiaxed course grained structure. K. Prasad Rao et al. observed the coating of low carbon steel over copper substrate by friction surfacing. The coating/substrate interface is continuous and free from defect. The coating hardness at the interface increases in comparison to the original base material because of grain refinement. The interfacial bond strength is found to be 105 MPa [215]. The Change of microhardness across the interface is shown in the Fig. 19.

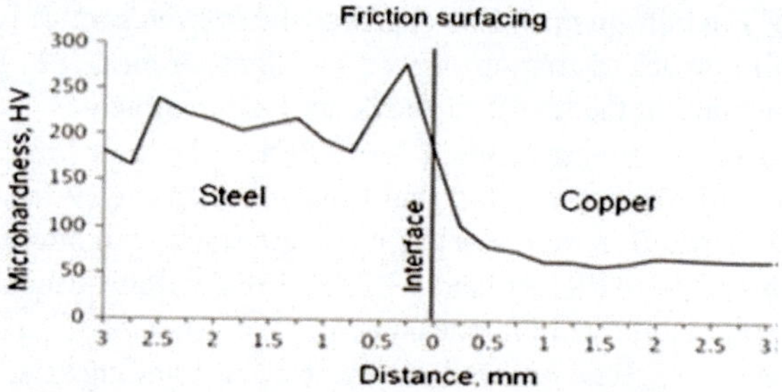

Fig. 19 Change of microhardness across the interface [215]

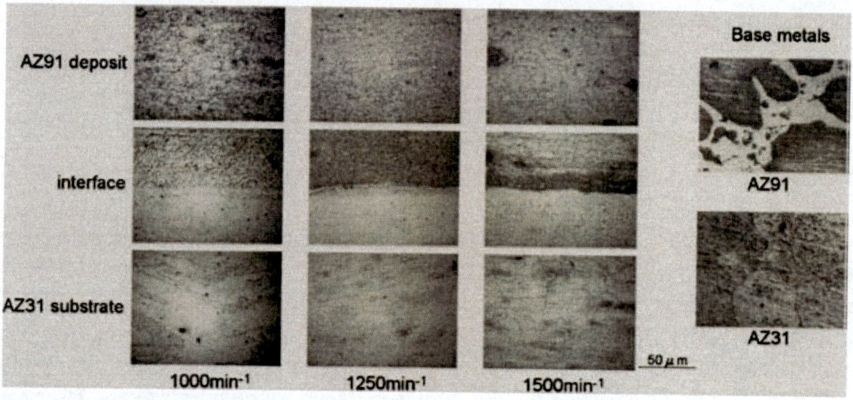

Fig. 20 Microstructures of deposit [220]

Magnesium Alloy

Nakama et al. [220] usded AZ91 magnesium alloy as coating materials and AZ31 magnesium alloy plate as substrate in case of friction surfacing. Grain structures of base metals, deposit, and interface are shown in Fig. 20.

The deposit exhibits finer structure than base metals. Also base metal of consumable rod showed cast solidified structure but deposit is not. In this case, it is observed that deposit material is not penetrated into the substrate. Also mechanically mixed layer which is observed for aluminum alloy is absent at the interface. The maximum temperature of surfacing in case of magnesium alloy is lower than that of aluminum alloy due to the difference of melting point.

Inconel 800 and Titanium Alloys

Chandrasekaran et al. [239] studied viability of deposit of Inconel 800. K. Prasad Rao et al. observed the deposition of low carbon steel over Inconel 800. The interface exhibits strong bonding having strength 144 MPa. Also there is no intermixing at the interface. Beyer et al. showed coating of Titanium 6.4alloy. The deposited layer indicating hardness of about 400 HV0.2. Also some defects are observed at the interface. Hence there is a possibility of brittle fracture of the coating [240].

2.5.3 Influence of Process Parameters

The measurement of coatings basically depends on thickness, width, bond strength, and process parameters (i.e., friction pressure, consumable rod rotational speed, and mechtrodetraverse speed). Also, the thermo-mechanical system is controlled by (i)substrate thickness, (ii) rod diameter (iii) material properties. The generation of heat due to friction and simultaneous cooling due to heat dissipation affects the friction surfaced coatings. The parameters regarding process, such as traverse speed, axial force and rotational speed controls the characteristics of the coatings developed by friction surfacing [241].

Generally vertical milling machine is used for friction surfacing. The force or pressure control with fixed axial load during deposition can be achieved by CNC machine equipped with proper instrument. Instead of using load control, traditional CNC machine can be used to set the rod feed rate [242].

The levels of process parameter selection depend on intermetallic compounds formation and physical properties like diameter, thermal conductivity and density and specific heat. The surface roughness depends on the selection of the levels of process parameters. It is observed that surface roughness is inversely proportional to frictional pressure, directly proportional to welding speed, combined effect of (i) frictional pressure and rotational speed and, (ii) inversely proportional to rotational speed and welding speed. The estimation of roughness or smoothness of coating surface depends upon the rate of material transfer in layers. Friction surfacing is controlled automatically by CNC machine and the process is reliable and repeatable [230–245]. Vitanov et al. [202, 246, 247] developed a model in order to make quick choice of process parameters. There is a limitation in increasing the rotational speed of mechtrode and friction pressure as the starting torque of the operation is more [248, 249]. The thickness of deposit is more conscious to substrate traverse speed than mechtrode rotational speed [250, 231]. The rotating speed of mechatrode is directly proportional to the time taken for deposition of metals. Also the traverse speed of substrate in mm/sec is inversely proportional to the thickness of deposition [209]. Vitanov [251] proposed for a constant rotational speed, a specific ratio between feed and traverse rates is to be maintained in order to get good bond. V. Pitchi Raju et al. proposed the width of the deposit is directly proportional to friction pressure and inversely proportional to rotational speed, combined effect of (i) friction pressure and rotational speed, (ii) friction pressure and welding speed, (iii) pressure, rotational speed, and welding speed [243]. The diffusivity of zinc atoms in copper is good. Friction surfacing can deposit a wide range of specification of materials, with a metallurgical bond, onto a variety of metal substrates. The process includes a hot forging action, and thereby microstructure of the deposition material refines. Friction stir surfacing was carried out using HSS flat tool instead of mechatrode to generate friction surfacing. Friction stir surfacing is a novel method to increase hardness and wear properties of a material without using any extra coating material. Friction stir surfacing process parameters can be optimized under different loads and speed. It was observed that zinc would be a useful material for protective coating on copper plate. A thick coating having range of thickness from 0.01 to 0.1 mm was deposited on a substrate [252, 221]. The input parameters take part a major role in coating and strength of bonding. It was observed that thinner coatings of stainless steel AISI 310 having high bonding strength is related with higher traverse speed. The relationship between input and output parameters are optimized by response surface methodology. The process parameters are studied by using newly developed theoretical regression model [253–255]. Sugandhi et al. [256] developed the empirical relations in order to predict coating thickness and coating width incorporating FS operating parameters. Scott F. Miller [237] deposited monel onto the steel in his experiment. The change in traverse speed can affect the thickness of the coating significantly. The thinnest coating (1.2 mm) can be obtained by highest traverse speed. However a regular coating was not found beyond the traverse speed of 5.6 mm/s, due to the discontinuity of dispersal of plasticized metal. But the coating thickness was found to be 3 mm for a lower traverse speed of 1.2 mm/s. The consequence of traverse speed on coating width was negligible in comparison to coating thickness [227]. Vitanov and Voutchkov [233] studied the flexibility for rapid changes of process parameters regarding numerous coating materials and geometries of substrate in a cost effective and reliable manner. They develop appropriate mathematical models for optimization techniques in order to achieve effective feedback due to flexible control.

3 Use of Additive Material in Medical Applications

Additive manufacturing technologies can be applied in the large area of the medical field. They transform medicine to rapid prototypes. They manufacture models of injured bone. AM printing methods allow scanning and developing a physical model of defective bones and thereby doctors are getting knowledge of bone and planning is made for saving cost and time [257].

The dentists can develop a plaster model of a patient's mouth or replace the teeth by additive manufacturing process like stereo lithography, SLS, and EBM [258, 259]. This technology will help patients suffering from loss of tissue in accidents, to recover quickly with better results. Also 3D cell printing technologies can be applied for printing artificial blood vessels which may be used in the bypass surgery or cardiovascular defects [260, 261].

Budzik et al. [262] proposed that the crown and roots of a tooth can be fabricated using the 3DP and FDM technologies. The results of checking fabricated models were compared with their CAD models and hence their accuracy is the result of errors of scanning, manufacturing, and the algorithm. AM is being used to produce a wide variety of personalized products including hearing aids, dental crowns, biomedical implantation for hard and soft tissues; customized casts, splints, orthotics, and prostheses (laser additive manufactured) in the medical and dental domain.

Atzeni and Salmi [131] analyzed two approaches for the production of dental restoration based on time compression techniques (i) the wax patterns for the investment casting of cobalt–chromium (CoCr) alloy (ii) the direct fabrication of metal restorations by means of selective laser sintering additive technique. After fabrication, the restoration surfaces were inspected by coordinate measuring machine (CMM). Commercially pure (CP) Titanium and the Ti-6Al-4V alloys are commonly used for orthopedic implants. Titanium implants acquire greater mechanical properties and stiffness in comparison to natural bone [29].

4 Conclusions

This study reviewed the recent research trend of the additive manufacturing from the research publication (2000–2017) and books. The regressive literature survey has been explored under the area of additive manufacturing process, coatings and the outcome of parameter on the coatings, effects of process parameters in production, hybrid additive manufacturing method, effect of microstructure of the additive materials in mechanical properties, friction surfacing as advanced additive manufacturing process, use of additive material in medical applications, and additive element having different combinations. The conclusions drawn from the vast literature survey are given below:

- Different AM processes are strongly dependent on the materials and process parameter. The process is controlled by material characteristic and process parameters. Large numbers of studies are made on different types of energy sources and characteristic of material powders. Laser is used as energy sources for most of the cases. The new technology 3D printing is developed by the researchers. Many researchers developed a dissimilar metal deposition system using wire and arc-based AM and powder and laser DED process. Also the combined use of x-ray computed tomography (XCT) and AM, developed by the researchers, become increasingly important in various fields of engineering. The thermo-mechanical Solid State Additive Manufacturing (SSAM) process provides a new and alternative path to AM.
- The researchers developed different types of coating materials and coating process for different applications. They developed the superficial properties of the coating and the bonding between the coating and the substrate material. They also developed composite coating materials. The researchers can control the thickness of the coating by optimizing or simulating the process parameter and also studied the post operation heat treatment.
- The production process, quality of product, time of production, cost of production, etc., are the production parameters. In case of high-precision manufacturing

processes the minimization of error is important. Different multilevel control concept and several sensors have the potential to reduce errors significantly.

- AM process may be modified to hybrid AM process which is an additive-subtractive process by the researchers to meet the need of advanced technology. During the last few years, a hybrid AM systems have been developed with the combination of LAM and CNC milling or multi-wire heads or different energy sources such as laser and electron beam by the reseaechers. Also Cold Spray (CS) and Friction Stir Processing (FSP) are the innovative hybrid process. Friction surfacing is a promising new technology in solid-state. Friction surfacing improves wear and corrosion resistance of the elements by coating deposition. Instead of depositing coatings, the process can be utilized for repair of defective or damaged metallic components during operation.
- AM printing technologies have been applied in the vast medical domain. Additive manufacturing is a very effective tool for dentists. It is used for the printing of artificial bloodvessels 263, 264, 265, 266, 267, 268, 269, 263, 270, 270.

References

1. Wong KV, Hernandez A (2012) A review of additive manufacturing. ISRN Mech Eng
2. Thompson MK, Moroni G, Vaneker T, Fadel G, Ian Campbell R, Gibson I, Bernard A, Schulz J, Graf P, Ahuja B, Martina F (2016) Design for additive manufacturing: trends, opportunities, considerations, and constraints. CIRP Ann—Manuf Technol 65:737–760
3. Gibson I, Rosen DW, Stucker BE (2010) Additive manufacturing technologies: rapid prototyping to direct digital manufacturing. Springer, New York
4. Dilip JS, Babu, VaradhaRajan S, Rafi KH, JanakiRam GD, Stucker BE (2013) Use of friction surfacing for additive manufacturing, materials and manufacturing processes 28:1–6
5. Grainger S, Blunt J (1998) Engineering coatings—design and application, 2nd edn. Abington Publishing, Cambridge, UK
6. Aydin M (2015) Additive manufacturing: is it a new era for furniture production? J Mech Eng Auto 5:338–347
7. Hopkinson N, Hauge RJM, Dickens PM (2006) Rapid manufacturing: an industrial revolution for the digital age. Wiley, West Sussex
8. Rafi KH, JanakiRam GD, Phanikumar G, Rao KP (2010) Friction surfaced tool steel (H13) coatings on low carbon steel: a study on the effectsof process parameters on coating characteristics and integrity. Sur Coat Tech 205:232–242
9. Puli R, Kumar EN, Janaki Ram GD (2011) Microstructural characterization of friction surfaced martensitic stainless steel AISI 410 coatings. Trans Indian Inst Metals 64(1–2):41–45
10. Bedford GM, Vitanov VI, Voutchkov II (2001) On the thermo-mechanical events during friction surfacing of high speed steels. Surf Coat Technol 141:34–39
11. Reddy GM, Prasad KS, Rao KS, Mohandas T (2011) Friction surfacing of titanium alloy with aluminum metal matrix composite. Surf Eng 27(2):92–98
12. Yamashita Y, Fujita K (2001) Newly developed repairs on welded area of lwr stainless steel by friction surfacing. J Nucl Sci Technol 38:896–900
13. Seefeld T, Theiler C, Schubet E, Sepold G (1999) Laser generation of graded metal-carbide components. Mater Sci Forum 308–311:459–466
14. Nikhilesh C, Krishan KC (2013) Metal matrix composites, 2nd edn. Springer, New York
15. Paulo Davim J (2012) Machining of metal matrix composites. Springer, NewYork

16. Tofaill SAM, Koumoulos EP, Bandyopadhyay A, Bose S, O'Donoghue L, Charitidis C (2017) Additive manufacturing: scientific and technological challenges, market uptake and opportunities. Mater Today 1–16 (Press)
17. Attaran M (2017) The rise of 3-D printing: the advantages of additive manufacturing over traditional manufacturing. Business Horizons (Press)
18. Gao W, Zhang Y, Ramanujan D, Ramani K, Chen Y, Williams CB, Wang CCL, Shin YC, Zhang S, Zavattieri PD (2015) The status, challenges, and future of additive manufacturing in engineering. Comput Aided Des 69:65–89
19. Gibson I, Rosen DW, Stucker B (2010) Additive manufacturing technologies-rapid prototyping to direct digital manufacturing. Springer
20. Gu DD, Meiners W, Wissenbach K, Poprawe R (2012) Laser additive manufacturing of metallic components: materials, processes and mechanisms J Int Mat Rev 57(3):133–164
21. Manthiram A, Bourell DL, Marcus HL (1993) Nanophase materials in solid free- form fabrication. JOM 45:66–70
22. Asgharzadeh H, Simchi A (2005) Effect of sintering atmosphere and carbon content on the densification and microstructure of laser-sintered M2 high-speed steel powder. Mater Sci Eng A 403:290–298
23. Santos EC, Shiomi M, Osakada K, Laoui T (2006) Rapid manufacturing of metal components by laser forming. Int J Mach Tool Manuf 46:1459–1468
24. Kruth JP, Wang X, Laoui T, Froyen L (2003) Lasers and materials in selective laser sintering. Assembly Autom 23:357–371
25. Chandrakar AK, Kachhawaha A (2015) Application of additive manufacturing on three dimensional printing. IJSR 4:2012–2016
26. Buswell RA, Soar RC, Gibb AGF, Thorpe A (2007) Freeform construction: mega-scale rapid manufacturing for construction. Autom Constr 16:224–231
27. Vaezi M, Seitz H, Yang SF (2013) A review on 3D micro-additive manufacturing technologies. Int J Adv Manuf Tech 67(5–8):1721–1754
28. Garg A, Lam JSM, Savalani MM (2015) A new computational intelligence approach in formulation of functional relationship of open porosity of the additive manufacturing process. Int J Adv Manuf Tech 80(1–4):555–565
29. Shahalia H, Jaggessara A, Yarlagaddaa PKDV (2017) Recent advances in manufacturing and surface modification of titanium orthopedic applications. Procedia Eng 174:1067–1076
30. Samarjy RSM, Kaplan AFH (2017) Using laser cutting as a source of molten droplets for additivemanufacturing: a new recycling technique. Mater Des 125:76–84
31. Fang X, Du J, Wei Z, Wang X, He P, Bai H, Wang B, Chen J, Geng R, Lu B (2016) Study on metal deposit in the fused-coating based additive manufacturing. Procedia CIRP 55:115–121
32. Ghazanfari A, Li W, Leu MC, Hilmas GE (2016) A novel extrusion-based additivemanufacturing process for ceramic parts. Austin, TX, USA, pp 1509–1529
33. Ghazanfari A, Li W, Leu MC, Watts JL, Hilmas GE (2017) Additive manufacturing and mechanical characterization of high densityfully stabilized zirconia. Ceram Int 43:6082–6088
34. Hu Y, Li Y, Cong W, Zhi L, Guo Z (2017) Additive manufacturing of alumina using laser engineered net shaping: effects of deposition variables. Ceram Int 43:7768–7775
35. Kumar BM, Panaskar NJ, Sharma A (2014) A fundamental investigation on rotatingtool cold expansion: numerical and experimental perspectives. Int J Adv Manuf Technol 73(5–8):1189–1200
36. Panaskar NJ, Sharma A (2014) Surfacemodification and nanocomposite layering of fastener-hole through friction-stir processing. Mater Manuf Process 29(6):726–732
37. Mishra RS, Ma ZY, Charit I (2003) Friction stir processing: a novel technique for fabrication of surface composite. Mater Sci Eng, A 341:307–310
38. Gandra J, Miranda R, Vilaca P, Velhinho A, PamiesT J (2011) Functionally gradedmaterials produced by friction stir processing. J Mater Process Technol 211:1659–1668
39. Miranda RM, Santos TG, Gandra J, Lopes N, Silva RJCC (2013) Reinforcement strategies for producing functionally graded materials by friction stir processing inaluminium alloys. J Mater Process Technol 213:1609–1615

40. Mojtaba S, Farnoush H, Mohandesi JA (2014) Fabrication and characterization of functionally graded Al–SiCnanocompositebyusing a novel multistep friction stir processing. Mater Des 63:419–426
41. Yao S, Wang T (2016) Improved surface of additive manufactured products by coating. J Manuf Process 24:212–216
42. Sharmaa A, Vijendra B, Ito K, Kohama K, Ramjia M, Sai BV (2017) A new process for design and manufacture of tailor-made functionally graded composites through friction stir additive manufacturing. J Manuf Process 26:122–130
43. Leitão, C, Galvão I, Leal RM, Rodrigues DM (2012) Determination of local constitutive properties of aluminum friction stir welds using digital image correlation Mater Des 33:69–74
44. Kruth JP, Leu MC, Nakagawa T (1998) Progress in additivemanufacturing and rapid prototyping. CIRP Ann-Manuf Technol 47(2):525–540
45. Mohri N, Takezawa H, Furutani K, Ito Y, Sata T (2000) A newprocess of additive and removal machining by EDM with a thinelectrode. CIRP Ann-Manuf Technol 49(1):123–126
46. Zhang H, Xu J, Wang G (2003) Fundamental study on plasma deposition manufacturing. Surf Coati Technol 171(1):112–118
47. Janaki Ram GD, Robinson C, Yang Y, Stucker BE (2007) Use of ultrasonic consolidation for fabrication of multi-materialstructures. Rapid Prototyping J 13(4):226–235
48. Yves-Christian H, Jan W, Wilhelm M, Konrad W, Reinhart P (2010) Net shaped high performance oxide ceramic parts by selectivelaser melting. Phys Procedia 5:587–594
49. Karunakaran KP, Suryakumar S, Pushpa V, Akula S (2010) Lowcost integration of additive and subtractive processes for hybrid layeredmanufacturing. Robot Comput-Integr Manuf 26(5):490–499
50. Melchels FP, Domingos MA, Klein TJ, Malda J, Bartolo PJ, Hutmacher DW (2012) Additive manufacturing of tissues and organs. Prog Polym Sci 37(8):1079–1104
51. Zhong W, Li F, Zhang Z, Song L, Li Z (2001) Short fibrereinforced composites for fused deposition modeling. Mater Sci Eng, A 301(2):125–130
52. Karalekas D, Antoniou K (2004) Composite rapid prototyping: overcoming the drawback of poor mechanical properties. J Mater Process Technol 153:2526–2530
53. Kumar S, Kruth JP (2010) Composites by rapid prototypingtechnology. Mater Des 31(2):850–856
54. Cerneels J, Voet A, Ivens J, Kruth JP (2013) Additive manufacturing of thermoplastic composites. Composites Week@ Leuven 1–7
55. Zhang Y, De Backer W, Harik R, Bernard A (2016) Build orientation determination for multimaterial deposition additive manufacturing with continuous fibers. Procedia CIRP 50:414–419
56. Gu DD et al (2012) Laser additive manufacturing of metallic components: materials, processes and mechanisms. Int Mater Rev 57(3):133–164
57. Hofmann DC et al (2014) Compositionally graded metals: a newfrontier of additive manufacturing. J Mater Res 29(17):1899–1910
58. Williams SW et al (2016) Wire + Arc additive manufacturing. Mater Sci Technol
59. Wanjara P, Brochu M, Jahazi M (2007) Electron beam freeformingofstainless steel using solid wire feed. Mater Des 28(8):2278–2286
60. Ding J et al (2014) A computationally efficient finite element model of wire and arc additive manufacture. Int J Adv Manuf Technol 70(1–4):227–236
61. Liu Q, Orme M (2001) High precision solder droplet printing technology and the state-of-the-art. J Mater Process Technol 115(3):271–283
62. Huang C, Orme M (1997) Phase change manipulation for droplet-based solid freeform fabrication. Trans AsmeSerie C J Heat Transf 119(4):818–823
63. Sames WJ et al (2016) The metallurgy and processing science of metal additive manufacturing. Int Mater Rev 1–46
64. Fang X, Du J, Wei Z, He P, Wang B, Geng R, Chen J, Lu B (2016) Experimental and analytical study offused-coating based metal additive manufacturing. In: Annual international solid freeform fabrication symposium

65. Brøtana O, Berg Å, Sørby K (2016) Additive manufacturing for enhanced performance of molds. Procedia CIRP 54:186–190
66. Lauwers B, Klocke F, Klink A, Tekkaya AE, Neugebauer R, Mcintosh D (2014) Hybrid processes in manufacturing. CIRP Ann—Manuf Technol 63:561–583
67. Panjan P, Urankar I, Navinšek B, Terčelj M, Turk, R, Čekada, M, Leskovšek V (2002) Improvement of hot forging tools with duplex treatment. Surf Coat Tech (151–152) 505–509
68. Lawrence J, Waugh DG (2015) Laser surface engineering: processes and applications. Woodhead Publishing Series in electronic and optical materials, p 65
69. Mahamood RM, Akinlabi ET, Shukla M, Pityana S (2012) Functionally gradedmaterial: an overview. In: Proceedings of the world congress on engineering, p 3
70. Shah K, ul Haq I, Khan A, Shah SA, Khan M, Pinkerton AJ (2014) Parametricstudy of development of inconel-steel functionally graded materials bylaser direct metal deposition. Mater Des 54:531–538
71. Gu DD, Meiners W, Wissenbach K, Poprawe R (2012) Laser additivemanufacturing of metallic components: materials, processes and mechanisms. Int Mater Rev 57(3):133–164
72. Fallah V, Corbin SF, Khajepour A (2010) Process optimization of Ti-Nballoycoatings on a Ti-6Al-4V plate using a fiber laser and blended elemental powders. J Mater Process Tech 2081–2087
73. Vilar R (2014) Laser powder deposition, comprehensive materials processing, p 10
74. Holzweissig MJ, Taube A, Brenne F, Schaper M, Niendorf T (2015) Micro structural characterization and mechanical performance of hot worktool steel processed by selective laser melting. Metall Mater Transac B 46
75. Klocke F, Arntz K, Teli M, Winands K, Stella Oliari MW (2017) State-of-the-art laser additive manufacturing for hot-work tool steels. Procedia CIRP 63:58–63
76. Silva RJ, Barbosa GF, Carvalho J (2015) Additive manufacturing of metal parts by welding. IFAC 48–3:2318–2322
77. Sansoucy E, Marcoux P, Ajdelsztajn L, Jodoin B (2008) Properties of SiC-reinforcedaluminium alloy coatings produced by the cold gas dynamic spraying process. Surf Coat Technol 202:3988–3996
78. Irissou E, Legoux JG, Arsenault B, Moreau C (2007) Investigation of Al–Al_2O_3 coldspray coating formation and properties. J Therm Spray Technol 16:661–668
79. Stoltenhoff T, Kreye H, Richter HJ (2002) An analysis of the cold spray process and its coatings. J Therm Spray Technol 11:542–550
80. Wang X, Feng F, Klecka MA, Mordasky MD, Garofano JK, El-Wardany T, Nardi A, Champagne VK (2015) Characterization and modeling of the bonding process in cold spray additive manufacturing. Addit Manuf 8:149–162
81. Taylor JS (2015) Physical processes linking input parametrs and surfacemorphology in additive manufacturing. Achieving precision tolerances in additive manufacturing. In: ASPE 2015 spring topical meeting 2015. ASPE
82. Buchbinder D, Schleifenbaum H, Heidrich S, Meiners W, Bültmann J (2011) Highpower, selective laser melting (HP SLM) of aluminium parts. Phys Procedia 12A:271–278
83. Martina F, Mehnen J, Williams SW, Colegrove P, Wang F (2012) Investigation of the benefits of plasma deposition for the additive layer manufacture of Ti–6Al–4V. J Mater Process Technol 212:1377–1386
84. Rangaswamy P, Griffith ML, Prime MB, Holden TM, Rogge RB, Edwards JM et al (2005) Residual stresses in LENS components using neutron diffraction and contourmethod. Mater Sci Eng A 399:72–83
85. Unocic RR, DuPont JN (2004) Process efficiency measurements in the laser engineered netshaping process. Metal Mater Trans B 35:143–152
86. Ding D, Pan Z, Cuiuri D, Li H (2015) Wire-feed additive manufacturing of metal components: technologies, developments and future interests. Int J Adv Manuf Technol l81:1465–1481
87. Dupont JN, Marder AR (1995) Thermal efficiency of arc welding processes. Weld J 406–416
88. Muller P, Mognol P, Hascoet J-Y (2013) Modelling and control of a direct laser powderdeposition process for functionally graded materials (FGM) parts manufacturing. J Mater Process Technol 213:685–692

89. Qu HP, Li P, Zhang SQ, Li A, Wang HM (2010) Microstructure and mechanical propertyof laser melting deposition (LMD) Ti/TiAl structural gradient material. Mater Des 31:574–582
90. Abe T, Sasahara H (2016) Dissimilar metal deposition with a stainless steel and nickel-basedalloy using wire and arc-based additive manufacturing. Precis Eng 45:387–395
91. Froes FHD, Eylon D, Bomberger H (eds) (1985) Titanium technology: present status and future trends. TDA, Dayton, OH
92. Francis FHS, Yau T-L, Weidenger HG (1996) Chapter 8, Materials science and technology—structure and properties of nonferrous alloys. In: Matucha KH (ed) VCH, Weinheim, FRG, p 401
93. Froes FHS (2000) Chapters 3.3.5a–3.3.5e encyclopaedia of materials science and engineering. In: subject Bridenbaugh P (ed) Elsevier, Oxford, UK
94. Froes FHS (2000) Chapter 8 of the handbook of advanced materials. In: Weasel JK (ed.) McGraw-Hill Inc., New York, NY
95. Froes FHS (2000) In: Ellis J (ed) Handbook of chemical industry economics, inorganic. Wiley, New York, NY
96. Boyer RR, Welsch G, Collings EW (eds) (1994) Materials properties handbook: titanium alloys. ASM Int., Materials Park, OH
97. Froes FH, Eylon D (1990) Powder metallurgy of titanium alloys. Int Mater Rev 35:162–184
98. Froes FH, Suryanarayana C (1993) Powder processing of titanium alloys. In: Bose A, German RM, Lawley A (eds) Reviews in particulate materials, vol 1. MPIF, Princeton, NJ, p 223
99. Froes FH (Sam) Powder metallurgy of titanium alloys. In: Chang I, Zhao Y (eds) (2013) Advances in powder metallurgy. Wood head Publishing, Philadelphia, USA, p 202
100. Froes FH, Imam MA, Fray D (eds) (2004) Cost affordable titanium. TMS, Warrendale, PA
101. Dutta B, Froes FHS (2017) The Additive Manufacturing (AM) of titanium alloys. Metal Powder Rep 72(2):96–116
102. Yin B, Ma H, Wang J, Fang K, Zhao H, Liu Y (2017) Effect of CaF2 addition on macro/microstructures and mechanicalproperties of wire and arc additive manufactured Ti-6Al-4V components. Mater Lett 190:64–66
103. Threadgill PL, Russell MJ (2007) Friction welding of near net shape performs in Ti-6Al-4V. In: 11th world conference on titanium (Ti-2007), (JIMIC-5). Kyoto, Japan, pp 3–7
104. Karl D, Sarat S (2016) Selective laser melting of duplex stainless steel powders: aninvestigation. Mater Manuf Process 31:1543–1555
105. Collins PC, Banerjee R, Banerjee S, Fraser HL (2003) Laser deposition of compositionallygraded titanium-vanadium and titanium-molybdenum alloys. Mater Sci Eng A 352:118–128
106. Zhang BC, Liao BC, Coddet C (2013) Selective laser melting commerciallypure Ti under vacuum. Vacuum 95:25–29
107. Wu GQ, Shi CL, Sha W, Sha AX, Jiang HR (2013) Microstructure and high cyclefatigue fracture surface of a Ti-5Al-5Mo-5V-1Cr-1Fe titanium alloy. Mater Sci Eng A 575:111–118
108. Bian L, Thompson SM, Shamsaei N (2015) Mechanical properties and micro structuralfeatures of direct laser deposited Ti-6Al-4V. JOM 67:629–638
109. Guo P, Zou B, Huang CZ, Gao HB (2017) Study on microstructure, mechanicalproperties and machinability of efficiently additive manufactured AISI 316Lstainless steel by high-power direct laser deposition. J Mater Process Technol 240:12–22
110. Li KB, Li D, Liu DY, Pei GY, Sun L (2015) Microstructure evolution and mechanicalproperties of multiple-layer laser cladding coating of 308L stainless steel. Appl Surf Sci 340:143–150
111. El Kadiri H, Wang L, Horstemeyer MF, Yassar RS, Berry JT, Felicelli S et al (2008) Phase transformations in low-alloy steel laser deposits. Mater Sci Eng A 494:10–20
112. Xue Y, Pascu A, Horstemeyer MF, Wang L, Wang PT (2010) Micro porosityeffects on cyclic plasticity and fatigue of LENS-processed steel. Acta Mater 58:4029–4038
113. Mickovski JK, Lazarev IJ, Lazarev J (2010) Microstructure case study of LENS processed-cylinder from AISI H13 steel. J Technol Plast 35:61–74
114. Wu D, Liang X, Li Q, Jiang L (2010) Laser rapid manufacturing of stainless steel 316 L/Inconel718 functionally graded materials: microstructure evolutionandmechanical properties. Int J Opt 1–5

115. Ganesh P, Kaul R, Sasikala G, Kumar H, Venugopal S, Tiwari P et al (2014) Fatiguecrack propagation and fracture toughness of laser rapid manufacturedstructures of AISI 316L stainless steel. Metallogr Microstruct Anal 3:36–45
116. Blackwell PL (2005) The mechanical and microstructural characteristics of laserdepositedIN718. J Mater Process Technol 170:240–246
117. Paul CP, Ganesh P, Mishra SK, Bhargava P, Negi JA, Nath K (2007) Investigating laser rapid manufacturing for Inconel-625 components. Opt LaserTechnol 39:800–805
118. Dinda GP, Dasgupta K, Mazumder J (2009) Laser aided direct metal deposition of Inconel 625 super alloy: microstructural evolution and thermal stability. Mater Sci Eng A 509:98–104
119. Ganesh P, Kaul R, Paul CP, Tiwari P, Rai SK, Prasad RC et al (2010) Fatigue andfracture toughness characteristics of laser rapid manufactured inconel 625 structures. Mater Sci Eng A 527:7490–7497
120. Ram GDJ, Stucker BE (2008) A feasibility study of LENS; deposition of CoCrMocoating on a titanium substrate. J Manuf Sci Eng 130:024503–024505
121. Janaki Ram GD, Esplin CK, Stucker BE (2008) Microstructure and wear propertiesofLENS deposited medical grade CoCrMo. J Mater Sci Mater Med 19:2105–2111
122. Liu Y, Yang Y, Mai S, Wang D, Song C (2015) Investigation into spatter behaviorduring selective laser melting of 5AISI6 316L stainless steel powder. Mater Des 87:797–806
123. Sander J, Hufenbach J, Giebeler L, Wendrock H, Kühn U, Eckert J (2016) Microstructureand properties of FeCrMoVC tool steel produced by selective lasermelting. Mater Des 89:335–341
124. Tolosa I, Garciandía F, Zubiri F, Zapirain F, Esnaola A (2010) Study of mechanicalproperties of AISI 316 stainless steel processed by selective laser melting, following different manufacturing strategies. Int J Adv Manuf Technol 51:639–647
125. Wang Q, Zhang S, Zhang CH, Wu C, Wang J, Chen J, Sun Z (2017) Microstructure evolution and EBSD analysis of a graded steel fabricated by laser additive manufacturing. Vacuum 141:68–81
126. Thompson A, Maskery I, Leach RK (2016) X-ray computed tomography for additive manufacturing: a review. Meas Sci Technol 27:072001 (pp 17)
127. Mankovich NJ, Cheeseman AM, Stoker NG (1990) The display of three-dimensional anatomy with stereo lithographic models. J Digit Imaging 3:200–203
128. Hirsch M, Patel R, Li W, Guan G, Leach RK, Sharples SD, Clare AT (2017) Assessing the capability of in-situ nondestructive analysis during layer based additive manufacture. Addit Manuf 13:135–142
129. Townsend A, Senin N, Blunt L, Leach RK, Taylor JS (2016) Surface texture metrology for metal additive manufacturing: a review. Precis Eng 46:34–47
130. Du J, Wang X, Bai H, Zhao G, Zhang Y (2017) Numerical analysis of fused-coating metal additive manufacturing. Int J Therm Sci 114:342–351
131. Atzeni E, Salmi A (2015) Evaluation of additive manufacturing (AM) techniques for the production of metal–ceramic dental restorations. J Manuf Process 20:40–45
132. Rao KP, Damodarama R, Khalid Rafia H, Janaki Rama GD, Reddy GM, Nagalakshmic R (2012) Friction surfaced Stellite6 coatings. Mater Charact 70:111–116
133. Norhafzan B, Aqida SN, Chikarakara E, Brabazon D (2016) Surface modification of AISI H13 tool steel by laser cladding with NiTipowder. Appl Phys A 122
134. Li WY, Jiang RR, Huang CJ, Zhang ZH, Feng Y (2015) Effect of cold sprayed Al coating on mechanical property and corrosionbehavior of friction stir welded AA2024-T351 joint. Mater Des 65:757–761
135. Spencer K, Zhang MX (2009) Heat treatment of cold spray coatings to form protective intermetallic layers. Scripta Mater 6:144–147
136. Rech S, Trentin A, Vezzu S, Legoux JG, Irissou E, Guagliano M (2010) Influence of preheatedAl 6061 substrate temperature on the residual stresses of multipassAlcoatings deposited by cold spray. J Therm Spray Technol 20:243–251
137. Kaczmarek L, Adamczyk-Cieślak B, Mizera J, Stegliński M, Kyzioł K, Miedzińska D, Kołodziejczyk L, Szymański W, Kozanecki M (2015) Influence of chemical composition of Ti/TiC/a-C: H coatings deposited on 7075 aluminum alloy on their selected mechanical properties. Surf Coat Technol 261:304–310

138. Huang C, Li W, Planche MP, Liao H, Montavon G (2017) In-situ formation of Ni-Al intermetallics-coated graphite/Al composite in a cold-sprayed coating and its high temperature tribologicalbehaviours. J Mater Sci Technol 33:507–515
139. Gorunov AI, Gilmutdinov AKh (2017) Investigation of coatings of austenitic steels produced by supersonic laserdeposition. Opt Laser Technol 88:157–160
140. Taylor JS (2015) Surface characteristics of additive-maufactured components. In: 15th international conference on metrology and properties of engineering surfaces
141. Chen C, Xie Y, Verdy C, Liao H, Deng S (2017) Modeling of coating thickness distribution and its application in offline programming software. Surf Coat Technol 318:315–325
142. Huang C, Li W, Zhang Z, Fub M, Planche M, Liao H, Montavon G (2016) Modification of a cold sprayed SiCp/Al5056 composite coating by friction stirprocessing. Surf Coat Technol 296:69–75
143. Wernick S, Pinner R, Draper R (1972) Surface treatment and finishing of aluminiumand its alloys, 1st edn., vols 1–2. Pergamon Books, Oxford, New York
144. Yang JYF (2009) Surface modification of aluminum alloy products for micro-arc oxidation processes
145. Xiang N, Song R, Zhuang J, Song R, Lu X, Su X (2016) Effects of current density onmicrostructure and properties of plasma electrolytic oxidation ceramic coatings formed on 6063 aluminum alloy. Trans Nonferrous Metals Soc China 26:806–813
146. Rama Krishna L, Somaraju KRC, Sundararajan G (2003) The tribological performance of ultra-hard ceramic composite coatings obtained through microarcoxidation. Surf Coat Technol 163–164, 484–490
147. You Q, Yu H, Wang H, Pan Y, Chen C (2014) Effect of current density on the microstructureand corrosion resistance of microarc oxidized ZK60 magnesium alloy. Biointerphases 9:31009
148. Jayaraj RK, Malarvizhi S, Balasubramanian V (2017) Optimizingthe micro-arc oxidation (mao) parameters to attaincoatings with minimum porosity and maximum hardness on thefriction stir welded AA6061 aluminum alloy welds. Defence Technol 13:111–117
149. Pohl M, Storz O, Glogowski T (2007) Effect of intermetallic precipitations on the properties of duplex stainless steel. Mater Charact 58(1):65–71
150. Karahan T, Emre HE, Tümer M, Kaçar R (2014) Strengthening of AISI 2205 duplexstainless steel by strain ageing. Mater Des 55:250–256
151. Cronemberger MER, Nakamatsu S, Della Rovere CA, Kuri SE, Mariano NA (2015) Effectof cooling rate on the corrosion behaviour of as-cast SAF 2205 duplex stainless steelafter solution annealing treatment. Mater Res 18:138–142
152. Sá Brito VRS, Bastos IN, Costa HRM (2012) Corrosion resistance and characterization of metallic coatings deposited by thermal spray on carbon steel. Mater Des 41:282–288
153. Wang SL, Cheng JC, Yi SH, Ke LM (2014) Corrosion resistance of Fe-based amorphous metallic matrix coating fabricated by HVOF thermal spraying. T Nonferr Metal Soc 24(1):146–151
154. Liang J, Srinivasan PB, Blawert C, Dietzel W (2009) Comparison of electrochemical corrosion behaviour of MgO and ZrO_2 coatings on AM50 magnesium alloy formed by plasma electrolytic oxidation. Corros Sci 51(10):2483–2492
155. Zhang P, Liu Z (2016) Physical-mechanical and electrochemical corrosion behaviours of additively manufactured Cr-Ni-based stainless steel formed by laser cladding. Mater Des 100:254–262
156. Pajukoski H, Nakki J, Thieme S, Tuominen J, Nowotny SP, Vuoristo P (2016) High performance corrosion resistant coatings by novel coaxial cold- and hot-wire laser cladding methods. J Laser Appl 28(1):012011–1–012011–12
157. Rezaei S, Wulfinghoff S, Reese S (2017) Prediction of fracture and damage in micro/nano coating systems using cohesive zone elements. Int J Solids Struct 121:62–74
158. Holmberg K, Laukkanen A, Ronkainen H, Wallin K (2013) Finite element analysis of coating adhesion failure in pre-existing crack field. Tribol Mater Surf Interfaces 742–751
159. Kurt A, Uygurb I, Cete E (2011) Surface modification of aluminum by friction stir processing. J Mater Process Technol 211:313–317

160. Zhang Y, Sahasrabudhe H, Bandyopadhyay A (2015)Additive manufacturing of Ti-Si-N ceramic coatings on titanium. Appl Surf Sci 346:428–437
161. Chen C, Xie Y, Verdy C, Liao H, Ren Z, Deng S (2017) Numerical investigation of transient coating build-up and heattransfer in cold spray. Surf Coat Technol
162. Bobbio LD, Otis RA, Borgonia JP, Dillon RP, Shapiro AA, Liu Z, Beese AM (2017) Additive manufacturing of a functionally graded material from Ti-6Al-4V to invar: experimental characterization and thermodynamic calculations. Acta Mater 127:133–142
163. Fang X, Du J, Wei Z, He P, Bai H, Wang X, Lu B (2017) An investigation on effects of process parameters in fused-coatingbased metal additive manufacturing. J Manuf Process (Press)
164. Arne R, Karina G, Matthias W, Florian B, Werner T (2016) Comparison of microstructureand mechanical properties of 316 L austenitic steel processed byselective laser melting with hot-isostatic pressed and cast material. Mater Sci Eng A 678:365–376
165. Zhang Q, Chen J, Tan, H, Lin X, Huang W (2016) Microstructure evolution and mechanical properties of Laser additive manufactured Ti−5Al−2Sn−2Zr−4Mo−4Cr alloy. Trans Nonferrous Met Soc China 26:2058–2066
166. Renken V, Albinger S, Goch G, Neef A, Emmelmann C (2017) Development of an adaptive, self-learning control concept for an additive manufacturing process. CIRP J Manuf Sci Technol (Press)
167. Wang ZQ, Palmer TA, Beese AM (2016) Effect of processing parameterson microstructure and tensile properties of austenitic stainless steel 304L made by directed energy deposition additive manufacturing. Acta Mater 110:226–235
168. Patil N, Pal D, Anam M, Gong H, Gu H, Dilip S et al (2014) Predictive modelingcapabilities for dimensional accuracy and surface finish in metal laser melting based additive manufacturing. Dimensional accuracy and surface finish in additive manufacturing. In: ASPE 2014 spring topical meeting. ASPE
169. Gong X, Lydon J, Cooper K, Chou K (2014b) Beam speed effects on Ti-6Al-4V microstructures in electron beam additive manufacturing. J Mater Res 29(17):1951–1959
170. Murr L, Gaytan S, Medina F, Martinez E, Hernandez D, Martinez L, Lopez M, Wicker R, Collins S (2009a) Effect of build parameters and build geometries on residual microstructuresand mechanical properties of Ti-6Al-4V components built by electron beam melting (EBM). In: Process of the solid freeform fabrication symposium. Austin, TX
171. Puebla K, Murr LE, Gaytan SM, Martinez E, Medina F, Wicker RB (2012) Effect of meltscan rate on microstructure and macrostructure for electron beam melting of Ti-6Al-4V. Mater Sci Appl 259(3)
172. Wang X, Gong X, Chou K (2015) Scanning speed effect on mechanical properties of Ti-6Al-4V alloy processed by electron beam additive manufacturing. Procedia Manuf 1:287–295
173. Romano J, Ladani L, Sadowski M (2015) Thermal modeling of laser based additive manufacturing processes within common materials. Procedia Manuf 1:238–250
174. Oyelola O, Crawforth P, Saoubi RM, Clare AT (2016) Machining of additively manufactured parts: implications for surface integrity. Procedia CIRP 45:119–122
175. Mishra RS, Ma ZY (2005) Friction stirs welding and processing. Mater Sci Eng R 50:1–78
176. Calignano F, Manfredi D, Ambrosio EP, Iuliano L, Fino P (2013) Influence of processparameters on surface roughness of aluminum parts produced by DMLS. Int J Adv Manuf Technol 67:2743–2751
177. Barari A, Jamiolahmadi S (2014) Estimation of surface roughness of additivemanufacturing parts using finite difference method. Dimensional accuracy and surface finish in additive manufacturing. In: ASPE 2014 spring topical meeting. ASPE
178. Poprawe R (2011) Tailored Light 2: laser application technology 2011
179. Gibson I, Rosen DW, Stucker B (2015) Additive manufacturing technologies: 3D printing, rapid prototyping, and direct digital manufacturing, 2nd edn.
180. Kaierle S, Barroi A, Noelke C, Hermsdorf J, Overmeyer L, Haferkamp H (2012) Review on laser deposition welding: from micro to macro. Phys Procedia 39:336–345
181. Lorenz KA, Jones JB, Wimpenny DI, Jackson MR (2015) A review of hybrid manufacturing. In: Solid freeform fabrication conference proceedings, p 53

182. Huang C, Li W, Feng Y, Xie Y, Planche MP, Liao H, Montavon G (2017) Micro structural evolution and mechanical properties enhancement of a cold-sprayed Cu-Zn alloy coating with friction stir processing. Mater Charact 125:76–82
183. Liu F, Ji Y, Meng Q, Li Z (2016) Microstructure and corrosion resistance of laser cladding and friction stir processing hybrid modification Al-Si coatings on AZ31B. Vacuum 133:31–37
184. Xiong Y, Qiu Z, Li R, Yuan T, Wu H, Liu J (2015) Preparation of ultra-fine grain Ni–Al–WC coating with interlocking bonding on austenitic stainless steel by laser clad and friction stir processing. Trans Nonferrous Met Soc China 25:3685–3693
185. Chew Y, Pang JHL, Bi G, Song B (2017) Effects of laser cladding on fatigue performance of AISI 4340 steel in the as-clad and machine treated conditions. J Mater Process Technol 243:246–257
186. Serres N, Tidu D, Sankare S, Hlawka F (2011) Environmental comparison of MESO-CLAD process and conventional machining implementing life cycle assessment. J Clean Prod 19:1117–1124
187. Salonitis K, D'Alvise L, Schoinochoritis B, Chantzis D (2015) Additivemanufacturing and post-processing simulation: laser cladding followed by high speed machining. Int J Adv Manuf Technol 85:2401–2411
188. Wang M, Xu B, Dong S, Zhang J, Wei S (2013a) Experimental investigations ofcutting parameters influence on cutting forces in turning of Fe-based amorphous overlay for remanufacture. Int J Adv Manuf Technol 65:735–743
189. Wang M, Xu B, Zhang J, Dong S, Wei S (2013b) Experimental observations onsurface roughness, chip morphology, and tool wear behaviour in machining Febasedamorphous alloy overlay for remanufacture. Int J Adv Manuf Technol 67:1537–1548
190. Nieslony P, Cichosz P, Krolczyk G, Legutko S, Smyczek D, Kolodziej M (2016) Experimental studies of the cutting force and surface morphology of explosivelyclad Tie steel plates. Measurement 78:129–137
191. Nieslony P, Krolczyk G, Zak K, Maruda R, Legutko S (2017) Comparativeassessment of the mechanical and electromagnetic surfaces of explosively cladTiesteel plates after drilling process. Precis Eng 47:104–110
192. Böß V, Denkena B, Wesling V, Kaierle S, Rust F, Nespor D, Rottwinkel B (2016) Repairing parts from nickel base material alloy by laser cladding and ball endmilling. Prod Eng 10:433–441
193. Zhang P, Liu Z (2017) On sustainable manufacturing of Cr-Ni alloy coatings by laser claddingand high-efficiency turning process chain and consequent corrosionresistance. J Cleaner Prod 161:676–687
194. Zahmatkesh B, Enayati MH (2010) A novel approach for development of surface nano composite by friction stir processing. Mater Sci Eng A 527:6734–6740
195. Triantou KI, Pantelis DI, Guipont V, Jeandin M (2015) Microstructure and tribological behavior of copper and compositecopper+alumina cold sprayed coatings for various alumina contents. Wear 336:96–107
196. Klopstock H, Neelands AR (1994) An improved method of joining and welding metals. Patent specifiction, No. 572789, U.K
197. Rafi HK, Balasubramaniam K, Phanikumar G, Rao KP, (2011) Thermal profilingusing infrared thermography in friction surfacing. Metall Mater Trans A 42:3425–3429
198. Bedford GM, Vitanov VI, Voutchkov II (2001) On the thermo-mechanical events during frictionsurfacing of high speed steels. Surf Coat Technol 141:34–39
199. Gandra J, Krohnband H, Mirandac RM (2014) Friction surfacing-a review. J Mater Process Technol 214(5):1062–1093
200. Mishra RS, Ma ZY (2005) Friction stir welding and processing. Mater Sci Eng 50(1–2):1–78
201. Chandrasekaran M, Batchelor AW, Jana S (1998a) Study of theinterfacial phenomena during friction surfacing of mild steel with tool steel and inconel. J Mater Sci 33:2709–2717
202. Vitanov VI, VoutchkovII BGM (2000) Decisionsupport system to optimize the Frictec (frictionsurfacing) process. J Mater Process Technol 107:236–242

203. Batchelor AW, Jana S, Koh CP, Tan CS (1996) The effect ofmetal type and multi-layering on friction surfacing. J Mater Process Technol 57:172–181
204. Reddy GM, Rao KS, Mohandas T (2009) Friction surfacing: novel technique for metal matrix composite coating on aluminium–silicon alloy. Surf Eng 25:25–30
205. Reddy GM, Prasad KS, Rao KS, Mohandas T (2011) Friction surfacing of titanium alloy with aluminium metal matrix composite. Surf Eng 27(2):92–98
206. Chandrasekaran M, Batchelor AW, Jana S (1997) Friction surfacing of metal coatings on steel and aluminum substrate. J Mater Process Technol 72:446–452
207. Tokisue H, Katoh K, Asahina T, Usiyama T (2006) Mechanical properties of 5052/2017 dissimilar aluminum alloys deposit by friction surfacing. Mater Trans 47(3):874–882
208. Madhusudhan RG, Srinivasa RK, Mohandas T (2009) Friction surfacing: novel technique for metal matrix composite coating on aluminum-silicon alloy. Surf Eng 25:25–30
209. Batchelor AW, Jana S, Koh CP, Tan CS (1996) The effect of metal type and multilayering on friction surfacing. J Mater Process Technol 57:172–181
210. Nixon GS, Mohanty BS, Barnabas SG, Manikandan S (2013) Factors affecting friction coating on stainless steel 304. Int J Sci Res Publ 3:3
211. Gandra J, Pereira D, Miranda RM, Silva RJC, Vilaça P (2013) Deposition of AA6082-T6 over AA2024-T3 by friction surfacing—mechanical and wear characterization. Surf Coat Technol 223:32–40
212. Janakiraman S, Bhat KU (2016) Formation of composite surface during friction surfacing of steel with aluminium. Adv Tribol 1–5
213. Rafi HK, Janaki Ram GD, Phanikumar G, Rao KP (2011) Microstructural evolution during friction surfacing of tool steel H13. Mater Des 32:82–87
214. Raoa KP, Veera SA, Rafi HK, Libin MN, Balasubramaniam K (2012) Tool steel and copper coatings by friction surfacing—a thermography study. J Mater Process Technol 212:402–407
215. Raoa KP, Sankar A, Rafi HK, Janaki Ram GD, Reddy GM (2013) Friction surfacing on nonferrous substrates: a feasibility study. Int J Adv Manuf Technol 65:755–762
216. Saravanakumar S, Gopalakrishnan S, Dinaharan I, Kalaiselvan K (2017) Assessment of microstructure and wear behavior of aluminum nitratereinforced surface composite layers synthesized using friction stirprocessing on copper substrate. Surf Coat Technol 322:51–58
217. Puli R, Nandha Kumar E, Janaki Ram GD (2011) Characterization of friction surfaced martensitic stainless steel (AISI 410) coatings. Trans IIM 64(1 & 2):41–45
218. Puli R, Janaki Ram GD (2012a) Corrosion performance of AISI 316L friction surfaced coatings. Corros Sci 62:95–103
219. Puli R, Janaki Ram GD (2012b) Microstructures and properties of friction surfaced coatings in AISI 440C martensitic stainless steel. Surf Coat Technol 207:310–318
220. Nakama D, Katoh K, Tokisue H, (2008) Some characteristics of AZ31/AZ91 dissimilar magnesium alloy deposit by friction surfacing. Mater Trans 49(5):1137–1141
221. Gopikrishna N, Sammaiah P, Kumar N S (2015) Effect of rotational speed on coating thickness during zinc deposition on copper by friction surfacing. Int J Eng Res Technol (IJERT) 4(01)
222. Khalid Rafi H, Janaki Ram GD, Phanikumar G, Rao KP (2010) Microstructure and properties of friction surfacedstainless steel and tool steel coatings. Mater Sci Forum 638–642:864–869
223. Barnabas S, Anantharam G, Shyamsundar V, Aravind BS, Prabhu T (2014) Friction surfacing in steel 304. Am J Eng Res (AJER) 3(4): 84–97
224. Barnabas G (2014) Parameters optimization in friction surfacing. Chem Mater Eng 2(6):127–136
225. Chandrasekaran M, Batchelor AW, Jana S (1998b) Study of the interfacial phenomena duringfriction surfacing of mild steel with tool steel and inconel. J Mater Sci 33:2709–2717
226. Miller SF (2013) New friction stir techniques for dissimilar materials processing. Manuf Lett 1:21–24
227. Sahoo DK, Mohanty BS (2016) Performance analysis of friction surfacing between two dissimilar materials. Int J Innov Eng Technol (IJIET) 7(4)
228. Gandraa J, Miranda RM, Vilac P (2012) Performance analysis of friction surfacing. J Mater Process Technol 212:1676–1686

229. KumarM V, Satish SV (2014) Friction stir surfacing of copper. Int J Sci Res (IJSR) 3(7):1414–1418
230. Raju VP, Hussain MM, Govardhan D (2015) Effect of process parameters on surface roughnessof tool steel M2 deposit over low carbon steelproduced by friction surfacing. Int J Mech Prod Eng 3(9)
231. Danish M, Usha S, Veerapandian R (2016) Characterization of aluminum alloy coated steel made using friction surfacing method. Int J Sci Res Dev 4(3)
232. Rafi HK, Balasubramaniam K, Phanikumar G, Rao KP (2011) Thermal profiling using infrared thermography in friction surfacing. In: The Minerals, Metals & Materials Society and ASM International 2011, vol 42A, p 3425
233. Vitanov VI, Voutchkov II (2005b) Process parameters selection for friction surfacing applications using intelligent decision support. J Mater Process Technol 159:27–32
234. Sekharbabu R, Rafi HK, Rao KP (2013) Characterization of D2 tool steel friction surfaced coatings over low carbon steel. Mater Des 50:543–550
235. Rafi HK, Kishore Babu N, Phanikumar G, Rao KP (2013) Microstructural evolution during friction surfacing of austenitic stainless steel AISI 304 on low carbon steel. Metall Mater Trans A 44A:345
236. Puli R, Janaki Ram GD (2012c) Wear and corrosion performance of AISI 410 martensitic stainless steel coatings produced using friction surfacing and manual metal arc welding. Surf Coat Technol 209:1–7
237. Raju VP, Manzoor HM (2016) Characterization of tool steel m2 friction surfaced deposit over low carbon steel. Int J Mater Sci 11(1):33–45
238. Raju VP, Manzoor Hussain M (2015) Experimental investigation of tool steel M2 coating on mild steel by friction surfacing. Int J Emerg Technol Adv Eng 5(4)
239. Chandrasekaran M, Batchelor AW, Jana S (1998c) Study of the interfacial phenomena during friction surfacing of mild steel with tool steel and inconel. J Mater Sci 33:2709–2717
240. Beyer M, Resende A, dos Santos JF (2003) Friction surfacing for multi-sectorial applications—FRICSURF. Institute for Materials Research, GKSS ForschungszentrumGeesthacht GmbH, Technical report
241. Gandra J, Miranda RM, Vilaca P (2012) Performance analysis of friction surfacing. J Mater Process Technol 212:1676
242. Vitanov VI, Javaid N (2010) Investigation of the thermal field in microfriction surfacing. Surf Coat Technol 204:2624–2631
243. Raju VP, Hussain MM, Govardhan D (2016) Effect of process parameters on the width of friction surfaced tool steel M2 deposit over low carbon steel. Int J Mater Sci 11:1–7
244. Rafi HK, Janaki Ram GD, Phanikumar G, Rao KP (2010) Friction surfaced tool steel (H13) coatings on low carbon steel: a study on the effects of process parameters on coating characteristics and integrity. Surf Coat Technol 205:232–242
245. Ashok Kumar U, Laxminarayana P (2014) Friction surfacing process of aluminum alloys. In: 5th international & 26th all india manufacturing technology, design and research conference (AIMTDR 2014) December 12–14th. IIT Guwahati, Assam, India
246. Vitanov VI, Voutchkov II (2005a) Process parameters selection for friction surfacing applications using intelligence decision support. J Mater Process Technol 159:27–32
247. Vitanov VI, Bedford GM (2001) Neurofuzzyapproach to process parameter selection for friction surfacing applications. Surf Coat Technol 140:256–262
248. Hidekazu S, Hiroshi T (2002) Mechanical properties of friction surfaced 5052 aluminium alloy. J Jpn Inst Light Met 52(8):346–351
249. Takeshi S, Sinya O (1995) Deposition of hard coating layer by friction surfacing. J Jpn Weld Soc 13(3):432–437
250. Raju VP, Hussain MM (2015) Experimental investigation of tool steel M2 coating on mild steel by friction surfacing. Int J Emerg Technol Adv Eng 5:4
251. Vitanov VI, Javaid N, Stephenson DJ. (2012) Application of response surface methodology for the optimization of micro friction surfacing process. Surf Coat Technol 204(21–22):3501–3508

252. Mubiayi MP, Akinlabi ET. (2014) Friction stir spot welding of dissimilar materials: an overview. In: Proceedings of the world congress on engineering and computer science 2014, vol II. WCECS 2014, San Francisco, USA
253. Navaneethakrishnan N, Loganatha VN (2016) Investigation of alloys and analysis of coating parameter in friction surfacing: a technical review 4(1)
254. Sakihama H, Tokisue H, Katoh K (2003) Mechanical properties of friction surfaced 5052 aluminum alloy. Mater Trans 44(12):2688–2694
255. Nixon GS, Barnabas SG, Edward A, Ezhilvannan R, Janani B, Sunddaram SM (2015) Microstructure analysis of SG iron friction surfacing. Aust J Basic Appl Sci 9(20):78–89
256. Sugandhi V (2012) Optimization of friction surfacing process parameters for AA1100 aluminum alloy coating with mild steel substrate using Response Surface Methodology (RSM) technique. Modern Appl Sci 6:2
257. James WJ, Slabbekoorn MA, Edgin WA, Hardin CK (1998) Correction of congenital malar hypoplasia usingstereolithography for presurgical planning. J Oral Maxillofac Surg 56(4):512–517
258. Van Noort R (2012) The future of dental devices is digital. Dent Mater 28(1):3–12
259. Hollister SJ (2005) Porous scaffold design for tissue engineering. Nat Mater 4(7):518–524
260. Conner M (2010) 3-D medical printer to print body parts. EDN 55(3):9
261. Zhao L, Lee V, Yoo S, Dai G, Intes X (2012) The integration of 3-D cell printing and mesoscopic fluorescence molecular tomography of vascular constructs within thick hydrogelscaffolds. Biomaterials 33(21):5325–5332
262. Budzik G, Burek J, Bazan A, Turek P (2016) Analysis of the accuracy of reconstructed two teeth models manufactured using the 3DP and FDM technologies. J Mech Eng 62(1):11–20
263. Spencer K, Fabijanic DM, Zhang MX (2009) The use of Al–Al2O3 cold spray coatings to improve the surface properties of magnesium alloys. Surf Coat Technol 204:336–344
264. Li WY, Zhang G, Liao HL, Coddet C (2008) Characterizations of cold sprayed TiNparticle reinforced Al2319 composite coating. J Mater Process Technol 02:508–513
265. Neef A, Seyda A, Herzo V, Emmelmann D, Schönleber C, Kogel-Hollacher M (2014) Low coherence interferometry in selective laser melting. Phys Procedia 56:82–89
266. Murr LE, Esquivel EV, Quinones SA, Gaytan SM, Lopez MI, Martinez EY, Medina F, Hernandez DH, Martinez E, Martinez JL, Stafford SW, Brown DK, Hoppe T, Meyers W, Lindhe U, Wicker RB (2009c) Microstructures and mechanical properties of electronbeam-rapid manufactured Ti-6Al-4V biomedical prototypes compared to wrought Ti-6Al-4V. Mater Charact 60(2):96–105
267. Tokisue H, Katohk K, Asahina T, Ushiyama T (2005) Structures and mechanical properties of multilayer friction surfaced aluminium alloys report of the research institute of industrial technology. Nihon University, vol 78, pp 1-13
268. Khalid Rafi H, JanakiRam GD, Phanikumar G, Rao KP (2010) Friction surfacing of austenitic stainless steelon low carbon steel: studies on the effects of traverse speed. In: Proceedings of the world congress on engineering, vol 2. London, U.K.
269. Mordfin L (1985) Nondestructive evaluation, materials and processes, Part B: processes. Marcel Dekker Inc.
270. Kumar R, Chattopadhyaya S, Ghosh A, Krolczyk GM, Vilaca P, Kumar R, Srivastava M, Shariq M, Triphathi R (2017) Characterization of friction surfaced coatings of AISI 316 tool over high-speed-steel substrate. Trans Famena 41(2):61–76
271. Rivera OG, Allison PG, Jordon JB, Rodriguez OL, Brewer LN, McClelland Z, Whittington D, Francis WR, Su J, Martens RL, Hardwick N (2017) Microstructures and mechanical behavior of Inconel 625 fabricated by solid-state additive manufacturing. Mater Sci Eng, A 694:1–9

Fabrication and Characterisation of Aluminium Matrix Composite (Al 2024) Reinforced with Zircon Sand and Flyash

Laxmikant Swain, Rabinarayan Sethi, A. K. Chaubey, and Silani Sahoo

Abstract Composites are becoming popular in advance engineering. Monolithic metals are replaced by metal matrix composites due to its low cost and improved properties like better strength, good wear resistance, hardness, etc. Aluminum Metal Matrix Composites (AMMCs) are used widely in aerospace, automobile, and marine industries. Among the Aluminum Alloys, Al-2024 as a copper-based aluminum alloy shows poor strength as well as poor wear resistance. In the present investigation, Al-2024 is reinforced with Zircon Sand and Fly Ash fabricated with the Liquid metallurgy route. Al-2024 is reinforced with reinforcements in different weight fractions like 0.25, 0.50, and 1 wt.%. The microstructural analysis is carried out with the help of FESEM. XRD analysis investigates the phase and structure of both matrix and dispersive phases. Wear test is conducted on Multiple Tribotester. It is found that with the increase in weight fraction of Ceramic Reinforcement, there is an increase in wear resistance. Reinforced AMMCs with 1 wt.% of Zircon Sand and Fly Ash in equal proportion have shown better results.

Keywords Al-2024 · Zircon sand · Fly ash · Micro hardness · Wear · Coefficient of friction

1 Introduction

Metal Matrix Composites (MMCs) are composed of two or more materials. MMCs are the multi-phase system that consists of a dispersive phase embedded in the matrix phase. These dispersive phases can help the matrix phase in improving its properties. Aluminum Metal Matrix Composites (AMMCs) are commonly available in Automotive, Marine, Aerospace, electronic packaging applications [1]. Aluminum is

L. Swain (✉) · R. Sethi
Department of Mechanical Engineering, I G I T, Sarang, Odisha 759146, India
e-mail: laxmikantswain93@gmail.com

A. K. Chaubey · S. Sahoo
CSIR-Institute of Minerals and Materials Technology, Bhubaneswar, Odisha 751013, India

© Springer Nature Singapore Pte Ltd. 2021

P. M. Pandey et al. (eds.), *Advances in Production and Industrial Engineering*,
Lecture Notes in Mechanical Engineering,
https://doi.org/10.1007/978-981-15-5519-0_20

commonly preferred with cost-effective reinforcement compensates the overall material cost in technological advancement [2]. Ceramic hard particles added to aluminum matrix composites have a low coefficient of thermal expansion (CTE) than Aluminum alloys [3]. Aluminum Alloys are lightweight and show better strength and stiffness [4]. But in the case of wear resistance, Aluminum alloys show very poor results. The addition of Ceramic Reinforcement increases the wear resistance as well as different mechanical properties like fatigue resistance, hardness, etc. This increases the efficiency of the material supplemented with reduced maintenance costs in Industrial applications. Composites reinforced with ceramic particles are more superior than unreinforced aluminum alloy in automotive, aerospace, and marine applications with improved mechanical and physical properties. These reinforced composites are also good in tribological properties [5].

Al 2 series are copper (3.8–4.9 wt.%) based alloying metals used in aircraft fittings, couplings, shaft, and gear materials [6]. Al-2024 is used in aeronautical sectors due to its high strength and light weight [7]. Al-2024 is having its disadvantages in cracking growth as well as defects like porosity and undercut [8, 9]. AMMCs are reinforced with various particulates like SiC, boron, TiC, SiC_3N_4, silica sand, MgO, mica, glass beads, and B_4N and Al_2O_3. Limited work has been investigated on Zircon Sand reinforced aluminum composites [10]. Zircon sand is preferred as a reinforcement because of its high hardness and high modulus of elasticity [11]. Zircon sand is also known as Zirconium silicate ($ZrSiO_4$). Zircon Sand is known for its high refractoriness and chemical inertness.

The density of the Ceramic reinforcement is very high that leads to an increase in the weight of the composites. The cost of the ceramic reinforcement is also very high that has become a problem in developing countries. The problem can be solved by multiple reinforcements in the aluminum matrix phase. The hybrid composites with improved mechanical properties can be manufactured with low cost and density than aluminum alloy. In recent work, Fly Ash as a secondary reinforcement is gaining more demand in the industrial application. Fly Ash is an Industrial waste product replaced by part of aluminum and magnesium which helps in energy savings [12]. Fly Ash is a byproduct of power plants. Fly Ash creates health hazards with air pollution and only 6% is utilized in India [13]. Fly Ash decreases the cost and density of the composite. The ceramic reinforcement is quite hard and brittle. The brittleness of the composites causes a threat to the machining of the composites. The addition of secondary reinforcement increases ductility [14]. AMMCs with small size reinforcements are generally preferred for low-speed applications [15].

In light of the above, the present work is to compare the study of unreinforced Al-2024 with up to 1 wt.% Zircon Sand and Fly Ash based Al-2024. The Wear properties are investigated and FESEM analysis is studied on worn surfaces of the specimen.

2 Materials

2.1 Matrix Material

In this study Al2024 with a theoretical density of 2770 kg/m^3 is taken as matrix material. It has better strength but low wear and corrosion resistance. The chemical composition of Al2024 is shown in Table 1.

In this present work, Zircon Sand and Fly Ash are used as reinforcement materials. Zircon Sand particle size is in a range of about 100 to 250 microns. The Zircon Sand is collected from Konark, Ramachandi, Odisha. The chemical composition of Zircon Sand is shown in Table 2.

The Particle size of Fly Ash as a secondary reinforcement ranges between 0.5 and 100-micron size. The Fly Ash is collected from NTPC, Talcher, Odisha. The Material composition of Fly Ash is shown in Table 3.

3 Hybrid Composite Sample Preparation

Hybrid Composite is fabricated through a liquid metallurgy route. Figure 1. shows the experimental setup of Vertical Chamber Pit Furnace with an automated stirrer. The Hybrid Composites samples are prepared at a different weight fraction of Ceramic reinforcement. Al-2024 blocks are kept inside Cast Iron crucibles which are aligned inside the furnace chamber with mold cavity. The whole metal flows through the mold cavity with the help of the ejector pin. Aluminum alloys are melted up to 750 °C. The molten metal is degassed by Hexachloroethane Tablet and preheated reinforcement kept at 200 °C is added to it. The stirring time is fixed at 8 min. and stirring speed is at 300 rpm. Four Casting Samples (Pure Al-2024, 0.25 wt.% $ZrSiO_4$

Table 1 Chemical composition of Al alloy used as matrix material in wt.%

Cu	Mg	Mn	Fe	Si	Zn	Ti	Cr	Others	Al
3.8–4.9	1.2–1.8	0.3–0.9	0.5	0.5	0.25	0.15	0.1	0.15	Bal

Table 2 Chemical composition of Zircon Sand (in wt. %)

ZrO_2	SiO_2	HfO_2	Al_2O_3	TiO_2	Fe_2O_3	FeO	P_2O_5
63.33	31.3	1.2	2.46	0.25	0.72	0.14	0.14

Table 3 Chemical composition of Fly Ash (in wt. %)

SiO_2	Al_2O_3	Fe_2O_3	Ca0	MgO	Na_2O	K_2O
63.9	29.8	3.9	1.01	0.12	0.07	0.05

Fig. 1 Stir casting set up

+ 0.25 wt.% Fly Ash, 0.5 wt.% $ZrSiO_4$ + 0.5 wt.% Fly Ash and 1wt.% $ZrSiO_4$ + 1wt.% Fly Ash) are prepared and pouring Temperature to the mold cavity is at 675 °C. The samples are machined with the EDM Wire machine.

Fig. 2 Machined wear samples

3.1 Microstructural Sample Preparation

FESEM analysis is carried out in ZEISS Microscopy. The samples are polished with 180, 220, 400, 600, 800, 1200, 1600, 2000 emery paper. The samples are polished again with diamond polish to get a better reflective surface. XRD analysis is observed in Rigaku Japan X-Ray Diffractometer. The dimension of the sample is 3 mm thick small block.

3.2 Micro Hardness Sample Preparation

The samples are resin-molded. These samples are polished with SiC paper in METCO Polishing machine. Polishing is done in a forward and backward direction to get a better reflective surface. As per ASTM E92, four samples are taken with 100 gf load in 10–15 s duration. The impression diameter is 0.404 mm.

3.3 Wear Sample Preparation

The Wear test is conducted on Multiple Tribotester TR-25. According to the ASTM G99 standard, the wear sample dimension is taken as 6.5*6.5*9 mm.

4 Results and Discussion

4.1 Microstructural Analysis

FESEM images reveal the presence of elements like oxygen, Zircon, Aluminum, Manganese, Magnesium, and Copper in 1 wt.% sample each of Fly Ash and Zircon Sand as shown in Fig. 3. The white dendritic structure shows the presence of Aluminum 2024 alloy constituents like Copper, Magnesium, Manganese, and Aluminum. The red circle marked near the dendritic structure shows the presence of a white globular structure of zirconium sourced from zircon sand. The presence of Oxygen indicates the presence of Alumina which is a major constituent of Fly Ash. The FESEM images give a view that the reinforcement is in less content but entrapped in the white dendritic structure of Al-2024.The less content of reinforcement may be due to less use of that reinforcement and depends upon the sensitivity of the sample.

XRD Phases of 1 wt. % each of Fly Ash and Zircon Sand Reinforced Composite is depicted in Fig. 4. It can be observed from the figure, the dominating peak represents the Al Matrix. Peaks corresponding to Zirconium Oxide and Silicon Oxide were

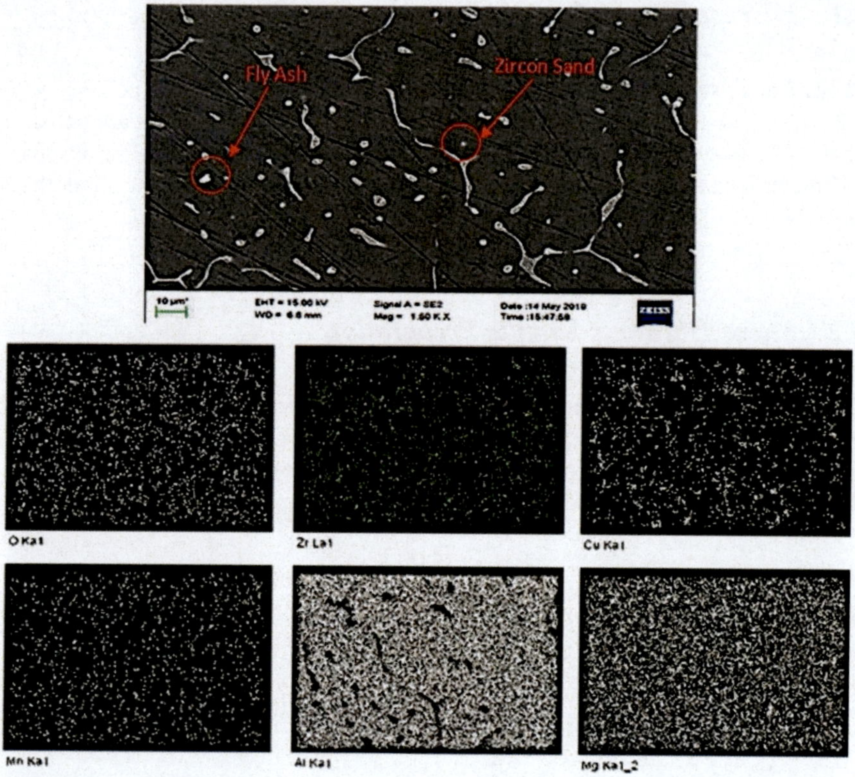

Fig. 3 FESEM analysis of 1 wt.% sample each of Fly Ash and Zircon Sand

not detected clearly due to the use of low content of that reinforcement. However, Small peaks correspond to Zirconium Oxide and Silicon Oxide are also detectable in the Fig. 4. The Al-matrix major peak is visualized at 2Θ value equal to 38.43, 44.83, and 78.31. The zirconium oxide and silicon oxide peak value visualized at 2Θ value equal to 38.90 and 47.38, respectively. The Crystal system of Zirconium Oxide shows the cubic structure and Silicon Oxide shows the tetragonal structure.

4.2 *Micro Hardness Test*

Five indentations are taken and mean value with standard deviation are reported in Table 4. The graph analysis of the Hardness value of different weight fraction samples shown in Fig. 5.

The results show that with the increase in sample concentration, there is an increase in hardness value. The increase in hardness value may be due to the presence of hard ceramic particles. The error is less in the case of 1 wt.% sample that exhibits

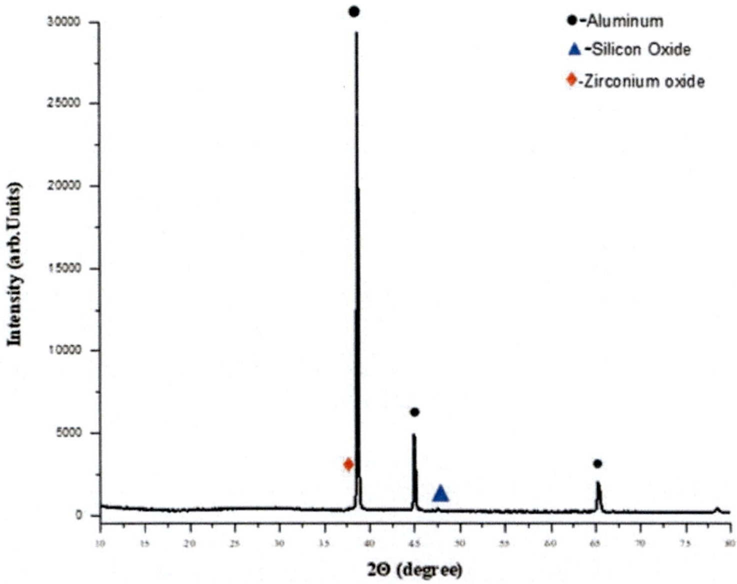

Fig. 4 XRD Analysis of 1 wt.% sample each of Fly Ash and Zircon Sand

Table 4 Micro vickers hardness test for all samples

Sample name	Mean hardness value with standard deviation (in HV)
Unreinforced Al-2024	102 ± 1.41
Al-2024 with 0.25 wt.% Fly Ash and Zircon Sand	105.5 ± 2.12
Al-2024 with 0.50 wt.% Fly Ash and Zircon Sand	114.5 ± 3.53
Al-2024 with 1 wt.% Fly Ash and Zircon Sand	116.5 ± 0.70

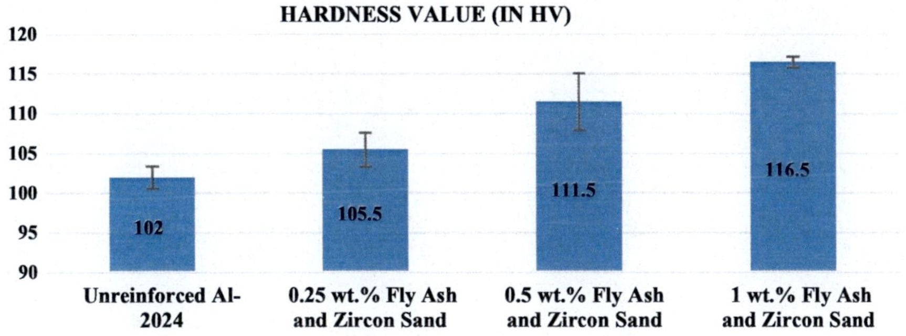

Fig. 5 MicroVicker hardness test for all samples

better hardness value than other samples and also indicates the good concentration of reinforcement. The better hardness value indicates the good interfacial bonding between reinforcement and base alloy and minimizes the porosity [4]. The presence of small concentration of silicon oxide and zirconium oxide is depicted from XRD analysis in Fig. 4 may be the cause of improvement in hardness property of reinforced aluminum 2024.

4.3 Wear Test

The input parameters necessary for wear test are load, sliding speed, and time. The time is fixed at 30 min. The combination of load and sliding speed is taken as 20 N–300 rpm.

The results of wear and coefficient of friction against variation of time are shown in Figs. 6 and 7.

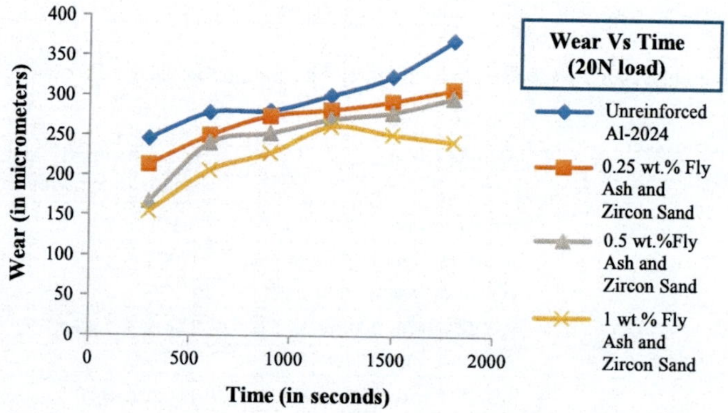

Fig. 6 Shows the wear versus time graph when 20 N load is applied

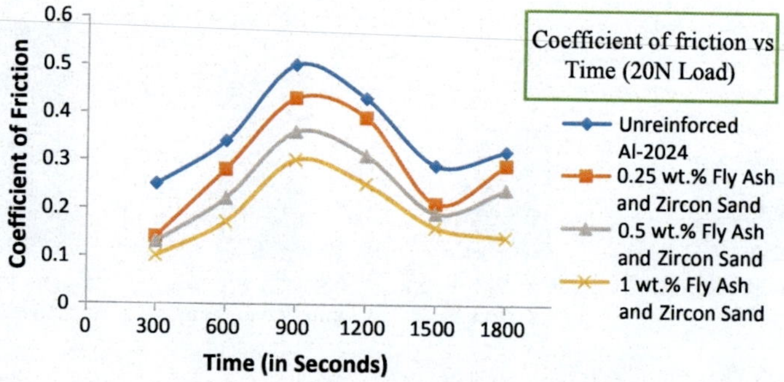

Fig. 7 Shows the coefficient of friction versus time graph when 20 N load is applied

The results show that there is an increase in wear resistance with the increase in reinforcement. The increase in wear resistance may be due to the presence of hard ceramic reinforcement. The wear resistance of 1 wt.% sample each of Fly Ash and Zircon Sand has increased may be due to the increase in overall hardness of the composite material as evident from Fig. 5. The samples with 0,0.25 and 0.5 wt.% have shown more wear than 1.0 wt. % samples. The base Al-2024 alloy has shown higher wear and 1.0 wt.% of sample shown lower wear. In the early stages of wear process, the addition of reinforcement acts as a load carrying agent and inhibits plastic deformation. In the later stages of wear process, these reinforcement forms wear debris along with base matrix material and worn particles. These wear debris fulfill the craters formed during the wear process and acts as a load bearing agent.

The Coefficient of Friction versus Time graph suggests that a decrease in coefficient of Friction with the increase in reinforcement. The average value of the coefficient of friction for base Al-2024 alloy is 0.35 but for 1 wt.% Sample each of Fly Ash and Zircon Sand is 0.22. The decrease in coefficient of friction of 1 wt.% samples of Fly Ash and Zircon Sand may be due to the presence of a higher percentage of hard Fly Ash and Zircon Sand particles that result in point contact between roller and Block.

5 Conclusion

Al-2024 is reinforced with Fly Ash and Zircon Sand through the stir casting method. FESEM images reveal the presence of aluminum, zircon, magnesium, manganese, copper, and oxygen which clears the presence of zirconium oxide. XRD analysis shows the lower content of zirconium oxide and silicon oxide in the aluminum matrix. The hardness value is maximum for 1 wt.% sample each of Fly Ash and Zircon Sand. The wear test reveals that 1 wt.% Sample each of Fly Ash and Zircon Sand having better wear resistant than base alloy Al-2024. This concludes that 1 wt.% sample has improved properties than base Al-2024 with better hardness property as well as tribological properties.

References

1. Ramgopal Reddy B, Srinivas C, (2018) Fabrication of silicon carbide and fly ash reinforced aluminum metal matrix hybrid composites. Mater Today: Proc 5:8374–8381
2. Karunakara S, Dinesh P (2018) Mechanical and tribological characterization of aluminum-hematite composites. Mater Today: Proc 5:2901–2906
3. Bhaskar KV, Sundarrajan S, Krishna MG, Ravindra K (2017) Microstructure and mechanical properties of fly ash/SiC particles reinforced AA2024 hybrid composites. Mater Today: Proc 5:7413–7419
4. Kumar HGP, Xavior MA (2017) Assessment of mechanical and tribological properties of Al-2024-SiC-graphene hybrid composites. Procedia Eng 174:992–999

5. Prakash Rao CR, Bhagyasekhar MS, Viswanath N (2014) Machining behavior of Al-6061-fly ash composites. Procedia Mater Sci 5:1593–1602
6. Khan MM, Dixit G (2017) Comparative study on erosive wear response of SiC reinforced and fly ash reinforced aluminum based metal matrix composite. Mater Today: Proc 4:10093–10098
7. Sirvent P, Garrido MA, Munez CJ, Poza P, Vezzu S (2018) Effect of higher deposition temperatures on the microstructure and mechanical properties of Al-2024 cold sprayed coatings. Surf Coat Technol 337:461–470
8. Ahn J, He E, Chen Li, Dear J, Shao Z, Davies C (2018) In-situ micro tensile testing of AA2024-T3 fibre laser welds with digital image correlation as a function of welding speed. Lightweight Mater Manuf 1:179–188
9. Chen Y, Ding H, Li J-Z, Zhao J-W, Fu M-J, Li X-H (2015) Effect of welding heat input and post-welded heat treatment on hardness of stir zone for friction welded 2024-T3 aluminum alloy. Trans Non Ferrous Metals Soc China 25:2524–2532
10. Kaur K, Pandey OP (2010) Microstructural characteristics of spray formed zircon sand reinforced LM 13 Composite. Alloy Compd 503:410–415
11. Das S, Das S, Das K (2007) Abrasive wear of Zircon sand and alumina reinforced Al-4.5%Cu alloy matrix composites—a comparative study. Compos Sci Technol 67:746–751
12. Vijaya Ramnath B, Elanchezhian C, Jaivignesh M, Rajesh S, Parswajinan C (2014) Evaluation of mechanical properties of aluminum alloy-alumina-boron carbide metal matrix composite. Mater Des 58:332–338
13. Satapathy LN (2000) A study on the mechanical, abrasion and microstructural properties of zirconia-flyash material. Ceram Int 26:39–45
14. Singh J, Chauhan A (2017) Fabrication characteristics and tensile strength of novel Al2024/SiC/red mud composites processed via stir casting route. Trans Non Ferrous Metals Soc China 27:2573–2586
15. Singh KK, Singh S, Shrivastava AK (2017) Comparison of wear and friction behavior of aluminum matrix alloy (Al-7075) and silicon carbide based aluminum metal matrix composite under dry condition at different sliding distance. Mater Today: Proc 4:8960–8970

Novel Technology on Recovery of Ceramic Materials from Partially Lateritised Khondalite Rocks-A Bauxite Mining Waste

Ranjita Swain, Sunita Routray, and R. Bhima Rao

Abstract Partially Lateritised Khondalite (PLK) rocks are found at the mine site of bauxite mining area. These materials cannot be processed for the production of alumina as these contain high reactive silica. But, the materials can be a raw material for other industrial application after processing. In view of this, a research work has been worked out on this problem. The main aim of the present investigation is to see the effect of calcinations of non-mag products for iron removal and to observe the improvement in brightness of the product. In view of this PLK rocks are calcined at 1473°K temperature in a kiln and subjected to Dry High Intensity Magnetic Separator (DHIMS) after crushing to below one mm size. It is observed that a product obtained from calcined non-magnetic product from kiln contains 70% brightness value which is a material for filler industries like paint, pigment, ceramic, pesticides, paper coatings, textile, etc.

Keywords PLK rocks · Calcination · Dry high intensity magnetic separator · Iron oxide · Filler/ceramic industries · Brightness

1 Introduction

The largest bauxite mining area is situated on Panchpatmali hills of Koraput district in Odisha [1, 2]. Partially Lateritised Khondalite (PLK) rocks are one of the mining wastes generated while mining the bauxite at Damanjodi. Whenever bauxite is being mined for alumina extraction, at the same time partially weathered khondalite rocks are also mined and dumped at mining area as a waste. It is also recorded that the present capacity for mining of bauxite is about 6.825 million tons per annum, from which 2.3 million tons per annum alumina is being produced. It is also observed that the same amount of partially weathered khondalite and lateritic rocks are being mined and kept at the site as it cannot be processed due to contain reactive silica. This

R. Swain (✉) · S. Routray · R. B. Rao
Chemical Engineering Department, C. V. Raman College of Engineering & CSIR-IMMT, Bhubaneswar, India
e-mail: ranjitaswain79@gmail.com

© Springer Nature Singapore Pte Ltd. 2021

P. M. Pandey et al. (eds.), *Advances in Production and Industrial Engineering*, Lecture Notes in Mechanical Engineering,
https://doi.org/10.1007/978-981-15-5519-0_21

mining waste mostly contains on an average of 30–56% of Al_2O_3, 0.1–35% of Fe_2O_3, 0.1–3% of TiO_2, 19–35% of SiO_2 and 16–31% of loss of ignition (LOI). As these rocks contain high alumina percentage still these are not suitable for metallurgical industry in extracting alumina due to presence of reactive silica or clay. But, these rocks can be used as a potential resource for different industrial applications such as ceramic, filler, paint, pigment, cosmetics and paper industries after removal of colouring impurities like iron using suitable mineral processing techniques. Demand for brighter and submicron particle size materials is increasing day by day due to higher demand for the same in filler, ceramic and refractory industries. As a result, natural reserves of such materials are depleting very fast due to increase in demand and hence it is necessary to explore the alternate resources especially from the mining and metallurgical wastes for industrial applications. Kaolin clay is one of the raw materials suitable for filler, ceramic and refractory industries. It occurs in patches throughout India. The total demand for kaolin is around one million tons per annum which are presently being met from indigenous sources for filler industries alone. Indian ceramic and filler industries also depend on other natural raw materials like pyrophillite, talc, calcite powder, etc., and also from derived products such as calcite precipitate, titanium dioxide, etc. In view of this, the PLK rocks can be a probable reserve to substitute kaolin clay, proper for ceramic and filler-grade resources after deletion of iron content by appropriate mineral dressing techniques. Several authors have investigated for removal of iron on calcined products and found that the reaction is faster using acids.

INDAL [3] is producing white, fine grained, high brightness and high purity kaolin clay from bauxite mines. The beneficiation process for recovery of white clay involves classification using hydrocyclone, centrifuging, high gravity mineral separation, bleaching, filtration, thickening into slurry, drying and calcinations methods to produce the desired product. Waite and Torikov [4] discussed the suspension of lepidocrocite (γ-FeOOH) and hematite (α-Fe_2O_3) from bauxite and also measured the apparent activation energy which is low for the dissolution of hematite than lepidocrocite. Gulfen et al. [5] have reported the optimum calcination conditions for bauxite as one hour at 1173°K followed by leaching with 8 M hydrochloric acid. It is also observed that dissolution of iron follows first-order reaction kinetics in shrinking core model with apparent activation energy of 41.38 kJ/mol. Goodarzi et al. [6] calcined the bauxite samples at temperatures between 873 and 1373°K and then leached with hydrochloric acid. Detailed leaching studies on sample calcined at 1173°K showed that leaching temperature is the most important parameter for iron dissolution. The optimum leaching time for iron removal was determined to be 120 min. Kinetics of iron removal from boehmitic and diasporic bauxite with acid followed first-order and shrinking core model reaction mechanism, respectively [7, 8]. The activation energy varied from 80 to 106 kJ/mol. It is also seen from the literature that the apparent activation energy varies from 19 to 80 kJ/mol [6–12].

Kaolin clay can be used as a value-added material after physical beneficiation study but some industries need very high quality and specialized kaolin clay.

Gajam and Raghavan [13] investigated the effect of calcination and the kinetics of dissolution of kaolinite clay in hydrochloric acid. The rate of dissolution of iron has been increased with calcination temperature, calcination time and leaching temperature. A first-order kinetic model is reported for the dissolution process. It is also reported that the acid leaching process removed the iron oxides associated with the clay [14].

Some of the work has been carried out by Swain et al. [15–17] on PLK rocks generated at bauxite mining place. Leaching experiments were carried out on different process parameters. This process and parameters are also optimized by using RSM application. After leaching with hydrochloric acid it can be used as a ceramic material for industrial application.

An attempt has been made that this work is to investigate the prospect of iron elimination from calcined sample of PLK rocks by using magnetic separation process and extend the new application of the sample in filler/ceramic industries.

2 Materials and Methods

2.1 Materials

Partially Lateritised Khondalite rock samples were collected during bauxite mining. These rock samples were calcined in an industrial kiln at 1473°K temperature for 2 h time, which shown in Fig. 1. The calcined rock was stage crushed to below

Fig. 1 **a** Natural and **b** calcined PLK rocks

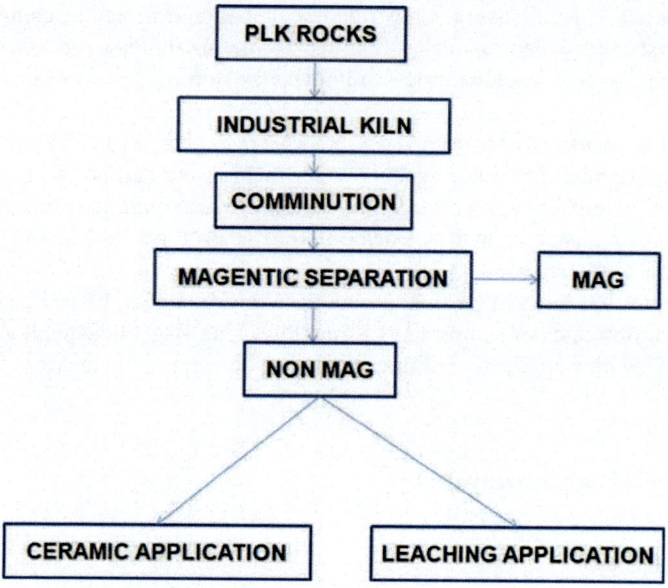

Fig. 2 Schematically presentation of raw materials and experimental conditions

1 mm size. The different coloured sample subjected to Dry High Intensity Magnetic Separator (DHIMS) at 0.6 T model no. ELB-YANTV, Ore Sorter Pty. Ltd., Australia. Details of the sample preparation for leaching studies are given below and also shown schematically in Fig. 2. The details of the raw materials and experimental conditions are also explained. The raw materials are calcined PLK rocks and magnetic separations are carried out at different conditions.

2.2 *Characterization and Evaluation*

The phase identification and quantification of mineralogy of crystalline compounds in rocks, solids and particulates using PANalytical X-Pert X-ray powder diffractometer with Mo-Kα radiation (λ = 0.709 Å) from 275 to 373°K scanning angle range resolution at a scanning rate of 273.02°K/s.

3 Results and Discussion

3.1 Characterization

The physical properties and chemical analysis of PLK rock is shown in Table 1. The data specify about the natural PLK rocks; contain 1.3 gm/cm^3 bulk density, 2.7 gm/cm^3 true density and angle of repose 35.9°. The data indicate that the rock contains 4.3% of Fe_2O_3 and 1.66% of TiO_2. Individual calcined rocks are also picked up and analysed.

The iron oxide analysis with respect to different sizes for calcined PLK rock samples is shown in Table 2. The data shown in Table 2 indicate that calcined kaolinized rocks contain less than 2% of Fe_2O_3 from 600 to 90 μm size whereas below 90 μm size Fe_2O_3% is increased. Hence, above 90 μm size sample fractions can be used directly as a refractory purpose. The sample fraction below 90 μm is subjected for grinding and leaching with organic or inorganic acid for filler industrial application.

XRD of PLK rock sample is made known in Fig. 3a indicates that the rock contains kaolinite, goethite, hematite, orthoclase, gibbsite, magnetite and quartz. The PLK rock shows a highly crystalline kaolinite phase along with other phases of gibbsite, goethite and quartz. The highest peak indicates that the high percentage of

Table 1 Chemical characterization of PLK rocks and calcined PLK rocks

Physical properties		Chemical properties	
Bulk density, g/cm^3	1.3	Al_2O_3, %	37.24
True density, g/cm^3	2.7	SiO_2, %	40.64
Porosity, %	51.9	Fe_2O_3, %	4.30
Angle of repose, °	35.9	TiO_2, %	1.66
		LOI, %	16.26

Table 2 Size and chemical analysis of calcined PLK rocks (White coloured sample)

Size, μm	wt, %	Fe_2O_3, %
+600	35.9	0.84
+420	13.3	0.84
+300	5.7	0.84
+210	15.0	0.84
+150	10.3	1.26
+90	6.4	1.39
+75	3.4	2.51
+45	1.8	2.65
−45	8.3	2.65
Total	100.0	1.15

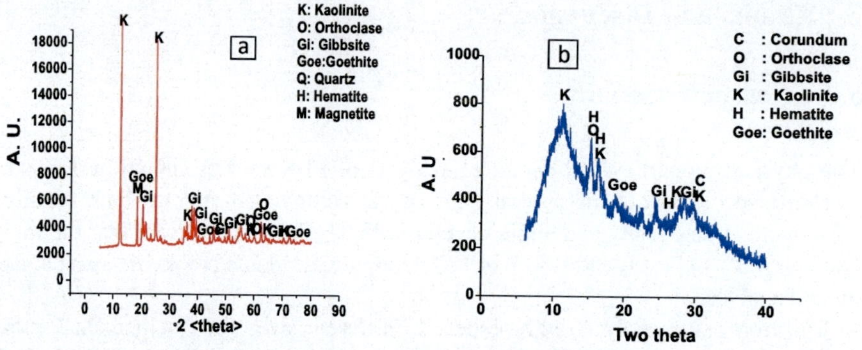

Fig. 3 XRD of **a** PLK rock sample, **b** calcined PLK rock sample

Fig. 4 XRD of calcined non-magnetic PLK rock sample

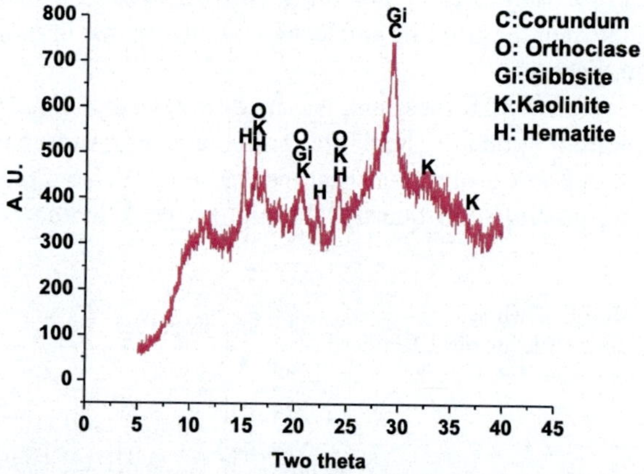

kaolinite is present in the PLK rocks. XRD of calcined PLK rock sample (Fig. 3b) contains corundum, kaolinite, orthoclase, gibbsite, goethite and hematite. The major crystalline phase is corundum due to the presence of Al_2O_3 in hexagonal system. Kaolinite shows its presence with a high peak. The iron content is present in the form of hematite and goethite. Some parts of hematite become magnetite also. In Fig. 4, the major phases are kaolinite, gibbsite, orthoclase and corundum. Some of the hematite phases is also present in the XRD figure.

3.2 *Effect Magnetic Separation on Reduction of Iron Content*

The individual size fractions of white coloured calcined PLK rock is subjected to DHIMS for reduction of iron content from the sample. The size and chemical analysis of the magnetic and non-magnetic fractions are tabulated in Table 3. The data indicate

Table 3 Size and chemical analysis of the samples after using DHIMS

Size, μm	Details	Wt., %	Fe_2O_3, %
+ 600	Mag	6.2	8.1
	Nonmag	93.8	0.6
	Total	100.0	1.06
+ 420	Mag	6.3	2.9
	Nonmag	93.7	0.2
	Total	100.0	1.07
+ 300	Mag	13.3	3.8
	Nonmag	86.7	0.3
	Total	100.0	1.37
+ 210	Mag	5.9	3.9
	Nonmag	94.1	0.25
	Total	100.0	0.99
+ 150	Mag	4.8	12.9
	Nonmag	95.2	0.4
	Total	100.0	1.0
+ 90	Mag	16.7	3.45
	Nonmag	83.3	0.56
	Total	100.0	1.0
+ 75	Mag	22.2	2.2
	Nonmag	77.8	0.84
	Total	100.0	1.15
+ 45	Mag	40	3.02
	Nonmag	60	0.9
	Total	100.0	1.74
+ 45	Mag	11.1	5.8
	Nonmag	88.9	1.2
	Total	100.0	1.71

that the non-magnetic fraction of the sizes from 600 to 90 μm contains less than 0.7% of Fe_2O_3. These samples can be suitable for ceramic industries as per ceramic standard for Fe_2O_3. But below 90 μm size fraction the samples contain more than 1% of Fe_2O_3.

3.3 *Characterization of the Product*

Chemical analysis of the product is given in Table 4 which indicates that the Fe_2O_3% is 0.7% and the brightness value is above 70%. The product is suitable in filler/ceramic

Table 4 Chemical characterization of calcined non-mag product

Sample	Al_2O_3, %	SiO_2, %	Fe_2O_3, %	TiO_2, %	LOI, %
Calcined non-mag product	43.67	46.34	0.7	0.78	9.01

Fig. 5 Photograph of calcined non-magnetic product for filler/ceramic purpose

industry. The product for filler/ceramic material is shown in Fig. 5. The non-magnetic product, the block made for firing and drying, plasticity test and the pellet for brightness testing are shown from left to right, respectively.

3.4 *Evaluation of Product*

The product is being validated with the standard ceramic property. The detail properties are tabulated in Table 5 which indicates that the product obtained from PLK

Table 5 Validation of material for ceramic property

Details	Product (calcined non-mag product)	Standard ceramic property*
Water plasticity	41.36%	32% (min.)
Shrinkage after drying	0.31%	7% (max.)
Shrinkage after firing	12.01%	14% (max.)
Colour after firing	Perfectly white	Perfectly white
Grit	Nil	2% (max.)
Fe_2O_3	0.70%	1% (max.)

*Standard ceramic property by Bureau of Indian Standard

rocks comparatively showing the best properties for ceramic industrial applications. The plasticity is more than desired and the shrinkage after firing and drying is in the range. Iron content is also less than 1%, no grit is found in the material too. The colour of the sample is also white after firing.

4 Conclusions

The PLK rock is a bauxite mining waste which is generated during mining of bauxite. The rock is calcined in an industrial kiln at 1473°K temperature for two hours. These rocks crushed and separated by using magnetic separator (DHIMS). There is two products are obtained such as magnetic fraction and non-magnetic fraction. The non-magnetic fractions are tested further for its suitability in ceramic industry. The product obtained from calcined non-magnetic product has 70% brightness. This product has low percentage of Fe_2O_3 that is 0.7%. The sample was also characterized and compared with BIS standard property of ceramics. The data obtained from the experiments are in the range and also found as a suitable material for filler/ceramic industries.

Acknowledgments One of the authors Dr. Ranjita Swain is thankful to NALCO for funding.

References

1. Mineral Commodity Summaries. US Geological Survey. (2006).
2. Retrived in 27 August 2019, https://en.wikipedia.org/wiki/National_Aluminium_Company.
3. INDAL Starts Calcined Kaolin Pilot Plant. Industrial Mineral News. World of Minerals. January (1999) 15
4. Waite TD, Torikov A (1987) Photo-assisted dissolution of colloidal iron oxides by thiol-containing compositions: 2. Composition of lepidocrocite (γ-FeOOH) and hematite (α-Fe_2O_3) dissolution. J Colloidal Interface Sci 119(1), 228–235
5. Gulfen G, Gulfen, M, Aydm AO (2006) Dissolution kinetics of iron from diasporic bauxite in hydrochloric acid solution. Ind J Chem Technol 13:386–390
6. Goodarzi MM, Rezaei B, Amini A (2007) Iron removal studies of low grade diasporic bauxite by using thermochemical treatment. Geosciences Spring 16(63):72–77
7. Zivkovic ZD, Strbac ND (1994) Kinetics and mechanism of leaching of low—quality boehmite bauxite with hydrochloric acid. Hydrometallurgy 36(2):77–90
8. Pattanaik N, Das B, Rao RB (1996) Removal of calcium and iron from bauxite for use in refractory industry. ERZMETALL 49(9):555–558
9. Reddy BR, Mishra SK, Banerjee GN (1999) Kinetics of leaching of a gibbsitic bauxite with hydrochloric acid. Hydrometallurgy 51(1):131–138
10. Patermarakis G, Paspaliaris Y (1994) The leaching of iron oxides in boehmitic bauxite by hydrochloric acid. Hydrometallurgy 23(1):77–90
11. Paspaliaris Y, Tsolakis Y (1987) Reaction kinetics for the leaching of iron oxides in diasporic bauxite from the Parnassus—Giona zone (Greece) by hydrochloric acid. Hydrometallurgy 19(2):259–266

12. Mergen A, Bilen M, Gunduz M (2002) Kinetics of leaching of boehmitic bauxite of doğankuzu-konya-turkey wıth hydrochlorıc acid. Silic Ind 67(7–8):75
13. Gajam SY, Raghavan S (1985) A kinetic model for hydrochloric acid leaching of kaolinite. Trans Inst Min Metall C 94:C115–C120
14. Ambikadevi VR, Lalithambika M (2000) Effect of organic acids on ferric iron removal from iron-stained kaolinite. Appl Clay Sci 16:133–145
15. Swain R, Routray S, Mohapatro RN (2018) Optimisation of process parameter using response surface methodology to recover ceramic materials from bauxite mining waste rocks. IOP Conf Ser: Mater Sci Eng 455. https://doi.org/10.1088/1757-899X/455/1/012012
16. Swain R, Routray S, Mohapatra A, Patra BR (2018) Industrial application of valuable materials generated from PLK rock-a bauxite mining waste. IOP Conf Ser: Mater Sci Eng 338:012049. https://doi.org/10.1088/1757-899X/338/1/012049
17. Swain R, Rao RB (2012) Kinetic study on leaching of iron in partially Laterised Khondalite rocks for ceramic industrial applications. Int J Min Process 112–113, 77–83

A Study on Heavy Mineral Distribution Pattern Along Brahmagiri Coast of Odisha, India and Its Beneficiation to Recover Industrial Minerals

Sunita Routray, Ranjita Swain, and R. Bhima Rao

Abstract In this research paper, it is attempted to study the grain size distribution of heavy minerals and their distribution pattern along with chemical characteristics of four samples collected from Bhrahmagiri coast of Odisha. The grain size distribution of as it is samples and heavy minerals are carried out. Size frequency mode of 300 and 212 μm are more significant for as it is samples. Whereas, size analysis of heavy minerals show size frequency mode of 212 μm. All the four samples are found to contain higher percentage of magnetic heavies.

Keywords Beach sand · Heavy minerals · Fluvial sand · Back dune · Frontal dune

1 Introduction

The country India has a coastline of around 6000 km, which hosts some of the largest and richest beach sand placer deposits. The beach and dune sands of this country contain heavy minerals like ilmenite, rutile, garnet, zircon, monazite and sillimanite. The total heavy mineral deposit of the World is about 2500 million tons (Mt), out of which India has about 270 Mt. In India, the resource contains about 463 Mt of ilmenite, 150 Mt of garnet, 29 Mt of zircon, 10 Mt of monazite and 190 Mt of sillimanite. The Indian resources of heavy mineral are: Indian resources constitute about 35% of world resource of ilmenite, 10% of rutile, 14% zircon and 71.4% of monazite [1]. Beach- dune sand deposits have some special features. These are environmentally favourable for mining and recovery of individual heavy minerals for industrial applications. The heavy minerals present in beach and dune sand have much industrial value and are used for different applications. Zircon finds its application in ceramics, refractory industries, foundry, TV glass, zirconia chemicals and other

S. Routray (✉) · R. Swain
C. V. Raman College of Engineering, Bhubaneswar, India
e-mail: sunitaroutray77@gmail.com

R. B. Rao
IMMT, Bhubaneswar, India

© Springer Nature Singapore Pte Ltd. 2021

P. M. Pandey et al. (eds.), *Advances in Production and Industrial Engineering*, Lecture Notes in Mechanical Engineering,
https://doi.org/10.1007/978-981-15-5519-0_22

applications [2]. Ilmenite is mainly used for manufacture of titanium dioxide (a white pigment) which is used in paints, paper, rubber, textile industries, etc. Garnet mineral is used in the manufacture of abrasives, grinding wheels for polishing glass/TV tubes, as sand blasting media, in water filtration, water jet cutting and in anti-skid surfaces. Sillimanite used as a raw material for high alumina refractory and insulators. Monazite is used as a source of thorium and rare earth compounds used in various chemical and electronic industries.

Hence, known resources are always explored for recovery of heavy minerals. The state Odisha has many resources of beach sand heavy minerals [3, 4]. In the present paper the study area chosen is Brahmagiri coast of state Odisha, which contains good source of heavy minerals as reported by AMD [5–8]. In this paper, an attempt is made to assess the occurrence of placer heavy minerals, their distribution, physical, chemical characteristics and recovery of heavy minerals by mineral separator. The mineral separator is widely recognized as a valuable tool of gravity concentration for quick and efficient separation of mineral grains of close specific density [9–11].

2 Study Area

In the present study area, i.e. Brahmagiri coast located in Puri district of the state Odisha, the district covers an area of 3050.8 km^2 and bounded by latitude 19°25′–20°25′N and longitudes 85°00′–86°20′E. The district is surrounded by Jagatsinghpur district in east, Ganjam district in west, Khurda district in north and Bay of Bengal in south. The district is drained by different rivers such as Kushabhadra, Daya, Bhargabi, Prachi, etc. Geologically, a small part in the extreme north of the area is represented by Khondalite gneiss and the remaining part is covered by a thick pile of Quaternary sediments. Department of Atomic Energy confirms the presence of heavy minerals along the Brahmagiri coast. Figure 1 shows the geographical location of the present study area.

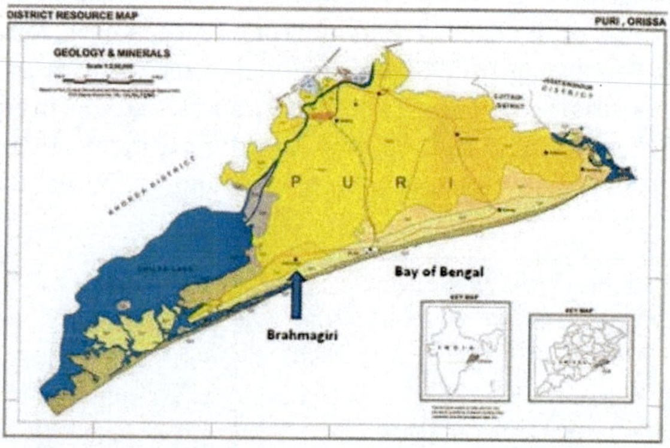

Fig. 1 Location map of Brahmagiri coast [12]

Fig. 2 Concentration of heavy minerals along the coast of Brahmagiri

3 Materials and Methods

Figure 2 shows the concentration of heavy minerals along the coast of Brahmagiri. Samples were collected from different locations along the coast. Representative samples of fluvial sand, beach sand, frontal dune sand and back dune sand samples were physically characterized after washing. Physical characterization such as angle of repose, bulk density, true density, sink-float analysis and size analysis using standard sieves were carried out. The sink portions obtained from sink-float analysis using bromoform was subjected to magnetic separation to get the magnetic and non magnetic heavy minerals. The non magnetic heavies are again subjected to diidomethane to separate non magnetic very heavy and light heavy minerals. Then all samples were sent for chemical analysis to know the elements present in it. Again, a composite sample was prepared from the samples collected. Gravity concentrator such as mineral separator was used to access the response of heavy minerals to recovery process.

4 Results and Discussion

4.1 Textural and Grain Size Distribution

The size frequency modes of fluvial sand, beach sand, frontal dune sand and back dune samples are presented in Fig. 3. It can be noted from the figure that size frequency mode of 300 and 212 μm are more significant for all samples. The bulk density, true density, angle of repose, d_{80} passing size and THM content of all the four samples is presented in Table 1.

The size analysis of the Total Heavy Minerals (THMs) obtained from the above samples through sink–float method is presented in Fig. 4. All samples show the 106 μm peak is significantly followed by either 150 or 212 μm peak.

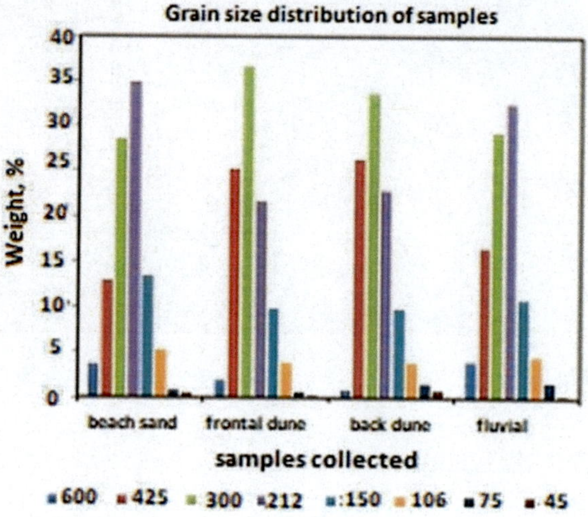

Fig. 3 Size frequency model of samples

Table 1 Physical characterization of all samples

Samples	Bulk density (g/cm^3)	True density (g/cm^3)	Angle of repose (°)	d_{80} passing size (μm)	THM (%)
Beach sand	1.625	2.7	34.41	415	9.75
Fluvial sand	1.78	2.85	30.56	395	10.4
Frontal dune sand	1.725	2.73	30.62	400	13.88
Back dune sand	1.74	2.73	31.26	390	10.6

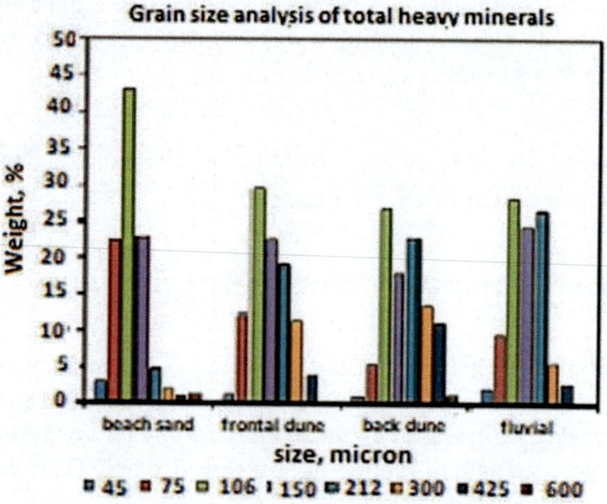

Fig. 4 Size analysis histogram of THMs

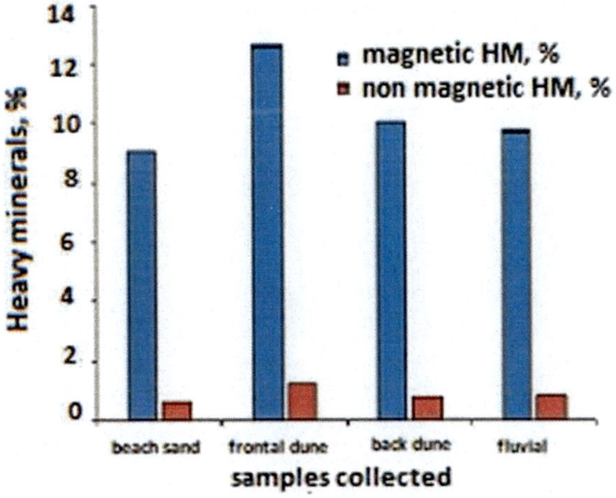

Fig. 5 Magnetic and non magnetic fraction in different samples

The THMs contain magnetic and non magnetic heavy minerals shown in Fig. 5. It can be seen from the figure that the percentage of magnetic minerals is much more than the non magnetic fraction. The frontal dune sand contains highest percentage of magnetic heavies among all samples. Figure 6 shows the fraction of heavy and light minerals in different samples. It can be concluded that more percentage of light minerals present in higher size fraction, whereas higher percentage of heavy minerals present in lower size fraction. It is quite obvious, because light minerals are coarser and heavy minerals are finer in size.

Table 2 presents the typical chemical analysis of sequential sink-float products of different samples. It can be observed from the table the TiO_2 (%) is much higher in magnetic heavy minerals and non magnetic very heavy minerals. It is due to the presence of ilmenite and rutile in the heavies, respectively. Ilmenite is magnetic and having specific gravity of 4.79 gm/cc. Rutile is non magnetic with specific gravity of 4.2 gm/cc. Fe_2O_3 (%) is also higher in all Magnetic heavy minerals due to presence of ilmenite. Al_2O_3 (%) is found to be higher in non magnetic light heavy minerals of all samples, which is due to the presence of sillimanite. Sillimanite is non magnetic and having specific gravity of 3.3 gm/cc is considered as a light heavy mineral. ZrO_2 (%) is found to be higher in non magnetic very heavy minerals as zircon is present in that. Zircon is non magnetic and also having high specific gravity (4.7 gm/cc). In all samples, presence of HfO_2 is very less.

X-ray diffraction data of total heavy minerals, magnetic heavies and non magnetic heavies are given in Fig. 7. The data indicate that the total heavy minerals contain mostly ilmenite followed by minor amounts of sillimanite, zircon, rutile, etc. The magnetic heavies significantly exhibit the presence of ilmenite intensity peak, whereas the non magnetic heavies significantly exhibit the presence of multiple mineral peak intensities such as sillimanite, zircon, rutile and quartz, etc.

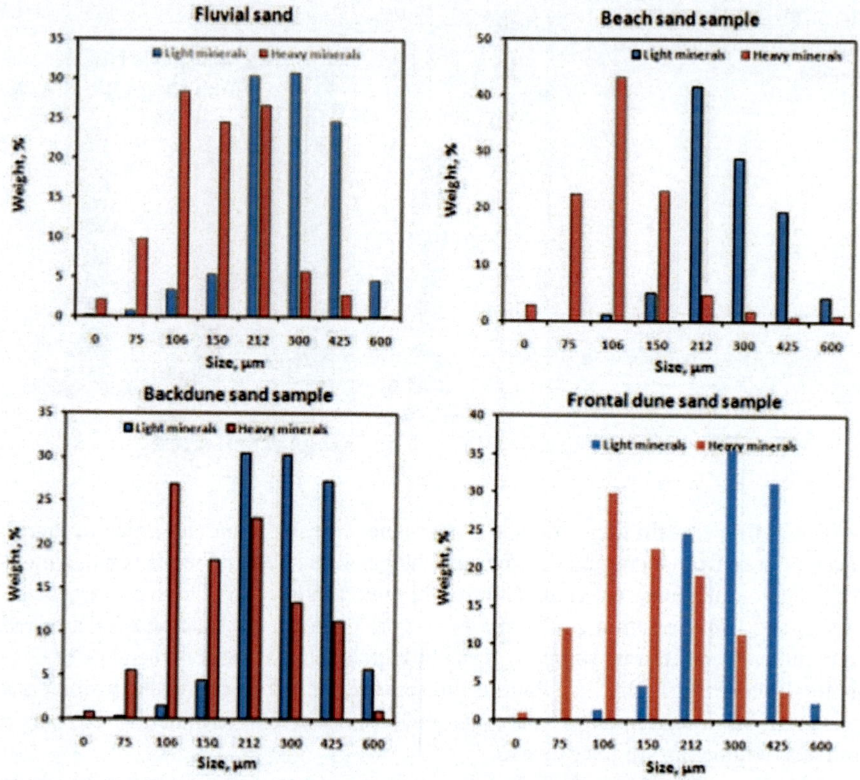

Fig. 6 Heavy and light mineral content in different samples

The composite bulk sample with 12.0% THM is subjected to mineral separator and the complete flowsheet with material balance is presented in Fig. 8. This study is carried out to observe the response of sand samples to gravity separation process. After five stages of operation, the final concentrate obtained shows a yield of 8.6% with grade and recovery of 67.4% and 48.3%, respectively.

The tailings obtained in all stages can be mixed and sent for further processing as it contains total heavy minerals of about 6.2%. Observing the results, it can be concluded that heavy minerals can be recovered from this deposit and scale up study can be done by using spiral concentrator.

5 Conclusions

The experimental results show that, for collected samples, size frequency mode is significant at 300 and 212 µm. Whereas, size analysis of heavy minerals show size frequency mode of 212 µm. It is also observed that, all the four samples contain higher

Table 2 Typical chemical analysis of sequential sink-float products of different samples *

Samples	Heavies	TiO_2 (%)	Fe_2O_3 (%)	Al_2O_3 (%)	SiO_2 (%)	ZrO_2(%)	HfO_2
Beach sand	MHM	38.3	38.4	5.2	10.8	1.3	658.6 ppm
	NMVH	50.6	5.4	4.2	11.3	22.1	0.654%
	NMLH	1.2	1.3	51.9	39.9	0.6	136.8 ppm
Frontal dune sand	MHM	35.3	36.8	5.6	14.7	0.7	434 ppm
	NMVH	47.8	4.2	3.1	11.9	24.7	0.715%
	NMLH	0.9	1.3	51.0	40.6	0.6	137.9 ppm
Back dune sand	MHM	31.9	37.1	7.1	17.2	0.8	403.2 ppm
	NMVH	26.7	3.4	1.5	5.5	10	0.3%
	NMLH	1.4	1.5	50.5	40.1	0.5	117.3 ppm
Fluvial sand	MHM	39.3	37.4	4.0	12.4	0.85	511.2 ppm
	NMVH	45.1	11.1	4.1	10.0	22.2	0.678%
	NMLH	1.2	1.5	50.5	40.2	0.645	138.5 ppm

*MHM: Magnetic heavy minerals, NMVH: Non magnetic very heavy mineral, NMLH: Non magnetic light heavy

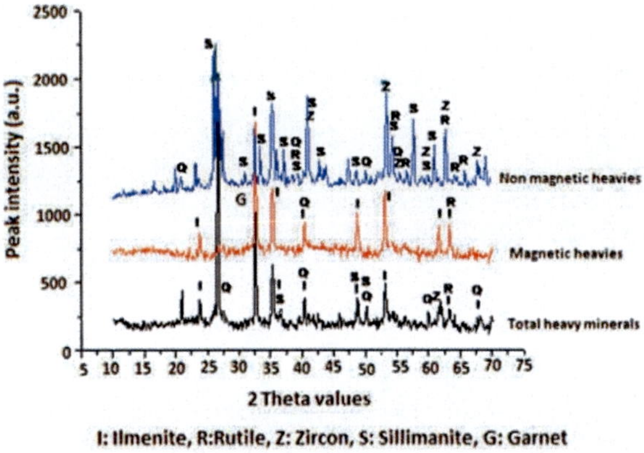

Fig. 7 X-ray diffraction data on feed heavies, magnetic heavies and non magnetic heavies

percentage of magnetic heavies. Chemical analysis of heavies shows that the heavy minerals are of suitable grade for industrial applications. The mineral separator study shows the samples collected from this coast are responding to the gravity separation process.

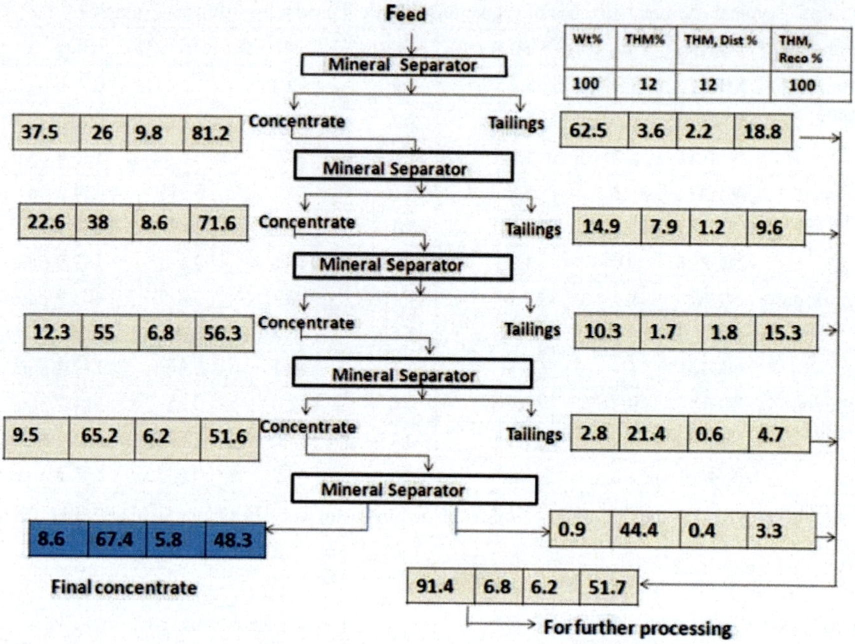

Fig. 8 Flowsheet for recovery of total heavy minerals

Acknowledgements Authors are thankful to SERB, Government of India for providing financial support to carry out the research work.

References

1. DAE Homepage (2011) Atomic minerals directorate for exploration and research. https://www.dae.gov.in. Accessed 07 Nov 2011
2. Murthy MVN (1969) Zircons. Miscellaneous Publication 9 (Geological Survey of India, Calcutta), 246.
3. Behera P (2003) Heavy minerals in beach sands of Gopalpur and Paradeep along Orissa coast line, East coast of India. Indian J Mar Sci 32(2):172–174
4. Acharya BC, Nayak BK, Das SK (2009) Heavy mineral placer sand deposits of Kontiagarh area, Ganjam district, Orissa, India. Resour Geol 59(4):388–399
5. Azam AM, Krishnan S, Banerjee DC (2001) Beach and inland heavy mineral sand investigations and deposits in India: an overview. Special Issue on Beach and Inland Heavy Mineral Sand Deposits of India. Explor Res Atom Miner 13:1–21
6. Jagannadha Rao K, Subramanyam AV, Kumar A, Sunil TC, Chaturvedi AK (2008) Discovery of heavy mineral-richsand dunes along the Orissa: Bengal coast of India using remote sensing techniques. J. Curr. Sci 94: 983–985
7. IMYB (2012) Indian Minerals Year Book. Indian Bureau of Mines. Chapter 27 (Ilmenite & Rutile)

8. IMYB (2013) Indian Minerals Year Book. Indian Bureau of Mines. Chapter 27 (Ilmenite & Rutile)
9. Routray S, Ranjita S, Bhimarao R (2017) Optimization of mineral separator for recovery of total heavy minerals of Bay of Bengal using central composite design. J Inst Eng India Ser D 98(1):119–130
10. Laxmi T, Rao RB (2011) Gravity concentration kinetics of red sediment placer minerals. AT Miner Process 52:35–49
11. Satpathy C, Routray S, Rao DS (2010) Heavy mineral recovery from beach and dune sands of Ganjam Coast, Orissa, India. World Metall ERZMETALL 63:5–13
12. District Resource Map (2009) Puri, Orissa. Published under the direction of DG, GSI, Kolkata

Webometric Study of Lean Manufacturing

Kritika Karwasra, Devesh Kumar, Gunjan Soni, and Surya Prakash

Abstract This paper contains info metric study of lean manufacturing. With the objective of reducing the cost TPS (Toyota Production System) came with the new practice called Lean Manufacturing. The goal of lean manufacturing is to eliminate the non-value added activities by reducing seven types of waste i.e. Overproduction, Over-processing, Rework, Inventory, Wait, Transportation, and Motion. In this paper summarization is done on the statistical data collected. The data has been collected over the years.and classified into various categories on the basis of their source of origin. The various sources of origin of the publications related to Lean Manufacturing are papers by various authors, publications through various international journals, publications among various fields of research and engineering, publications from various countries, and publications from various conferences. By using this data, webometric has been used to explore the impact of the field of Lean Manufacturing, the impact of set of researchers that has been publishing over the years.

Keywords Lean manufacturing · Webometric

K. Karwasra (✉) · D. Kumar · G. Soni
Mechanical Engineering, Malaviya National Institute of Technology, Jaipur, Rajasthan, India
e-mail: mekritika28@gmail.com

D. Kumar
e-mail: deveshkumar1993@gmail.com

G. Soni
e-mail: gsoni.mech@mnit.ac.in

S. Prakash
BML Munjal University, Haryana, India
e-mail: suryayadav8383@gmail.com

© Springer Nature Singapore Pte Ltd. 2021

P. M. Pandey et al. (eds.), *Advances in Production and Industrial Engineering*,
Lecture Notes in Mechanical Engineering,
https://doi.org/10.1007/978-981-15-5519-0_23

1 Introduction

Lean manufacturing is a combination of the craft production system, in which highly skilled manufacturers manufacture a product according to specifications given by the customer and mass production system, in which a large number of products are manufactured on an assembly line [1]. The goal of lean manufacturing is to reduce the Muda (waste) in labor effort, inventory, time to market and manufacturing space to become highly responsive to customer demand while producing world-class quality products in the most efficient and economical manner. Lean manufacturing is not a set of a processor to follow; it is continuous improvement to reduce waste [2]. Lean focuses on eliminating non-value added activities and waste (muda) in industry. The seven wastes are: motion, overproduction, overprocessing, lead time, rework, inventory, and defects. In addition, two more types of waste have appeared in literature that is underutilization of people's creativity and environmental waste. The major tools used earlier for lean manufacturing was (i) Value Stream Mapping [3], (ii) Kanban [4], (iii) Just in time [5], (iv) Kaizen [6], (v) Cellular manufacturing [7], (vi) Single Minute Exchange Die [8], (vii) 5S [9], etc.

With time it was observed, lean still contains some challenges such as fundamental shift required in an organization's culture. Therefore the concept of merging lean with six sigma arrived. Six Sigma is a methodology that improves business processes by using statistical analysis. It is a data-driven and highly disciplined approach that eliminates the number of defects in an organization by using DMAIC (Define, Measure, Analyze, Improve, and Control) approach. So, integration of these two approaches gives the organization more efficiency and affectivity and helps to achieve superior performance faster than the implementation of each approach in isolation [10].

Lean Six Sigma is defined as a business strategy and methodology that boosts process performance, and develops customer satisfaction, leadership, and bottom line results by improving quality, speed, and costs [11].

Now, with the rise of environment and social consciousness, the definition of lean has been expanded to incorporate concepts of economic, social, and environmental sustainability. The USA's Environmental protection Agency defined aims of lean as: Develop the quality products, at the lowest cost with the shortest lead times by systematically and continuously eliminating waste, while respecting people and the environment. In order to respond to the growth of customer demands for products and services that are environmentally sustainable and comply with governmental environmental regulations, companies have now been forced to rethink these objectives and hence how they manage their operations and processes.

The green paradigm is operationalized through green initiatives [12], supported by methods and tools that include environmental operations management also known as green operations [13], green manufacturing, green supply chains, reverse logistics, eco-design, design for environment or green building, sustainable value stream mapping, and life cycle assessment.

In general the enterprises, nowadays, are driven by market demand. The presence of fierce competition, ever-expanding business with rapid pace and time schedules that are supposed to be kept up to the mark and compressed are the factors by which market demand is characterized these days. On the other hand, the production cycles in manufacturing are being shortened with reduction batch size with depending on the need of the hour. The dynamic nature of product demand and its customization is also inevitable and experienced frequently by the organizations. This also poses a challenge for leading organizations in industrial sector which in turn creates a need to improve optimization and process efficiency. Hence in order to maintain and improve competitive advantages, the possible initiative which should be undertaken by the companies is Lean Manufacturing. The term lean manufacturing was introduced by Womack to address the working methodology implemented by Japanese companies, particularly Toyota. This methodology relies on the elimination of wastes and nonproductive processes, in order to focus on value-added operations and produce high-quality products, at the customers demand pace, with ideally no waste [14]. Manufacturers around the world have always been inclined toward enhancing their competitiveness. The idea and the concept behind lean manufacturing is the best way to achieve their aspirations. With the methodology behind lean manufacturing, it would not be wrong to believe that this process will also result in waste control and regulation of essential processes related to production [15].

Lean manufacturing brought an industrial revolution in world that's why; a statistical analysis is done to understand it more intricately. This paper presents a bibliometric analysis of lean manufacturing. Historical data over the years is collected from Scopus and has been classified into various categories on the basis of their various authors, various international journals, fields of research, and countries.

2 Statistical Analysis

2.1 Statistics by year

With the keyword of "Lean Manufacturing" 3950 research publications were found from Scopus database (accessed on May 6, 2019). The general direction in which lean manufacturing is developing is shown in Fig. 1. As can be seen from Fig. 1, the first research article was published in 1991 as the concept of lean manufacturing came in the late 1980s. The number of research papers increase from 1995 to 19 then falls to 10 in 1996, then tremendous increase in publications was observed from 42 in 2001 to 227 in 2005. After this, the number of publications was decreased and remained to an average of 197 till 2012. Then there was a sudden rise of 255 publications in the year 2013 and kept increasing till the year 2018 with the maximum publication of 378. Then there was a huge fall down in the year 2019, in which 86 publications happened. The trend in Fig. 1 shows that the number of publications will decrease to 1 in 2020, as other manufacturing practices are arising like genetic algorithm, etc.

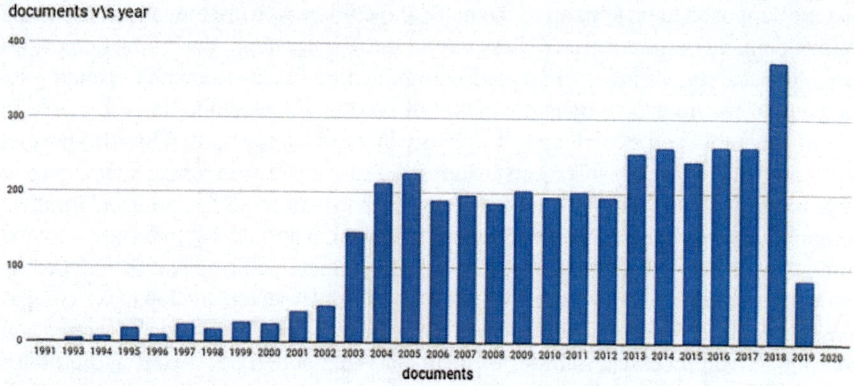

Fig. 1 Number of publications by year

2.2 *Statistics by Field*

Publications discussed in the previous section are majorly from "Engineering," "Business, management and accounting, and Computer Science." Total 2694 documents are in Engineering field, 1257 in Business, management and accounting, and 767 in Computer science, as seen from Fig. 2. Sum of these publications exceeds the total number of publications in previous section as one publication can be designated in more than one field. All three fields have maximum number of publications in the year 2018, i.e., 254 in Engineering field, 171 in Business, management and accounting,

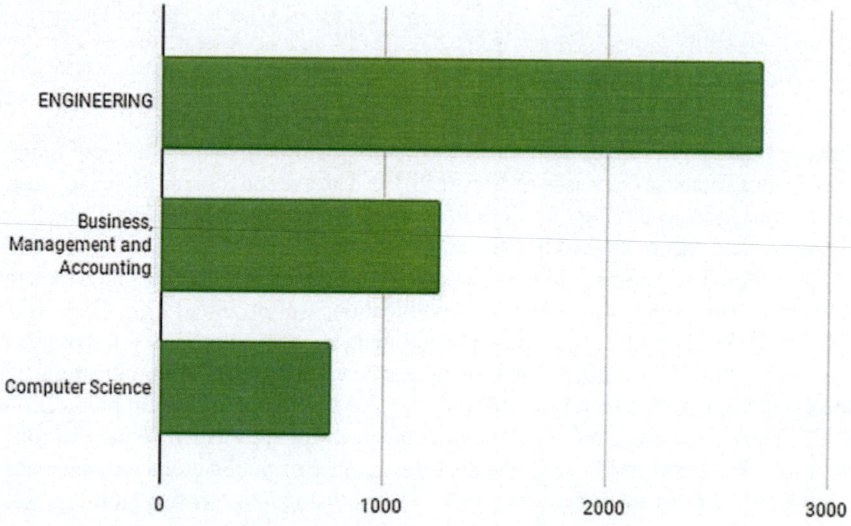

Fig. 2 Number of paper in top three field

and 119 in Computer science. As seen from Figs. 3, 4, and 5 the fashion of publication is the same in all three fields. The first publication under Engineering field and

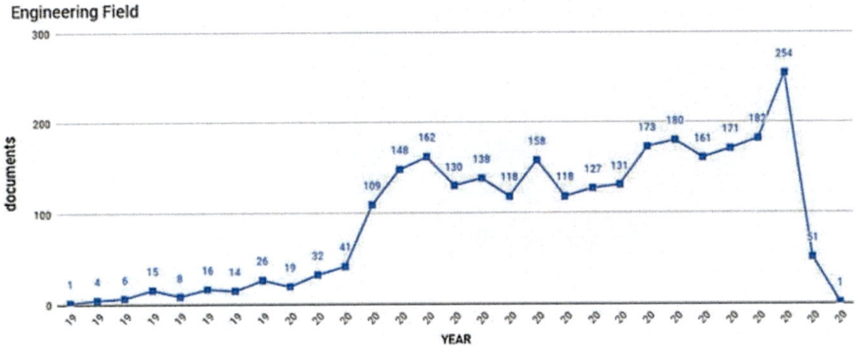

Fig. 3 Distribution of publications by year of engineering field

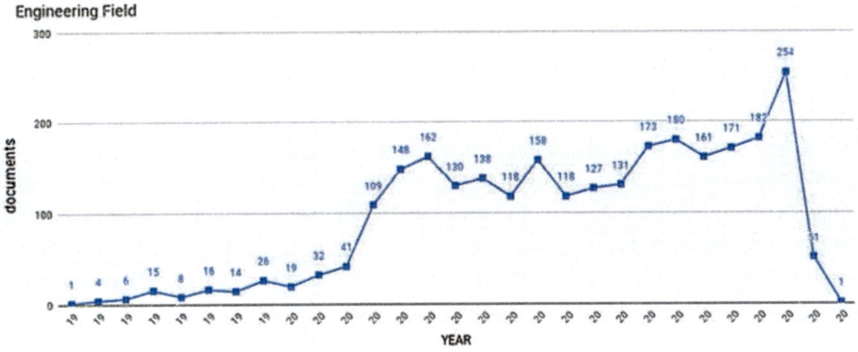

Fig. 4 Distribution of publications by year of business, management and accounting field

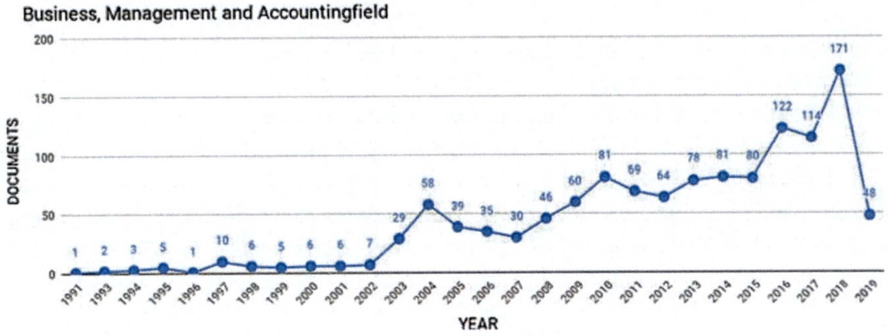

Fig. 5 Distribution of publications by year of computer science field

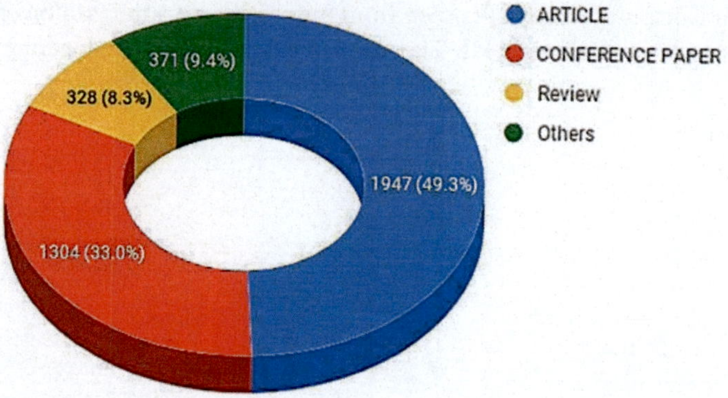

Fig. 6 Number of publications in different document type in lean manufacturing

Business, management and accounting field was in 1991 and the first publication under Computer science field was in 1994.

2.3 *Statistics by Document Type*

Majority of publications of Lean manufacturing are articles, conference papers, and review papers with 1947 publications are articles, 1304 are conference papers, and 328 are research papers. Figure 6 shows the distribution of different papers in each category. Another category consists of all other document types, i.e., book, book chapter, note, report. As shown in Fig. 6 around 91% papers are of articles, conference papers, and review type and rest 9% belongs to the other category.

2.4 *Statistics by Country*

Top three countries with the highest number of publications are the United States, India, and the United Kingdom with 1109, 401, and 233 publications, respectively. Figure 7 shows the year-wise distribution of publications in each country. The first paper was published in 1991 in the United Kingdom, while it was published in United States in 1993 and in India in 2004, still the United Kingdom has less number of publications as compared to India and United States. Around 44% of all papers are published by these countries. Trend of each country is seen in Figs. 8, 9, and 10, United States has maximum number of publications in 2006, i.e., 90 India and United Kingdom has maximum number of publications in the year 2018 that are 82 and 24, respectively. The number of publications continuously increased from 1991 to 2006 and then there were on average 73 papers per year till 2012 but then decreased continuously from 2013 to 2019 in the United States. In India, trend was

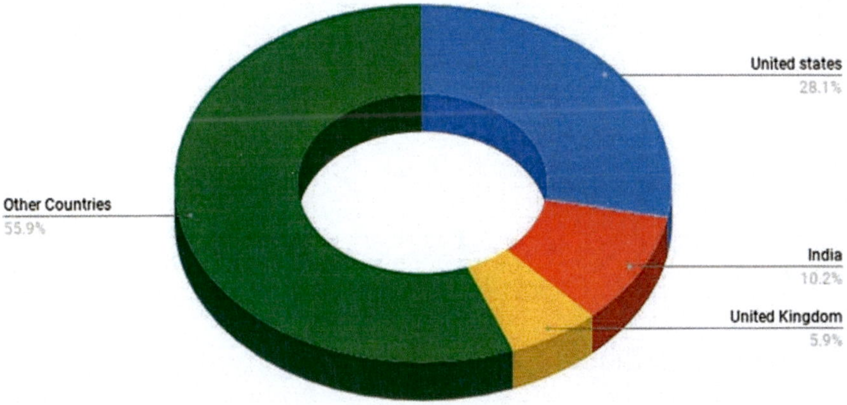

Fig. 7 Number of publications in top three country in lean manufacturing

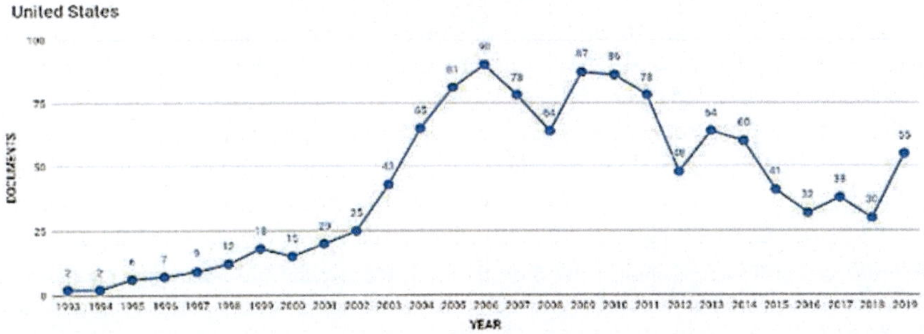

Fig. 8 Distribution of publication in United States by year

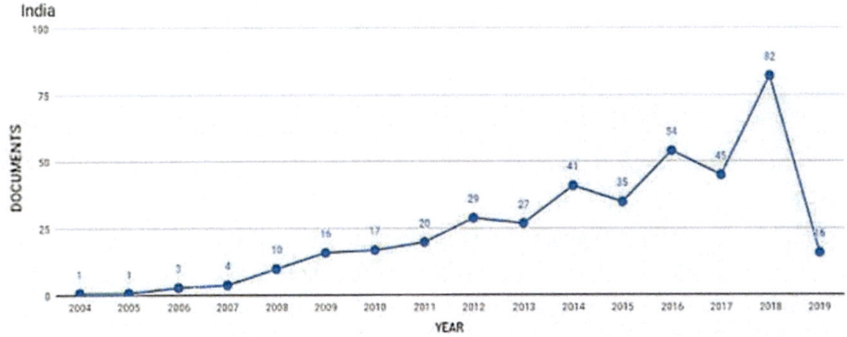

Fig. 9 Distribution of publication in India by year

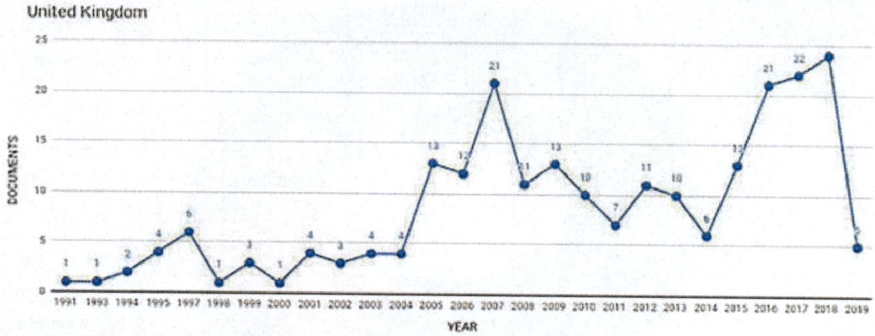

Fig. 10 Distribution of publication in United Kingdom by year

different it continuously increased from 2004 to 2012 then remained around 40 per year till 2017 then it increased to 82 in 2018 and decreased to 18 in 2019. The United Kingdom has a very less number of publications, it has on average 3 publications per year from 1991 to 2004, and then this number increased to 12 publications per year from 2005 to 2015, then increased to 24 in 2018 and decreased to 5 in 2019.

It has been observed that India does less work in Lean or India lack in Lean practices. Individual barriers of Lean Manufacturing implementation has been identified as lack of management focus, lack of urge to create sense of urgency, lack of management support, lack of long-term vision, lack of labor resources, lack of capital fund, lack of communication, lack of idea innovation, mediocre consultants, lack of time, lack of training, lack of understanding about Lean, lack of implementation know-how, conflicts with other Initiatives like TQP, TPM, JIT, disparate manufacturing environments, demand volatility, conflicts with ERP Implementations, company culture, employee's resistant to change, middle management resistance, no direct financial advantage, not recognizing financial benefits, no financial targets, past experience of failure and sliding back to previous state in the absence of staying power [16].

2.5 *Statistics by Journal*

Top 3 journals with the highest number of papers are International Journal of Six Sigma Lean, International journal of Production Research, and International Journal of Advanced Manufacturing Technology with 48, 85, and 39 publications, respectively as shown in Fig. 11. International journals have been a solid medium to convey research to the community of scientists, researchers, and learned people. Thus medium has also proved its importance to the lean manufacturing. Past years data show that Lean Manufacturing's first-ever paper in International Journal of Six Sigma Lean was published in 2010. Since 2010, every year International Journal of Six Sigma Lean have been publishing in this topic with an increase of 7 in 2016 from just 1 in 2010. The year 2018 experienced the highest number of research

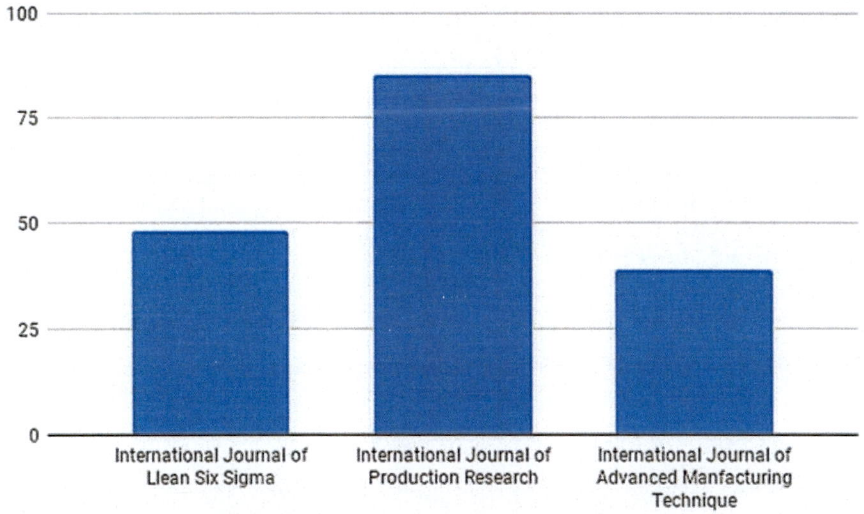

Fig. 11 Number of publications in top three journal in lean manufacturing

papers published which is 16 as shown in Fig. 12. In case of International Journal of Production Research, it has been a warm topic of interest for researchers as the publications are being published since 1991. The number has also increased in subsequent years with highest number of publications in 2017, i.e., 10. From 2001 to 2007, the average number of publications of international journals production research was 3. In the following years from 2007, trends show that the average increased from 3 to 7 as shown in Fig. 13. In case of International Journal of Advanced Manufacturing, the first publication was published in 2006 which was only 1 publication that year. According to past data, only 1 journal every year was published till 2008. There has been increased number of publications since 2008 as shown in Fig. 14. In the present year of 2019 there has been only 1 publication in Advance Manufacturing while

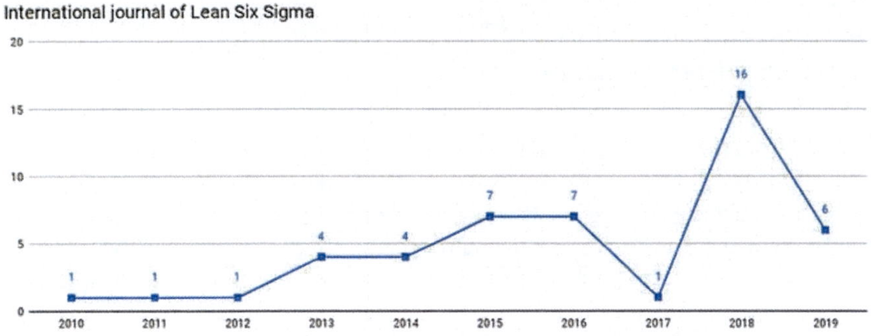

Fig. 12 Distribution of publications in international journal of lean six sigma by year

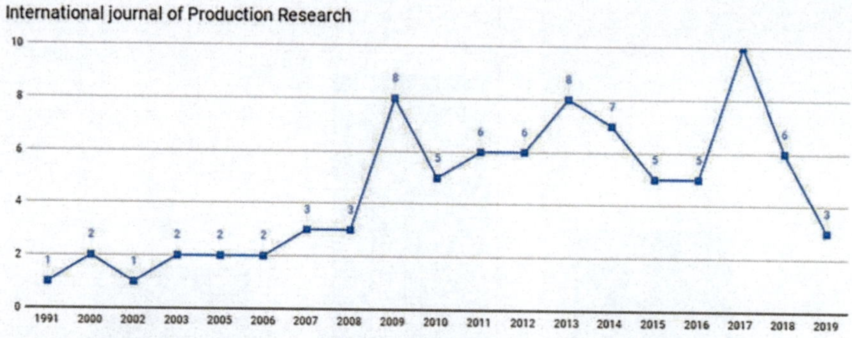

Fig. 13 Distribution of publications in international journal of production research by year

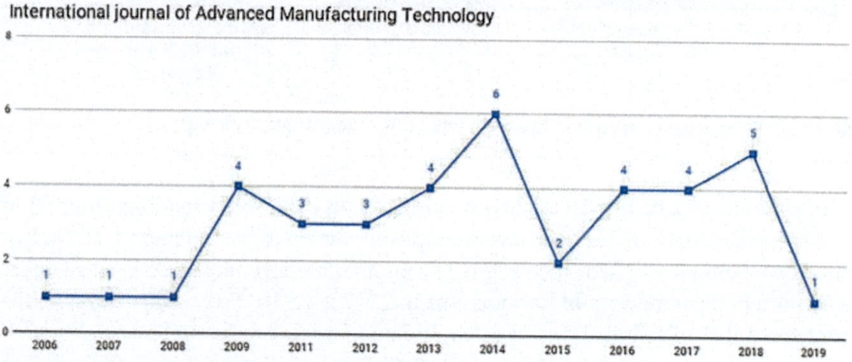

Fig. 14 Distribution of publications in international journal of advanced manufacturing technology

in Lean Six Sigma and Production Research there have been 6 and 3 publications, respectively.

2.6 *Statistics on the Basis of Various Publications from Different Authors*

Statistics from the past shows that three authors who have been forefathers of major publications from time to time are A. Weber, R. Kodali, and Vinod S as shown in Fig. 15. As shown from Fig. 17 it is observed that, the first-ever publication from Vinod S came out in 2008 while R. Kodali wrote his first publication in 2007. Out of these three, the first-ever publication was written by A. Weber in 2003. Data shows that he wrote 2 other publications as well in that same year. From that year all the three authors have been frequently active in publishing their works from time to time every year. The maximum publications by anyone of these authors in a particular year are 8 which are published by S. Vinod in 2008. In year 2013 there was no publication by

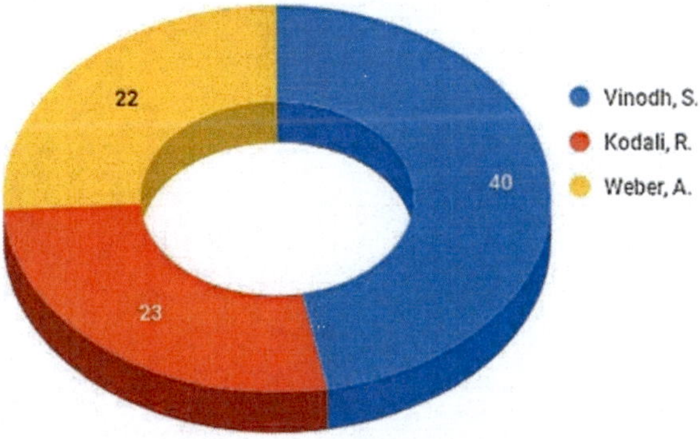

Fig. 15 Number of publications by top three author in lean manufacturing

R. Kodali. Similarly, after publishing 5 publications in year 2005, author A. Weber came out with his next publication in 2013. He has been continuously active in qualifications since then. In the present year, i.e., 2019, the only publication till now is 1 which is the work of author Vinod S.

3 Results and Discussion

Lean manufacturing is an organized approach introduced by Japanese after World War II to reduce waste in Industries. The first research document was published in 1991 as seen in Fig. 1, a total of 3950 papers are published till May 6, 2019. Most of the publications are in "Engineering" field and in "Business management and accounting" field, as seen from Fig. 2. In addition, majority of publications are of article type and conference paper type as can be seen from Fig. 6. The United States has published maximum number of documents in this field, i.e., 1109 since 1991. India and the United Kingdom are the other two countries with top number of publications, as seen from Fig. 7, and also the first document was published in the United Kingdom as seen from Fig. 16. The trend shows that there were less than 100 papers from 1991 to 2002, after this year number of documents was increased to an average of 200 papers per year till 2017 as seen from Fig. 1.

Journal with highest number of publications are International Journal of Six Sigma Lean (48), International Journal of Production Research (85), and International Journal of Advanced Manufacturing Technology (39) as seen from Fig. 11. In the end, top discussion is done between the top 3 authors, as shown in Fig. 17. Author Vinodh S. publishes his first research work in 2008 and from then he publishes his work every year in this field. Author Kodali R. published his first document in 2007 and published continuously every year till 2016, after 2016 there is no publication by him. Instead author Weber A. published his first publication in 2003 and published every year till 2005, after 2005 he publishes his next document in the year 2013.

Statistical analysis by document type is discussed in which top 3 document types are article, review paper, and conference paper. Year-wise distribution of all three data

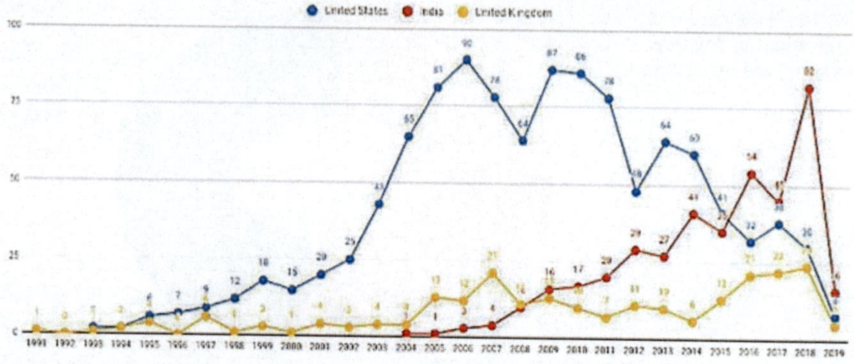

Fig. 16 Number of publications of lean manufacturing by top three countries

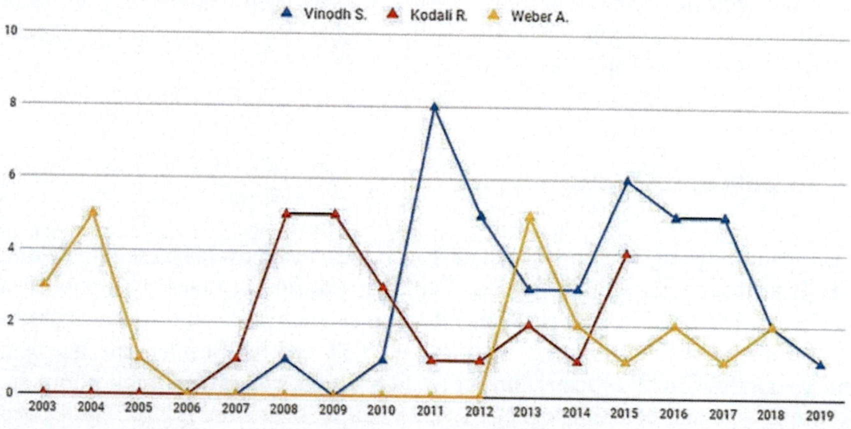

Fig. 17 Number of publications of lean manufacturing by top authors

types is shown in Fig. 18. The first publication of article was in 1991, of conference paper was in 1993, and of review paper was in 1997. As seen from Fig. 18 article has the highest rate of publishing as compared to review and conference paper. Article and conference paper have the same trend but conference paper has different pattern as seen from Fig. 18. The number of article publications and conference paper publications was increased from 1993 to 2018, but review paper publications had different trend it had maximum papers in 2005 and after that there was only an average of 7 papers per year. No document type has more than 200 papers.

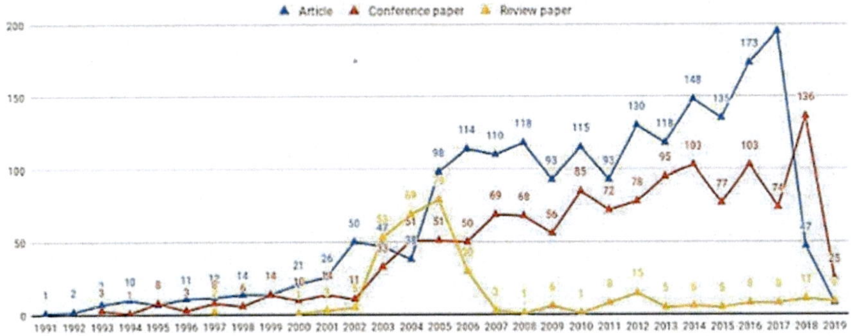

Fig. 18 Number of publications of lean manufacturing in top three document type

4 Conclusion

This paper informetric study of lean manufacturing. Historical data over the years is collected from Scopus and has been classified into various categories on the basis of their various authors, various international journals, and fields of research and engineering, countries, and from various conferences. There are a total of 3950 papers from year 1980 to 2019. We figure out that the maximum number of publications was in 2018, i.e., 378, but it dropped down to 86 in 2019 till May. The top three authors with maximum number of publications are Vinodh S., Kodali R., and Weber A. with 40, 23, and 22 number of publications simultaneously. Engineering field has the highest number of publications, i.e., 2694 and the other top two fields Business Management and Accounting has 1257 and Computer Science has 767 publications. Major publications of lean manufacturing are of article type, around 49.3% of publications are of article type, 33% are conference paper, and 8.3 % of them are of review paper type. Article and Conference paper has the majority of publications in 2018 but then review paper has the highest publication in 2005, in addition there is no more than 80 review paper in any year, instead article and conference paper has some years with more than 100 publications. The United States has the highest number of research document till May 2019. Fraction of publication of top 3 countries is, 28.1% from the United States, 10.2 % from India, and 5.9 % of documents from the United Kingdom. The United States publishes on an average of 74 documents from 2004 to 2013 per year, but India and the United Kingdom have the majority of documents published in 2018. The first document in India was published in 2004 however United kingdom had its first publication in 1991 and the United States had in 1993. The First document was published in the International Journal of Production Research in 1991 and this journal published maximum number of documents, i.e., 85. In 2006 International Journal of Advanced Manufacturing Technology published its first certificate and has 39 publications. The International Journal of Lean Six Sigma published its first publication in 2010 and has 48 documents.

In future more expanded study should be done on the basis of different tools used in lean manufacturing.

References

1. Womack JP, Jones DT, Roos D (1990) The machine that changed the world. Macmillan, New York
2. Pavnaskar SJ, Gershenson JK, Jambekar AB (2003) Classification scheme for lean manufacturing tools. Int J Prod Res 41(13):3075–3090
3. Rahani AR, al-Ashraf M (2012) Production flow analysis through value stream mapping: a lean manufacturing process case study. Procedia Eng. 41:1727–1734
4. Rahmana NAA, Sharifb SM, Esac MM (2013) Lean manufacturing case study with Kanban system implementation. Procedia Econ. Finance 7:174–180
5. Goyal SK, Deshmukh SG (1992) A critique of the literature on just-in-time manufacturing. Int J Oper Prod Manag 12(1):18–28
6. Modarress B, Ansari A, Lockwood DL (2007) Kaizen costing for lean manufacturing: a case study, pp 1751–1760
7. Pattanaik LN, Sharma BP (2009) Implementing lean manufacturing with cellular layout: a case study. Int J Adv Manuf Technol 42(7–8):772–779
8. Cakmakci M (2009) Process improvement: performance analysis of the setup time reduction-SMED in the automobile industry. Int J Adv Manuf Technol 41(1–2):168–179
9. Bayo-Moriones A, Bello-Pintado A, Merino-Díaz de Cerio J (2010) 5S use in manufacturing plants: contextual factors and impact on operating performance. Int J Qual Reliab Manag 27(2):217–230
10. Salah S, Rahim A, Carretero J (2010) The integration of Six Sigma and Lean management. Int J Lean Six Sigma 1(3):249–274
11. Snee RD (2010) Lean Six Sigma-getting better all the time. Int J Lean Six Sigma 1:9–29
12. Digalwar AK, Tagalpallewar AR, Sunnapwas VK (2013) Green manufacturing performance measures: an empirical investigation from India manufacturing industries. Meas Bus Excell 4:59–75
13. Nunes B, Bennett D (2010) Green operations initiatives in the automotive industry: an environmental reports analysis and benchmarking study. Benchmark Int J 3:396–420
14. D'Antonioa G, Bedollaa JS, Chiabert P (2017) A novel methodology to integrate manufacturing execution systems with the lean manufacturing approach. Procedia Manuf 11:2243–2251
15. Kumar M, Vaishya R, Parag (2018) Real-time monitoring system to lean manufacturing. Procedia Manuf 20:35–140
16. Kumar R, Kumar V (2014) Barriers in implementation of lean manufacturing system in Indian industry: a survey. Int J Latest Trends Eng Technol 4(2)

Investigation of the Criticality of Flux Leakage of a Magnetizer Assembly of a MFL Tool for Oil and Gas Pipelines Inspection

Surya Prakash, Bhuvanesh Kumar Sharma, Chandra Prakash, Shubham Saini, Vikas, and Ankur Sharma

Abstract Almost all pipelines need constant monitoring and frequent inspections. This paper presents the design of a magnetizer assembly and experimental simulation investigation of flux leakage. The simulations carried out in a well known multiphysics software to get insights for the designed magnetizer assembly based MFL tool. Defects of different size and depth are modeled and simulated so that a better design aspect of the magnetizer assembly can be conceived. Several simulations were carried out, and magnetic field density plots of the pipe cross-section were obtained. It was observed that as the thickness of the pipe is decreasing the magnetic flux density in that region increases. The center of the pipe, where the metal erosion is maximum, the flux density was also found to be increased and gradually decreases as we move toward the edges of the metal erosion. These findings helped in selecting the second chamber of the tool consisting of a magnetic section and optimal magnets arrangement for a given material of pipeline. The findings will help in selecting the chamber of the tool consisting of a magnetic section having magnets that are arranged in a crisscross pattern to provide stability to the tool.

Keywords Magnetic flux leakage · Inline inspection · Oil and gas · MFL

S. Prakash (✉) · S. Saini · Vikas
Department of Mechanical Engineering, BML Munjal University, Gurgaon, India
e-mail: suryayadav8383@gmail.com

B. K. Sharma · A. Sharma
VDT Pipeline Integrity Solutions Pvt. Ltd., IIM Lucknow-Noida Campus, Noida, Uttar Pradesh, India

C. Prakash
Department of Computer Engineering & Applications, GLA University, Mathura, Uttar Pradesh, India

© Springer Nature Singapore Pte Ltd. 2021

P. M. Pandey et al. (eds.), *Advances in Production and Industrial Engineering*, Lecture Notes in Mechanical Engineering,
https://doi.org/10.1007/978-981-15-5519-0_24

1 Introduction

Pipelines are one of the key infrastructures of any nation as they are the most economical and useful mode of transporting oil, gas, and other petrochemical products. The hydrocarbon products that are transported in pipelines are vital for running everyday life and industries. Any unwanted event or accident in these infrastructures may cause great damages to life and assets. The buried pipes can pass through anywhere such as crowded city, desert, or even a water body. They are cost-effective, convenient, and are environmental friendly. Pipelines have a lot more advantages than other means, but at the same time, it needs huge capital investment and regular maintenance, monitoring, and inspection to prevent the risk. These risks involve degradation or crack which may cause hazardous damage, accident, or danger to life and environment. Hence, pipelines need constant monitoring, frequent inspections, and regular adherence to the norms.

There are numerous methods available for monitoring and inspections of pipelines such as nondestructive testing, e.g., X-rays, magnetic particle test, eddy current test, magnetic flux leakage (MFL) [1] method and ultrasonic. In this paper, the design of a magnetizer assembly of an MFL geometry inspection (GI) tool for in-line inspection of oil and gas pipelines is presented and flux leakage behavior is analyzed.

The magnetizer assembly is a vital component of an MFL tool and we have designed a kit for experimenting the behavior of flux leakage under certain conditions. The scope of the paper is limited to design a magnetizer assembly and test the experimental abilities through simulation of flux leaks. The simulations carried out in COMSOL multiphysics software to get insights for its proposed magnetizer assembly of the MFL tool. Defects of different size and depth are modeled and simulated so that a better design aspect of the magnetizer assembly can be conceived.

2 Literature Review

Broadly, the pipeline integrity management deals with the concept of failure prevention, inspection, and care [2]. Kishawy and Gabbar have [2] reviewed the pipeline integrity management practices in general. In their study, the authors have given significant attention to designing phase of inspection, maintenance, and monitoring. Some other excellent review studies are [2–4] which provide in-depth information about research on intelligent pipeline MFL detection. Corrosion is also a major defect. Tawancy et al. [5] provided the analysis of corroded segments of pipes. Feng et al. [3] provide excellent theory and application summary of in-line inspection methods for oil and gas pipelines. One more study by Usarek and Warnke [6] did some experimental study on inspection of gas pipelines using MFL method. It was also observed that debris also affect the defect detection. The debris in pipeline may have detrimental effect on MFL data which was studied by Yang et al. [7]. To improve the capabilities of MFL tools, the effects of apparatus parameters, like magnet height, concentrator, backing iron, etc., are chosen through OED method was there in the study of Song [8]. A very interesting study by Narang et al. [9] presented the simulation of defects and analyzed the quantification of the dimension of defect. It seems that there is a rich literature related to MFL investigation in oil and gas pipeline sector. However, there is a need to investigate the multiphysics aspects of various pipeline materials. In the present study, the simulations are carried out in COMSOL multiphysics software to get insights for its proposed magnetizer assembly of the MFL tool. Defects of different size and depth are modeled and simulated so that a better design aspect of the magnetizer assembly can be conceived. The next section provides the theory of the pipelines and major concepts required for MFL applications in tool design or criticality related to magnetizer assembly.

2.1 *Pipeline Integrity Design*

In general, the design of pipes is mostly done based on utility. The pipe systems can be broadly classified into two following categories: *high-pressure pipes* that are used for transporting fiery fluids like oil/gases. The precise estimate of the failure in such pipelines is still a key issue [10]. There are many issues such as, the effect of temperature change, thermal expansion/contraction, mechanical forces, etc. which make pipes prone to failure. On the other side, the *low-pressure pipes* focus on the rigidity of pipe, soil type, and pipe's interaction with soil if they are underneath. In these aspects, the designing is more intensive on external load rather than the internal, and the weight of the earth becomes the most critical design factor in such condition.

Material Selection: The pipe material is selected based on utility and type of fluid which they carry. The common standards are laid down by American Petroleum Inst. (API), American Soc. for Testing and Materials (ASTM), ASME, and ANSI. For example, B16.5 steel pipe flanges have seven pressure classes as per the ANSI standard [11]. It is advised that the user should follow these standards. Proper care should be taken while designing for allowable stress, hoop stress, axial stress calculation. These parameters affect the pipeline inspection before the operation or during maintenance.

2.2 *Principle of Flux Leakage and Detection*

Some examples of the inspection tools are geometry inspection tools (multichannel), leakage detection, magnetic flux leakage (MFL), ultrasonic tools, EMAT, etc. The MFL tool deals with the inspection of pipeline defects and works on the principle of magnetic flux leakage [12]. On the areas where there are corrosion, metal loss, dent, porosity, pinhole, and cracks, there can be a variation in magnetic field lines under a given magnetic field that causes the leakage of flux. In a typical MFL tool, a magnetic detector is placed between the poles of the magnet to detect the leaked

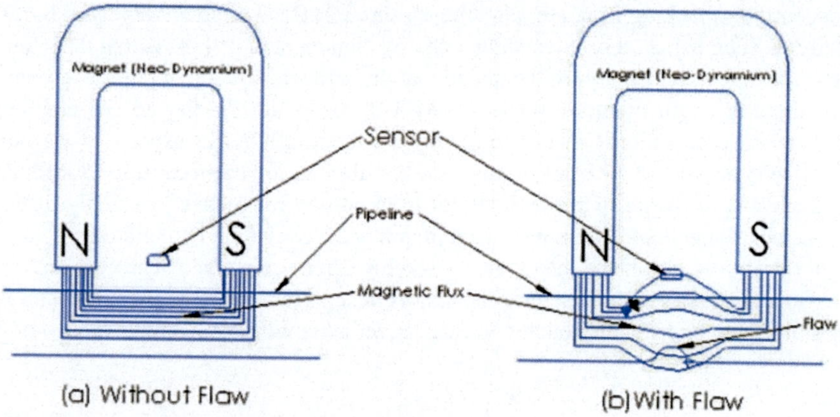

Fig. 1 Principle of flux leakage and detection through magnets

field, which is undoubtedly a kind of nondestructive testing. MFL tools are used to spot if a corrosion anomaly is interior or exterior in the wall of pipeline and also determine the location of the region of metal loss anomaly [13]. Figure 1 shows the principle of flux leakage and detection through magnets. The magnitude of the defects is measured by estimating the flux density leaked from the external of the ferromagnetic material. If there is a defect on the pipes, some of the flux lines will break the exterior of the components. Hall sensor or mobile induced coils could detect the leaked part of the flux [4, 14]. The magnets produce a strong magnetic field in the steel pipe, and sensors present in the tool record the disturbance in the magnetic field pattern. The recorded magnetic flux data gives us an indication of where the magnetic field is deflected, providing all related conditions of the pipeline. The magnetizer section is the assembly of these magnets in a particular manner so that all points on the circumference of the pipe can be covered. The details of this part are presented in the next section for context making.

2.3 Magnetic Flux Leakage (MFL) Tool

In a pipeline, the areas where there are corrosion, metal loss, dent, porosity, cracks MFL detects the leakage or variation in magnetic field lines [12]. Our study first presents a CAD model of MFL tool and provides specific attention to magnetizer assembly. The tool is designed for ferromagnetic material as the MFL tool magnetizes the pipeline so that the field lines can pass through them. A typical MFL tool incorporates several arrangements: drive system, magnetic system, sensor system, data conditioning and recording system, and power system [4]. Figure 2 shows the general design of an instrumented pipeline inspection gauge (MFL based). The magnetizer

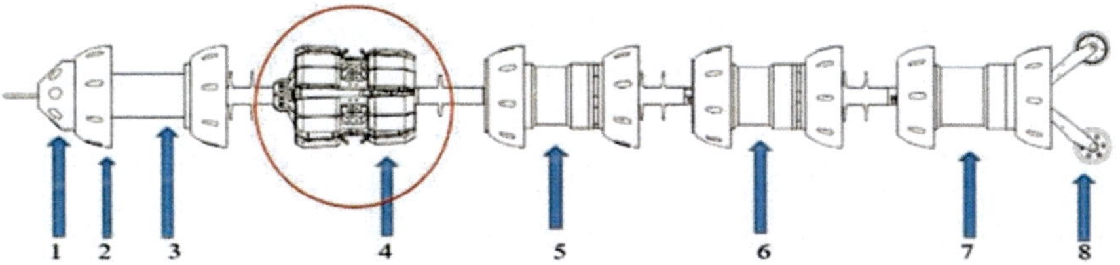

Fig. 2 Design of an instrumented pipeline inspection gauge (MFL based) (*Note* 1—Nose Ccone, 2—Pu cups, 3—Drive section, 4—Magnetic section, 5—CPU section, 6—Bettery section, 7—Transmitter section, 8—Odometer)

assembly is a vital component of an MFL tool, and we aim to design it for experimenting the behavior of flux under certain conditions so that flux leakage can be detected and approximated for a pipeline abnormality. The tool captures the data of the pipeline structural features and links it up with the real-time clock data/GPS. Specialized software is used for analyzing the data obtained from the tool after running in pipelines. The software predicts the correct location of each dent and corrosion. The magnetic section also contains some flappers which are small PCB sections which consist mainly of sensors for the tool such as the proximity sensor and the hall sensor. The hall sensor is used to detect the presence of the magnetic field in the environment, and the proximity sensor provides the distance between the flapper and the inner surface of the metal pipeline. The proximity sensor is primarily used to detect whether a defect detected by the tool is internal or external, i.e., whether a dent or pipeline lies on the inner surface or the outer surface of the pipeline thus working as an ID-OD (internal defect and outer defect indicator). The hall sensor as mentioned previously can detect the presence of a magnetic field in the environment. The data of the hall sensor is recorded, and then it is calibrated using it in such a way that it considers the data as a reference point and any deviations from it observed by the tool is considered as a defect by the software. Machine learning as a new approach that can be used to avert false data and thus make the final report more reliable. The inertial measurement unit can be used to detect the orientation of the tool and hall sensor number and pipe position.

The magnetizer section has the following parts, magnet bar, anchor block front, anchor block rear, end shaft, linkage arm, and mag shaft, wear plate front, and wear plate rear. In the designed tool, we are using a strong magnet, Neodymium N52 having Remanence of 1.437 T and 14.35 kGs. The magnetizer section has six bars and each bar has four magnets. In between the four magnets, four flappers (assembly of hall sensors and proximity sensors in PCB) are used. The magnetized field is formed, and sensors are installed between the magnetized fields so that when the drive section pulls the inspection tool, a continuous flow of magnetic field established. When there is a defect at any point of the inspection tool, this magnetic field gets deflected. For this deflection of the magnetic field, the installed sensors will collect the field deflection data and store. In the present design, the sensors used for data storage are mainly of two types, hall sensors, and customized proximity sensors. Hall sensors detect the severity and type of defect in the tool and proximity sensors detect whether the defect is inside or outside the pipeline. Figure 3a shows the magnetizer section and exploded view of magnetizer subassembly. Figure 3b shows the proposed arrangement of magnet, sensor, and flapper in magnetizer section for experiment.

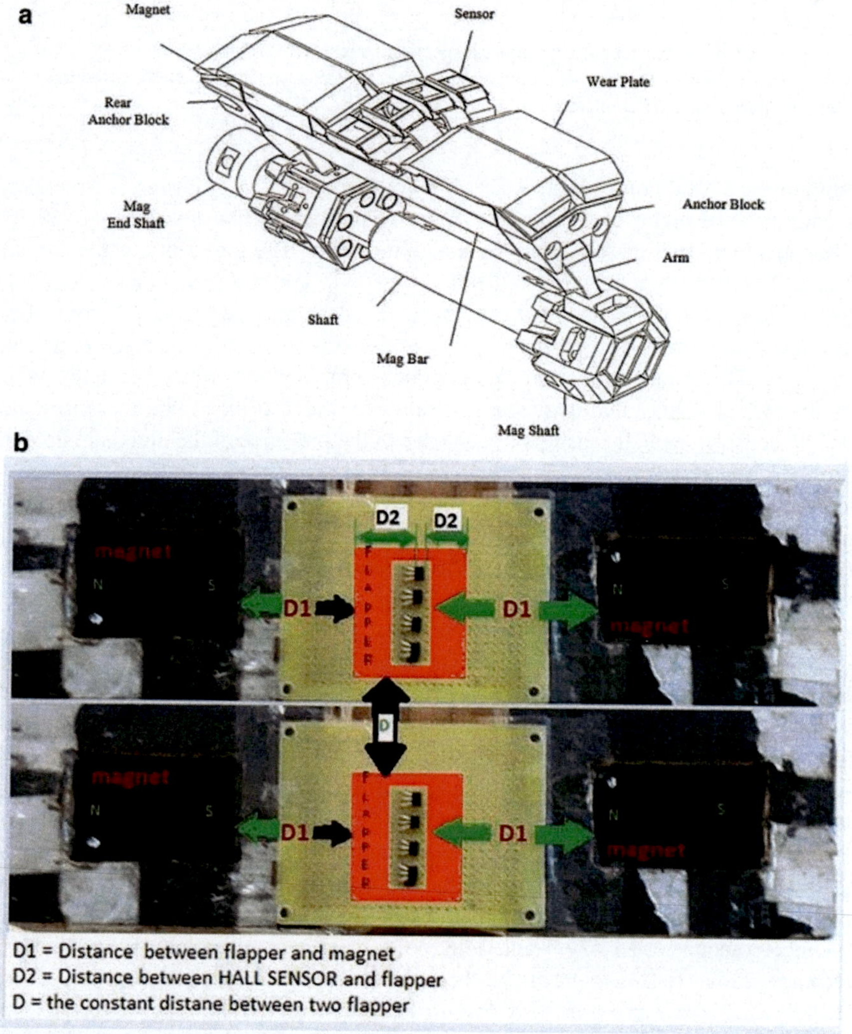

Fig. 3 **a** Magnetizer section and exploded view of magnetizer sub-assembly. **b** Arrangement of magnet, sensor, and flapper in magnetizer section for experiment

2.4 Parts and Functioning of Magnetizer:

- Magnetiser bar—They used to hold all the parts of the magnetizer, which mainly includes magnets and sensors.
- Sensors—They will be mounted on PU strips.
- Linkage arm—Used to connect the mag arm and center shaft (mag shaft).
- Spring on linkage arm—Used to make the constant surface contact of the magnetizer to the surface of the pipeline.
- Anchor Block Front and Rear—They provide support to the magnets and will be connected to mag shaft via a linkage arm.
- Wear Plate Front and Rear—Reduces the wear out of material at contact points.
- Magnetizer Nominal Diameter for $8' = 210$ mm, length of MFL = 2300 mm and weight = 85 kg.

3 Simulation and Method

As stated earlier, the scope of the paper is limited to design a magnetizer assembly and test the experimental abilities through simulation of flux leaks. The simulations carried out in COMSOL multiphysics software to get insights for its proposed magnetizer assembly of the MFL tool. Defects of different size and depth are modeled and simulated so that a better design aspect of the magnetizer assembly can be conceived. After collecting the information from design, the focus was shifted to creating a simulation model for COMSOL. In the initial study, a small portion of the pipe is taken. For primary analysis on material, the flat surface is selected. The design of the tool-pipe assembly was made in NX 10 (Fig. 4 of the CAD model of Pipe-tool assembly) and CAD file was transferred to COMSOL 5.0 Multiphysics on magneto-static studies. Further, the FEM analysis was carried out as per [12]. It is worth noting that we are here presenting a very fundamental simulation study apart from design. There are different modules present in the software for different studies, the magnetic field study which was present in the AC/DC module, and the study was done assuming the system to be stationary. Different materials were assigned to their respective part according to the data given which can be a good indicator [15]. The following details were useful for simulation; pipe material—low carbon steel 1010; tool material—aluminum 6061 t6; magnet material—neodymium iron boron; brushes—soft iron; medium was air. The following additional assumptions are made while simulating the behavior of magnetic flux leakage.

- The system was assumed stationary.
- We did not consider any losses like friction loss, hydraulic loss, and viscous force.
- We assumed temperature 293.15 K and pressure to be assumed 1 atm.
- We considered some magnetic vector potential.
- Brushes are considered as solid.

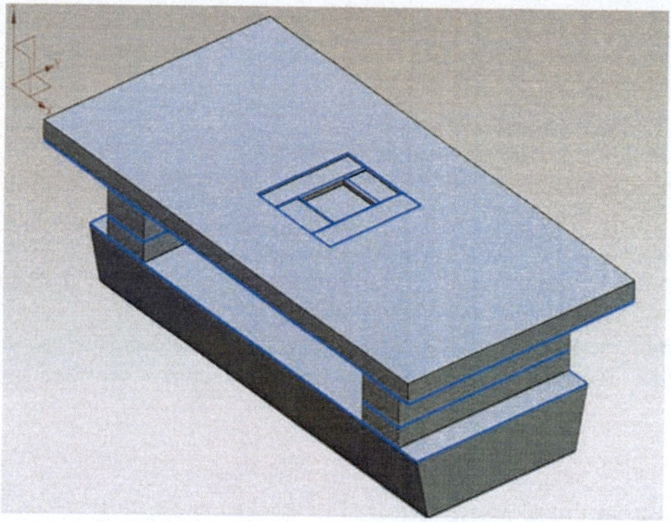

Fig. 4 CAD model of pipe-tool assembly

To start with, a rectangular cross-section of pipe is made of dimension 200 mm * 400 mm. A 40 mm * 40 mm was made on both interior and exterior of the pipe with depth 4 mm and 3 mm, respectively. Several other offsets were made on the interior section with varying depths. Beneath the pipe, there is the tool assembly consisting of magnets, brushes (to hold the magnets), and a backplate to assemble all components. Figure 5 shows the mesh generation for pipe-tool assembly. A magnetic field was

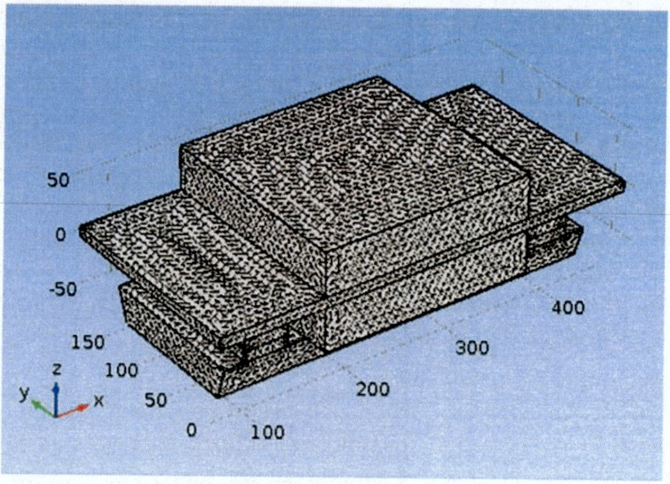

Fig. 5 Mesh generation on pipe-tool assembly

generated through neodymium iron boron magnets which produced a magnetic vector potential of 1000 wb/m.

In the simulations, Neodymium Iron Boron (NdFeB) magnets were taken to perform the FEM analysis, which has a coercivity of 827,600 A/m. Two mediums were defined on the interior and exterior of the pipe. Materials were assigned to different components, as discussed above. The governing equations for the analysis are $\Delta \times H = Je$ and $B = \Delta \times A$ Where, "A" is vector potential, "B" is Magnetic flux density, and "J" is current density.

3.1 Results and Findings

A number of simulations were carried out, and magnetic field density plots of the pipe cross-section were obtained. It was observed that as the thickness of the pipe is decreasing the magnetic flux density in that region increases. High magnetic field densities were observed where the pipe thickness is reduced. At the center of the pipe, where the metal erosion is maximum the flux density is also higher and gradually decreases as we move towards the edges as the metal erosion while offset decreases as shown in Figs. 6 and 7.

The placement of the sensor was done, and simulations were carried out. The critical aspects identified are: sensor safety is the topmost priority, because of the hydrocarbon environment if placed on the proximity of surface plates. Placement of the proximity sensor is critical because it will affect the accuracy of the defect

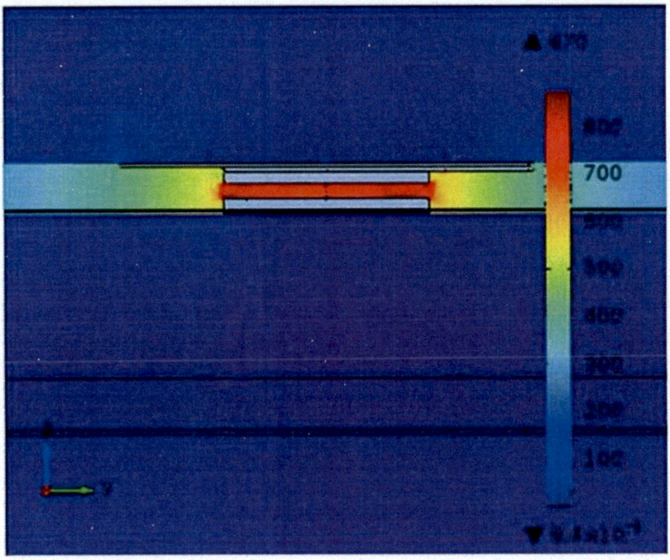

Fig. 6 Magnetic flux density

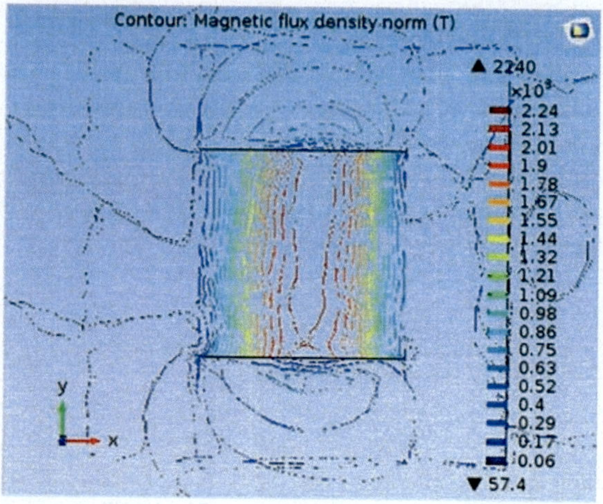

Fig. 7 Magnetic contours on the defect region

detected. The use of proper steel is recommended, such as SS 304 because of its nonmagnetic, noncorrosive behavior in trials. In actual practice, sufficient data rate is achieved for storing the information of the sensors, thus making it possible for the tool to be moved at a higher velocity and make the scanning process quicker.

4 Conclusion

The paper presents the design of a magnetizer assembly and experimental abilities through simulation of flux leaks. The simulations carried out in COMSOL multiphysics software to get insights for its proposed magnetizer assembly of the MFL tool. Defects of different size and depth are modeled and simulated so that a better design aspect of the magnetizer assembly can be conceived. Several simulations were carried out, and magnetic field density plots of the pipe cross-section were obtained. It was observed that as the thickness of the pipe is decreasing the magnetic flux density in that region increases. High magnetic field densities were observed where the pipe thickness is reduced. At the center of the pipe, where the metal erosion is maximum, the flux density is also greater and gradually decreases as we move toward the edges as the metal erosion. These findings helped in selecting the second chamber of the tool consists of a magnetic section which may consist of a large number of Neodymium N52 grade magnets, and the magnets are arranged in a crisscross pattern to provide stability to the tool.

References

1. Bubenik TA, Nestlroth JB, Eiber RJ, Saffell BF (1997) Magnetic flux leakage (MFL) technology for natural gas pipeline inspection. NDT E Int 1(30):36
2. Kishawy HA, Gabbar HA (2010) Review of pipeline integrity management practices. Int J Press Vessels Pip 87(7):373–380
3. Feng Q, Li R, Nie B, Liu S, Zhao L, Zhang H (2017) Literature review: theory and application of in-line inspection technologies for oil and gas pipeline girth weld defection. Sensors 17(1):50
4. Yang L, Fong H, Wong Y (2001) Research on intelligent pipeline magnetic flux leakage detector. In: Conference proceedings of 10th Asia-Pacific conference on non-destructive testing, Australia, pp 17–21. Last accessed on https://www.ndt.net/article/apcndt01/papers/11922/11922.htm
5. Tawancy HM, Al-Hadhrami LM, Al-Yousef FK (2013) Analysis of corroded elbow section of carbon steel piping system of oil-gas separator vessel. Case Stud Eng Fail Anal 1:6–14
6. Usarek Z, Warnke K (2017) Inspection of gas pipelines using magnetic flux leakage technology. Adv Mater Sci 17(3):37–45
7. Yang L, Zhang G, Liu G, Gao S (2008) Effect of lift-off on pipeline magnetic flux leakage inspection. In: Proceedings of 17th world conference on non-destructive testing, China, pp 25–28
8. Song Q (2011) Effects of apparatus parameters on MFL signals using orthogonal experimental design. Appl Mech Mater 44–47:3524–3528
9. Narang R, Chandrasekaran K, Gupta A (2017) Experimental investigation and simulation of magnetic flux leakage from metal loss defects. J Fail Anal Prev 17(3):595–601
10. Liu H (2017) Pipeline engineering, 1st ed. CRC Press, US. https://doi.org/10.1201/9780203711019
11. Pipeline design consideration and standards. https://www.pngrb.gov.in/OurRegulation/ppp-GSR-Infra.html. Last accessed on 28 May 2019
12. Al-Naemi FI, Hall JP, Moses AJ (2006) FEM modelling techniques of magnetic flux leakage-type NDT for ferromagnetic plate inspections. J Magn Magn Mater 304(2):790–793
13. Varela F, Yongjun TM, Forsyth M (2015) An overview of major methods for inspecting and monitoring external corrosion of on-shore transportation pipelines. Corros Eng Sci Technol 50(3):226–235
14. Umeadi BBN, Jones KG (2008) The development of an intelligent sensor for the monitoring of pipeline system integrity. Soc Pet Eng. https://doi.org/10.2118/115503-MS
15. Raghav V (2017) Finite element analysis of rectangular defects in inspection of steel pipelines using magnetic flux leakage technique. Mater Today Proc 4(2):2665–2672

DIC Correlation with Analysis Under Impact of Fiber Metal Laminates

S. K. Abhishek, G. Sunil Kumar, and R. Ramesh Kumar

Abstract Impact analyses are performed on fiber metal laminates (FMLs) by free fall of a mass at low to medium velocities and time-out of plane displacement responses are compared with following Digital Image Correlation (DIC) test procedure involving random pattern image projection on the specimen and high-speed camera. CARALL 1 consists of five aluminium layers which are reinforced with twelve embedded quasi-isotropic carbon fibre laminates and CARALL 2 involves three layers of aluminium reinforced with two embedded carbon fibre laminates. The thin plate undergoes predominant elastic deformation and explicit dynamic analysis shows good agreement with experimental results than CARALL 1 that undergoes both elastic and plastic deformation. Both test and analyses show good agreement on the assessment of response time of the FML laminates. The behaviour of deflection under impact loading up to low velocity is found to be linear and beyond which, a non-linear behaviour is observed.

Keywords Fiber metal laminates · Digital image correlation · Carbon fibre reinforced aluminium laminate

1 Introduction

One of the critical issues in the design of composite structures in the aerospace industry against impact is to enable the structure to absorb more impact energy. Impact loading occurs mainly under high velocity or low velocity. High velocity impact loading induces a localised form of damage that may not affect the structural integrity unlike in the case low velocity impact [1]. Hence most of the studies are confined to low velocity impact, say up to 15 m/s. The impact resistance of the CFRP

S. K. Abhishek (✉) · R. Ramesh Kumar
Govt. Engineering College Barton Hill, Thiruvananthapuram, India
e-mail: skabhishek4747@gmail.com

G. Sunil Kumar
Vikram Sarabhai Space Centre, Thiruvananthapuram, India
e-mail: gsunilkumar008@gmail.com

© Springer Nature Singapore Pte Ltd. 2021

P. M. Pandey et al. (eds.), *Advances in Production and Industrial Engineering*, Lecture Notes in Mechanical Engineering,
https://doi.org/10.1007/978-981-15-5519-0_25

can be improved by embedding metal constituents in between the CFRP laminates [2]. These are called fiber metal laminates (FML), which have improved impact resistance and damage tolerance than metal plates and plain composites. FML is generally based on either GLARE, that stands for glass laminate reinforced aluminium epoxy or ARALL and CARALL represented by Aramid and Carbon reinforced aluminium laminates, respectively [3]. Study on aluminium/carbon-epoxy fiber metal laminates on the inter-relation between the damage zone area, maximum depth of the deformation, the layup configurations and energy levels were reported and it was observed that Al/CFRP laminates have high impact damage resistance for low velocity [4]. The FML with (0/90) ply sequence in the carbon fibre was observed to exhibit the best behaviour. Mishra and Nayak simulated non-linearity due to contact and observed that contact force variation increased with impact velocity [5]. In most of the experimental investigations on the low velocity impact on composite plate was performed by falling weight, in which mass of indenter was kept high [6]. A comparison with experimental studies was based on the observation of damage zone size with respect to indenter diameter with finite element analysis. Carrillo performed a low velocity impact on aramid-reinforced polypropylene and obtained force–time variation and absorbed energy for different impact velocities. Results showed that FML configuration is based on 3 Aluminium and 4 composite layers, exhibited highest specific absorbed energy for first damage and perforation threshold when compared to other laminates [7]. The impact angle played a significant influence on the dynamic response of the GLARE. The impact process became slow as the impact angle which in turn reduced the plastic deformation, maximum impact contact force and energy absorption. So 90° impact is found as the most dangerous condition. Compared with the matrix failure and delamination of composite layers, the plastic deformation and fracture of aluminium layers were the two effective energy-absorbing modes which slowed down the failure progress of composite layers [8].

In this work, CARALL plate consisting of the aluminium layer with quasi-isotropic carbon laminates are considered for the impact analysis following material non-linearity for aluminium layer and digital correlation experimental investigation is performed using random pattern projection to capture out of plane displacements. Comparison of time displacement–response of CARALL is made to verify finite element model for the design of FML under impact velocities. Moreover study is focused to understand any failure of embedded composite layers.

2 Experimental Investigation Using DIC

The experimental study is carried out to generate test data on displacement–time response of the CARALL 1 and CARALL 2 specimens of 250 × 250 mm size with two different thickness of 4.34 and 1.22 mm to compare with explicit dynamic analyses considering material non-linearity.

Table 1 Details on FML

Nomenclature	Stacking sequence
CARAL 1	Al/0°/60°/−60°/Al/−60°/60°/0°/Al/0°/−60°/60°/Al/60°/−60°/0°/Al 17 layers each with 0.18 mm thickness
CARALL 2	Al/0°/Al/90°/Al 5 layers each with 0.18 mm thickness

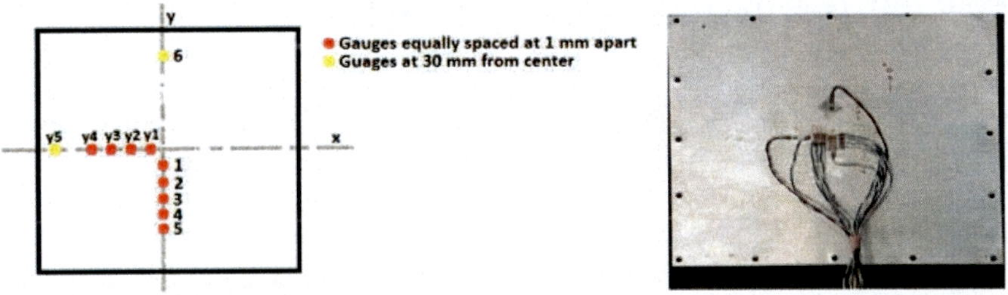

Fig. 1 Location of strain gauges

2.1 *Geometry of Target Plate*

The size of the CARALL specimen is 250 x 250 mm with thickness of 0.18 mm for both aluminium and CFRP layers. The layup sequence and orientation of the specimens are given in Table 1.

2.2 *Strain Gauge Location*

In order to measure the strain of the specimen strain gauge is bonded at the bottom of the specimen. Strain gauge parallel in numbers with gauge factor 2.065 are bonded as shown in Fig. 1.

Five gauges are bonded in X direction to measure strain in Y direction and six gauges are placed in Y direction to capture strain in X direction. The numbering of gauges is done to identify the position of each gauge.

2.3 *Indenter Geometry and Mass and Velocity*

The impactor is modelled as an available spherical ball of mass 66.6 g. The diameter of the spherical ball is 25.3 mm and is modelled as a rigid body. The impact velocities are 11.29, 13.65 and 16.86 m/s and its associated heights are 6.5, 9.5 and 14.25 m.

Table 2 Material properties of Aluminium AA-2024-T3

Material	Density	Young's modulus	Poisons ratio	Yield stress	Tangent modulus
AA 2024-T3	2700 kg/m^3	70 GPa	0.3	300 MPa	1.3 MPa

Table 3 Overall orthotropic properties of CFRP

Material	E1	E2	E3	G12	G23	G13	μ_{12}	μ_{23}	μ_{13}
CFRP	112 GPa	112 GPa	112 GPa	15.49 GPa	4 GPa	15.49 GPa	0.32	0.3	0.32

2.4 Material Properties

Bilinear kinematic model is used for aluminium. The material properties of aluminium alloy Al-2024-T3 are shown in Table 2.

Linear orthotropic material model is applied as a material model for CFRP. The overall orthotropic material properties of the CFRP laminates are presented in Table 3.

3 Method of Approach

3.1 Numerical Simulation of CARALL

ANSYS LS-DYNA is used for the numerical simulation of CARALL as shown in Fig. 2.

Eight noded solid elements are used to model the FML. The element is applicable for bilinear kinematic and plastic kinematic models. The specimen is modelled as per the layup sequence given in Table 1. Each layer is modelled as solid with thickness of 0.18 mm and the impactor is modelled as per the geometry mentioned in Sect. 2.2. Hexahedral element of size 12.5 × 12.5 mm each in the X–Y plane is used. Now two

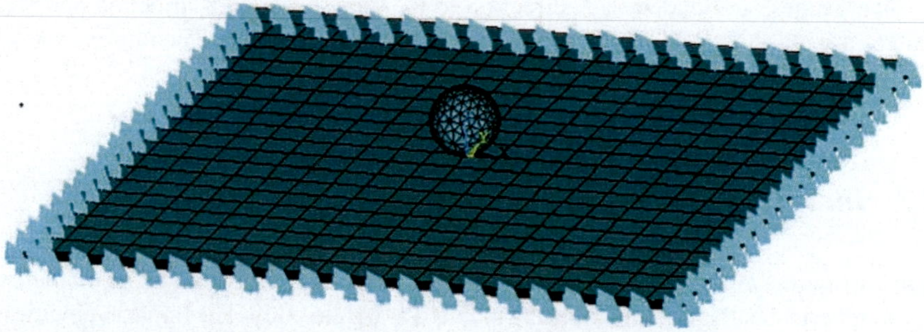

Fig. 2 Geometry of CARALL

components are created for defining the contact between impactor and specimen. The interaction between spherical ball and top layer is defined by automatic surface-to-surface contact. All the nodes in the periphery of the CARALL are constrained for all degree of freedom. The spherical ball is restricted from rotation in all directions and is allowed to move only in Z direction (out of plane direction of plates). The ball is given an impact velocity as planned (11.29, 13.65 and 16.86 m/s) in the thickness direction. The above problem is analysed and time response variation of the CARALL is obtained.

3.2 *Experimental Investigation*

All the low velocity impact tests are conducted using the free-fall drop test setup. A schematic sketch of the impact test setup is shown in Fig. 3. The specimen is placed on the fixture and clamped from all edges and the fixture is held rigidly on a supporting frame to avoid shocks after impact. Out of plane displacement of the specimens is captured using a combination of two-dimensional digital image correlation and pattern projection. This method employs a high-speed camera, which is aligned perpendicular to the specimen surface. The high-speed camera is capable of capturing 10,000 frames per second.

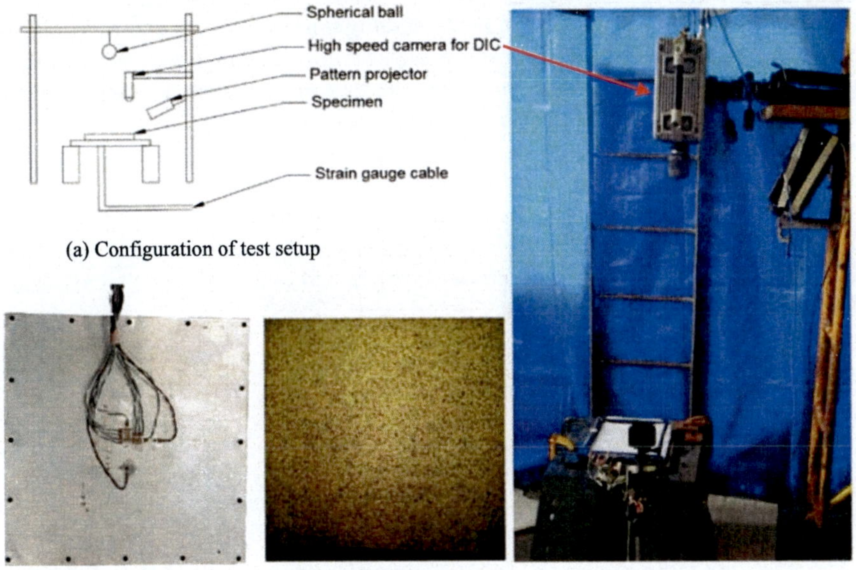

(a) Configuration of test setup

(b) Strain gauge (c) Random patterns (d) Test components

Fig. 3 Experimental setup

Table 4 Test plan

Test series	Specimen	Height of drop (m)	Touch down velocity of ball (m/s)
1.a	CARALL 1	6.5	11.29
1.b	CARALL 2	6.5	11.29
2.a	CARALL 1	9.5	13.65
2.b	CARALL 2	9.5	13.65
3.a	CARALL 1	14.5	16.86
3.b	CARALL 2	14.5	16.86

In this technique, the surface of the specimen is coated with a white paint and random patterns are projected on to the surface using a pattern projector. The high-speed camera captures the speckle patterns during the impact and calculates the deflection based on the changes in position of the speckles. The images of the speckles before undergoing deformation and after deformation are captured. Now computer algorithms are used to calculate the out of plane displacements. The impact test on the specimens is carried out by dropping a spherical ball of mass 66.6 g on to it. There will be three test series as shown in Table 4. First test series is performed at a height of 6.5 m, the height from which the first drop is to be performed is marked and the setup for tieing the ball is fixed at that point. Now the position on the FML, where the ball is to be dropped is determined by using a plumb bob, the ball is tied accordingly, and the remaining tests are performed as per the test plan.

For measuring the displacement, DIC test setup is used. This consists of placing a high-speed camera at two metres above the specimen as discussed and a pattern projector is placed in position to project the random patterns for displacement measurement. The focus of the camera is adjusted and the position where the ball will impact is located. Finally, the ball is released from the desired height using a ball detachment mechanism. Now the impact occurs and the deflection and strains are captured. After the impact, the specimen is removed and the next specimen is placed. Test is repeated on the second specimen and observations are noted. Now the height of drop is changed and impact is performed as per the test plan to obtain deflection and strains for each test.

4 Results and Discussion

Post-test observation of the test specimens is made. Displacement–time response of CARALL 1 specimen (Al/0°/60°/–60°/Al/–60°/60°/0°/Al/0°/–60°/60°/Al/60°/–60°/0°/Al) following the explicit dynamic analyses and test data for an impact velocity of 11.29 m/s is given in Fig. 4. Similar studies on comparison for different impact velocities of 13.6 and 16.86 m/s are presented in Figs. 5 and 6. In the case of CARALL 2

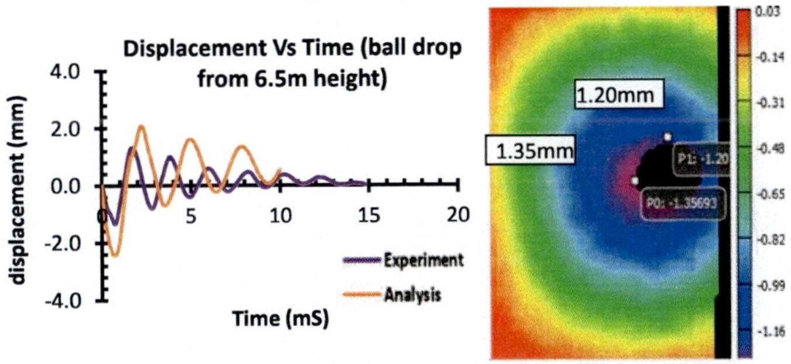

Fig. 4 Out of plane displacement response between DIC and analysis for CARALL 1 (11.29 m/s)

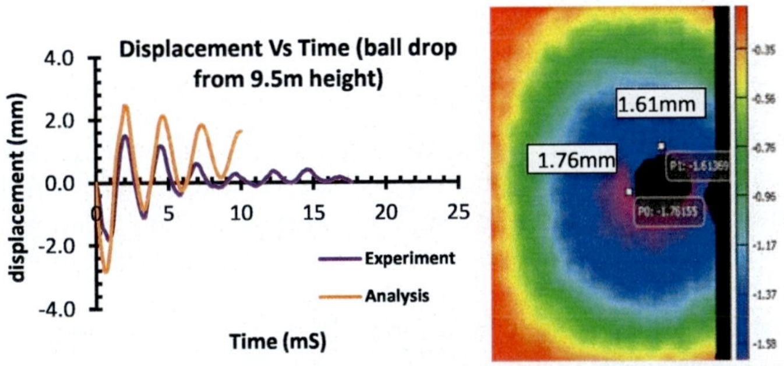

Fig. 5 Out of plane displacement response between DIC and analysis for CARALL 1 (13.65 m/s)

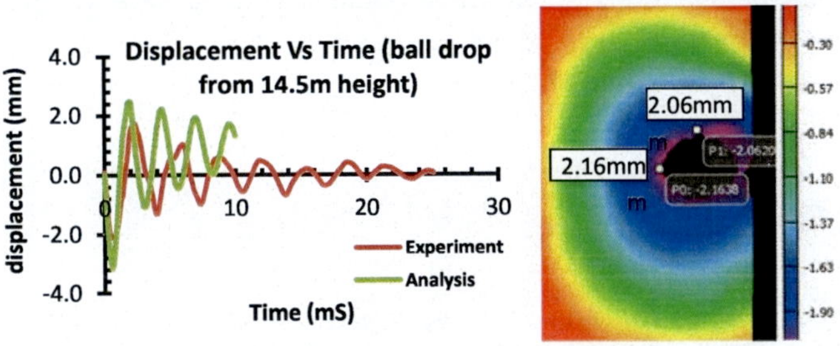

Fig. 6 Out of plane displacement response between DIC and analysis for CARALL 1 (16.86 m/s)

(Al/0°/Al/90°/Al) that has low overall thickness of 0.9 mm, the dynamic displacement–time response are given for the three impact velocities and given in Figs. 7, 8 and 9. Tables 5 and 6 compare the peak response time and associated deflections from both the approaches. The influence of the impact velocities on the behaviour of peak response deformation is given in Fig. 10.

4.1 Post-test Observations

1. It has been found that the strain gauges are damaged after the first test.
2. Coin tapping NDT is done on both the specimens. The dent or deformation of the specimen where the impact occurred is examined. It is found that in the case of CARALL 1, local dent at the impact point has been seen and the dent size increases with increase in impact velocities. Besides a damage zone of one-inch diameter around the centre of plate, the plate has been bend permanently about an axis parallel to zero degree fibre direction over the full length of 250 mm.
3. In the case of CARALL 2 only very minor localised damage at the impact points may be seen.

4.2 CARALL 1

For the CARALL 1 specimen, impacted at a point that happened to be in the periphery of a circle of 10.1 mm radius from the centre of the specimen, the maximum deflection obtained from analysis is 2.42 mm at 0.76 ms whereas the DIC value of deflection is 1.35 mm at 0.8 ms, beyond which the impact response reduces to zero at 15 ms (Fig. 4.). The post-test observation has shown an overall permanent bend of the specimen with local dent at the point of impact. It may be noted that the test values are found lower than analysis. Such trends are observed even for higher velocities (Figs. 5 and 6). In other words, there is a considerable deviation between the prediction and the test for the deformations but for the response time. One of the reasons for the deviation may be due to the difference in the thickness considered in the analysis (3.06 mm) when compared to actual thickness that has been enhanced by more adhesive layers during curing (4.34 mm).

4.3 CARALL 2

It is observed from Fig. 7 that the FML specimen for an impact velocity of 11.29 m/s shows a maximum deflection of 3.65 mm at 1.58 ms from the analysis while from

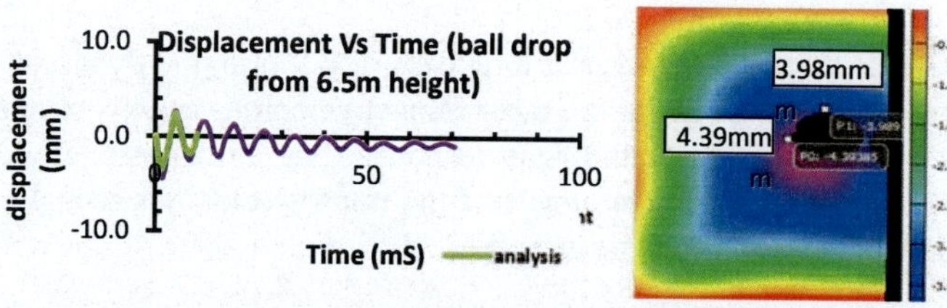

Fig. 7 Out of plane displacement response between DIC and analysis for CARALL 2 (11.29 m/s)

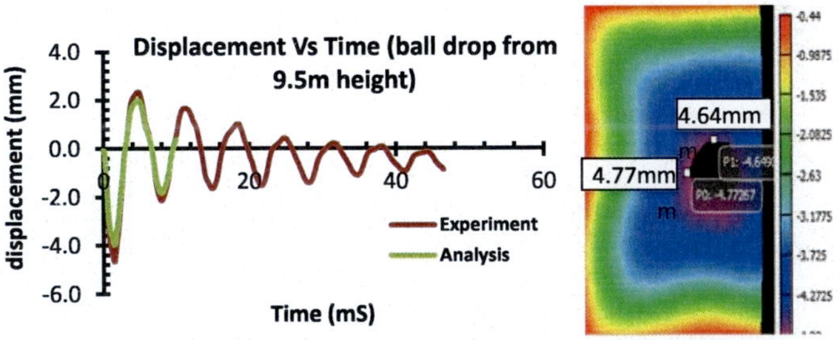

Fig. 8 Out of plane displacement response between DIC and analysis for CARALL 2 (13.65 m/s)

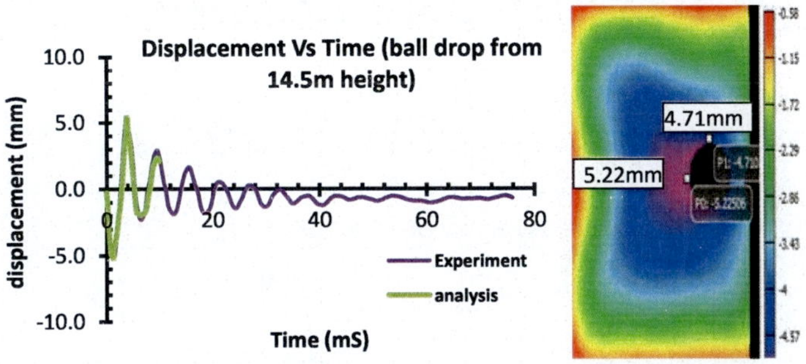

Fig. 9 Out of plane displacement response between DIC and analysis for CARALL 2 (16.86 m/s)

experimental results the values are in close agreement with 4.3 mm at 1.7 ms, respectively. As the impact velocity increases, the gap between the two approaches narrows down (Figs. 8 and 9).

From Figs. 7, 8 and 9, it is obvious that both response time and deformations are in good agreement for the case CARALL 2 under three impact velocities. The post-test observation did not indicate either any local dent causing plastic deformation or any overall permanent bend. Therefore, it is concluded that CARALL 2 has predominant elastic deformation.

It is interesting to note that in the case of CARALL 1 (Al/0°/60°/−60°/Al/ …) the DIC pattern is in a circular form while in the case of CARALL 2, it is elongated in one direction (transverse to zero degree fibre direction). This is because the former has a quasi-isotropic layup sequence and latter is orthotropic (Al/0°/Al/90°/Al).

4.4 Comparison of FML Under Low and Medium Velocities

From Fig. 10, a linear variation is observed up to 15 m/s, beyond which a non-linear trend is seen for the medium velocity.

Table 5 Comparison of response time and deflection for CARALL 1

Impact velocity (m/s)	Response time		Deflection	
	FEM (ms)	Test (ms)	FEM (mm)	Test (mm)
11.29	0.76	0.8	2.42	1.35
13.65	0.64	0.7	2.82	1.76
16.86	0.59	0.6	3.18	2.16

Table 6 Comparison of response time and deflection for CARALL 2

Impact velocity (m/s)	Response time		Deflection	
	FEM (ms)	Test (ms)	FEM (mm)	Test (mm)
11.29	1.58	1.7	3.65	4.3
13.65	1.61	1.6	4.13	4.77
16.86	1.41	1.4	5.56	5.22

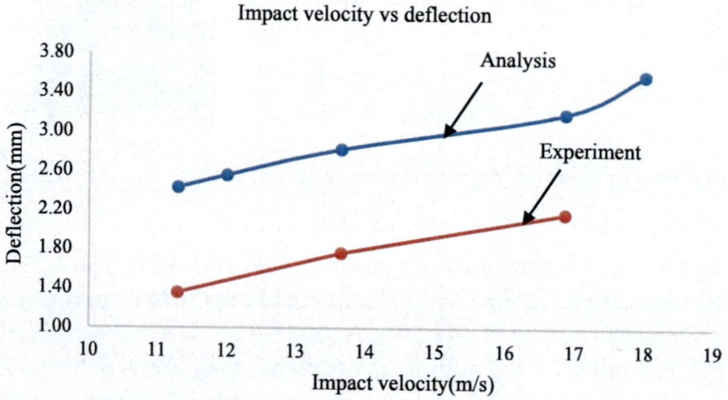

Fig. 10 Variation of peak displacement with impact velocity

Stiffness variation of FML plates between the test and analysis and post-test NDT has given in Sec. 4 clearly shows that composite layers have not undergone failure (but for close to the impact point) as the test results are in a lower bound even though aluminium layers have undergone overall plastic deformation for the CARALL 1.

5 Conclusions

Experimental investigation under impact loading has been performed on fiber metal laminates by free fall of a mass of 66.6 g at three different velocities of 11.29, 13.65 and 16.86 m/s. Explicit dynamic analysis results have shown good agreement on response time-deflection variation for thin FML that has undergone predominant elastic deflection with little plastic deformation. Out of plane displacement and response time due to impact have been captured with very good accuracy following present experimental study using DIC technique. In the case of CARALL 1, post-test observation clearly indicated an overall permanent bend that assisted much absorptivity besides local plastic deformations. The present study has also shown a non-linear behaviour of dynamic response under medium impact velocity when compared to linear variation for the case of low velocity. It is concluded based on stiffness variation of FML plates between the two approaches that composite laminates have not undergone failure even though the FML has undergone overall permanent deformation over whole width.

References

1. Cantwell WJ, Morton J (1989) Comparison of low and high velocity impact response of CFRP. Composites 20:545–551
2. Sinmazcelik T, Avcu E, Bora MO, Coban O (2011) A review: fiber metal laminates, background, bonding types and applied test methods. Mater Des 32:3671–3684
3. Chai GB, Manikandan P (2014) Low velocity impact response of fibre-metal laminates–a review. Compos Struct 107:363–380
4. Bienias J, Jakubczak P (2012) Low velocity impact resistance of aluminium/carbon-epoxy fiber metal laminates. Polish Soc Compos Mater 12:193–197
5. Mishra A, Naik NK (2010) Failure initiation in composite structures under low-velocity impact: analytical studies. Compos Struct 92:436–444
6. Muslim Ansari MD, Chakrabarti A, Ashraf Iqbal M (2016) Dynamic response of laminated GFRP composite under low velocity impact: experimental and numerical study. In: 11th international symposium on plasticity and impact mechanics
7. Carrillo JG, Gonzalez Canche NG, Flores Johnson EA, Cortes P (2019) Low velocity impact response of fibre metal laminates based on aramid fibre reinforced polypropylene. Compos Struct 220:708–716
8. Li L, Sun L, Wang T, Kang N, Cao W (2019) Repeated low-velocity impact response and damage mechanism of glass fibre aluminium laminates. Aerosp Sci Technol 84:995–1010

Evaluation of Tensile Strength Behaviour of Friction Stir Welding Joints of Aluminium Alloy with Interlayer

Avtar Singh, Vinod Kumar, and Neel Kanth Grover

Abstract In the present study, the influence of interlayer is evaluated on friction stir welding of AA6082 aluminium alloy. Zn and Cu alloying elements in proportion mixed with adhesive are used as an interlayer. Interlayer is preplaced in a gap of 0.2 mm between the abutting plates. Friction stir welding is carried out at various rotational and welding speeds. Thereafter joints are evaluated for tensile properties. It is found that FSW joints prepared with interlayer possess higher tensile strength than joints produced without interlayer.

Keywords Friction stir welding · Mechanical properties · Interlayer · Gap · Defect

1 Introduction

Friction stir welding is solid state process in nature [1]. In this process material does not melt completely but softens due to heat generated by friction between rotating tool and workpiece [2]. Rotating tool consists of a shoulder and a pin profile. Tool pin plunge into workpiece to generate required heat and mixing of soften material. The tool shoulder generates the additional heat and also prevents escaping of soften material outside [3]. It is evident that workpiece material is under the influence of high heat and plastic deformation. Subsequently material gets soften due to recrystallization and dissolution of strengthening precipitates in similar material welding [4]. Similarly, brittle intermetallic compounds are formed in dissimilar materials. The softening of material and brittle intermetallic compound significantly decreases

A. Singh (✉)
Yadavindra College of Engineering, Talwandi Sabo, Bathinda, PB 151302, India
e-mail: avtardhanjal@gmail.com

V. Kumar
Department of Mechanical Engineering, Punjabi University, Patiala, PB 147002, India

N. K. Grover
Department of Mechanical Engineering, IKG Punjab Technical University, Kapurthala, PB 144603, India

© Springer Nature Singapore Pte Ltd. 2021

P. M. Pandey et al. (eds.), *Advances in Production and Industrial Engineering*, Lecture Notes in Mechanical Engineering,
https://doi.org/10.1007/978-981-15-5519-0_26

the mechanical properties of the joints [5]. These problems can be mitigated to some extent by adding the alloying elements interlayer at the interface of abutting plates during welding. Interlayer elements prevent the softening of material by formation of new solid solution/high strength intermetallic compounds [4]. During friction stir welding of aluminium alloys, nugget zone experienced softening due to high heat and dissolution of strengthening precipitates. It is attributed that joints under tensile loading are failed from the softened zone [6]. Therefore, in this experiment study an attempt has been made to strengthen the FSW joint nugget zone using interlayer of alloying elements.

Various researches have been conducted recently on similar and dissimilar materials to study the impact of interlayer on mechanical properties of FSW joints. Cam et al. [7] studied the used of high strength interlayer on friction stir welding of AA6061-T6 aluminium alloy. A slice of AA7075-T6 aluminium alloy was inserted at the interface of abutting plates followed by friction stir welding. Thereafter FSW joints were characterised for microstructural and mechanical analysis. The microstructural analysis revealed that joints' chemistry alter due to insertion of interlayer. It was also found that interlayer significantly enhanced the hardness of the nugget zone. But no considerable effect showed on improvement of tensile properties. Mokabberi et al. [4] conducted an investigation to evaluate the effect of Zn, Cu and brass interlayer on FSW joint integrity. It was found from the results that Zn and Cu interlayer showed no significant improvement in joint efficiency. Whereas joint prepared with brass interlayer showed approximately 30% improvement in joint efficiency. The microstructural observations indicated that interlayer particles were shuttered and distributed into the nugget zone of fiction stir welding joint. It was also observed that Zn and Cu formed solid solution with aluminium that significantly reduced the grain size for 20–10 μm. Similarly various investigations were carried out with interlayer such as brass [4], aluminium oxide [8], Silicon carbide [9], Titanium carbide [10], Cu [11–13] and Zn [14–16] demonstrated the improvements in joint efficiency.

In the present experimental investigation, the influence of Zn and Cu mixture with adhesive is used as interlayer; and tensile behaviour of FSW joint is evaluated.

2 Experimental Procedure

In this study AA 6082-T6 aluminium alloy plates of size of 200 × 75 × 6 mm were selected. The composition of material used for welding is given in Table 1. Edge preparation was carried out using shaping machine and dirt was removed using

Table 1 Composition of welded material

Major element	Si	Mg	Fe	Al
%age	1.215	1.175	0.259	Rest

acetone solution. To perform friction stir welding, the plates were firmly clamped into special fixture by maintaining a 0.2 mm gap between abutting plates as shown in Fig. 1. Pure Zn and Cu particles in the ratio of 30:70 are mixed with adhesive that is used as interlayer. Interlayer was preplaced into a gap of abutting plates. Vertical milling machine of 3 HP, owing rotational speed in the range of 600–4600 rpm was used to carry out friction stir welding. Taper cylindrical tool of shoulder, pin diameter and pin depth of 18 mm, 6 mm and 5.7 mm, respectively, was used for FSW welding as shown in Fig. 2. The diameter of pin was 1 mm tapering towards bottom side of pin. Rotational speed and welding speed were varied keeping tool tilt angle and plunge depth constant to 2° and 0.2 mm, respectively, as listed in Table 2.

Fig. 1 Schematic of FSW

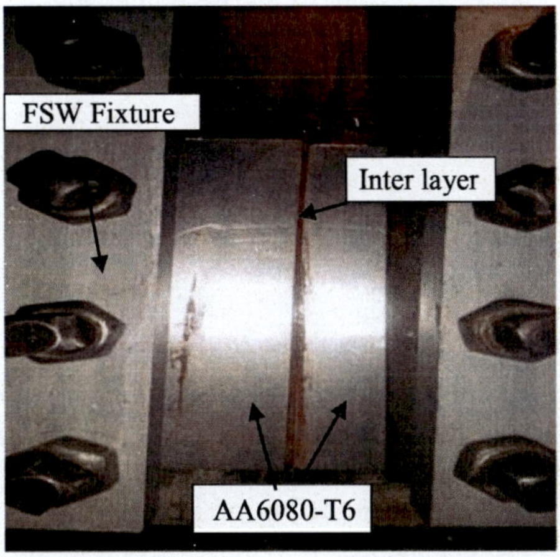

Fig. 2 FSW Tool used

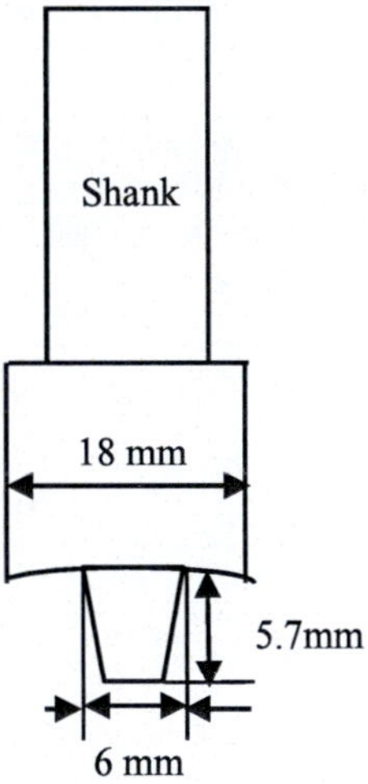

Table 2 Process parameters of FSW

S. No.	Rotational speed (RPM)	Welding speed (mm/min)	Tool tilt angle (degree)	Plunge depth (mm)
1	1200	30	2	0.2
2	1540	50	2	0.2

Three tensile strength specimens were extracted from each FSW joint as per ASTM E-8 using wire cut electrical discharge machine. Further tensile strength of the joints was evaluated using a tensile testing machine of Tinius Olsen make: Model H50KS at strain rate of 1 mm/min. SEM of make Jeol 6510 was used to reveal the fracture behaviour of tensile specimens.

3 Results and Discussion

It is evident that upper surface of FSW joints without interlayeris free from any major defect. Figure 3 shows that there is no flash; only smooth joint line with concentric circle is visible.

Figure 4 shows FSW joints with interlayer, rough surface with concentric circles on the top surface [17]. Flash is also present on the surface of FSW joints. It is due to improper consolidation of material on the top surface of the joints. It seems that interlayer particles in loose condition at the interface of abutting plates interfere in proper consolidation of material. Figure 4b illustrates that concentric circles' intensity enhances with increase of rotational and welding speed [18]. It is attributed that high heat input reduces flow stresses in plasticized material, therefore tool loses control over soften material resulting in higher intensity of concentric circles.

Fig. 3 Top surface of FSW without Interlayer at rotational and welding speeds

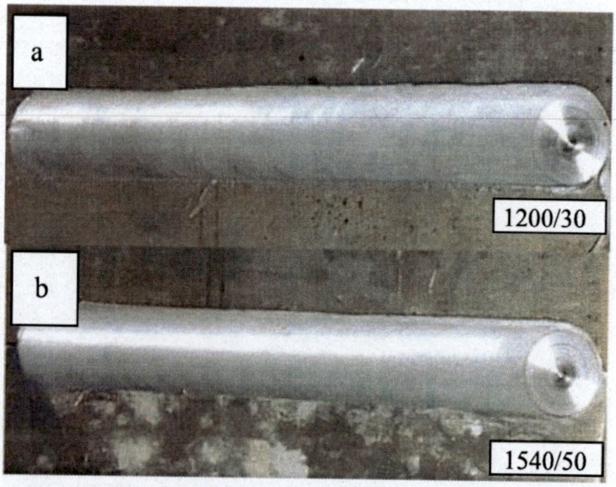

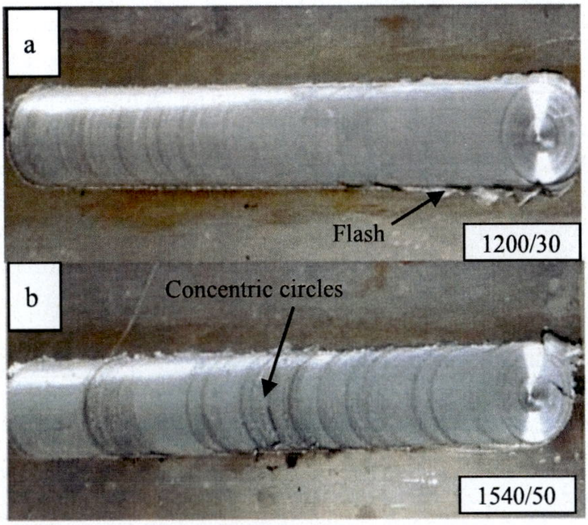

Fig. 4 Top surface of FSW with Interlayer at different rotational and welding speeds

The visual inspection revealed no defect in cross section of FSW joints with and without interlayer as shown in Figs. 5 and 6. It is confirmed that sufficient heat input experiences by material and proper consolidation succeed in the nugget zone. Figure 6 also demonstrates interlayer particles in the nugget zone. But joint thinning is observed in FSW joints with interlayer. It is because of the material used to fill the gap between abutting plates.

Fig. 5 Cross section of FSW joints without interlayer at various rotational and welding speeds

Fig. 6 Cross section of FSW joints with interlayer at various rotational and welding speeds

3.1 Tensile Strength

Tensile stength results of joint with and without interlayer are shown in Table 3. It is observed that FSW joints with interlayer owing high tensile strength in comparison to the joint without interlayer as shown in Fig. 7. The joints with interlayer have 20% more efficiency than joint without interlayer. But increase in rotational speed and welding speed reduces the tensile strength in both the cases. It has happened due to high heat input which is responsible for lower tensile strength.

Elongation experiences inverted effect of interlayer. FSW joint with interlayer owns lower elongation than joints without interlayer as shown in Fig. 8. It is also confirmed that increase in rotational and welding speed reduced elongation of the friction stir welding joints. The high localised deformation has significant effect on elongation of the joints. It is concluded that interlayer has negative effect on elongation of the joints.

Table 3 Tensile strength of FSW joints with and without interlayer at different process parameters

S. No.	Rotational Speed (rpm)	Welding Speed (mm/min)	Average Ultimate Tensile Strength (MPa)	Elongation (%age)	Average Ultimate Tensile Strength (MPa)	Elongation (%age)
			Without Interlayer		*With Interlayer*	
1	1200	30	147.23	8.60	176.12	6.91
2	1540	50	139.58	4.91	163.54	4.21

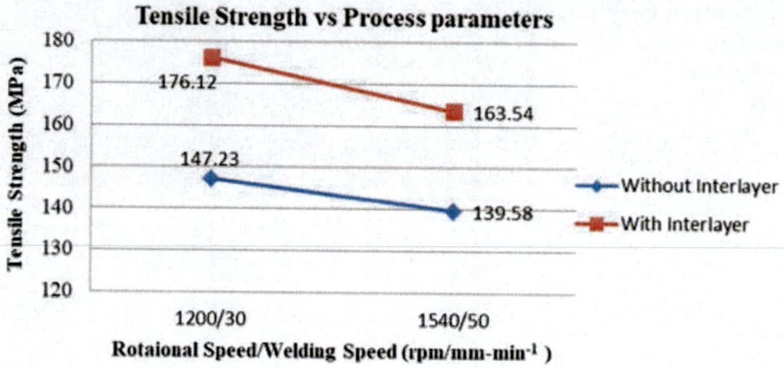

Fig. 7 Comparison of Tensile Strength of FSW joints with and without interlayer at various rotational and welding speeds

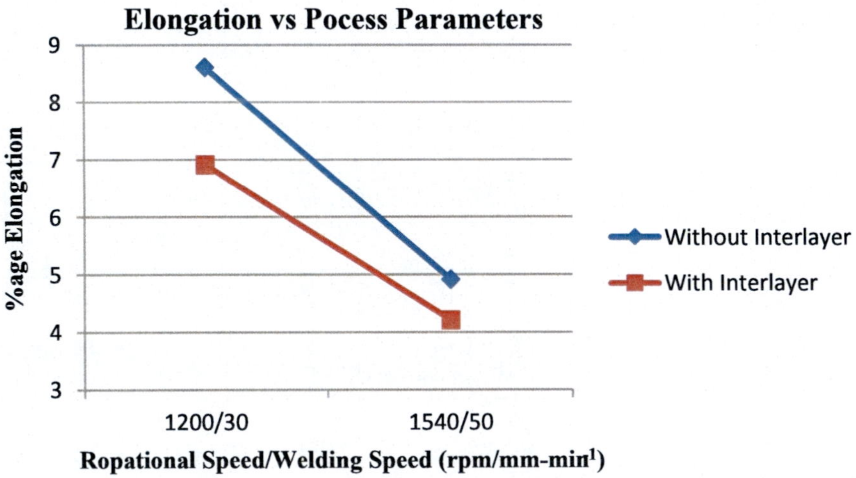

Fig. 8 Comparison of elongation of FSW joints with and without interlayer at various rotational and welding speeds

3.2 Fracture Behaviour

It is depicted from Figs. 9 and 10 that under tensile loading, the joints get fractured at or near the interface of nugget zone and towards retreating side in both the cases. The failure in all the joints occurred nearly vertical to tensile loading [19]. The micrograph shown in Fig. 11a of the fracture surface reveals numerous and deep dimples. But shallow dimples are noticed in the fracture surface of the joint produced at higher rotational speed as shown in Fig. 11b. It indicates that joint produced at lower rotational speed receives appropriate heat input resulting in proper consolidation of material that lead to good ductile behaviour. Voids and shallow dimples are reported

Fig. 9 Fractured Tensile specimens with interlayer

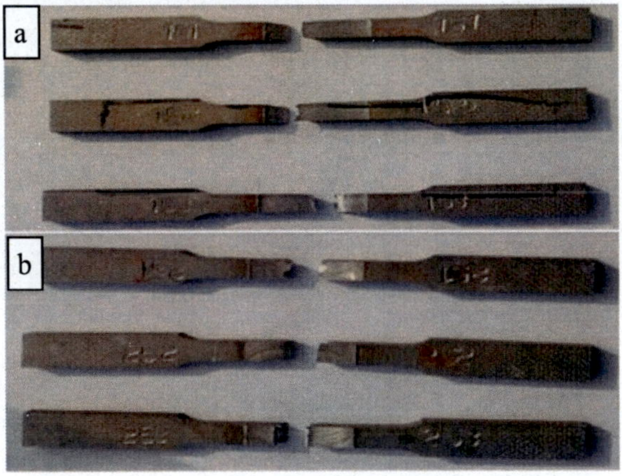

Fig. 10 Fractured Tensile specimens without interlayer

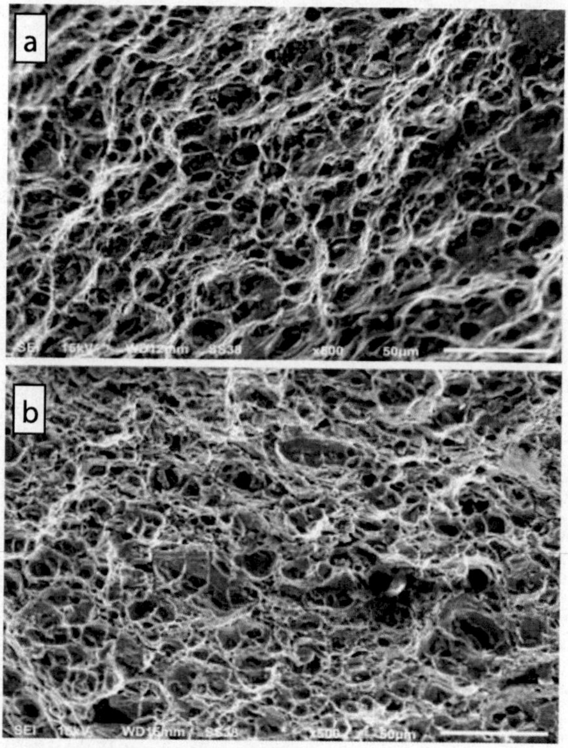

Fig. 11 Fracture surface of joints without interlayer

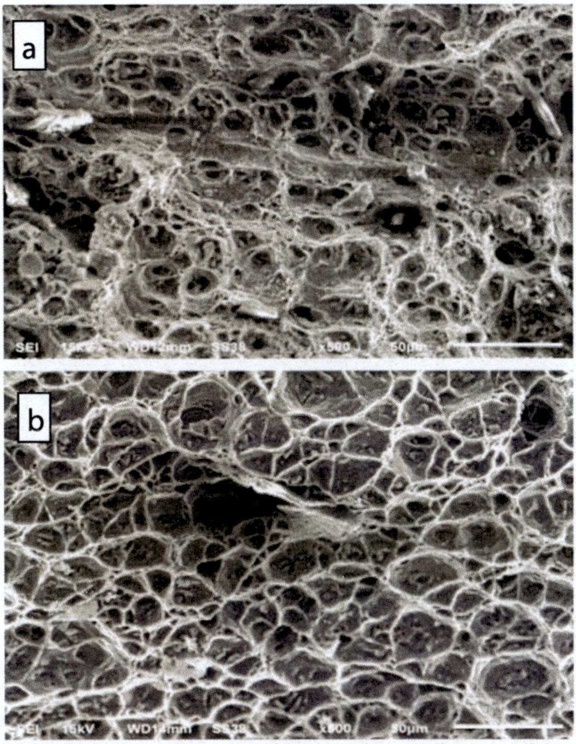

Fig. 12 Fracture surface of joints with interlayer

on fractured surfaces of FSW joint with interlayer as shown in Fig. 12a. It is observed that deep and larger size dimples are the evidences of ductile behaviour in the joints with interlayer from Fig. 12b.

4 Conclusion

On the basis of experimental research carried out in the present study on welded joints of AA6082 aluminium alloy, realized according to FSW process at various rotational and welding speeds, the following conclusions can be reached:

- Tensile strength of the joints with interlayer increased about 20% in comparison to FSW joint without interlayer.
- Tensile strength decreased with increase of rotational and welding speed. With increase of rotational speed, heat input enhanced resulting in dissolution of strengthening precipitates of the material.
- FSW joints with interlayer possess lower elongation in compression to the joint without interlayer due to recrystallization of material. Hence grain refinement takes place.
- All joints in both cases get fractured from the nugget zone as softening of material possess by this zone.
- FSW joint with and without interlayer exhibited ductile fracture behaviour.

References

1. Aonuma M, Nakata K (2012) Dissimilar metal joining of ZK60 magnesium alloy and titanium by friction stir welding. Mater Sci Eng B: Solid-State Mater Adv Technol. 177:543–548. https://doi.org/10.1016/j.mseb.2011.12.031
2. Elangovan K, Balasubramanian V (2008) Influences of tool pin profile and welding speed on the formation of friction stir processing zone in AA2219 aluminium alloy. J Mater Process Technol 200:163–175. https://doi.org/10.1016/j.jmatprotec.2007.09.019
3. Sakthivel T, Sengar GS, Mukhopadhyay J (2009) Effect of welding speed on microstructure and mechanical properties of friction-stir-welded aluminum. Int J Adv Manuf Technol 43:468–473. https://doi.org/10.1007/s00170-008-1727-7
4. Mokabberi SR, Movahedi M, Kokabi AH (2018) Effect of interlayers on softening of aluminum friction stir welds. Mater Sci Eng, A 727:1–10. https://doi.org/10.1016/j.msea.2018.04.093
5. Zhang HM, Chao YJ, Luo Z (2017) Effect of interlayer on microstructure and mechanical properties of Al–Ti ultrasonic welds. Sci Technol Weld Joining 22:79–86. https://doi.org/10.1080/13621718.2016.1193386
6. Mahoney MW, Rhodes CG, Flintoff JG, Bingel WH, Spurling RA (2007) Properties of friction-stir-welded 7075 T651 aluminum. Metall Mater Trans A. 29:1955–1964. https://doi.org/10.1007/s11661-998-0021-5
7. Çam G, İpekoğlu G, Tarık Serindağ H (2014) Effects of use of higher strength interlayer and external cooling on properties of friction stir welded AA6061-T6 joints. Sci Technol Weld Joining 19:715–720. https://doi.org/10.1179/1362171814Y.0000000247
8. Kumar, RA., T.M.: Reinforcement with alumina particles at the interface region of AA6101-T6 and AA1350 alloys during friction stir welding. Mater Res Express. 5, 046521 (2018)
9. Uzun H (2007) Friction stir welding of SiC particulate reinforced AA2124 aluminium alloy matrix composite. Mater Des 28:1440–1446. https://doi.org/10.1016/j.matdes.2006.03.023
10. Karakizis PN, Pantelis DI, Fourlaris G, Tsakiridis P (2018) Effect of SiC and TiC nanoparticle reinforcement on the microstructure, microhardness, and tensile performance of AA6082-T6 friction stir welds. Int J Adv Manufact Technol. 95:3823–3837. https://doi.org/10.1007/s00170-017-1446-z

11. Razal Rose A, Manisekar K, Balasubramanian V (2011) Effect of axial force on microstructure and tensile properties of friction stir welded AZ61A magnesium alloy. Trans Nonferrous Met Soc China 21, 974–984. https://doi.org/10.1016/S1003-6326(11)60809-1
12. Inada K, Fujii H, Ji YS, Sun YF, Morisada Y (2010) Effect of gap on FSW joint formation and development of friction powder processing. Sci Technol Weld Joining 15:131–136. https://doi.org/10.1179/136217109X12568132624244
13. Abnar B, Kazeminezhad M, Kokabi AH (2014) The effect of Cu powder during friction stir welding on microstructure and mechanical properties of AA3003-H18. Metall Mater Trans A. 45:3882–3891. https://doi.org/10.1007/s11661-014-2310-5
14. Lenina AWA, Periyasamyb N, Georgec L (2016) Influence of interlayer thickness (Zn) on the properties of Al 7020 FSW joints. Mater Res. 19:817–823. https://doi.org/10.1590/1980-5373-MR-2016-0163
15. Shamsipur A, Anvari A, Keyvani A (2018) Improvement of microstructure and corrosion properties of friction stir welded AA5754 by adding Zn interlayer. Int J Miner, Metall, Mater. 25:967–973. https://doi.org/10.1007/s12613-018-1646-z
16. Xu RZ, Ni DR, Yang Q, Liu CZ, Ma ZY (2015) Influence of Zn interlayer addition on microstructure and mechanical properties of friction stir welded AZ31 Mg alloy. J Mater Sci 4160–4173 (2015). https://doi.org/10.1007/s10853-015-8841-3
17. Meshram SD, Reddy GM, Rao AV (2016) Role of threaded tool pin profile and rotational speed on generation of defect free friction stir AA 2014 aluminium alloy welds. Defence Sci J. 66:57–63. https://doi.org/10.14429/dsj.66.8566
18. Singh A, Kumar V, Grover NK (2019) Influence of tool pin profiles on friction stir welding with a gap for AA6082-T6 aluminium alloy. Mater Res Express. 6:086543. https://doi.org/10.1088/2053-1591/ab18cc
19. Ren SR, Ma ZY, Chen LQ (2007) Effect of welding parameters on tensile properties and fracture behavior of friction stir welded Al-Mg-Si alloy. Scripta Mater 56:69–72. https://doi.org/10.1016/j.scriptamat.2006.08.054

Effect of Machining Parameters and MQL Parameter on Material Removal Rate in Milling of Aluminium Alloy

Kamaljeet Singh, Anoop Kumar Singh, and K. D. Chattopadhyay

Abstract In the current work, Taguchi technique is applied on aluminium alloy (Al-6061) during face milling to optimize the machining parameters and minimum quantity lubrication (MQL) parameter for increasing the material removal rate (MRR). The MQL method is applied during face milling of Al-6061using coated carbide inserts. Coolant flow rate (CFR), cutting velocity, feed per revolution, and depth of cut are selected as input parameters. Taguchi L_9 (3^4) orthogonal array is chosen for four input parameters with three levels for experimentation. Signal-to-noise (*S*/*N*) ratio is calculated based on "larger the better" for each experiment. Analysis of means is used to evaluate the optimum settings of input parameters. Significance and contribution of each input parameter is evaluated by using analysis of variance (ANOVA). Results of ANOVA show that depth of cut has a major contribution of 80.287% on material removal rate.

Keywords Face milling · MQL · ANOVA · *S*/*N* ratio · CFR · MRR

1 Introduction

Cutting fluids are used in machining processes for achieving desired machining performances (low surface roughness, high material removal rate, etc.) [1–4]. As a result of the latest environmental protection laws, the use of cutting fluids should be less during machining. So, to overcome this problem, MQL was used as an alternative with respect to wet or dry cooling system. MQL technique combines the effect of dry and wet machining. MQL benefits in many areas, i.e., production costs, legal laws, environmental laws, and human wellness [5–8].

MQL technique was applied on aluminium alloy during reaming process with coated carbide tool inserts and shows comparable results in comparison to dry machining [9]. Authors have investigated that MQL technique could be used in respect of flood machining without much affecting the machining results [10]. MQL

K. Singh (✉) · A. K. Singh · K. D. Chattopadhyay
Chitkara University Institute of Engineering and Technology, Chitkara University, Punjab, India
e-mail: kamaljeet.singh@chitkara.edu.in

© Springer Nature Singapore Pte Ltd. 2021
P. M. Pandey et al. (eds.), *Advances in Production and Industrial Engineering*, Lecture Notes in Mechanical Engineering,
https://doi.org/10.1007/978-981-15-5519-0_27

was used for turning of steel with a mixture of air and lubricant oil in machining show good results with respect to flood cooling system [11]. The effect of MQL, cutting velocity, and feed rate on the brass specimen under different lubricant conditions environment concludes that MQL lubrication technique performance is very much comparable to flood lubrication [12]. Author has found that the different machining performance improves by 10% with the use of MQL in comparison with flood cooling system [13].

This study is based on the analysis of machining parameters and CFR on MRR in milling of aluminium alloy (Al-6061) using carbide tool on vertical machining center (HURCO VM10). Four input parameters, i.e., cutting velocity, feed per revolution, depth of cut, and CFR are evaluated at three different levels by using L_9 (3^4) orthogonal array. Taguchi method is used to find the optimum setting of input parameters by using *S/N* ratio values. Further contribution of each input process parameter is evaluated by using ANOVA [14, 15].

2 Methodology

In this Study, face milling is performed on Al-6061 using MQL cooling technique. CFR is varied; with fixed nozzle distance of 150 mm and nozzle angle of 45°. MOTUL Lubricant—SAFCO RUBRIC SZ 32—20L is used as a coolant to make mist with constant compressed air at pressure of 3 bar. Suction is generated by passing the high velocity air through venturi to supply the mixture of air and lubricant oil for machining. Further CFR is controlled by air-lubricant control valves which are fixed on venturi as shown in Fig. 1(b).

2.1 *Taguchi Method*

Taguchi method statistically designed the experiments with the help of different orthogonal arrays to attain better results with less number of experiments and also minimize the cost and time of experimentation. Further, ANOVA is used to study the response variation by using *S/N* ratio and to measure the percentage contribution of each input parameter. *S/N* ratio is categorized into three types: smaller the better, larger the better, and nominal the best [16, 17]. Highest value of *S/N* ratio among different levels is considered as an optimum level of a particular input process parameter which is applicable to all the three types of *S/N* ratio. This paper provides the best possible combination of cutting velocity, feed per revolution, depth of cut, and CFR to improve the MRR of Al-6061 alloy with minimal use of coolants without affecting the environment and human health.

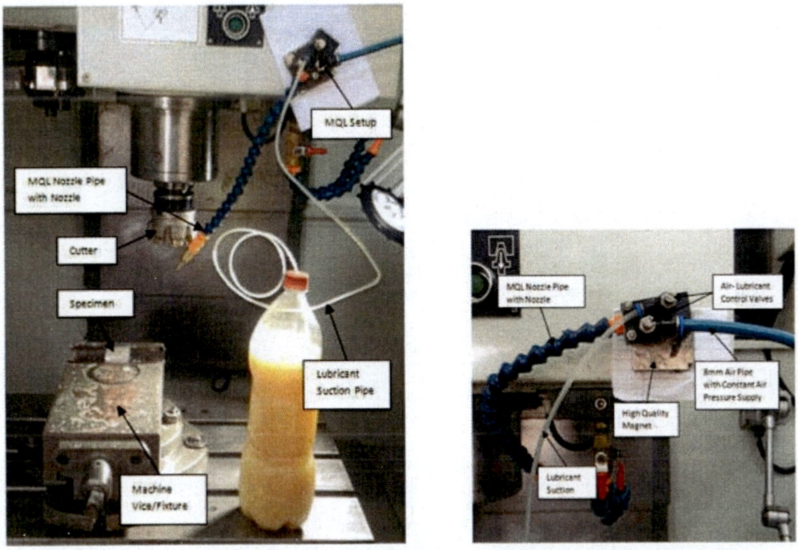

(a) VMC (HURCO-VM10) machine used for Experimental Work (b) Details of MQL setup

Fig. 1 **(a)** VMC (HURCO-VM10) machine used for experimental work **(b)** details of MQL set-up

2.2 Design of Experiments

In this experiment, Taguchi method is used with four factors at three levels of each input parameter. According to this, nine experiments are performed with L_9 orthogonal array [18]. All the experiments are performed three times, and these experiments are carried out at random to avoid the error. In this orthogonal array, each input parameter has three levels which are represented by "1", "2", or "3" as given in Table 1.

3 Experimentation

3.1 Experimental Setup

In this study, Aluminium alloy (Al-6061) is chosen as substrate material because of its vast applications in the manufacturing and automobile industry. Alloying elements of Al-6061 are presented in Table 2 [19]. VMC (HURCO-VM10) machine is used for face milling operation as shown in Fig. 1a and the details of MQL setup are represented by Fig. 1b with the help of carbide insert. The cutter of 80 mm diameter with equally spaced seven numbers of cutting inserts is used for face milling. Specimens of size 36 mm × 36 mm × 12 mm are prepared as per L_9 orthogonal array (with

Table 1 L_9 orthogonal array [18]

Trial No	A	B	C	D
1	1	1	1	1
2	1	2	2	2
3	1	3	3	3
4	2	1	2	3
5	2	2	3	1
6	2	3	1	2
7	3	1	3	2
8	3	2	1	3
9	3	3	2	1

A, B, C, D—Input Parameters

Table 2 Alloying elements of Al-6061 (wt%) [19]

Material	Si	Fe	Cu	Mg	Cr	Other impurities	Al
Al-6061	0.4–0.8	Max. 0.7	0.15–0.40	0.8–1.2	0.04–0.35	0.15–0.25	Balance 95.8–98.6

three replications) as shown in Fig. 2 (Before machining and after machining). The experiments are performed at different cutting velocity, feed per revolution, depth of cut, and coolant flow rate.

(a) Specimens (Before Machining) (b) Specimens (After Machining)

Fig. 2 Aluminium alloy-6061 specimens **(a)** before machining **(b)** after machining

Table 3 Details of input parameters with levels

Input parameters	Code	Unit	Levels		
			1	2	3
Cutting velocity	C_V	m/min	300	400	500
Feed per revolution	F_{PR}	mm/rev	0.14	0.28	0.42
Depth of cut	D_{OC}	mm	0.3	0.4	0.5
Coolant flow rate	C_{FR}	ml/h	80	115	150

3.2 *Identification of Input Factors with levels.*

The four input parameters Cutting velocity (C_V), Feed per revolution (F_{PR}), Depth of cut (D_{OC}), and Coolant flow rate (C_{FR}) are considered with three levels and their related orthogonal array $L_9(3^4)$ is chosen, and they are are tabulated in Table 3. As per the experimental design, the specimens have been machined in vertical milling machine (HURCO-VM10) by using various levels of the input parameters. The milling operation is performed under different coolant flow rate of MQL (Lubricated with MOTUL—Lubricant—SAFCO RUBRIC SZ 32—20L).

3.3 *Material Removal Rate (MRR).*

Material removal rate (MRR) is taken as an output which is calculated with the help of digital weighing scale of specimens (before machining and after machining) along with machining time as shown in Fig. 3a and Fig. 3b and their related equation for calculating MRR is shown in Eq. (1).

(a) Initial Weight (Before Machining) (b) Final Weight (After Machining)

Fig. 3 Weight measurement

Table 4 Results of experimentation

Exp. no	C_V	F_{PR}	D_{OC}	C_{FR}	Avg. MRR (gms/min)	*S/N* ratio (dB)
1	1	1	1	1	7.893	17.945
2	1	2	2	2	7.513	17.516
3	1	3	3	3	10.013	20.011
4	2	1	2	3	9.367	19.431
5	2	2	3	1	10.917	20.761
6	2	3	1	2	7.017	16.922
7	3	1	3	2	11.843	21.469
8	3	2	1	3	7.957	18.014
9	3	3	2	1	8.413	18.499

$$\text{MRR} = (\text{Initial weight in gms} - \text{Final weight in gms})/\text{Machining Time (in min)} \tag{1}$$

For each specimen, an average of three different observations is calculated to avoid the measurement error. The results of the average MRR and their related *S/N* ratio are given in Table 4.

As per Taguchi philosophy, response output is measured in signal-to-noise ratio (*S/N* ratio). Instead of response average values, *S/N* ratio considers the variability in the results of experiments which improves the accuracy and precision of data analysis.

4 Result with Discussion

As MRR is to be maximized, so "larger the better type" quality feature is selected for evaluating the *S/N* ratio of MRR and is determined by using Eq. (2) [20, 21]:

$$\frac{S}{N}\text{ratio} = -10 * \log\left(\frac{1}{n}\sum_{i=1}^{n}\frac{1}{y_i}^2\right) \tag{2}$$

where y_i is the response value for *i*th experiment and n is the number of replications.

An investigation is carried out to evaluate the effect of input parameters on material removal rate. The obtained *S/N* ratio from Eq. (2) and the result of material removal rate is given in Table 4.

The result of an analysis of means in the main effect plot of input parameters for MRR is presented in Fig. 4. Higher *S/N* ratio value is the best suitable level for the particular parameter. Cutting velocity with level 3, feed per revolution with level 1, depth of cut with level 3, and coolant flow rate with level 3 reveal higher *S/N* ratio.

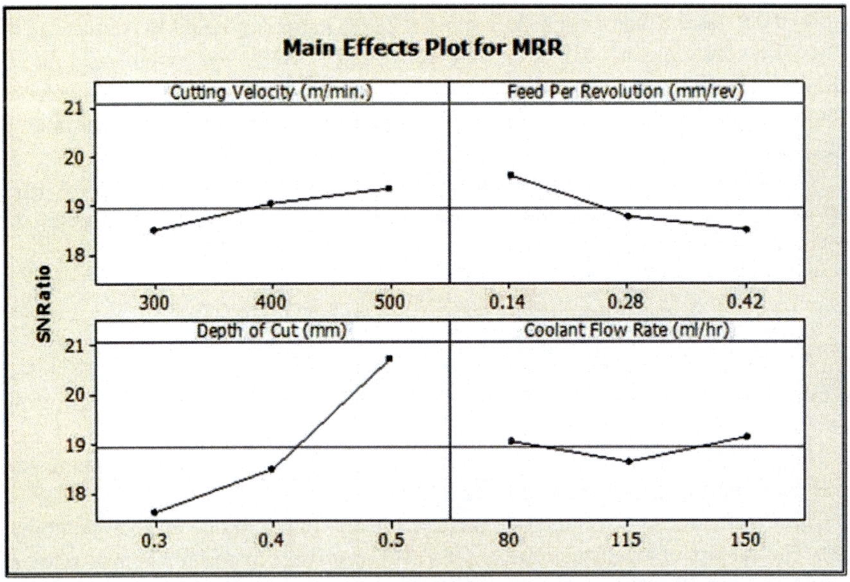

Fig. 4 Main effect plot of input parameters on MRR

Depth of cut has maximum range between highest and lowest *S*/*N* ratio value shown in main effect plot. It means that depth of cut has maximum contribution in variance. So, as per the maximum *S*/*N* ratio values of all input parameters, the optimum settings for the current study are $C_{V3}F_{PR1}D_{OC3}C_{FR3}$. Therefore, the optimum levels of input parameters are cutting velocity (C_{V3}) with 500 m/min, feed per revolution (F_{PR1}) with 0.14 mm/rev, depth of cut (D_{OC3}) with 0.5 mm, and coolant flow rate (C_{FR3}) with 150 ml/h.

Table 5 represents the result of ANOVA for MRR. Significance of parameter is checked on the basis of critical value (4.41) of *F*-Test at 95% confidence level. If *F*-ratio value of particular parameter is greater than the critical value then it will be considered a significant parameter. Results of Table 5 show that all the parameters are significant with *F*-ratio greater than 4.41. Outcomes of ANOVA show that the

Table 5 Results of ANOVA

Input parameters	DOF	SS	Variance	F-ratio	Percentage contribution
C_{V3}	2	1.084	0.542	73.014	5.553
F_{PR}	2	2.101	1.050	141.559	10.839
D_{OC}	2	15.466	7.733	1042.156	80.287
C_{FR}	2	0.461	0.231	31.067	2.319
Error	18	0.134	0.007		1.002
Total	26	19.245	0.740		100

depth of cut has the major contribution of 80.287% in maximizing MRR followed by feed per revolution of 10.839%, cutting velocity of 5.553%, and coolant flow rate of 2.319%. ANOVA table also indicates that only 1.002% error contribution is present; which clearly indicates that the effect of interactions between the input parameters is insignificant.

After finding out the optimum values of input parameters ($C_{V3}F_{PR1}D_{OC3}C_{FR3}$), further it is required to evaluate predicted optimum value of MRR through the equation given below [20]:

$$\Psi\text{opt} = \Psi\text{m} + \sum\nolimits_{k=1}^{n} (\Psi\text{m}_i - \Psi\text{m}) \tag{3}$$

where Ψopt is the predicted optimum MRR, Ψm is the overall mean of MRR, Ψm_i is the mean of optimum level and n is the number of input parameters.

The predicted *S/N* ratio of MRR by using Eq. (3) for optimum input parameter level combination is 21.986 dB and 12.573 gms/min.

Confirmation tests have been performed for optimum parameter level combination. The results of confirmation tests for *S/N* ratio and material removal rate are 21.871 dB and 12.403 gms/min. Confirmation test reveals that an improvement of 4.510 gms/min (in terms of MRR) and 36.35% (in terms of percentage) can be attained by using optimum values of input parameters ($C_{V3}F_{PR1}D_{OC3}C_{FR3}$) in comparison with the initial settings ($C_{V1}F_{PR1}D_{OC1}C_{FR1}$).

In this study, it is summarized that cutting velocity at level 3, feed per revolution at level 1, depth of cut at level 3, and CFR at level 3 is required to increase the MRR values.

5 Conclusion

In the current research work, optimization of material removal rate during face milling of aluminium alloy (Al-6061) is evaluated using Taguchi technique. L_9 orthogonal array is used to perform the experiments. The optimal quantity of coolant flow rate and the most suitable depth of cut, cutting velocity, and feed per revolution are determined by using analysis of means and their significance is measured through ANOVA. The conclusions of this study are given below:

- The main effect plot indicates that optimum cutting velocity with 500 m/min, feed per revolution with 0.14 mm/rev, depth of cut with 0.5 mm, and coolant flow rate with 150 ml/h are the required settings to optimize MRR.
- The depth of cut (80.287%) is the main contributor followed by feed per revolution (10.839%), cutting velocity (5.553%), and coolant flow rate (2.319%) to optimize the MRR.
- As the error contribution is 1.002%, the effect of interactions is neglected in this study.

- An improvement of 36.35% is achieved by using optimum input parameters settings ($C_{V3}F_{PR1}D_{OC3}C_{FR3}$) in comparison with the initial settings ($C_{V1}F_{PR1}D_{OC1}C_{FR1}$).

References

1. Naidu GG, Vishnu AV, Raju GJ (2014) Optimization of process parameters for surface roughness in milling of EN-31 steel material using Taguchi robust design methodology. Int J Mech Prod Eng **2(9)**, (2014).
2. Vishnu AV, Tilak KBG, Naidu GG, Raju GJ (2015) Optimization of different process parameters of aluminium alloy 6351 in CNC milling using Taguchi method. Int J Eng Res Gen Sci. **3(2)**, (2015).
3. Ross PJ (2005) Taguchi techniques for quality engineering, 2nd edn. Tata McGraw Hill
4. Thakre AA (2013) Optimization of milling parameters for minimizing surface roughness using Taguchi's approach. Int J Emer Technol Adv Eng. 3(6):226–230
5. Vishnu AV, Naidu GG, Srikanth P, Chowdary SV, Sharan SS (2016) Experimental study on the high speed machining of steel alloy. Int J Sci Res Dev 30–34
6. Gupta MK, Sood P (2017) Machining comparison of aerospace materials considering minimum quantity cutting fluid: a clean and green approach. Proc Inst Mech Eng, Part C: J Mech Eng Sci. 231(8):1445–1464
7. Sarikaya M, Gullu A (2015) Multi-response optimization of minimum quantity lubrication parameters using Taguchi-based grey relational analysis in turning of difficult-to-cut alloy Haynes 25. J Cleaner Prod. 91:347–357
8. Braga DU, Diniz AE, Miranda GWA, Coppini NL (2002) Using a minimum quantity of lubricant (MQL) and a diamond coated tool in the drilling of aluminium–silicon alloys. J Mater Process Technol 122(1):127–138
9. Lugscheider E, Knotek O, Barimani C, Leyendecker T, Lemmer O, Wenke R (1997) Investigations on hard coated reamers in different lubricant free cutting operations. Surf Coat Technol 90(1–2):172–177
10. Fratila D, Caizar C (2011) Application of Taguchi method to selection of optimal lubrication and cutting conditions in face milling of AlMg3. J Cleaner Prod. 19(6–7):640–645
11. Dhar NR, Islam MW, Islam S, Mithu MAH (2006) The influence of minimum quantity of lubrication (MQL) on cutting temperature, chip and dimensional accuracy in turning AISI-1040 steel. J Mater Process Technol 171(1):93–99
12. Davim JP, Sreejith PS, Silva J (2007) Turning of brasses using minimum quantity of lubricant (MQL) and flooded-lubricant conditions. Mater Manuf Process 22(1):45–50
13. Kumar S, Singh D, Kalsi NS (2017) Analysis of surface roughness during machining of hardened AISI 4340 steel using minimum quantity lubrication. Mater Today: Proc. 4(2):3627–3635
14. Taguchi G (1990) Introduction to quality engineering. Asian Productivity Organization, Tokyo
15. Montgomery DC (1991) Design and analysis of experiments. Wiley, Singapore
16. Kwak JS (2005) Application of Taguchi and response surface methodologies for geometric error in surface grinding process. Int J Mach Tools Manuf 45(3):327–334
17. Phadke MS (1989) Quality engineering using robust design. Prentice-Hall, Englewood Cliffs. NJ
18. Unal R, Dean EB (1990) Taguchi approach to design optimization for quality and cost: an overview. In: Proceedings of the international society of parametric analyst 13th annual, pp 21–24
19. Muraca RF, Whittick JS (1972) Materials data handbook: aluminium alloy 6061, 2nd edn. Western Applied Research and Development, Inc.

20. Tosun N, Huseyinogly M (2010) Effect of MQL on surface roughness in milling of AA7075-T6. Mater Manuf Process 25(8):793–798
21. Phadke MS (1995) Quality engineering using robust design. Prentice Hall, PTR

Influence of Cutting Force and Drilling Temperature on Glass Hole Surface Integrity During Rotary Ultrasonic Drilling

Ankit Sharma, Atul Babbar, Vivek Jain, and Dheeraj Gupta

Abstract The surface integrity of the machined surface is having a significant impact on the application of the glass and ceramic materials. Surface integrity is of paramount importance to the industries to prevent the loss of useful life of critical components. However, conventional machining processes are unable to fulfill the demands of the present scenario but hybrid machining processes such as rotary ultrasonic drilling process has shown its potential for enhanced surface integrity. Nowadays, researchers are focused on getting superior machined glass quality because of the hardness and esthetic look. Therefore, the authors performed an experimental investigation to estimate the surface roughness while drilling a hole in glass specimen using rotary ultrasonic machining (RUD). It is concluded that the best drilling condition to get the least value of average surface roughness (167.702 nm) is the usage of coolant with 5000 rpm of spindle speed. At 5000 rpm with coolant supply, the cutting force and drilling temperature have been reduced. It is noticed that the least value of drilling temperature is 54.7 °C with a cutting force of 26.65 N. Finally, the improved surface integrity has been achieved noteworthy using RUD process. It would directly enhance the life span of the glass component and increase its functional usage.

Keywords Float glass · Rotary ultrasonic machining · Surface roughness

1 Introduction

The applications of the glass (brittle material) have been rapidly increasing worldwide which has attracted the attention of numerous researchers. Surface integrity is of paramount importance to the industries to prevent the loss of useful life of critical components. However, conventional machining processes are unable to fulfill the

A. Sharma (✉)
Chitkara College of Applied Engineering, Chitkara University, Rajpura, Punjab, India
e-mail: ankit.sharma@chitkara.edu.in

A. Babbar · V. Jain · D. Gupta
Department of Mechanical Engineering, Thapar Institute of Engineering and Technology, Patiala, India

© Springer Nature Singapore Pte Ltd. 2021
P. M. Pandey et al. (eds.), *Advances in Production and Industrial Engineering*, Lecture Notes in Mechanical Engineering,
https://doi.org/10.1007/978-981-15-5519-0_28

demands of the present scenario but hybrid machining processes such as rotary ultrasonic drilling process has shown its potential for enhanced surface integrity [1]. Low material removal rate, less surface finish, high wear, and geometrical inaccuracies are always associated with the conventional machining process [2]. On another side, RUD utilizes the material removal mechanism of the conventional drilling and vibrational drilling which leads to enhanced surface integrity and high material removal rate [3]. The cutting force developed during the drilling process is always a concern as it may affect the surface integrity [4]. Kumar and Singh [5] investigated the effect of machining parameters on surface roughness during rotary ultrasonic drilling of BK-7 glass. Parameters such as rotational speed, feed rate, and ultrasonic power have been evaluated at three different states. It was found that the feed rate has more influence on the output characteristics followed by rotational speed and ultrasonic power. Micrographs highlight the plastically deformed regions when drilling is performed at a low rotational speed. Wang et al. [6] proposed a model to predict the edge chipping during rotary ultrasonic drilling (RUD) considering the cutting forces and subsurface cracks which can significantly affect the surface roughness. The experiments were conducted on optical K9 glass followed by confirmatory trials. The relationship was established for ultrasonic amplitude, material properties, and cutting force. Both theoretical and experimental results were in good agreement with each other. In another study, reduction in the cracks was observed during RUD of brittle material in comparison to the conventional drilling process [7]. Liu et al. [8] outlined that the cutting parameters during rotary ultrasonic machining can be optimized by knowing the cutting forces. A mechanistic model was presented assuming the brittle fracture as the main contribution towards material removal. The relationship between numerous machining parameters such as speed, feed rate, frequency, amplitude, abrasives concentration, and size of abrasives was established with respect to cutting force. The predicted results were further validated and a positive relationship was found with experimental findings. The effect of rotary ultrasonic machining has been explored by past researchers for numerous processing parameters [3, 9–17]. The investigations of the effect of drilling force and temperature produced during drilling have been scarce and need further studies to enhance the surface integrity during rotary ultrasonic drilling for various industry-oriented applications. Therefore, the present study uses a novel RUD process, which was recently developed in the last decade to explore the effect of cutting force and temperature.

2 Materials and Methods.

The complete experimental setup has been installed to make the hole in the workpiece. The setup consists of rotary ultrasonic arrangements like CNC milling machine, rotary ultrasonic horn assembly, power generator, dynamometer, infrared camera, abrasive coated hollow drill tool, and coolant arrangement. The complete experimentation set up and the instruments used for the characterization are illustrated as shown in Fig. 1. The rotary ultrasonic machine is having a vibration frequency of 20 kHz along with its machine could displace in all three (X–Y–Z) axis directions.

The float glass is selected as a workpiece specimen (Fig. 1). It contains the composition of silica sand, dolomite, limestone, broken cullets, soda ash, and coloring agent, etc. A hollow abrasive coated drill tool is used for the study. The tool outer and inner diameter is 10 and 8 mm approximately. The abrasives that are coated over the tool lateral and end face are having mesh number around 120. The tool length is taken as 50.75 mm.

In the pilot study of experimentation, the fixed parameters/working conditions, which are used throughout the study has been mentioned in Table 1.

During this pilot study, the range of spindle speed is taken as 1000, 2000, 3000, 4000, and 5000 rpm. Overall, 10 holes have been created. Out of them, five holes are created at with-coolant condition and other five holes are drilled at without-coolant condition. It would help to visualize the dual experimentation study that incorporates the effect of various range of cutting speed and coolant on/off conditions. The other parameters remain fixed as stated in Table 1.

In situ, the drilling holes, the dynamometer (Kistler Dynoware type 9272), and infrared camera (Keysight Technologies-U5855A) have been installed to measure the cutting force and drilling temperature simultaneously. So that the instant effect of cutting force and drilling temperature over the drilled hole surface integrity has

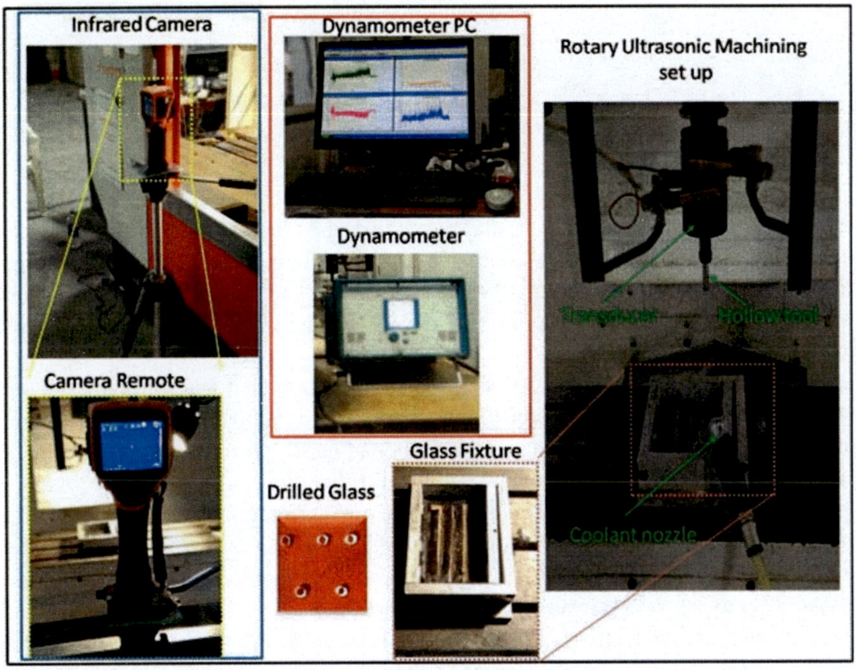

Fig. 1 Experimentation setup and the instruments used for the characterization

Table 1 Illustration of fixed working condition used for drilling

S. no	Parameters/Working conditions	Detail/Remarks
1	Feed rate	6 mm/min
2	Vibration amplitude	20 μm
3	Hollow abrasive coated tool dimensions	O.D.—10 mm I.D.—8 mm
4	Coolant	Water
5	Float glass specimen size	Length—148 mm Breadth—80 mm Thickness—5 mm

been evaluated. Here, the dynamometer is used to measure the cutting force in the longitudinal direction of the glass specimen. The cutting force data are saved in the commercial software which is further used for the analysis. Along with it, an infrared camera is used to evaluate the maximum drilling temperature for each drilled hole. The thermal resolution and temperature range of the camera are 320 × 240 pixels and 0–350 °C, respectively. A total of 400 frames are captured with 8 frames per second. The 'TrueIR Analysis and Reporting tool' software has been used for capturing the thermal images. The temperature is measured at a fixed drilling region. In this study, atomic force microscopy (AFM) is used to evaluate the drilled hole surface roughness. The sample position range is 5 × 5 mm.

3 Results and Analysis

During drilling operation (Fig. 2), it is noticed that coolant plays a crucial for any spindle speed. It is noticed that whenever the coolant is in on condition the cutting force is significantly declined from beginning to the end of drilling. Label diagram (Fig. 2) represents the cutting force (F_c) versus Time (t) graph for minimum (1000 rpm) and maximum (5000 rpm) spindle speeds at with-coolant and without-coolant conditions.

Maximum cutting force (98.23 N) is noticed at 1000 rpm and without-coolant condition and the least cutting force (26.65 N) is found at 5000 rpm and with-coolant condition. After achieving the maximum values of cutting force, the force declined dramatically after that second. It shows that the drilled hole is created.

After analyzing the cutting force, drilled hole surface integrity has been evaluated. The surface roughness is the key focus to work on surface integrity. Atomic force microscopy technique has been selected to get the 3D topography of the drilled glass surface. Figure 3 shows the surface roughness profile of the drilled hole region for a float glass specimen. The high-resolution 3D image shows the roughness profile at spindle speed 5000 rpm with coolant on. This combination of drilling conditions is observed as the best parameter combination for drilling an 8-mm hole in float

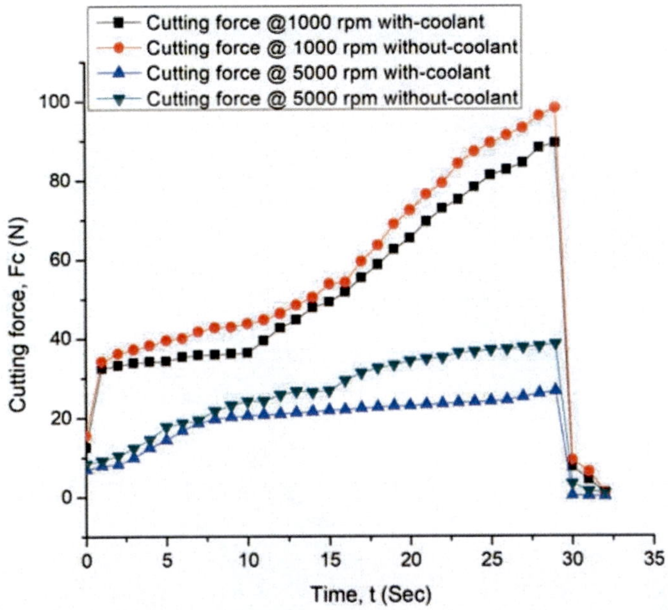

Fig. 2 Cutting force (F_c) versus time (t) for minimum and maximum spindle speeds at with-coolant and without-coolant conditions

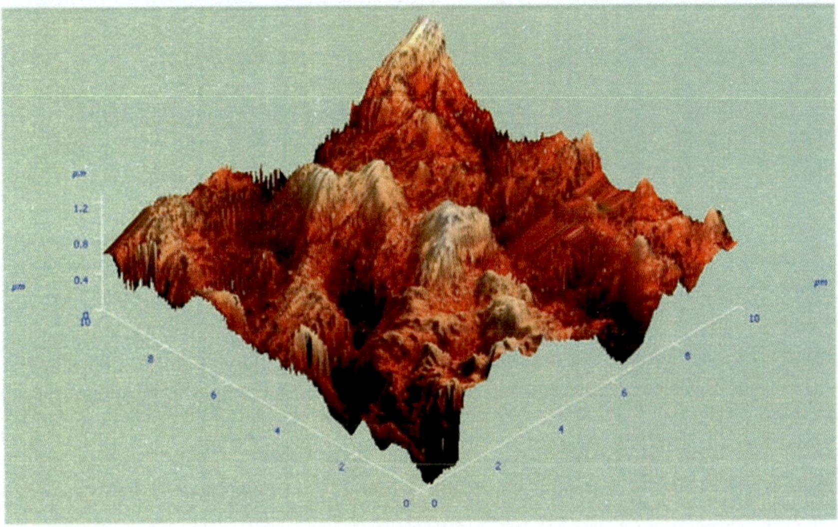

Fig. 3 Surface roughness profile of the drilled hole region for a float glass specimen at spindle speed 5000 rpm with coolant on condition using a tool of 8 mm diameter

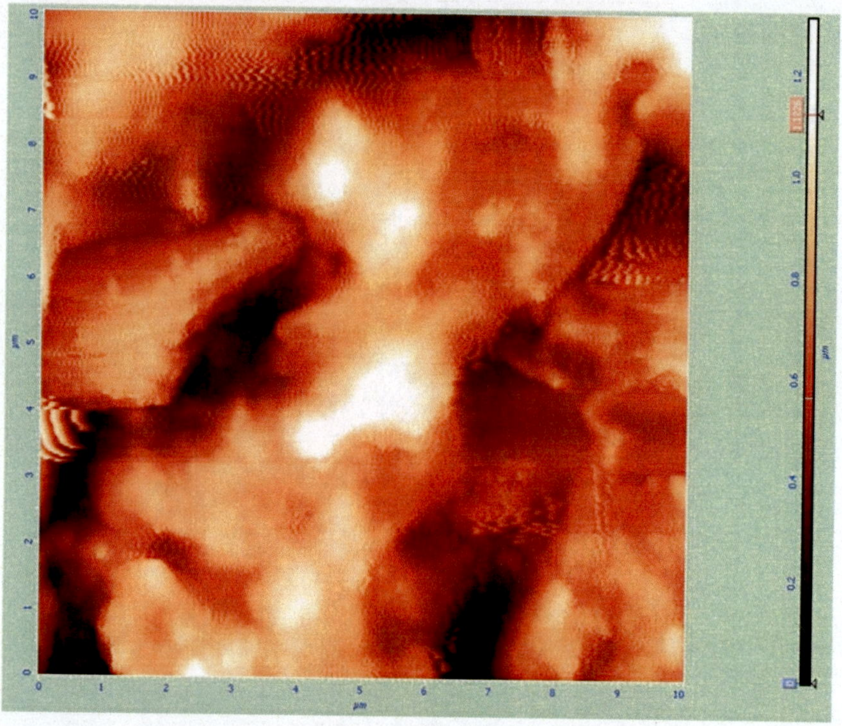

Fig. 4 3D topography of the drilled glass hole surface at best drilling condition

glass. At the best drilling condition, the maximum peak region value of surface roughness (R_{max}) and the average surface roughness (R_a) are reported as 1309.75 nm and 167.702 nm, respectively. Figure 4 represents the 2D view of the roughness profile of the drilled glass surface at best drilling condition. The surface roughness image and cutting force chart (Fig. 2) shows that the cutting force having a key effect on the surface roughness. Because the least values of cutting force are observed at 5000 rpm, similarly the least value of average surface roughness is noticed at 5000 rpm. Along with it, with-coolant drilling is suitable to achieve the least cutting force. Therefore, it is stated that the drilled hole surface roughness is having a considerable prominence by the cutting force.

Table 2 shows the various roughness factors at the best parameter arrangement. It is found that the average roughness (R_a) value and root mean square roughness (R_q) values are found as 167.702 nm and 212.77 nm, respectively. Other major roughness factors are illustrated in Table 2.

Finally, the influence of coolant presence at a spindle speed of 5000 rpm has been investigated by figuring out the in situ temperature that occurred in s fixed drilling region (as shown in Fig. 5). Figure 5a, b shows the infrared images at best parameter (i.e. spindle speed-5000 rpm): (a) At with-coolant condition and (b) At without-coolant condition. The infrared images help to visualize the influence of

Table 2 Illustration of various surface roughness factors at spindle speed 5000 rpm and with-coolant condition

S. no.	Roughness type	Roughness value(s)
1	Measuring range	10 × 10 μm
2	Average roughness, R_a	167.702 nm
3	Mean height of roughness in ten points, R_z	646.97 nm
4	Surface roughness, R_{max}	1309.75 nm
5	Root Mean Square (RMS) roughness, R_q	212.77 nm

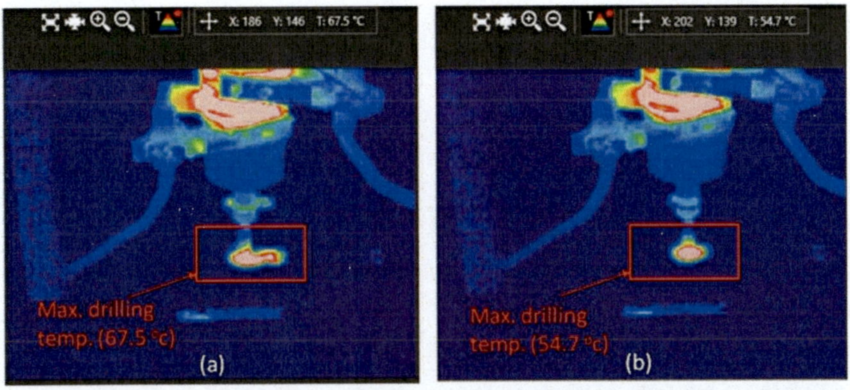

Fig. 5 Infrared images at the best parameter (spindle speed—5000 rpm) to achieve least surface roughness values: **a** at with-coolant condition, **b** at without-coolant condition

cutting temperature on surface roughness. It is revealed that the maximum temperature reached is up to 54.7 °C during float glass drilling while using coolant, whereas, without-coolant conditions, the drilling temperature rises to 12.8 °C.

For both ('with-coolant' and 'without-coolant') the drilling conditions, the values of average surface roughness (R_a) at the drilled hole region has been presented. The drilling parameters: feed rate, vibration amplitude, and tool diameter, etc. (refer: Table 1) has been kept constant in the study except for spindle speed. Thus, the effect of spindle speed is explained thoroughly. Figure 6 shows the graphical representation of the effect of 'Spindle speed' with respect to 'Average surface roughness' without-coolant condition. Figure 7 depicts the illustration of the 'Spindle speed' versus 'Average surface roughness' with-coolant condition.

It is observed visually in the graphical images (Figs. 6 and 7) that the surface roughness value has been mitigated by increasing the spindle rotation speed from 1000 to 5000 rpm. The cognition behind the employ of higher spindle speed is that it directly diminished the effective cutting force (refer Fig. 2). Because the cutting force reduction leads to a decrease in the drilling temperature that offers by-product as superior hole surface integrity [9]. The same drilling temperature trend is noticed in thermal infrared images (Fig. 5) that the with-coolant condition, lesser drilling

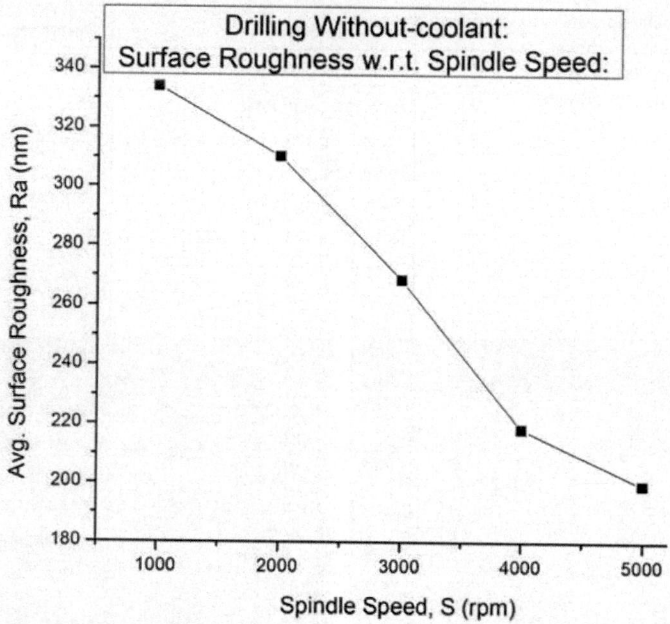

Fig. 6 Graphical representation of the effect of 'Spindle speed' with respect to 'Average surface roughness' without-coolant condition

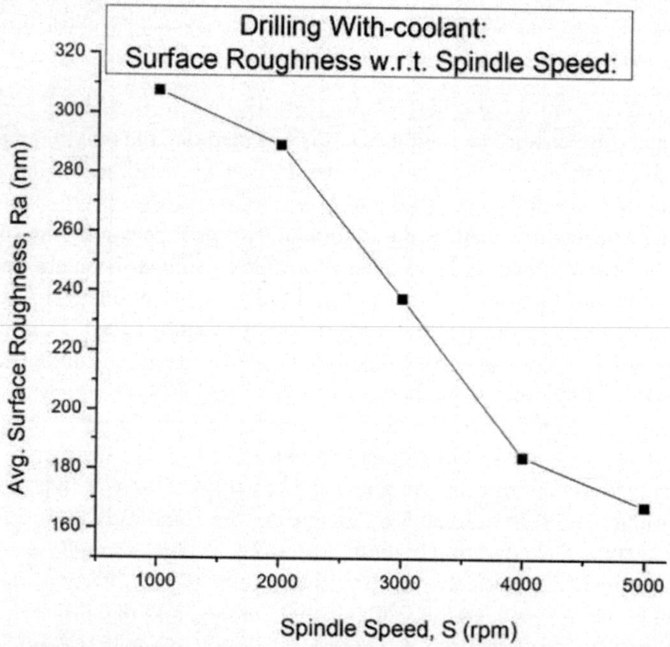

Fig. 7 The effect of 'Spindle speed' versus 'Average surface roughness' with-coolant condition

temperature is observed visually. This reduction in temperature prominence, owing to the reduction in friction between the tool and workpiece drilled to be region [18]. Finally, the propagation of stress is diminished that subsequently enhanced the overall drilled hole quality.

The maximum and minimum values of surface roughness (R_a) at the time of 'without-coolant' drilling are observed 333.892 nm at 1000 rpm and 199.314 nm at 5000 rpm, respectively, where the maximum and minimum roughness (R_a) values during 'with-coolant' drilling are reported 307.375 nm at 1000 rpm and 167.702 nm at 5000 rpm, respectively. Hence, the authors recommend the employ of 'with-coolant condition' to get the least surface roughness. Overall, it is stated that cutting force and drilling temperature have a significant influence on the glass hole integrity using RUD process. It would directly enhance the life span of the glass component and increase its functional usage.

4 Conclusion

During rotary ultrasonic drilling of float glass specimen, the effect of drilling force and cutting temperature has been presented while observing the hole surface integrity. The key accomplished outcomes such as the best drilling condition are revealed as 5000 rpm of cutting speed and with-coolant conditions to generate the least value of average surface roughness (R_a), i.e., 167.702 nm. It has been investigated that the usage of coolant with 5000 rpm of cutting speed has decreased the cutting force by 11.88 N and the drilling temperature is also reduced by 12.8 °C. The ultimate cause of mitigating the cutting force and drilling temperature is stress reduction at the time of drilling. Therefore, the fine surface integrity has been achieved noteworthy using the rotary ultrasonic drilling process.

References

1. Debnath K, Singh I, Dvivedi A (2015) Rotary mode ultrasonic drilling of glass fiber-reinforced epoxy laminates. J Compos Mater 49(8):949–963. https://doi.org/10.1177/0021998314527857
2. Singh RP, Singhal S (2016a) Rotary ultrasonic machining: a review. Mater Manuf. Process 31:1795–1824
3. Li ZC, Jiao Y, Deines TW (2005) Rotary ultrasonic machining of ceramic matrix composites: feasibility study and designed experiments. Int J Mach Tools Manuf 45(12–13):1402–1411. https://doi.org/10.1016/j.ijmachtools.2005.01.034
4. Zhang C, Zhang J, Feng P (2013) Mathematical model for cutting force in rotary ultrasonic face milling of brittle materials. Int J Adv Manuf Technol 69:161–170. https://doi.org/10.1007/s00170-013-5004-z
5. Kumar V, Singh H (2019) Optimization of rotary ultrasonic drilling of optical glass using Taguchi method and utility approach. Eng Sci Technol Int J 22(3):956–965. https://doi.org/10.1016/j.jestch.2019.02.004

6. Wang J, Feng P, Zhang J (2016) Modeling the dependency of edge chipping size on the material properties and cutting force for rotary ultrasonic drilling of brittle materials. Int J Mach Tools Manuf 101:18–27. https://doi.org/10.1016/j.ijmachtools.2015.10.005
7. Wang J, Zha H, Feng P, Zhang J (2016) On the mechanism of edge chipping reduction in rotary ultrasonic drilling: a novel experimental method. Precis Eng 44:231–235. https://doi.org/10.1016/j.precisioneng.2015.12.008
8. Liu D, Cong WL, Pei ZJ, Tang Y (2012) A cutting force model for rotary ultrasonic machining of brittle materials. Int J Mach Tools Manuf 52(1):77–84. https://doi.org/10.1016/j.ijmachtools.2011.09.006
9. Sharma A, Babbar A, Jain V, Gupta D (2018) Enhancement of surface roughness for brittle material during rotary ultrasonic machining. MATEC Web Conf 249:01006. https://doi.org/10.1051/matecconf/201824901006
10. Sharma A, Jain V, Gupta D (2018) Characterization of chipping and tool wear during drilling of float glass using rotary ultrasonic machining. Measurement 128:254–263. https://doi.org/10.1016/j.measurement.2018.06.040
11. Babbar A, Singh P, Farwaha HS (2017) Regression model and optimization of magnetic abrasive finishing of flat brass plate. Indian J Sci Technol 10:1–7. https://doi.org/10.17485/ijst/2017/v10i31/113860
12. Babbar A, Jain V, Gupta D (2019) Neurosurgical bone grinding. In: Biomanufacturing. Springer International Publishing, Cham, pp 137–155
13. Babbar A, Sharma A, Jain V, Jain AK (2019) Rotary ultrasonic milling of C/SiC composites fabricated using chemical vapor infiltration and needling technique. Mater Res Express 6:085607. https://doi.org/10.1088/2053-1591/ab1bf7
14. Singh D, Babbar A, Jain V (2019) Synthesis, characterization, and bioactivity investigation of biomimetic biodegradable PLA scaffold fabricated by fused filament fabrication process. J Brazilian Soc Mech Sci Eng 41:121. https://doi.org/10.1007/s40430-019-1625-y
15. Fernando P, Zhang M, Pei Z, Cong W (2017) Intermittent and continuous rotary ultrasonic machining of K9 glass: an experimental investigation. J Manuf Mater Process 1:20. https://doi.org/10.3390/jmmp1020020
16. Singh RP, Singhal S (2016b) Rotary ultrasonic machining: a review. Mater Manuf Process 31:1795–1824. https://doi.org/10.1080/10426914.2016.1140188
17. Zeng WM, Li ZC, Pei ZJ, Treadwell C (2005) Experimental observation of tool wear in rotary ultrasonic machining of advanced ceramics. Int J Mach Tools Manuf 45:1468–1473. https://doi.org/10.1016/j.ijmachtools.2005.01.031
18. Deng G, Lu Z, Yao G, Liu J, Li Z, Zhang D (2017) Cutting temperature and resulting influence on machining performance in rotary ultrasonic elliptical machining of thick CFRP. Int J Mach Tools Manuf 123:160–170

Analysis on Development of Beeswax as Phase Change Material for Thermal Energy Storage

Durgesh Kumar Mishra, Sumit Bhowmik, and Krishna Murari Pandey

Abstract The current review article is focused on the development of beeswax as phase change material (PCM) for thermal energy storage. Beeswax is an organic non-paraffin PCM, which is suitable for heat energy storage. But the main hamper of the beeswax during energy storage is less thermal conductivity and leakage during phase transformation. So, the researchers have been started attacking to resolve this kind of problem. During this fashion they were started to add high thermal conductivity particles with beeswax to improve thermal conductivity and at the same time leakage prevention porous material to prevent leakage during phase transformation. During the review, it is established that expanded graphite, graphene, carbon nanotube, expanded perlite carbon fiber, copper oxide, polymer, dammer gum, and tallow were used for conductivity enhancement and leakage preventing agent.

Keywords Beeswax · Phase change material · Thermal conductivity · Leakage prevention · Heat storage capacity

1 Introduction

As time is moving, the demand for energy is increasing continuously. So, for fulfilling the gap between demand and supply researchers have to focus on the energy management. Phase change material (PCM) is one of the most utilizing materials for thermal energy management. It is utilized for storage and management of waste heat in solar panel, solar water system, battery thermal management, electronic system, etc. There are three different kinds of PCM, which are classified as organic, inorganic, and eutectic; where organic is divided into paraffin and non-paraffin, inorganic is categorized as metal, alloy, salt, etc., and eutectic is divided into the combination of organic + inorganic and inorganic + organic [1–6]. PCM stores energy in the form of sensible heat and latent heat, but the latent heat storage is most utilizing. Latent

D. K. Mishra · S. Bhowmik · K. M. Pandey (✉)
Department of Mechanical Engineering, National Institute of Technology Silchar, Silchar 788010, Assam, India
e-mail: kmpandey2001@yahoo.com

© Springer Nature Singapore Pte Ltd. 2021

P. M. Pandey et al. (eds.), *Advances in Production and Industrial Engineering*, Lecture Notes in Mechanical Engineering,
https://doi.org/10.1007/978-981-15-5519-0_29

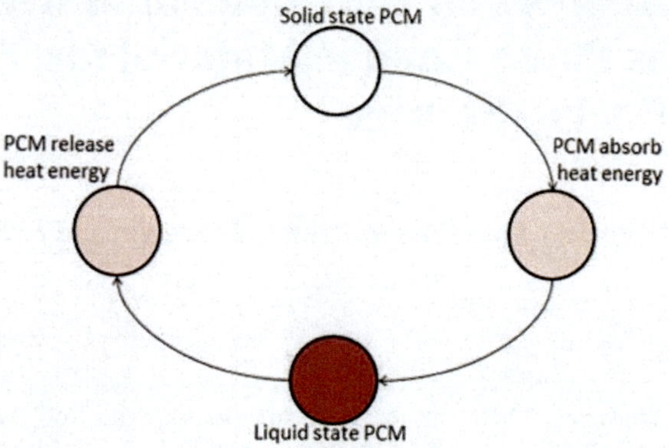

Fig. 1 Working cycle of PCM

heat accumulates energy at constant temperature condition. PCM works in a cyclic manner which is given in Fig. 1, it stores energy and becomes liquefied after that at less temperature zone PCM release energy and solidified [7–12]. Beeswax is an organic non-paraffin PCM with the empirical formula as $C_{15}H_{31}COOC_{30}H_{61}$ which is produced with metabolic process of bees, and wax is extracted through the abdominal segments of bees [13–15]. It is utilized as PCM because of high heat storage capacity. But the main hamper of the beeswax during energy storage is less thermal conductivity and leakage during phase transformation [16]. For removing this huddle, various researchers incorporate properties gaining materials like expanded graphite (EG), graphene, expanded perlite (EP), and multiwall carbon nanotubes (MWCNTs) with beeswax. In the current article, it is reviewed how the beeswax is developing as an enhanced property PCM for thermal energy storage.

2 Development of Beeswax Composite PCM

In the recent year, beeswax is developing as PCM because it is naturally available and contains the properties which are suitable for the application in solar water heater and building given in Tables 1 and 2. The explanation of the researcher for development of beeswax as PCM is discussed in the next section.

2.1 *With Incorporation of EG*

Dinker et al. [17] developed a novel kind of composite PCM for energy storage with a combination of beeswax and EG. The composite was fabricated through melt mixing method where EG is mixed in molten beeswax. During the experimental analysis, it was found that thermal conductivity of the composite enhanced from 0.29 to 0.63 W/m-K that reduced the melting time from 1020 to 900 min. From the

Table 1 Thermal properties of beeswax and their composites

References	Materials composition	Thermal conductivity(W/m-K)		Heat storage capacity (J/g)		Melting temperature (°C)	
		Beeswax	Composite	Beeswax	Composite	Beeswax	Composite
[17]	Beeswax + EG (10%)	0.2998	0.9741	214	210		59.89
[19]	Beeswax + Graphene (0.05%, 0.1%, 0.15%, 0.2%, 0.25%, 0.3%)	0.25	1.15, 1.5, 1.88, 2.1, 2.3, 2.89	141.49	152.66, 154.25, 163.23, 163.40, 180.07, 186.74	62.28	62.05, 62.59, 62.59, 61.57, 62.12, 62.42
[20]	Beeswax + tetradecanol + Carbon Fiber (2%) + EP	0.24	1.245	194.5	178.7	63.1	34
[21]	(i) Beeswax + Acid treated CNTs (5%, 20%)	0.25	0.46, 0.58	153.2	115.5, 91.6	62.3	60.2, 59.8
[23]	Beeswax + CuO (0.05%, 0.1%, 0.15%, 0.2%, 0.25%)	0.25	0.53, 1.55, 1.91, 1.97, 2.07	359.29	315.12, 207.43, 170.07, 199.67, 176.49	64.22	63.62, 63.59, 63.66, 63.19, 62.45
[24]	Beeswax + Polymer + Silica	–	–	155.3	84.1	62.4	61.7

Table 2 Journal name and outcome of research

Reference	Journal	Outcome of the research
[17]	Experimental Heat Transfer	1. It is shown that no chemical reaction between beeswax and EG 2. Thermal conductivity of the composite was enhanced by 117% 3. Melting point temperature of the composite was slightly lower than pure beeswax 4. Decrement in the heat storage capacity 5. Utilized in heating water unit
[19]	Applied Thermal Engineering	1. There was no agglomeration of graphene within beeswax 2. New compounds did not form after synthesis 4. Thermal conductivity on the composite has increased with mass fraction 5. Latent heat of the composite was also increased as the mass ratio increased 6. Used in building energy management
[20]	Composites: Part A	1. Found that there is only physical interaction between the materials no chemical interaction 2. It revealed that EP is not only utilized for mechanical strength but also for leakage prevention 3. Melting temperature of composite reduced drastically 4. Latent heat of composite reduced slightly 5. Used for solar energy utilization
[21]	Journal of Energy Storage	1. Acid treated CNTs are more suitable for leakage prevention and thermal conductivity 2. It is found that 132% increment in thermal conductivity 3. Melting temperature and latent heat of composite reduced slightly 4. Used for solar energy utilization in buildings
[23]	International Journal of Technology	1. Modified MWCNTs is used for shape stability 2. Thermal conductivity of the composite increased gradually according to mass fraction 3. Melting temperature and latent heat of composite reduced slightly 4. The authors suggested using this composite for building application

(continued)

Table 2 (continued)

Reference	Journal	Outcome of the research
[24]	Advanced Materials Interfaces	1. Polymer and Silica are utilized for encapsulation 2. Melting temperature decreased slightly 3. Heat storage capacity of the PCM also decreased 4. The authors suggested to utilize the material for coating in electronics items as well as in building internal and external structure

physical characterization, it is explained that EG formed regular network inside the beeswax and few of the beeswax was absorbed in the porous network. Experimental results also revealed that there is no chemical reaction occurred within the composite. Thermal analysis of the material revealed that melting point of the composite was slightly lesser than the pure beeswax. The authors also studied the performance of thermal energy storage material with the help of shell and tube heat storage unit. The prepared material was filled in the shell and hot water passed through the pipe as a heat transfer fluid.

Dinker et al. [18] scrutinized beeswax composite with EG as heat energy storage material in shell-tube type thermal energy storage unit. During the fabrication, 10% of the EG to be incorporated with the beeswax and it is found that the increment of thermal conductivity by 117%, at the same time melting temperature of the composite changed slightly. The author studied the thermal performance of thermal energy storage unit at 0.25, 0.50, and 1.0 LPM flow rate and 60, 70, and 80 °C inlet temperature. It was found that with the addition of EG the charging time of the composite reduced at high flow rate and high inlet temperature.

It was also suggested that with low flow rate thermal storage efficiency enhanced to maximum and at high temperature and low flow rate combination charging time reduced by 33% when it was comparing through beeswax. Thermal storage efficiency improved from 84 to 87% when comparing pure beeswax and composite beeswax. The authors were also inspired to use beeswax because low cost and naturally availability. The improved thermal efficiency of storage unit filled with composite material is due to its enhanced thermal conductivity and reduced charging time. Results of the study on beeswax–EG composite material are boosting and it can be utilized for thermal storage applications.

2.2 *With Incorporation of Graphene*

Amin et al. [19] experimentally investigated the physical and thermal properties of beeswax/graphene composite PCM to decrease the energy feeding in the buildings. Ultrasonic method is used for synthesis of composite. It was found from the results that complete dispersion of graphene into the beeswax and there was no chemical reaction occurred within the material. Results revealed that with the incorporation of graphene latent heat, thermal conductivity and viscosity has been increased linearly. Thermal conductivity and heat storage capacity of the composite increased from 0.25 to 2.89 W/m-K and 141.49 to 186.74 J/g, when the mass fraction of graphene increased from 0.05 to 0.3%.

2.3 *With the Incorporation of EP*

Cheng et al. [20] studied a composite PCM which is prepared after the mixing of beeswax/tertradecanol/carbon fiber/expanded perlite with the help of magnetic stirrer and vacuum impregnation. Microstructure, chemical compatibility, and thermal properties of the fabricated composite were analyzed. It was found that no chemical interaction between the raw materials but there was only physical combination. Beeswax/tetradeconal/carbon fiber was sufficiently absorbed into the porous EP structure which could prevent the leakage during molten state. The melting temperature of the composite reduced drastically as compared with pure beeswax and latent heat was also slightly reduced. From the thermal cycling test, it is indicated that after 200 charge–discharge cycle the composite has sufficient stability. Thermal conductivity of the composite PCM was also increased from 0.24 to 1.245 W/m-K. Finally, the authors suggested that the prepared form-stable composite PCM has more suitable thermal properties and thermal stability for solar energy utilization in buildings.

2.4 *With the Incorporation of MWCNTs*

Putra et al. [21] discussed the physical, chemical, and thermal properties of beeswax/CNT composites for thermal storage. The authors tried to discuss the two main drawbacks of PCM, first one is low thermal conductivity and second is reduction in volume during phase transition. For the above drawbacks, a shape stabilized composite PCM has been developed which prevents leakage at the time of phase transition. Here three kinds of MWCNTs were used as a supporting material, named as: ball-milled MWCNTs, pristine MWCNTs, and acid-treated MWCNTs. The results indicated that acid-treated MWCNTs were more useful in the case of thermal conductivity as well as for shape stability. 5 and 20% CNTs were mixed with the beeswax and the samples were examined after 0 cycles, 150 cycles, and 300 cycles. Results indicated that the thermal conductivity of the composite increased by 132% in the case of 20% of CNTs but melting temperature and latent heat have been decreased. The authors suggested utilizing this kind of composite in the field of building thermal regulation.

Rawi et al. [22] experimentally investigated physical, chemical, and thermal properties of beeswax/MWCNTs composite as shape stabilized PCM. For the experimentation, two samples were prepared, first one was pure beeswax and second Beeswax/MWCNTs composite. MWCNTs were used to produce shape stable material, which was modified and then added with beeswax. It was seen that with the addition of 5 wt% modified CNTs thermal conductivity of the composite increased by 84%. Results also indicated that the composite contained good physical and thermal compatibility. Thermal characterization of the composite revealed that melting temperature and latent heat was 59.79 °C and 91.64 J/g. The authors suggested that the proposed composite PCMs were used for the building energy management.

2.5 *With the Incorporation of CuO*

Putra et al. [23] examined the behavior of beeswax and CuO as composite PCM for thermal energy storage. Five samples were prepared with the 0.05, 0.1, 0.15, 0.2, and 0.25 wt% of CuO and found that the melting temperature of samples was 63.62 °C, 63.59 °C, 63.66 °C, 63.19 °C, and 62.45 °C, respectively. Differential scanning calorimetry was used for finding out melting temperature and thermal capacity. Results also revealed that there is no chemical reaction between the materials. The presence of CuO nanoparticles improved thermal conductivity of PCM at the same time slightly reduced the heat-storing capacity. Nonetheless, the change in heat storage capacity caused no significant effects in the performance of beeswax/CuO. The results also showed that the melting rate of PCM increased and melts PCM faster.

2.6 *With the Incorporation of Polymer*

Naderizadeh et al. [24] investigated the PCM used for coating applications in electronics items and building interior or exterior part, which fulfills the thermal energy management. During the current research, authors used beeswax as PCM and polymer and silica particles for the properties improving materials. The fabrication of coating material was through emulsification of beeswax in commercial polymer dispersion. The silica nanoparticles were spread during emulsion at different concentrations for superhydrophobicity. Results showed that due to emulsification beeswax encapsulated by the polymer as a nanothin shell that prevents the leakage during phase transformation. It was also suggested by the authors that the present coating was ecofriendly.

2.7 *With the Incorporation of Dammer Gum and Tallow*

Umar et al. [25] investigated the properties of concrete contain beeswax as a PCM. During the current study beeswax, dammar gum, and tallow were used to make enhanced properties composite PCM. With the help of T-history method author try to find out thermal conductivity, melting temperature, and enthalpy. Results indicated that beeswax and tallow exhibit great potential for latent heat, whereas dammer gum has been utilized for thermal conductivity enhancement. It was also seen that as the PCM containing in the concrete increased, the compressive strength of the concrete decreased drastically. From this study, it could be said that PCM has been utilized as thermal energy storage material for buildings rather than construction material.

3 Conclusions and Future Scope

After reviewing recent papers it can be concluded that.

- Beeswax is suitable PCM for solar energy system and building energy management when incorporating with properties gaining material like EG, EP, MWCNTs, graphene, CuO, polymer, dammer gum, and tallow.
- Thermal conductivity of the composite increases gradually when the mass fraction of the properties gaining material increases.
- Leakage prevention during the phase change also occurs, when porous material is utilized to prepare composite.

In the future scope, it is suggested that finding new combination of material which can be utilized as properties gaining material and leakage preventing agent for various engineering applications.

Acknowledgements The authors are thankful to TEQIP-III of National Institute of Technology, Silchar for their financial support to present the review paper at an international conference.

References

1. Pielichowska K, Pielichowski K (2014) Phase change materials for thermal energy storage. Prog Mater Sci 65:67–123
2. Ibrahim NI, Al-Sulaiman FA, Rahman S, Yilbas BS, Sahin AZ (2017) Heat transfer enhancement of phase change materials for thermal energy storage applications: a critical review. Renew Sustain Energy Rev 74:26–50
3. Bhowmik C, Bhowmik S, Ray A, Pandey KM (2017) Optimal green energy planning for sustainable development: a review. Renew Sustain Energy Rev 71:796–813
4. Safari A, Saidur R, Sulaiman FA, Xu Y, Dong J (2017) A review on super cooling of phase change materials in thermal energy storage systems. Renew Sustain Energy Rev 70:905–919

5. Li M, Wu Z (2012) A review of intercalation composite phase change material: preparation, structure and properties. Renew Sustain Energy Rev 16(4):2094–2101
6. Khan Z, Khan Z, Ghafoor A (2016) A review of performance enhancement of PCM based latent heat storage system within the context of materials, thermal stability and compatibility. Energy Convers Manag 115:132–158
7. Chandel SS, Agarwal T (2017) Review of current state of research on energy storage, toxicity, health hazards and commercialization of phase changing materials. Renew Sustain Energy Rev 67:581–596
8. Lin Y, Jia Y, Alva G, Fang G (2018) Review on thermal conductivity enhancement, thermal properties and applications of phase change materials in thermal energy storage. Renew Sustain Energy Rev 82:2730–2742
9. Zhang P, Xiao X, Ma ZW (2016) A review of the composite phase change materials: fabrication, characterization, mathematical modeling and application to performance enhancement. Appl Energy 165:472–510
10. Mohamed SA, Al-Sulaiman FA, Ibrahim NI, Zahir MH, Al-Ahmed A, Saidur R, Yılbaş BS, Sahin AZ (2017) A review on current status and challenges of inorganic phase change materials for thermal energy storage systems. Renew Sustain Energy Rev 70:1072–1089
11. Quanying Y, Chen L, Lin Z (2008) Experimental study on the thermal storage performance and preparation of paraffin mixtures used in the phase change wall. Sol Energy Mater Soar Cells 92(11):1526–1532
12. Bashirnezhad K, Kebriyaee SA, Moosavi A (2018) The experimental appraisement of the effect of energy storage on the performance of solar chimney using phase change material. Sol Energy 169:411–423
13. Xiao X, Zhang P, Li M (2013) Preparation and thermal characterization of paraffin/metal foam composite phase change material. Appl Energy 112:1357–1366
14. Jackson MA, Eller FJ (2006) Isolation of long-chain aliphatic alcohols from beeswax using lipase-catalyzed methanolysis in supercritical carbon dioxide. J Supercrit Fluids 37(2):173–177
15. Dinker A, Agarwal M, Agarwal GD (2017a) Experimental study on thermal performance of Beeswax as thermal storage material. Mater Today Proc 4(9):10529–10533
16. Sharma A, Tyagi VV, Chen CR, Buddhi D (2009) Review on thermal energy storage with phase change materials and applications. Renew Sustain Energy Rev 13(2):318–345
17. Dinker A, Agarwal M, Agarwal GD (2017b) Preparation, characterization, and performance study of beeswax/expanded graphite composite as thermal storage material. Exp Heat Transf 30(2):139–150
18. Dinker A, Agarwal M, Agarwal GD (2018) Experimental performance analysis of beeswax/expanded graphite composite for thermal energy storage in a shell and tube unit. Int J Green Energy 15(11):585–595
19. Amin M, Putra N, Kosasih EA, Prawiro E, Luanto RA, Mahlia TM (2017) Thermal properties of beeswax/graphene phase change material as energy storage for building applications. Appl Therm Eng 112:273–280
20. Cheng F, Wen R, Zhang X, Huang Z, Huang Y, Fang M, Liu YG, Wu X, Min X (2018) Synthesis and characterization of beeswax-tetradecanol-carbon fiber/expanded perlite form-stable composite phase change material for solar energy storage. Compos A Appl Sci Manuf 107:180–188
21. Putra N, Rawi S, Amin M, Kusrini E, Kosasih EA, Mahlia TMI (2019) Preparation of beeswax/multi-walled carbon nanotubes as novel shape-stable nanocomposite phase-change material for thermal energy storage. J Energy Storage 21:32–39
22. Rawi S, Amin M, Kusrini E, Putra N (2018) Characterization of shape-stabilized phase change material using beeswax and functionalized multi-walled carbon nanotubes. In: IOP conference series: earth and environmental science, vol 105, p 012042

23. Putra N, Prawiro E, Amin M (2016) Thermal properties of beeswax/CuO nano phase-change material used for thermal energy storage. Int J Technol 7(2):244–253
24. Naderizadeh S, Heredia-Guerrero JA, Caputo G, Grasselli S, Malchiodi A, Athanassiou A, Bayer IS (2019) Superhydrophobic coatings from Beeswax-in-water emulsions with latent heat storage capability. Adv Mater Interfaces 6(5):1801782
25. Umar H, Rizal S, Riza M, Mahlia TMI (2018) Mechanical properties of concrete containing beeswax/dammar gum as phase change material for thermal energy storage. AIMS Energy

Evaluation of Material Handling Using MCDM Techniques: A Case Study

Pardeep Kumar Verma, Raman Kumar, and Gyanendra Singh Goindi

Abstract Choosing the best possible Material Handling Equipment (MHE) is a significant assignment in MSEs due to the extensive capital involved. There are numerous substantial and impalpable factors affecting the selection process of appropriate MHE. Multi-criteria decision-making (MCDM) has been observed to be a helpful way to deal with and break down these clashing components. The assessment of MHE alternatives from different criteria and the weight of these different criteria are normally communicated in various ways. This paper proposes an MCDM strategy for assessment and selection of MHE type for small-scale industry in Punjab. Two-phase is used for getting weights to different criteria and for their ranking enabling us to get more reliable results required for the selection of suitable MHE. Analytical Hierarchy Process (AHP) and Technique for Order Preference by Similarity to Ideal Solution (TOPSIS) are utilized to compute the priority weightage and ranking of alternates, respectively. Selection of the most appropriate MHE is our objective while taking into account the tangible and intangible criteria.

Keywords MHE selection · Ranking · Decision-making

1 Introduction

Raw material, work in process and even finished product flows in the premises of facility such as industry from one working station to another, from final working station to packing, from packing to store to complete its intended purpose in the facility. Material flow is the description of the transportation of material as a flow of entities, which is accomplished through human power or other equipment types. It consists of the lifting, movement and dropping off material while caring to avoid its pilferage and minimum time consumption and cost-effective. So, the problem of

P. K. Verma (✉) · R. Kumar · G. S. Goindi
Department of Mechanical Engineering, Chandigarh University, Gharuan, Punjab, India
e-mail: gne.ldh99@gmail.com

© Springer Nature Singapore Pte Ltd. 2021

P. M. Pandey et al. (eds.), *Advances in Production and Industrial Engineering*, Lecture Notes in Mechanical Engineering,
https://doi.org/10.1007/978-981-15-5519-0_30

MHE selection may become single or multi-objective, multi-criteria. While considering these factors, selections of MHE become a decision to be taken by the management. The decision-maker plays a vital role to decide the approach leading to making the decision. Material handling is responsible for 25–87% in personnel, space, time related to production and product cost [1]. Our purpose is to identify the most appropriate MHE from the available options of its category for the desired level of performance ranging from low to high with the objectives to minimum investment, minimization of operational cost, material handling and time related to it, minimum number of equipment used for material handling and maximization of compatibility between part and its handling equipment.

High-level selection considers MHE from entirely different categories, such as cranes to conveyors robots to rail, intermediate level takes up selection within the category such as powered hand truck and forklift truck, and the low-level selection debates on the model of MHE within the types such as an appropriate model of AGV.

When choosing what MHE to utilize, it is essential to consider the general attributes of the equipment types available in the market. At that point, decision-maker needs to decide which MHE matches better to the required application.

In general categories of MHE used in industries are Static Storage Systems such as Storage Racks, Drawer Storage, Mezzanines, Powered and Non-powered Industrial Trucks, a fixed Path Conveyor such as gravity conveyors, powered conveyors, roller conveyors, vertical conveyors, a variable Path Conveyor such as Automatic Guided Vehicles (AGV), monorails, Work Positioning Systems such as cranes, electric hoists, pallet trucks [2].

Different approaches are used to solve the questione raised regarding the appropriate MHE. Majorly, these approaches can be categorized into four, i.e. Multi-Criteria Decision-Making (MCDM), Artificial Intelligence (AI), optimization and Hybrid. Some of the most popular MCDM methods are Analytic Hierarchy Process (AHP), Analytic Network Process (ANP), utility models, goal programming, Data Envelopment Analysis (DEA), Simple Multi-Attribute Ranking (SMART), outranking methods, TOPSIS and disaggregate–aggregate approaches [3–7].

In a manufacturing unit, selecting an appropriate MHE assumes a significant role in expanding the viability and effectiveness of the complete unit [8]. Engineers are normally looked with the challenge of picking the correct MHE without essential experience or recognition. The guidance from MSE providers is free, yet providers need to influence clients to purchase their items and their recommendation. Therefore, it has low dependability as they offer their own products in which they are dealing and have a good margin. Interestingly, systematic models dependent on MCDM are less utilized for MHE selection in MSEs [9].

As of date, every consumer has access to a variety of MHE, all of them possess different aspects along with pros and cons, different price ranges, making the selection procedure hard. Decision-maker involved needs to consider different quantitative (e.g., volume and weight of the load, distance to be moved, cost, etc.) and subjective (e.g., shape of load, load type, maintenance of MHE) criteria. Along these lines, we can examine the selection of MHE as the MCDM problem [10].

The selection of a materials handling system (MHS) is a cumbersome task in a flexible manufacturing system (FMS) for its dependency on multiple factors. ANP is fit for considering both the internal and external dependence factors up to 35 or even greater than that. As the criterion grows, complication arises in calculations making it less advisable to solve shop floor issues [11].

The AHP technique was used to find the weightage of factors while using AHP and MC simulation model for MHE selection. The managerial guidelines were provided for the evaluation of MHE [12].

Usage of the Fuzzy Multi-Attribute Selection of MHE (FUMAHES) to select the best conveyor from four available alternatives and six selection criteria by developing a Decision Support System (DSS) can be helpful for decision-makers [13]. Modified Grey Relational Analysis (M-GRA) was employed for AGV seletion. Utility index of AGV was calculated by using M-GRA method [14]. MHE selection problem was solved by using the Weighted Utility Additive (WUTA) method in conveyor systems [15]. Fuzzy VIseKriterijumska Optimizacija I Kompromisno Resenje (FVIKOR) approach was used to select optimal MHE through weighing 4 and 20 criteria and sub-criteria, respectively, by the voting method [16]. AHP is a viable instrument that can be used by decision-makers for better decision-making in complex situations having multiple criteria and options [17].

Mostly in MHE selection, the level is high, such issues under multi-criteria when a number of options are less applied in MCDM approach. When the decision is to be taken by human and the options are limited, MCDM technique works well. While reviewing the literature, it exhibits that the MCDM approach can possibly take care of the issues related to the selection of appropriate MHE selection problems issue in vagueness.

The organization of the remaining paper is as per the following: Section 2 describes the present work and the last segment finishes up results and scope for further research.

2 Present work

A. Profile of Case Company

The case unit is engaged in the manufacturing of a variety of hand tools. The company is located in the northern part of India which has been chosen to conduct the industrial case study. Approximately 100 employees have been working in the various sections of the company. The leading manufacturing industry is facing problems in delays of material movement from one work station to another, accumulation of Work In Process (WIP), as small bins are used for the storage of WIP on different workstations. Battery-operated forklift, pallet jacks, hand trucks, manual forklifts, die loader and overhead cranes are used for material movement within the premises. A cross-functional team (CFT) of five employees including shop incharge, supervisor and operators from case company was involved in brainstorming and group decision-making for selecting the attributes and alternates. In this model, three criteria, i.e. operational controllability, maintenance and compatibility and nine sub-criteria were implicated as these criteria were assumed to be major inputs for the selection of MHE in MSEs. The MCDM problem bifurcated into smaller constituent in three-level hierarchy diagram and the lowest levels depict the alternatives showing the relationships among objectives and alternatives through links in Fig. 1.

Transforming linguistic expressions into quantitative values [18] describes fuzzy method that reflects the exact linguistic descriptions in terms of crisp scores. Consequently, it gives a better guess of linguistic descriptions that are generally utilized. The 5-point scale is considered to exhibit the transformation of fuzzy numbers into crisp scores as given in Table 1.

Implementation of Analytical Hierarchy Process:

Step 1: The level of inclination or decision-maker's intensity in the decision of each pair-wise correlation has been utilized in this model is measured on a scale of 1–9, where even numbers represent the compromises among the preferences above as shown in Table 2

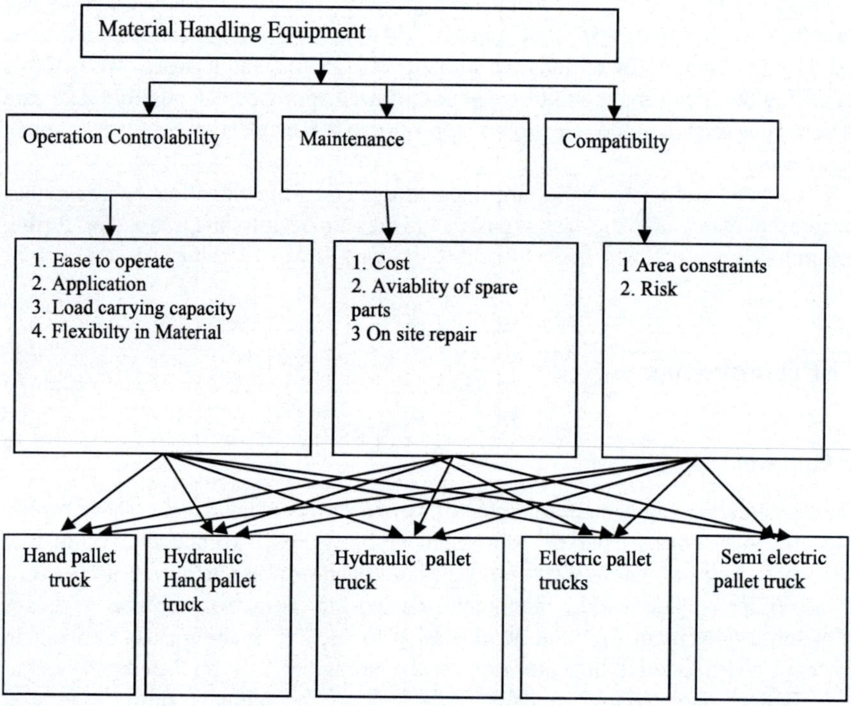

Fig. 1

Table 1 Conversion of fuzzy numbers into crisp scores

Linguistic term	Fuzzy number	Crisp score
High	M1	0.895
Above average	M2	0.695
Average	M3	0.495
Below average	M4	0.295
Low	M5	0.115

Table 2 Nine-point intensity of importance scale and its description

Definition	Intensity of importance
Equally important	1
Moderately more important	3
Strongly more important	5
Very strongly more important	7
Extremely more important	9
Intermediate values	2, 4, 6, 8

Step 2: Screening of different pallet trucks on certain basic criteria and sub-criteria fulfilling the requirements of the manufacturing unit. The selected criteria and sub-criteria are listed in Table 3

Step 3: Pair-wise comparison of different sub-objectives

The significance of *i*th sub-objective is contrasted and *j*th sub-objective is determined Pair-wise observation of different sub-objectives. In the analysis of sub-criteria, i.e. ease to operate is slightly more important than its application, that's why it was assigned two numbers during brainstorming of cross-functional team aforementioned as shown in Table 4.

Step 4: Normalized matrix

Table 3 Criteria selected

Criteria	Sub-criteria
Operational controllability	Ease to operate
	Application
	Load-carrying capacity
	Power required
	Flexibility in material
Maintenance	Cost
	Availability of spare parts
	On-site repair
Compatibility	Area constraints
	Risk
	Distance to be moved

Table 4 Pairwise comparison matrix

Ease to operate	Application	Load-carrying capacity	Flexibility in material	Cost	On-site repair	Area constraints	Availability of spares part	Risk
1.00	2.00	3.00	4.00	0.33	5.00	5.00	7.00	1.00
0.50	1.00	2.00	5.00	0.25	3.00	4.00	5.00	0.50
0.33	0.50	1.00	2.00	0.20	5.00	4.00	4.00	0.33
0.25	0.20	0.50	1.00	0.20	3.00	4.00	5.00	0.25
3.00	4.00	5.00	5.00	1.00	3.00	5.00	5.00	2.00
0.20	0.33	0.20	0.33	0.33	1.00	2.00	2.00	0.50
0.20	0.25	0.25	0.25	0.20	0.50	1.00	2.00	0.50
0.14	0.20	0.25	0.20	0.20	0.50	0.50	1.00	0.25
1.00	2.00	3.03	4.00	0.50	2.00	2.00	4.00	1.00

The estimations of normalized matrix r_{ij} are determined as given in the below-mentioned formulae, i.e. by dividing each value in a column by the sum of squares of values in that particular column

$$r_{ij} = a_{ij} \Big/ \sum_{i=1}^{n} a_{ij}^2$$

where a_{ij} refers to the inputs received through brainstorming session among the CFT which was further simplified using the conversion of fuzzy numbers into a crisp score. Normalized matrix of different sub-objective is shown in Table 5.

The approximate priority weight ($W_1, W_2, \ldots\ldots, W_j$) for each attribute is obtained as shown in Table 6 and in graphical format in Appendix 1

$$nW_j = GM_j \Big/ \sum_{j=1}^{n} GM_j$$

3 Topsis Method

The TOPSIS (Technique for Order Performance by Similarity to Ideal Solution) was first developed [19] and it explains the best alternatives which are nearest to the positive ideal solution [20].

Step 1: The objective is to evaluate the 11 factors and the factors are ease to operate, application, load-carrying capacity, power required, flexibility in material,

Table 5 Normalized matrix of different sub-objective

	Ease to operate	Application	Load-carrying capacity	Flexibility in material	Cost	On-site repair	Area constraints	Availability of spares part	Risk
Hand pallet truck	0.894	0.295	0.204	0.251	0.245	0.558	0.595	0.546	0.095
Hydraulic Hand pallet truck	0.115	0.534	0.618	0.591	0.551	0.558	0.595	0.546	0.408
Hydraulic pallet truck	0.295	0.414	0.480	0.591	0.153	0.433	0.462	0.546	0.573
Electric pallet trucks	0.115	0.534	0.480	0.421	0.612	0.308	0.196	0.232	0.573
Semi-electric pallet truck	0.295	0.414	0.342	0.251	0.489	0.308	0.196	0.232	0.408

Table 6 Priority weights of attributes

Sub-criteria	Priority weights
Ease to operate	0.183
Application	0.127
Load-carrying capacity	0.094
Flexibility in material	0.068
Cost	0.259
On-site repair	0.050
Area constraints	0.040
Availability of spares parts	0.030
Risk	0.150

cost, availability of spare parts, on-site repair, area constraints, risk, distance to be moved.

Step 2: Formation of a decision matrix.

Step 3: Calculate the values of normalized matrix and priority weight. These values were calculated in AHP part.

Step 4: Calculation of weighted normalized decision matrix by multiplying the normalized decision matrix by its associated weights, which will enable us to get the positive ideal value and negative ideal value for each sub-criterion. The weighted normalized decision matrix is shown in Table 7.

$$Vij = Wj \times Rj$$

Step 5: Determination of best solution:

The positive-ideal (best) and negative-ideal (worst) solutions are determined as the most favourable values and least favourable values for each sub-criterion. The positive-ideal (best) and negative-ideal (worst) solutions can be expressed as given below and values are shown in Table 8 and graphically in Appendix 2;

$$V^{+} = \{(\overset{\max}{\underset{i}{\sum}} V_{ij} / j \in J), (\overset{\min}{\underset{i}{\sum}} V_{ij} / j \in J') / i = 1,2,3,\ldots\ldots,N\},$$

$$= \{V_1^{+}, V_2^{+}, V_3^{+}, V_4^{+}, \ldots\ldots V_M^{+}\}$$

$$V^{-} = \{(\overset{\min}{\underset{i}{\sum}} V_{ij} / j \in J), (\overset{\max}{\underset{i}{\sum}} V_{ij} / j \in J') / i = 1,2,3,\ldots\ldots,N\},$$

$$= \{V_1^{-}, V_2^{-}, V_3^{-}, V_4^{-}, \ldots\ldots V_M^{-}\}$$

Step 6: Calculations of separation measures:

Calculate the separation measures, using the *n*-dimensional Euclidean distance. The separation of each alternative from the positive-ideal solution and the negative-ideal solution is given by the following equations.

Table 7 Weighted normalized decision matrix

	Ease to operate	Application	Load-carrying capacity	Flexibility in material	Cost	On-site repair	Area constraints	Availability of spares part	Risk
Hand pallet truck	0.163	0.038	0.019	0.017	0.063	0.028	0.024	0.016	0.014
Hydraulic Hand pallet truck	0.021	0.068	0.058	0.040	0.143	0.028	0.024	0.016	0.061
Hydraulic pallet truck	0.054	0.053	0.045	0.040	0.040	0.021	0.018	0.016	0.086
Electric pallet trucks	0.021	0.068	0.045	0.029	0.158	0.015	0.008	0.007	0.086
Semi-electric pallet truck	0.054	0.053	0.032	0.017	0.127	0.015	0.008	0.007	0.061

Table 8 Positive-ideal (best) and negative-ideal (worst) solutions

Sub-factors	Positive-ideal	Negative-ideal
Ease to operate	0.163	0.021
Application	0.068	0.038
Load-carrying capacity	0.058	0.019
Flexibility in material	0.04	0.017
Cost	0.04	0.158
On-site repair	0.028	0.015
Area constraints	0.024	0.008
Availability of spares part	0.007	0.016
Risk	0.014	0.086

Table 9 Separation measures of attributes

	Positive separation measure S_J+	Negative separation measure S_J-
Hand pallet truck	0.060	0.187
Hydraulic hand pallet truck	0.182	0.065
Hydraulic pallet truck	0.133	0.130
Electric pallet trucks	0.201	0.043
Semi-electric pallet truck	0.165	0.165

$$S_1+ = \left\{ \sum_{j=1}^{M} (V_{ij} - V_j+)2 \right\} 0.5 \quad \text{and} \quad S_1- = \left\{ \sum_{j=1}^{M} (V_{ij} - V_j-)2 \right\} 0.5$$

The values are shown in Table 9 and graphically in Appendix 3;

Step 7: Calculations of relative closeness to the ideal solution and rank the performance order. The relative closeness of the alternative A_j can be expressed as follows (Table 10):

Table 10 Calculate the relative closeness

	Relative closeness values	Ranking
Hand pallet truck	0.756	1
Hydraulic Hand pallet truck	0.263	4
Hydraulic pallet truck	0.493	3
Electric pallet trucks	0.175	5
Semi-electric pallet truck	0.500	2

$$C_J = S_J - /(S_J - + S_J +)$$

4 Conclusions and Future Scopes

In present research work, MCDM methods have been implemented on multi-criteria MHE using AHP and TOPSIS methods. AHP and TOPSIS methods have been used for computing the weightage of sub-criteria and ranking alternatives, respectively. This two-phase methodology includes a hierarchical structure, pair-wise comparisons of attributes and consistency test following this final ranking of the attributes are achieved, and so final ranking is more rational and authentic. Hand Pallet truck is placed at rank one followed by semi-electric pallet truck using relative closeness value. However, a great deal of points of interest, there are restrictions in this approach as all the potential criteria couldn't include this model and positioning of characteristics might be changed if another property is included. The priority weighting by AHP method is subjective as its value depends upon the judgement of decision-makers. Modified TOPSIS and PROMETHEE techniques can be employed for the extension of the proposed framework. Mathematical modelling can be combined with this methodology for improvement. This framework can be replicated in other similar kinds of manufacturing units.

Appendix "1": Priority Weights of Attributes

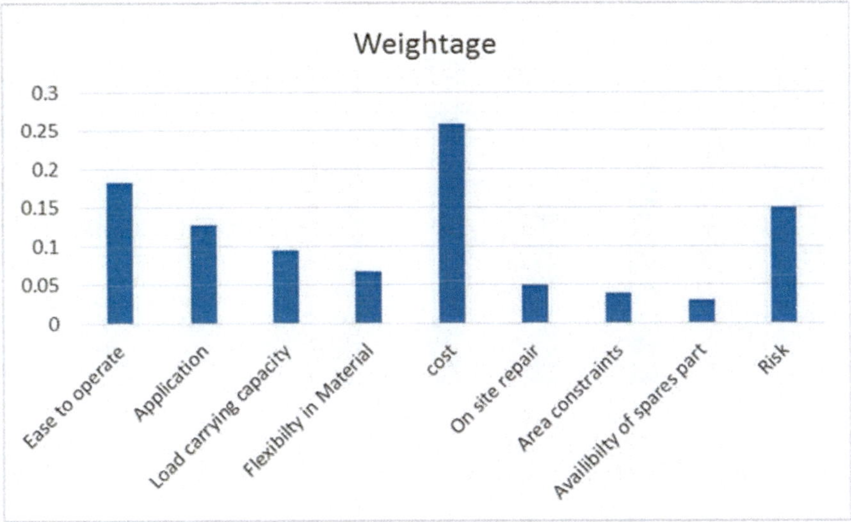

Appendix "2": Positive-Ideal (best) and Negative-Ideal (worst) Solutions

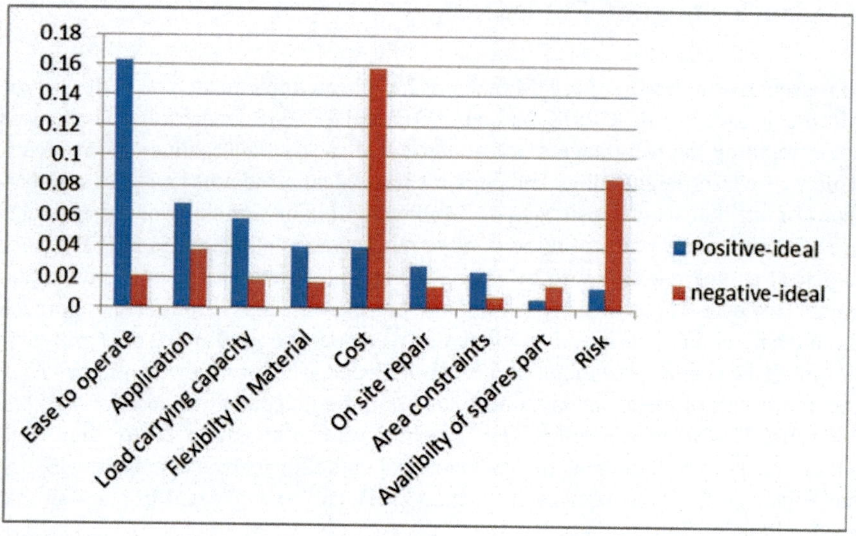

Appendix "3": Separation Measures of Attributes

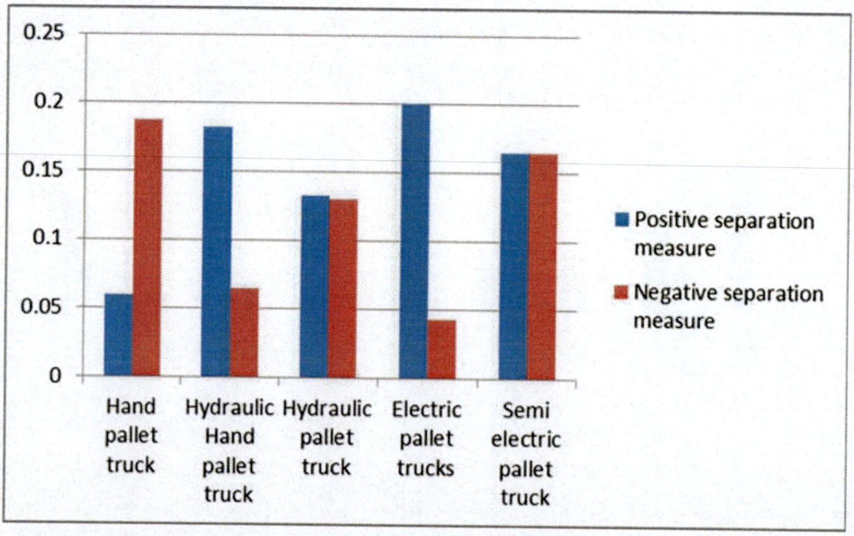

References

1. J. A. Tompkins. J. A. White, Y. A. Bozer. E. H. Frazelle, J. M. Tanchoco, and J. Trevino, " Facilities Planning", 3rd edition, New York:Wiley, pp. 137–287, 2003.
2. Chen FTS, Ip RWL, Lau H (2001) Integration of expert system with analytic hierarchy process for the design of material handling equipment selection system. J Mater Process Technol 116:137–145
3. Onut S, Kara SS, Mert S (2009) Selecting the suitable material handling equipment in the presence of vagueness. Int J Adv Manuf Technol 44:818–828
4. Tuzkaya G, Gülsün BC, Kahraman C, Özgen D (2010) An integrated fuzzy multi-criteria decision making methodology for material handling equipment selection problem and an application. Expert Syst Appl 37(4):2853–2863
5. Lashgari A, Yazdani-Chamzini A, Fouladgar MM, Zavadskas EK, Shafiee S, Abbate N (2012) Equipment selection using fuzzy multi criteria decision making model: key study of Gole Gohar iron min. Eng Econo 23(2):125–136
6. Komljenovic D, Kecojevic V (2009) Multi-attribute selection method for materials handling equipment. Int J Ind Syst Eng 4(2):151–173
7. Singh H, Kumar R (2018) Optimization of production planning in manufacturing sector: a case study. Ind Eng J 11(5):16–19
8. Fonseca DJ, Uppal G, Greene TJ (2004) A knowledge-based system for conveyor equipment selection. Expert Syst Appl 26(4):615–623
9. Hadi-Vencheh A, Mohamadghasemi A (2014) A new hybrid fuzzy multi-criteria decision making model for solving the material handling equipment selection problem. Int J Comput Integr Manuf 28(5):534–550
10. Chakraborty S, Banik D (2006) Design of a material handling equipment selection model using analytic hierarchy process. Int J Adv Manuf Technol 28:1237–1245
11. Anand G, Kodali R, Kumar BS (2011) Development of analytic network process for the selection of material handling systems in the design of flexible manufacturing systems (FMS). J Adv Manag Res 8(1):123–147
12. Rosenbloom ES (1996) A probabilistic interpretation of the final rankings in AHP. Eur J Oper Res 96:371–378
13. Kulak O (2005) A decision support system for fuzzy multi-attribute selection of material handling equipments. Expert Syst Appl 29(2):310–319
14. Maniya KD, Bhatt MG (2011) A multi-attribute selection of automated guided vehicle using the AHP/M-GRA technique. Int J Prod Res 49(20):6107–6124
15. Chakraborty S, Banik DS (2013) Design of a material handling equipment selection model using analytic hierarchy process. Int J Adv Manuf Technol 28(11–12):1237–1245
16. Vencheh AH, Ghasemi AM (2015) A new hybrid fuzzy multi-criteria decision making model for solving the material handling equipment selection problem. Int J Comput Integr Manuf 28(5):534–550
17. Saaty TL (1980) The Analytic Hierarchy Process. McGraw-Hill, New York
18. Rao RV (2013) Decision making in the manufacturing environment using graph theory and fuzzy multiple attribute decision making methods. Adv Manuf 2:8–39
19. Hwang C, Yoon K (1981) Multiple attribute decision making: method and application. Springer Publications
20. Shyur HJ, Shih HS (2006) A hybrid MCDM model for strategic vendor selection. Math Comput Model 44(7–8):749–761

Application of Machine Learning Technique for Demand Forecasting: A Case Study of the Manufacturing Industry

Arvind Jayant, Anshul Agarwal, and Vaibhav Gupta

Abstract The objective of this work is to develop a machine learning-based Support Vector Machine (SVM) demand forecasting model and its application in supply chain management. The proposed SVM model will predict future demand with high accuracy as compared to the conventional forecasting methods. To demonstrate the effectiveness of the present model, demand forecasting issue was investigated in a piston-manufacturing industry as a real-life case study. In this proposed research, an SVM model is developed using radial basis kernel function and sigmoid function to forecast monthly piston demand for Bajaj Discover motorbikes. Various factors that affect the product demand such as produced units, inventory, sales cost, and the number of competitors have been taken into consideration in the development of the model. A comparative analysis of the SVM model and various traditional forecasting methods used in the company like exponential smoothing, moving average, and autoregressive model has been done and the best demand forecasting model has been recommended to the case company.

Keywords Support vector machine (SVM) · Demand forecasting · Exponential smoothing method · Moving average method · Autoregressive model · Python

1 Introduction

Machine learning is like learning from experience that comes naturally to animals and humans. Machine learning algorithms work without depending on the fixed model or equation, rather they work by learning directly from data. The performance of the algorithm improves as the data available for learning increases. Machine learning algorithms help in making better predictions and decisions by generating useful

A. Jayant (✉) · A. Agarwal · V. Gupta
Department of Mechanical Engineering, Sant Longowal Institute of Engineering & Technology, Longowal, Sangrur 148106, Punjab, India
e-mail: arvindjayant@gmail.com

A. Agarwal
e-mail: anshul.cri15@gmail.com

© Springer Nature Singapore Pte Ltd. 2021

P. M. Pandey et al. (eds.), *Advances in Production and Industrial Engineering*, Lecture Notes in Mechanical Engineering,
https://doi.org/10.1007/978-981-15-5519-0_31

insights from data by finding patterns in the data. These algorithms are frequently used to make significant decisions in energy load forecasting, stock trading, medical diagnosis, etc [16, 17]. There are two types of machine learning algorithms: Supervised learning and unsupervised learning. In supervised learning, the model is trained from input and output data that are already known, and after training the model, the same model can be used to predict future outputs through the unknown data. In unsupervised learning, the algorithm finds a hidden pattern in the data as the data are unknown and make predictions based on those intrinsic patterns found in input data.

If supply chain operations are not properly managed, then it may lead to poor customer service, increasing costs, potential revenue losses, and, most importantly, loss of customer trust, which causes loss of profits [1, 2]. Supply Chain process has many uncertainties such as follows:

- Inventory shortage to meet the demands
- Uncertainty in logistics and shortage in supply
- Increase in order backlog
- Fluctuations in demand
- No proper communication among stakeholders
- Quality of inventory levels may vary

It is very difficult to forecast with pre-set rules as the reasons for these variations change on a daily basis; therefore, machine learning algorithms could help in reducing these uncertainties and predict precisely.

- Machine learning algorithms help in a better understanding of these challenges and allow real-time analysis and forecasting in advance.
- As there are a lot of data involved in these predictions, so ML algorithm helps in processing large volumes of data, which is not possible manually.
- Models developed using machine learning help in reducing operational inefficiencies, minimizing loss in revenue, and ensures on-time delivery, which improves the level of customer satisfaction.
- Methodology developed by ML is equipped to handle huge data and work in real time and can provide solutions proactively rather than after the problem occurs.

Thus, machine learning is an asset for supply chain management since they help reduce uncertainties thereby increasing profits and minimizing losses.

Demand forecasting plays an important role in managing sales operations in any business organization. In such an environment of uncertainties and competitiveness, forecasting proves to be an integral part of decision-making functions for the people involved in management. It provides specific guidelines and instructions that must be performed by the management for better prospects of any organization. It is applicable to all trading firms, manufacturers, and distributors involved in any business unit. Forecasting, where demand uncertainty is high, is more difficult rather than in a situation where demand uncertainty is low. Accurate forecasting helps in keeping inventory in check, as it helps in reducing excess and preventing the shortage of inventory and thus improving profitability. When there is a case of overestimated demand the manufacturer tries to increase the production and this may lead to extra stock on the one hand and, on the other hand, when there is underestimated demand, it leads to lost sales, unfulfilled orders, decrease in service levels, and lost opportunities [18]. In both the cases, the supply chain becomes inefficient and may cause loss to the company. Therefore, accurate demand forecasting poses as a challenge to any organization.

Methods and techniques to improve accuracy in forecasting have become a major topic of discussion in industries and this can help to increase competitiveness in the market. The technique to forecast demand based on past data plays a key role in an organization and an individual decision-making. In time series forecasting, the technique tries to find a pattern, and based on this, predictions are done for complex systems. For a successful supply chain management, forecasting is very important. There are various demand forecasting techniques which include traditional as well as some latest methods. Traditional methods have some vital demerits which can affect the accuracy of forecasting. Machine learning is an advanced technique that could be used for demand forecasting. There are a number of machine learning algorithms that are used for making predictions and classification [3]. One such algorithm is Support Vector Machine (SVM) which can be applied to develop a framework for demand forecasting. SVM algorithm helps in a better understanding of the challenges such as fluctuations in demand, uncertainties in the market, etc. Models developed using SVM allow real-time analysis and help in reducing operational efficiencies, minimizing loss in revenue, and ensures on-time delivery, which improves the level of customer satisfaction. A model is developed for demand analysis using the SVM algorithm in a piston manufacturing industry as part of real-world case study. This machine learning algorithm is applied using an advanced computer language Python.

The outstanding portion of the paper is organized as follows: Sect. 2 presents a brief review of the literature available on-demand forecasting using machine learning. Section 3 describes the problem to be evaluated. Section 4 discusses the various demand forecasting techniques including the advanced and the traditional methods and performance indices to evaluate the forecasting models. Section 5 presents the forecasting model development using SVM. Section 6 provides the results obtained from different demand forecasting models. Section 7 shows the comparison of these models and their validity. Finally, Sect. 8 presents the conclusion and future research directions which concludes the case study.

2 Literature Review

Over the past few years, researchers and academicians have focused on the area of artificial intelligence and its application in various fields. Machine learning is a vast emerging area in which a lot of research is going on. It is being applied in different areas of supply chain management. Demand forecasting is an area of SCM which has a lot of potential for application of machine learning algorithms. A brief description of the application of machine learning algorithm SVM in demand forecasting in SCM is as follows.

Villegas et al. [3] proposed a novel framework selection method that combines diverse criteria using a SVM for demand forecasting. Chouksey et al. [4] identified "various critical success factors and their overall and individual impact on sales forecasting on the basis of real-time primary data and secondary data of the automobile firm by applying regression analysis." Shabani et al. [5] proposed a SVM model using polynomial kernel function to forecast monthly water demand of a city and to assess the use of phase space reconstruction prior to the design of models' input variables combinations. Fan et al. [6] used "a sentiment analysis method, the Naive Bayes algorithm, to find the sentiment index from the text of each online review and incorporate it into the imitation coefficient of the Bass/Norton model to increase the forecasting accuracy." Amirkolaii et al. [7] presented "a survey of existing forecasting methods used in service and non-service supply chains to select best performing AI methods and performance measures using ABC classification and subsequently neural networks and mean square error are modelled using supply chain data from aircraft spare parts industry." Kochak and Sharma [8] modeled a demand forecasting technique by artificial intelligence approaches using artificial neural networks. Jaipuria and Mahapatra [14] developed an integrated method of artificial neural network (ANN) and discrete wavelet transform (DWT) analyses, denoted as DWT–ANN, to predict the demand of a product. Jayant and Dhillon [15] established "an extensive evaluation hierarchy model of automobile short-term demand to prevent the disadvantages of previous models mainly for single time series." Yolcu et al. [10] and Sheremetov et al. [11] focused on forecasting models, which are time series, based on the context of petroleum engineering and oil production in the future is predicted in diverse situations in a changeable time window. Jayant et al. [16, 17] proposed a novel neural network model, which consists of both linear and nonlinear structures, for time series forecasting. Carbonneau et al. [12] developed a model to forecast biased demand at the end of a supply chain (bullwhip effect) and then a comparison is made with traditional methods of forecasting. Yue et al. [13] employed "the technique of Support Vector Machine (SVM) for demand forecasting and various factors that affect the product demand such as seasonal and promotional factors have been taken into consideration in the model also a comparison with approaches such as Statistical Model, Winter Model and Radius Basis Function Neural Network (RBFNN) is made."

3 Problem Description

Product demand is always prone to fluctuation thus making the supply chain inefficient and ineffective. Therefore, demand forecasting is an important and crucial part of the downstream activity of any supply chain and is the process of predicting future sales of a product. The fluctuation of product demand causes uncertainty in the supply chain. Thus, the accuracy of sales forecast of a product in a supply chain is

Table 1 Company profile

Business characteristics	Case company
Year of establishment	1954
Total income (INR) as of March 2018	136221.3 lakhs
Products manufactured	Pistons, piston rings, cylinder liners, light metal castings, sintered metal products
Type of business	Manufacturer, supplier

certainly an important key to competitiveness. Therefore, there is a need to develop an efficient and precise model for forecasting future demand.

In the present work, the demand forecasting issue has been investigated in a piston-manufacturing company as a real-world case study. Two SVM models with kernel functions: Radial Basis system (RBF) and sigmoid has been developed using Python. Nine years supply chain data of piston for Bajaj Discover motorbikes have been collected from an Indian automobile parts company (Refer Table 1) and demand for the subsequent years has been computed.

4 Supply Chain Demand Forecasting Techniques

- Support Vector Machine
- Exponential Smoothing Method
- Moving Average Method
- Autoregressive Method

Out of these, the first technique is a machine learning algorithm and is an advanced method of demand forecasting. The remaining techniques are considered to be traditional methods of demand forecasting.

4.1 Support Vector Machine

Support Vector Machine (SVM) first came onto existence in 1992, brought by way of Boser, Guyon, and Vapnik in Computational Learning Theory (COLT) convention with the paper. SVM is an algorithm of supervised learning methods used for regression and categorization. In this approach, every data object as a factor is plotted in an n-dimensional area (in which n is a variety of features) with the cost of every function being the value of a selected coordinate. SVM is a kind of technique, identified by the absence of local minima, usage of kernels, sparseness of the solution, and potential manage acquired through acting on the number of support vectors or on the margin, etc. (Refer Fig. 1) [1].

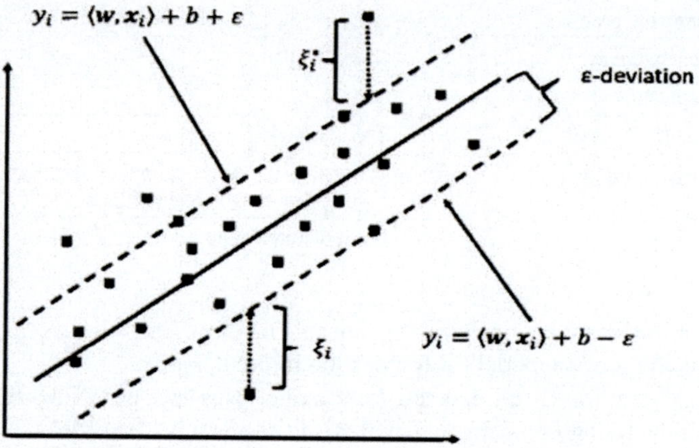

Fig. 1 Diagrammatic representation of working of SVM algorithm

SVM may be implemented to the categorization problems in addition to the case of regression. Some terms that are used in support vector regression are given below:

- **Kernel**: This function is used to map data that is in a lower dimension into a data of higher dimension.
- **Hyper Plane**: In SVR it is the plane that is used to predict the target value or continuous value with the help of given inputs.
- **Boundary line**: These are the lines on both sides of the hyperplane and it creates a margin that decides the data points that are support vectors.
- **Support vectors**: These are those data points that are nearest or within the boundary line and distance of the points is least or minimum.

The algorithm tries to fix a margin by taking a decision boundary at a certain distance of 'ε' around both sides of the original hyperplane. This is done in such a manner that the support vectors are nearest to the hyperplane or the data points are within the margin. In simple words, the points which have error rate minimum are taken into consideration and this provides a better fitting model.

4.2 Exponential Smoothing Method

This technique calls for the present data and to give the next prediction the forecast value for the present time is used. It is a changed form of WMA and loads are assigned in exponential decreasing order. The assigned weights are exponentially diminished from the latest data to earlier data. The latest statistics is assigned to maximum weight and the weight assigned to older data exponentially decreased [14].

$$F_t = F_{t-1} + \alpha(D_{t-1} - F_{t-1})$$

$$\alpha = \frac{2}{(n+1)}$$

where F_t is the present forecast, F_{t-1} is the past forecast, D_{t-1} is the past demand, α is smoothing constant, and n is the number of periods of moving average.

4.3 Moving Average Method

It is also called a rolling average or simple moving average. This approach used past information and calculate the rolling average for a steady period. A fresh average is computed at the culmination of each duration by way of adding real demand records for the latest period and deleting facts for the older duration (Chongli et al. 2007). In this technique as information modifies for each duration so it is known as moving average method.

$$\text{Moving average} = (n_1 + n_2 + n_3 + \cdots)/n$$

where n is the number of periods of a simple moving average.

4.4 Autoregressive Model (Zou et al. 2018)

"A time series is a sequence of measurements of the same variable(s) made over time. Usually, the measurements are made at evenly spaced times—for example, monthly or yearly. Let us first consider the problem in which we have a y-variable measured as a time series. As an example, we might have y a measure of global temperature, with measurements observed each year. To emphasize that we have measured values over time, we use t as a subscript rather than the usual i, i.e. y_t means y measured in time period t. An autoregressive model is when a value from a time series is regressed on previous values from that same time series. For example, y_t on y_{t-1}

$$y_t = \beta_0 + \beta_1 y_{t-1} + \epsilon_t$$

In this regression model, the response variable in the previous time period has become the predictor and the errors have our usual assumptions about errors in a simple linear regression model. The order of an autoregression is the number of immediately preceding values in the series that are used to predict the value at the present time. So, the preceding model is a first-order autoregression, written as AR(1)."

4.5 Performance Evaluation Indices

For evaluating the performance and accuracy of different demand forecasting models, two performance indices have been used in this research work: Mean Absolute Deviation (MAD) and Mean Absolute Percentage Error (MAPE). The formulae for both these indices are as follows:

- Mean Absolute Deviation (MAD):

$$\text{MAD} = \frac{\sum_{i=1}^{n} |D_i - F_i|}{n}$$

- Mean Absolute Percentage Error (MAPE):

$$\text{MAPE} = \frac{\sum_{i=1}^{n} \left| \frac{D_i - F_i}{D_i} \times 100 \right|}{n}$$

where D_i is the actual value for period i, F_i is the forecasted value for period i and n is the number of periods.

The smaller is the values for both these indices, the better is the prediction model.

5 Development of SVM Model for Demand Forecasting

Two models have been developed using Python computer language with the SVM algorithm on two kernel functions 'RBF' and 'SIGMOID.' The data used in developing the model are taken from an Indian Automobile parts industry. The monthly supply chain data of the last 9 years, i.e. from the year January 2010 to March 2019, for the piston of Bajaj Discover bikes has been collected and is used to train and test the developed model. A detailed graphical representation of the dataset is given in Fig. 2.

5.1 Steps in Building a Machine Learning Model (As shown in Fig. 3)

- Access and explore the data.
- Preprocess the data and extract features.
- Train the model
- Develop predictive models.
- Optimize the model.
- Deploy analytics to a production system.

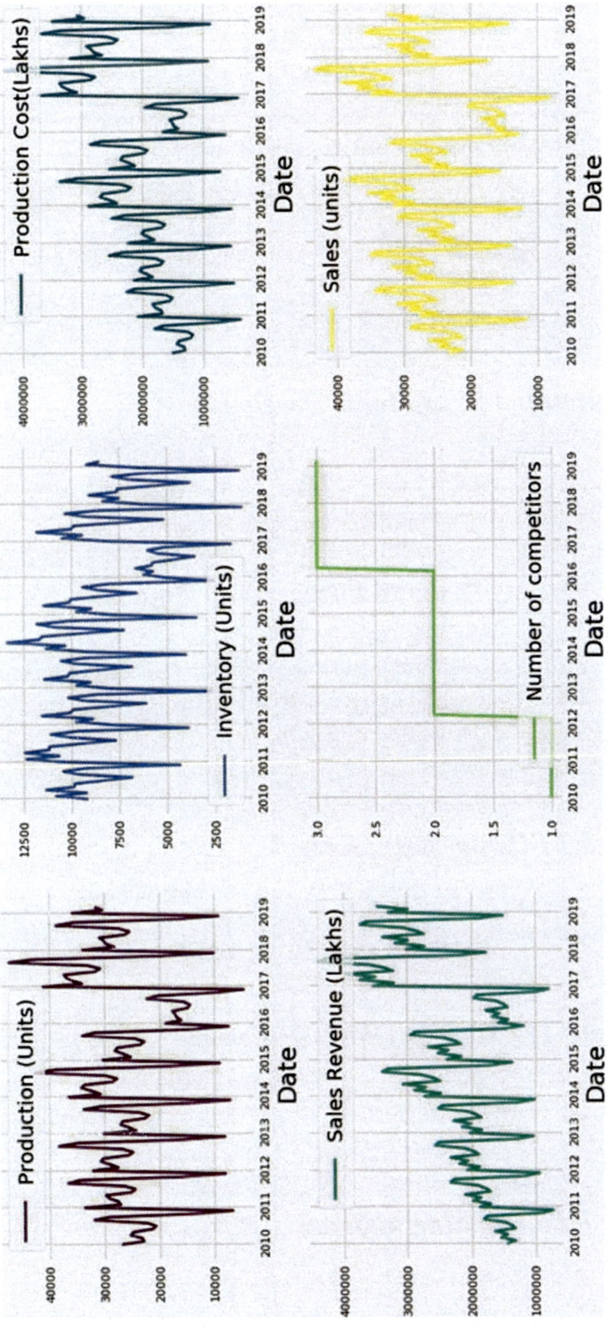

Fig. 2 Graphical Representation of all features of a dataset

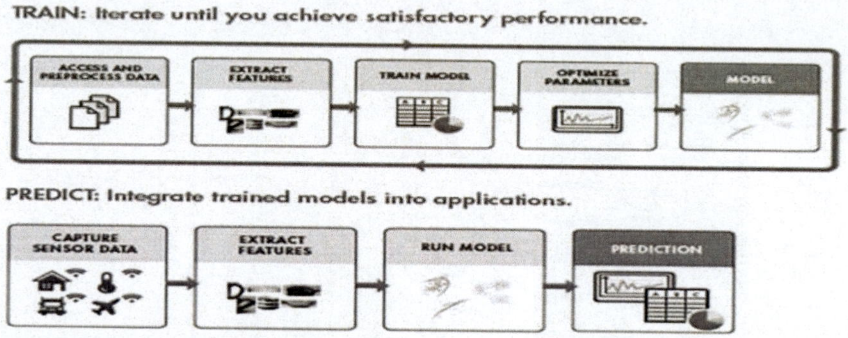

Fig. 3 Steps in the development of ML model

5.2 *Flowchart of the SVM Model Developed*

See Fig. 4.

6 Results of Various Demand Forecasting Models

This section presents the results obtained from different forecasting techniques on the given dataset. The same dataset has been used for all the techniques and the same data are used for testing the model for validation.

6.1 *SVM Model I Using 'RBF' Kernel*

See Fig. 5.

6.2 *SVM Model II Using 'SIGMOID' Kernel*

See Fig. 6.

6.3 *Exponential Smoothing (alpha = 0.2)*

See Fig. 7.

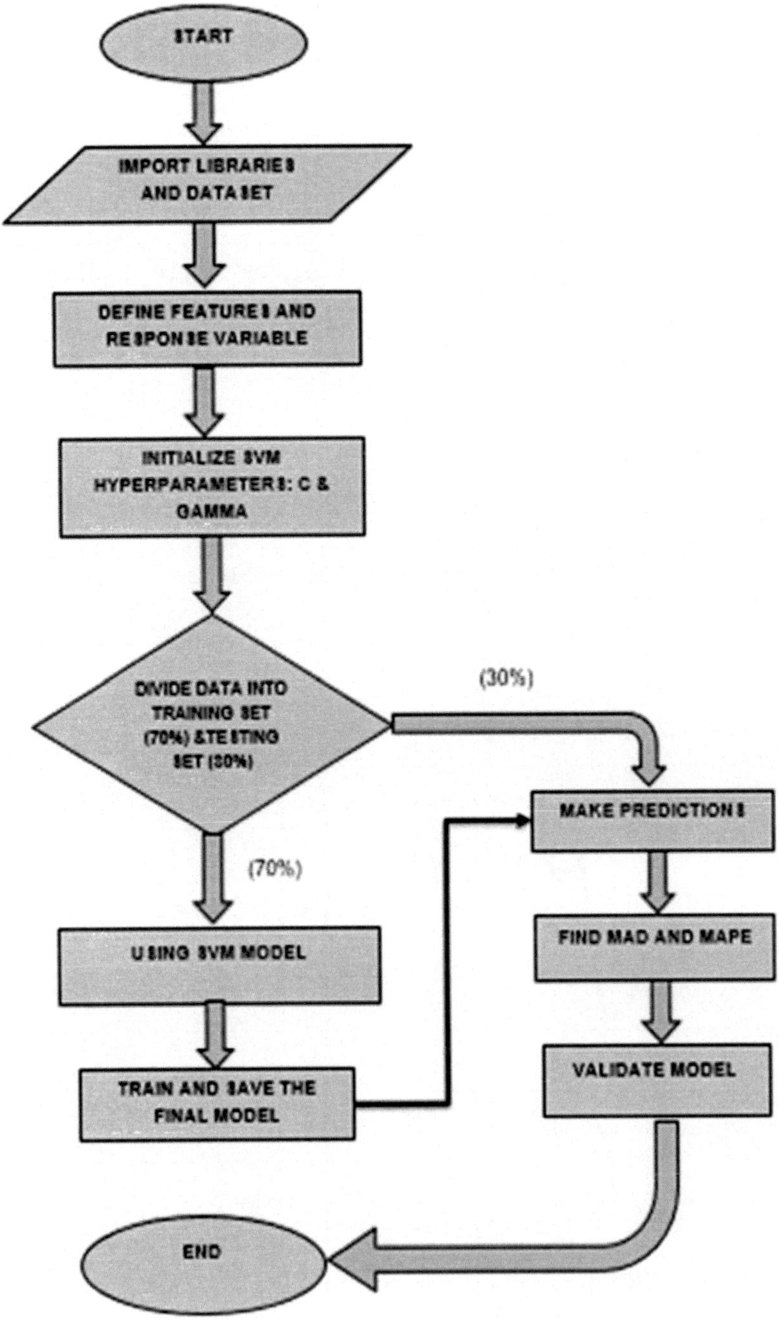

Fig. 4 Flowchart of SVM Model

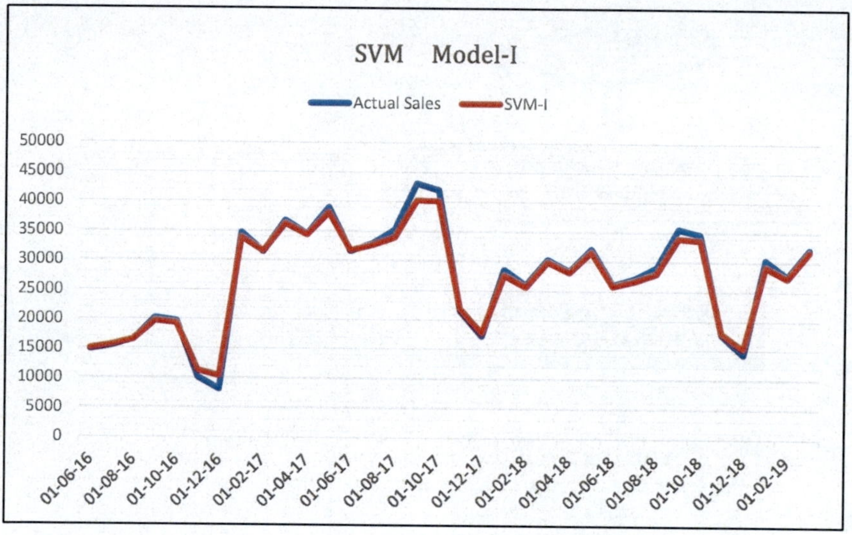

Fig. 5 Graph showing actual and forecasted values of SVM Model-I

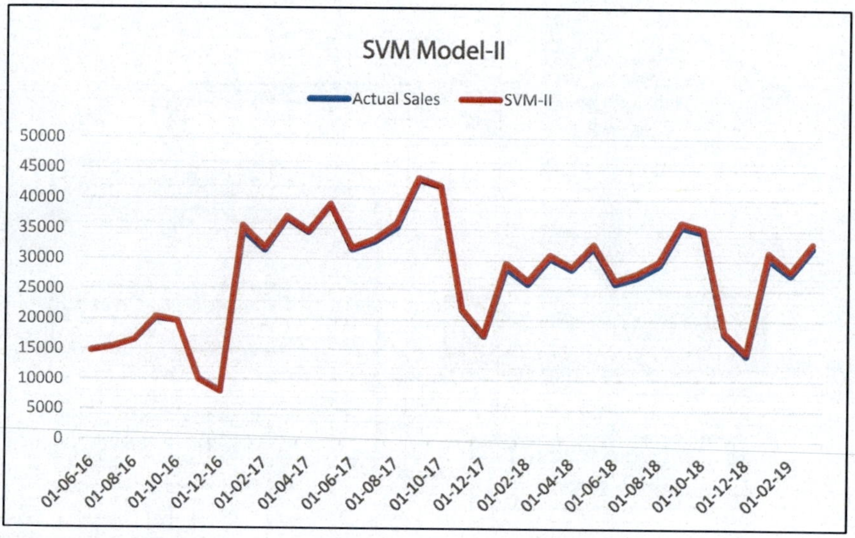

Fig. 6 Graph showing actual and forecasted values of SVM Model-II

6.4 Exponential Smoothing (alpha = 0.5)

See Fig. 8.

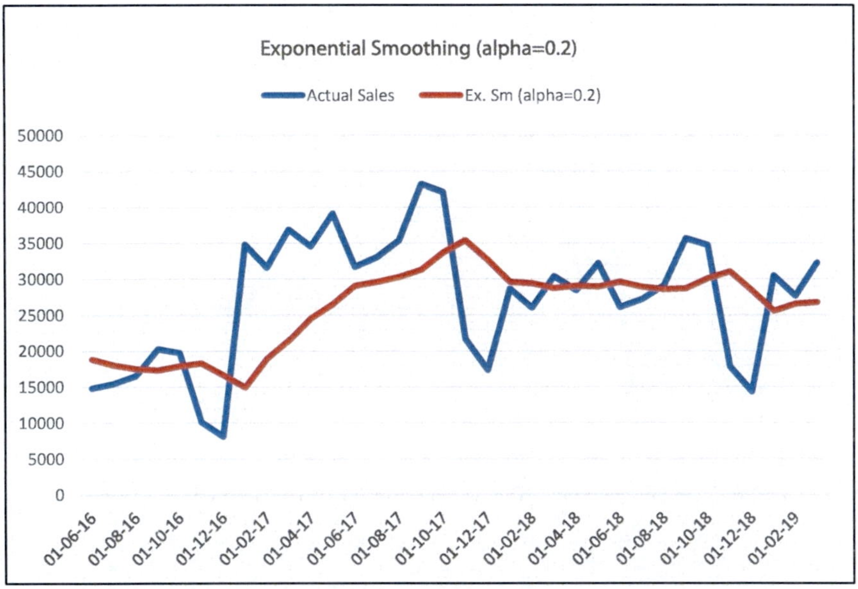

Fig. 7 Graph showing actual and forecasted values of Exponential Smoothing ($\alpha = 0.2$)

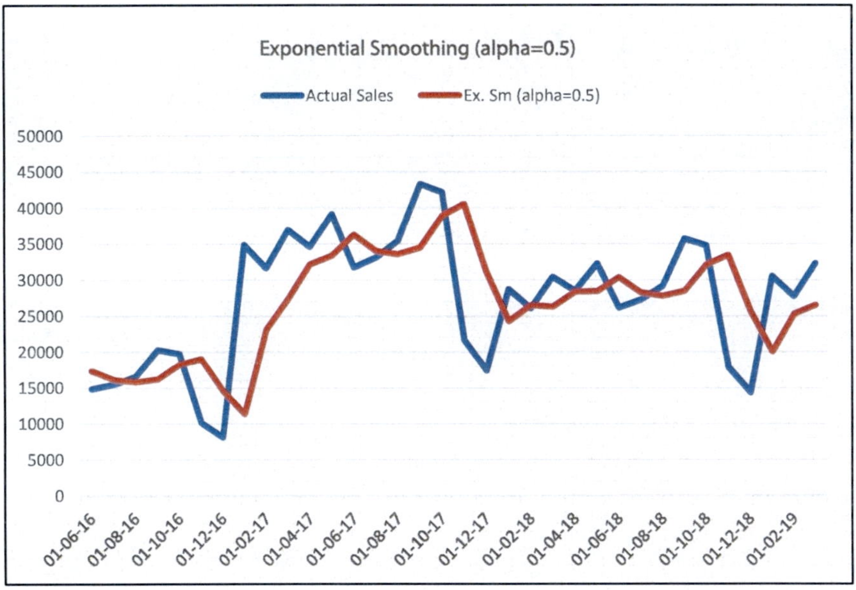

Fig. 8 Graph showing actual and forecasted values of Exponential Smoothing ($\alpha = 0.5$)

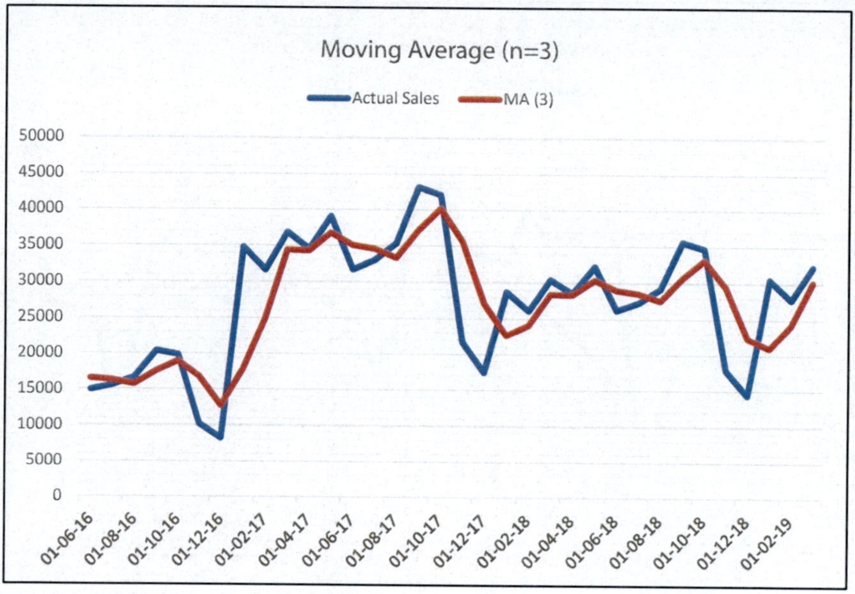

Fig. 9 Graph showing actual and forecasted values of Moving Average Method ($n = 3$)

6.5 *Moving Average Method* (n = 3)

See Fig. 9.

6.6 *Moving Average Method* (n = 6)

See Fig. 10.

6.7 *Autoregressive Model*

See Fig. 11.

7 Comparison of Results and Models Validity

A comparison of the results of the advanced and traditional forecasting techniques has been shown in Table 2. The comparison has been made based on performance indices discussed in Sect. 4.5. The results show clearly that forecasting models developed

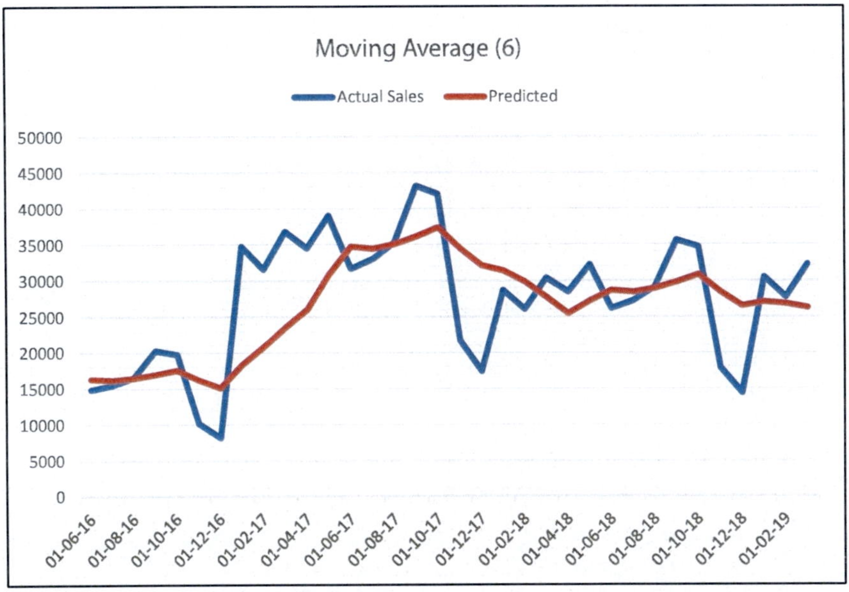

Fig. 10 Graph showing actual and forecasted values of Moving Average Method ($n = 6$)

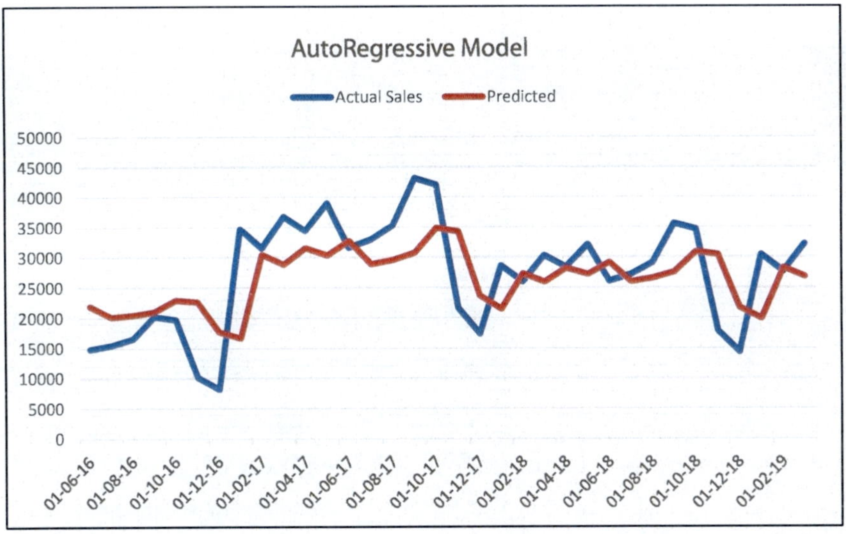

Fig. 11 Graph showing actual and forecasted values of Autoregressive Model

Table 2 Comparison of the performance (MAD, MAPE) of forecasting techniques

Technique	Mean Absolute Deviation (MAD)	Mean Absolute Percentage Error (MAPE)
Exponential smoothing (alpha = 0.2)	6633.46	28.97
Exponential smoothing (alpha = 0.5)	5918.24	26.43
Moving average (n = 3)	4248.72	19.27
Moving average (n = 6)	5474.30	23.95
Autoregressive model	5979.72	27.56
SVM model-I	685.78	3.15
SVM model-II	409.96	1.47

using machine learning techniques SVM are far better than the traditional methods (Figs. 12 and 13).

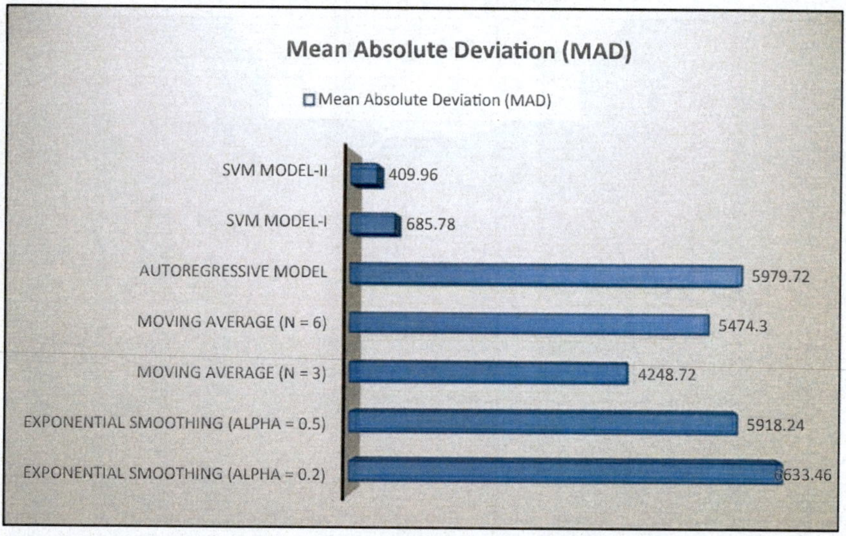

Fig. 12 Graph showing MAD for various techniques of demand forecasting

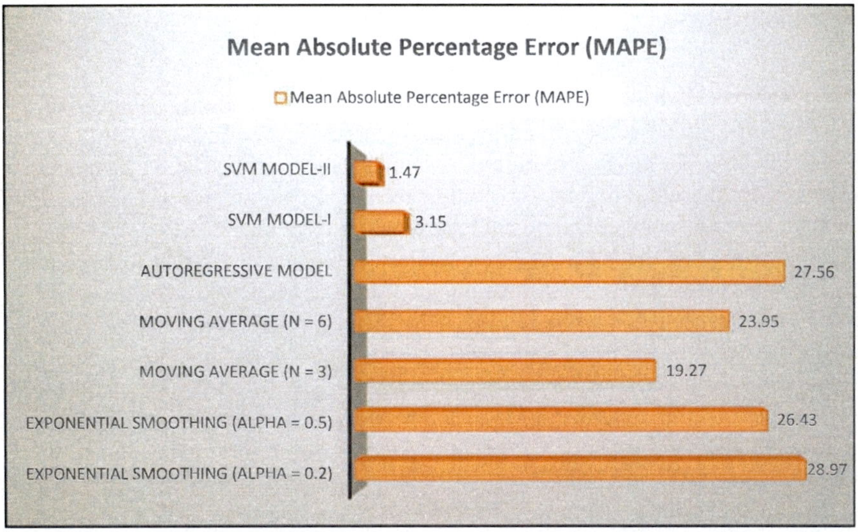

Fig. 13 Graph showing MAPE for various techniques of demand forecasting

8 Managerial Implication

In the present research, SVM models with RBF and SIGMOID kernel function have developed using Python v3.5 for forecasting the future demand of the product and these models are further validated by comparing the results of SVM models with the traditional forecasting methods like Exponential smoothing method, Moving average method, and Autoregressive model. SVM models developed, forecast the demand with high accuracy, and thus reduces the fluctuation of the supply chain, as the demand forecasting is an integral part of the supply chain.

The objective of this study is to find out the importance of SVM models in forecasting the future demand of the product and to compare the SVM models with the traditional forecasting methods. Predicting the future demand with the traditional forecasting methods yields low accuracy and greater fluctuation. Thus, it is necessary to develop some models, which will forecast the demand with high accuracy. SVM models developed in the present work predict the demand with high accuracy and reduces the fluctuation of the supply chain. Managers can use these proposed ANN models for predicting the future demand of their company's product and make their supply chain stable and reliable.

9 Conclusions

Two SVM models based on RBF and SIGMOID kernel functions have been developed by using the Python v3.5 for forecasting future demand using the data of piston for Bajaj Discover motorbikes. Demand forecasted by the SVM models is validated by the conventional forecasting techniques like exponential smoothing method, moving average method, and Autoregressive model. MAD and MAPE are used for estimating the accuracy of the models. Results revealed that forecasting of demand, in case of RBF kernel function, the values of MAD and MAPE are 685.78 and 3.15%, respectively, while in case of SIGMOID kernel function, the values of MAD and MAPE are 409.96 and 1.47%, respectively. Thus, the SVM model with Sigmoid kernel function is far better and more accurate than the SVM model with RBF kernel function in terms of MAD and MAPE.

In relation to the traditional forecasting methods for forecasting the demand, the values of MAD and MAPE are 5918.24 and 26.43%, respectively, in case of Exponential smoothing method with alpha = 0.5, while the values of MAD and MAPE are 4248.72 and 19.27%, respectively, in case of Moving average method and the values of MAD and MAPE are 5979.72 and 27.56%, respectively, in case of Autoregressive model. Thus, the Moving average method is best suited for forecasting future demand as compared to the Exponential smoothing method and Autoregressive model in our case study. It can be concluded that the SVM models outperform the traditional forecasting model in predicting the demand and forecast the demand with greater accuracy.

References

1. Jayant A, Gupta P, Garg SK (2014) Reverse logistics network design for spent batteries: a simulation study. Int J Logist Syst Manag 18(3):343–365
2. Jayant A, Paul V, Kumar U (2014) Application of analytic network process (ANP) in business environment: a comprehensive literature review. Int J Res Mech Eng Technol 4(3): 29–43
3. Villegas MA, Pedregal DJ, Trapero JR (2018) A support vector machine for model selection in demand forecasting applications. Comput Ind Eng 121:1–7
4. Chouksey P, Deshpande A, Agarwal P, Gupta RC (2017) Sales forecasting study in an automobile company—a case study. Ind Eng J
5. Shabani S, Yousefi P, Naser G (2017) Support vector machines in urban water demand forecasting using phase space reconstruction. Procedia Eng 186:537–543
6. Fan ZP, Che YJ, Chen ZY (2017) Product sales forecasting using online reviews and historical sales data: a method combining the Bass model and sentiment analysis. J Bus Res 74:90–100
7. Amirkolaii KN, Baboli A, Shahzad MK, Tonadre R (2017) Demand forecasting for irregular demands in business aircraft spare parts supply chains by using artificial intelligence (AI). IFAC-PapersOnLine 50(1):15221–15226
8. Kochak A, Sharma S (2015) Demand forecasting using neural network for supply chain management. Int J Mech Eng Robot Res 4(1):96–104
9. Zhu Z, Chen T, Shen T (2014) A new algorithm about market demand prediction of automobile. Int J Mark Stud 6(4):100

10. Yolcu U, Egrioglu E, Aladag CH (2013) A new linear & nonlinear artificial neural network model for time series forecasting. Decis Support Syst 54(3):1340–1347
11. Sheremetov LB, González-Sánchez A, López-Yáñez I, Ponomarev AV (2013) Time series forecasting: applications to the upstream oil and gas supply chain. IFAC Proc Vol 46(9):957–962
12. Carbonneau R, Laframboise K, Vahidov R (2008) Application of machine learning techniques for supply chain demand forecasting. Eur J Oper Res 184(3):1140–1154
13. Yue L, Yafeng Y, Junjun G, Chongli T (2007) Demand forecasting by using support vector machine. In: Third international conference on natural computation (IEEE). vol 3, pp 272–276, Aug 2007
14. Jaipuria S, Mahapatra SS (2014) An improved demand forecasting method to reduce bullwhip effect in supply chains. Expert Syst Appl 41(5):2395–2408
15. Jayant A, Dhillon MS (2014) Use of analytic hierarchy process (AHP) to select welding process in high pressure vessel manufacturing environment. Int J Appl Eng Res 10(8):586–595
16. Jayant A, Gupta P, Garg SK (2011) Design and simulation of reverse logistics network: a case study. In: The proceedings of world congress on engineering, London, U.K.
17. Jayant A, Singh A, Patel V (2011) An AHP based approach for supplier evaluation and selection in supply chain management. Int J Adv Manuf Syst 2(1):1–6
18. Jayant A (2013) Evaluation of 3PL Service provider in supply chain management: an analytic network process approach. Int J Bus Insights Transform (IJBIT) 6(2):78–82

A Robust Hybrid Multi-criteria Decision-Making Approach for Selection of Third-Party Reverse Logistics Service Provider

Arvind Jayant, Shweta Singh, and Tanmay Walke

Abstract Environmental awareness has universally driven the move for sustainable supply chain management. Accordingly, manufacturing companies or organizations try to seek sustainable business strategies to respond to market pressure toward corporate social responsibility (CSR). Sustainable reverse logistics service provider selection is one of the practical strategies for competitive organizations. With the large-scale development of the automotive products industry, sustainable reverse logistics service provider evaluation method is the key for decision authority when dealing with big data information and possible risks of unstructured data. For instance, the choice of decision authority possibly may responsible for a misleading decision, thus leading to undesirable waste of less available resources and time. Therefore, the objective of present work is to apply the integrated multi-criteria decision methods using the "MOORA and WASPAS" approaches in the evaluation of third-party logistics service providers (3PRLSPs). It also incorporates the significance weight provided by SWARA technique and helps decision-makers for efficient decision-making. The proposed model is to evaluate, and criteria weight is determined using the step-wise weight assessment ratio (SWARA) approach and then ranking of the alternatives was decided by MOORA and WASPAS. The automotive parts manufacturing company may be benefited by their commitment toward environmental safety, economic, and corporate social responsibility (CSR) leading to improved brand value and sustainable business development.

Keywords WASPAS · Sustainability · MOORA · MCDM · 3PRLSPs

1 Introduction

Organizations are working towards sustainable development by incorporating sustainability aspects in their business operations. "Sustainability had remained as

---

A. Jayant (✉) · S. Singh · T. Walke
SLIET Deemed to be University, Longowal, Sangrur, Punjab, India
e-mail: arvindjayant@sliet.ac.in

© Springer Nature Singapore Pte Ltd. 2021

P. M. Pandey et al. (eds.), *Advances in Production and Industrial Engineering*,
Lecture Notes in Mechanical Engineering,
https://doi.org/10.1007/978-981-15-5519-0_32

part of many cultures but formal introduction of sustainability as a concern for businesses can be traced to the Brundtland report published in a book entitled 'Our Common Future'. According to World Commission on Environment and Development (1987) sustainability is more accepted which states that: "Sustainable development is the development that gains the need of present generation without causing any effect to the future." Over the years, "sustainability supported by Agenda 21 adopted at United Conference on Environment and Development at Rio de Janeiro, 1992" has emerged as a motivator for global operations to improve their mutual understanding about sustainable development approach and paradigm [1, 2] "propose several reasons as to why growing interest has raised up in reverse logistics worldwide: It is an effectual means to deal with the large load of returned material, particularly for the industries that experience high return rates, at times over 60% of sales."

Reuse or secondary and international level business are increasingly fast growing and hence provide companies with a chance to promote their business sales by secondary or discarded parts and products. There is major difference in management of forward logistics and reverse logistics operation as shown in Table 1. Latest rules and regulations gradually evolved in developed nations in terms of EOL take-back guidelines in the past decade and thus requires OEM to fully manage the complete life cycle of their parts and products.

According to Jayant et. al. [3], "reverse logistics emphasized green logistics in traditionally, that means added environmentally into logistics strategies, including product return, recycling, waste disposal, refurbishing, repair, and remanufacturing." Many industries have recognized the financial impact on RL operations and effective RL could improve industry performance outcomes and secondary level business competitiveness.

The remaining portion of the research work is organized as follows: Sect. 2 explains review of the literature. Section 3 defines the case study problem to be evaluated. Section 4 discusses the proposed research framework including the procedure for implementation of SWARA, MOORA, and WASPAS methods to solve the proposed problem. Section 5 presents the discussion with proposed solutions. Finally,

Table 1 Difference between forward and reverse logistic

Forward logistics	Reverse logistics
"Forecasting relatively straightforward" "One-to-many transportation" "Product quality uniform" "Product packaging unbroken" "Destination/routing clear" "Consistent channel" "Prominence of speed recognized" "Inventory management reliable" "Product life cycle manageable"	"Forecasting more difficult" "Many-to-one transportation" "Product quality not uniform" "Product packaging often broken" "Purpose/routine unclear" "Exception driven" "Temperament not clear" "Pricing dependent on many factors" "Speed often not considered a priority"

Fig. 1 Reverse logistics

Sect. 6 presents the conclusion and future research directions which concludes the case study (Fig. 1).

2 Literature Review

Important criteria for 3PRLSPs are identified through extensive literature review. There are three main criteria of sustainability which further divided into 17 sub-criteria. Further, the 17 very important sustainable supply chain selection and evaluation criteria identified in the case study and criteria's were validated with the help of subject experts' inputs and categorized into three dimensions of sustainability (Social, economic, and environment). The identified criteria are shown with the help of flow diagram below in Table 2.

3 Problem Description

3.1 Company Profile

The case company XYZ Limited started in July 1983. The company has plant situated in Haryana (India). XYZ is a private limited company, manufactured various parts of automotive industry. The strength of the company is its technology, people at work, widest network, accolades, etc. (Tables 3, 4 and 5).

Table 2 Selection and evaluation criteria for sustainable 3PLRSP

Main criteria	Code	Sustainable supplier selection sub-criteria	References
Economic factors	E1	Ordering and logistic cost	Marcin Stępień et al. (2016)
	E2	Custom and insurance cost	Syed A. M. Tofail et al. (2017)
	E3	Quality management	CostacheRusu (2016)
	E4	On time delivery rate	Ricarda Schäfer et al. (2016)
	E5	Transportation	Christopher Hendrickson (2002)
	E6	Delivery and service	Jie Yu et al. (2014)
Social factors	S1	Occupational health and safety program	Kwesi Amponsah-Tawiah and Justice Mensah (2016)
	S2	Operation	Christoph Teller et al. (2018)
	S3	Wages	Yanting Chen, and Qijun Liu (2018)
	S4	Prevention and risk control program	A. Romero Barriuso et al. (2018)
	S5	Flexible working facilities	Suzanne R. Dhaini et al. (2018)
Environmental factors	EN1	Air emission	Yee Van Fan et al. (2018)
	EN2	Wastewater	Nikolay Makisha (2016)
	EN3	Use of harmful material	Nikolay Makisha et al. (2018)
	EN4	Use of environment friendly material and technology	Jiao Chen et al. (2018)
	EN5	Recycle	Huaidong Wang et al. (2018)
	EN6	Reuse	S. Arden and X. Ma (2018)

Table 3 Company profile

Year of establishment	1983
Turnover of company	119628 million
Employees strength	200
Production capacity	1.5 million
Type of parts manufactured	Automobiles, hydraulic brakes. Rubber pads, suspension system, automotive components
Type of business	Manufacturer, supplier

Table 4 Name of the company customers

1	KINGAS	6	Hi-Pad Auto parts Co. Ltd.
2	JBM Group	7	GALCO group
3	Ion Exchange Limited	8	Canara Standard Keys
4	IZEST SDN BHD		
5	Hinduja Foundries		

Table 5 List of suppliers

1	Alex machine tools Pvt. Ltd.
2	Dawn motors Pvt. Ltd.
3	Paras Industry
4	Power India Wheels Co.
5	Metaforge Engineering Pvt. Ltd.
6	Kew Industries Limited

3.2 *Industrial Survey*

Data collected from ABC automobile industry for 3PRLSPs selection considering the criteria are considered and then ranking of different service providers accordingly. Weight age of the criteria selected on a scale of triangular fuzzy (Table 6).

10 Alternative and 17 criteria are under consideration. These seventeen criteria are E1 Ordering and Logistics cost; E2 Custom and Insurance Cost; E3 Quality management; E4 On time delivery rate; E5 Transportation; E6 Delivery and service; S1 Occupational health and safety program; S2 Operations; S3 Wages; S4 Prevention & risk control program; S5 Flexible working facility; EN1Air emission; EN2 Wastewater; EN3 Use of harmful material; EN4 Use of environment friendly technology and material; EN5 Recycle; EN6 Reuse.

Weight age of each criterion on every alternative is resolute by using MCDM techniques and finally the result gives us the ranking of alternatives. Ranking could be concluded with the help of fuzzy methods so that the problem becomes more structured.

4 Research Framework

See Figs. 2 and 3.

Table 6 The fuzzy scale

Linguistic scale	Response scale
"Equally important"	"(1, 1, 1)"
"Moderately less important"	"(2/3, 1, 3/2)"
"Less important"	"(2/5, 1/2, 2/3)"
"Very less important"	"(2/7, 1/3, 2/5)"
"Much less important"	"(2/9, 1/4, 2/7)"

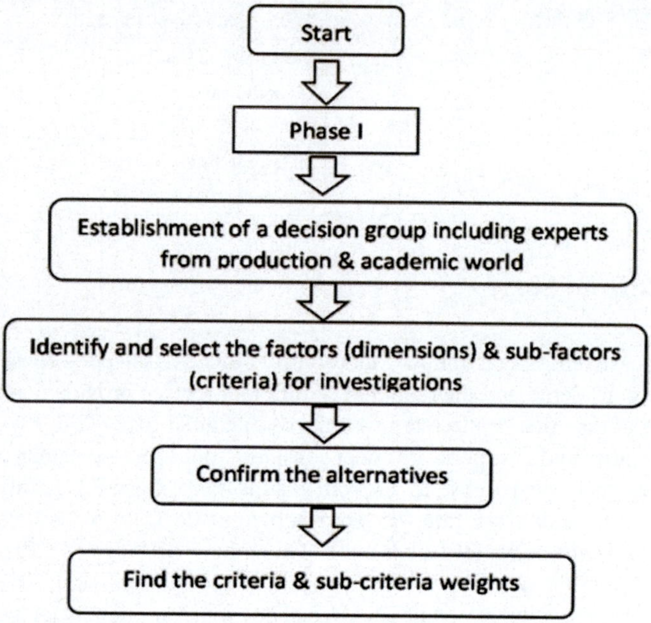

Fig. 2 Phase I of research framework including SWARA method

4.1 Evaluation of Reverse Logistic Provider Selection by Step-Wise Weight Assessment Ratio (SWARA) Method

Since conventional MADM methods cannot meritoriously handle problems with such inexact information, "therefore, fuzzy MADM methods have been developed owed to the inaccuracy in measuring the relative importance of attributes and the performance ratings of alternatives with respect to attributes. Hence, present work aims to extend SWARA to fuzzy SWARA. This assumes that all criteria are independent."

The process of decisive the relative weights of criteria by fuzzy SWARA is as same as the SWARA such as the following steps:

Step 1 Sort the evaluation factors in descending order of expected significance (Tables 7 and 8).

Step 2 According to Table 9, "state the relative importance of the factor j in relation to the previous $(j-1)$ factor, which has higher importance, and follow to the last factor. After determining all relative importance scores by all experts, to aggregate their judgments, the geometric mean of corresponding scores was obtained. Kersuliene et al. term this ratio as the comparative importance of average value Sj."

Step 3 Determine the coefficient Kj as follows:

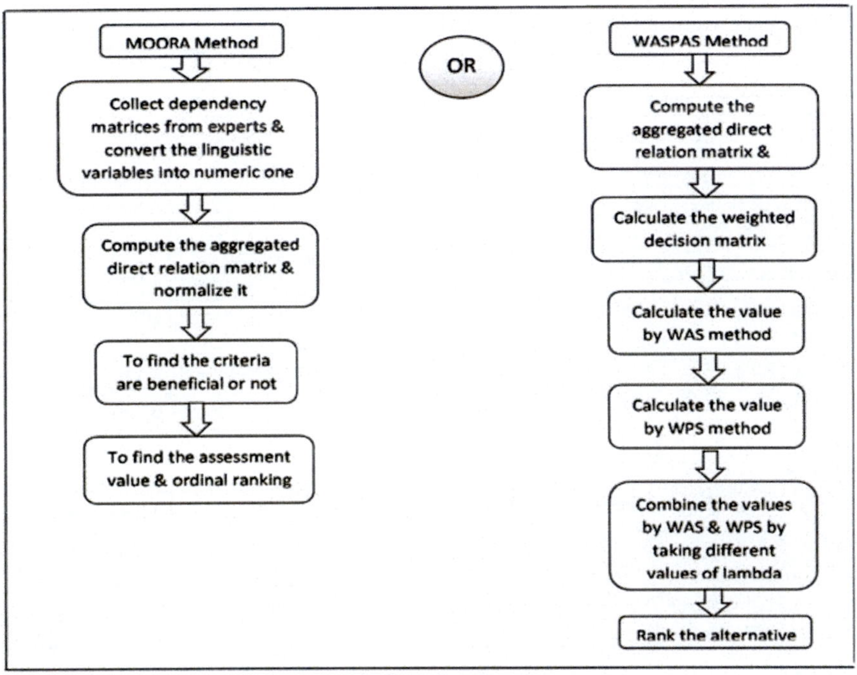

Fig. 3 Phase II of research framework including MOORA and WASPAS

Table 7 The fuzzy scale

Linguistic scale	Response scale
"Equally important"	"(1, 1, 1)"
"Moderately less important"	"(2/3, 1, 3/2)"
"Less important"	"(2/5, 1/2, 2/3)"
"Very less important"	"(2/7, 1/3, 2/5)"
"Much less important"	"(2/9, 1/4, 2/7)"

$$Kj = \begin{cases} 1 & j = 1 \\ Sj + 1 & j > 1 \end{cases}$$

Step 4 Determine the fuzzy weight qj as follows (Table 10):

$$qj = \begin{cases} 1 & j = 1 \\ \frac{Kj-1}{kj} & j > 1 \end{cases}$$

Step 5 The relative weights of the evaluation criteria are determined as follows:

$$wj = \frac{qj}{\sum_{k=1}^{n} qk}$$

Table 8 Comparative importance of average value *Sj*

Criteria	Comparative importance of average value *Sj*		
E1: Ordering and logistic cost	0.286	0.333	0.400
E2: Custom and insurance cost	0.222	0.250	0.286
E3: Quality management	0.667	1.000	1.500
E4: On time delivery rate	0.400	0.500	0.667
E5: Transportation	0.667	1.000	1.500
E6: Delivery and services	1.000	1.000	1.000
S1: Occupational health and safety program	0.667	1.000	1.500
S2: Operations	0.286	0.333	0.400
S3: Wages	0.400	0.500	0.667
S4: Prevention and risk control program	0.222	0.250	0.286
S5: Flexible working facilities	0.400	0.500	0.667
EN 1: Air emission	1.000	1.000	1.000
EN 2: Wastewater	0.667	1.000	1.500
EN 3: Use of harmful material	0.400	0.500	0.667
EN4: Use of env. friendly tech. and material	1.000	1.000	1.000
EN5: Recycle	0.667	1.000	1.500
EN 6: Reuse	0.400	0.500	0.667

"where *wj* denotes the relative weight of criterion *j*" (Table 11).
where "$wj = (w_l, w_m, w_u)$ is the relative fuzzy weight of the *j*th criterion and n shows the number of evaluation criteria".
Basic arithmetic operations on triangular fuzzy numbers "$A1 = (l_1, m_1, u_1)$," where "$l_1 \leq m_1 \leq u_1$ and $A2 = (l_2, m_2, u_2)$".
where "$l_2 \leq m_2 \leq u_2$" is done as follows:

- "Fuzzy addition":
  "$A1 \oplus A2 = (l_1 + l_2, m_1 + m_2, u_1 + u_2)$"
- "Fuzzy subtraction":
  "$A1 \ominus A2 = (l_1 - U_2; m_1 - m_2; U_1 - l_2)$"
- "Fuzzy multiplication":
  "$A1 \otimes A2 = l_1 l_2; m_1 m_2; u_1 u_2$"
- "Fuzzy division":
  "$A1 \oslash A2 = l_1/U_2; m_1/m_2; U_1/l_2$"

Step 6 Using above steps find the fuzzy weight for criteria (Table 12).
Step 7 Final weights for each sub-criteria (Table 13 and Fig. 4).

Table 9 Value of coefficient *Kj*

Criteria	Coefficient $Kj = Sj + 1$		
E1: Ordering and logistic cost	1.000	1.000	1.000
E2: Custom and insurance cost	1.222	1.250	1.286
E3: Quality management	1.667	2.000	2.500
E4: On time delivery rate	1.400	1.500	1.667
E5: Transportation	1.667	2.000	2.500
E6: Delivery and services	2.000	2.000	2.000
S1: Occupational health and safety program	1.667	2.000	2.500
S2: Operations	1.286	1.333	1.400
S3: Wages	1.400	1.500	1.667
S4: Prevention and risk control program	1.222	1.250	1.286
S5: Flexible working facilities	1.400	1.500	1.667
EN 1: Air emission	2.000	2.000	2.000
EN 2: Wastewater	1.667	2.000	2.500
EN 3: Use of harmful material	1.400	1.500	1.667
EN4: Use of env. Friendly tech. and material	2.000	2.000	2.000
EN5: Recycle	1.667	2.000	2.500
EN 6: Reuse	1.400	1.500	1.667

4.2 *Evaluation of Reverse Logistic Provider Selection by MOORA Method*

Brauers and Zavadskas introduced "fuzzy MOORA in a privatization-themed study in subsistence economy as a MCDM method. There are three different approaches for solving problems with fuzzy MOORA: fuzzy ratio method, reference point approach, and full multiplicative form. In this paper, we use the fuzzy ratio method of Mavi et. al. [4]."

Step 1 Construct the decision matrix using triangular fuzzy numbers

$$\begin{bmatrix} (x^l_{11k}, x^m_{11k}, x^u_{11k}) & \cdots & (x^l_{1nk}, x^m_{1nk}, x^u_{1nk}) \\ \vdots & \ddots & \vdots \\ (x^l_{1mk}, x^m_{1mk}, x^u_{1mk}) & \cdots & (x^l_{mnk}, x^m_{mnk}, x^u_{mnk}) \end{bmatrix}$$

where "*m* is the number of alternatives, *n* is the number of criteria, and *xmnk* presents the judgment of decision-maker *k* ($k = 1; 2; \ldots; K$) about the performance of alternative *i* in criterion *j*. Fuzzy numbers (*xijki*, *xijkm*, *xijku*) are assigned to each alternative based on Table 14."

Step 1–1: Obtain the aggregated decision matrix, $\tilde{X}$;

Table 10 Recalculated fuzzy weight *qj*

Criteria	Coefficient $Kj = Sj + 1$			Recalculated weight *qj*		
E1: Ordering and logistic cost	1.000	1.000	1.000	1.0000	1.0000	1.0000
E2: Custom and insurance cost	1.222	1.250	1.286	0.7770	0.8000	0.8180
E3: Quality management	1.667	2.000	2.500	0.3110	0.4000	0.4900
E4: On time delivery rate	1.400	1.500	1.667	0.1860	0.2660	0.3500
E5: Transportation	1.667	2.000	2.500	0.0740	0.1330	0.2100
E6: Delivery and services	2.000	2.000	2.000	0.0370	0.0660	0.1050
S1: Occupational health and safety program	1.667	2.000	2.500	0.0140	0.0330	0.0620
S2: Operations	1.286	1.333	1.400	0.0100	0.0250	0.0480
S3: Wages	1.400	1.500	1.667	0.0063	0.0160	0.0340
S4: Prevention and risk control program	1.222	1.250	1.286	0.0049	0.0130	0.0270
S5: Flexible working facilities	1.400	1.500	1.667	0.0029	0.0080	0.0190
EN 1: Air emission	2.000	2.000	2.000	0.0014	0.0040	0.0090
EN 2: Wastewater	1.667	2.000	2.500	0.0006	0.0020	0.0053
EN 3: Use of harmful material	1.400	1.500	1.667	0.0004	0.0014	0.0037
EN4: Use of env. friendly tech. and material	2.000	2.000	2.000	0.0002	0.0007	0.0018
EN5: Recycle	1.667	2.000	2.500	0.0001	0.0003	0.0011
EN 6: Reuse	1.400	1.500	1.667	0.0000	0.0002	0.0007
Total				2.42572	2.7686	3.1846

$$“\begin{bmatrix} (x^l_{11k}, x^m_{11k}, x^u_{11k}) & \cdots & (x^l_{1nk}, x^m_{1nk}, x^u_{1nk}) \\ \vdots & \ddots & \vdots \\ (x^l_{1mk}, x^m_{1mk}, x^u_{1mk}) & \cdots & (x^l_{mnk}, x^m_{mnk}, x^u_{mnk}) \end{bmatrix}”$$

where

$$“x^l_{ij} = \frac{\sum_{k-1}^{k} x^l_{ijk}}{k}, x^m_{ij} = \frac{\sum_{k-1}^{k} x^m_{ijk}}{k}, x^u_{ij} = \frac{\sum_{k-1}^{k} x^u_{ijk}}{k}”$$

Step 2 “Normalize the aggregated initial decision matrix to form a more comparable structure”.

As $\widetilde{r_{ij}} = (r^l_{ij}, r^m_{ij}, r^u_{ij})$

Table 11 Relative weights of evaluation criteria

Criteria	Fuzzy weight *qj*			Weight *wj*		
E1: Ordering and logistic cost	1.0000	1.0000	1.0000	0.314011	0.361193	0.41225
E2: Custom and insurance cost	0.7770	0.8000	0.8180	0.243987	0.288955	0.33722
E3: Quality management	0.3110	0.4000	0.4900	0.097657	0.144477	0.20200
E4: On time delivery rate	0.1860	0.2660	0.3500	0.058406	0.096077	0.14429
E5: Transportation	0.0740	0.1330	0.2100	0.023237	0.048039	0.08657
E6: Delivery and services	0.0370	0.0660	0.1050	0.011618	0.023839	0.04329
S1: Occupational health and safety program	0.0140	0.0330	0.0620	0.004396	0.011919	0.02556
S2: Operations	0.0100	0.0250	0.0480	0.003140	0.009030	0.01979
S3: Wages	0.0063	0.0160	0.0340	0.001978	0.005779	0.01402
S4: Prevention and risk control program	0.0049	0.0130	0.0270	0.001539	0.004696	0.01113
S5: Flexible working facilities	0.0029	0.0080	0.0190	0.000911	0.002890	0.00783
EN 1: Air emission	0.0014	0.0040	0.0090	0.000440	0.001445	0.00371
EN 2: Wastewater	0.0006	0.0020	0.0053	0.000185	0.000722	0.00218
EN 3: Use of harmful material	0.0004	0.0014	0.0037	0.000110	0.000506	0.00153
EN4: Use of env friendly tech. and material	0.0002	0.0007	0.0018	0.000053	0.000253	0.00074
EN5: Recycle	0.0001	0.0003	0.0011	0.000022	0.000108	0.00045
EN 6: Reuse	0.0000	0.0002	0.0007	0.000013	0.000072	0.00029

$$\text{“}r_{ij}^{l} = x_{ij}^{l} / \sqrt{\sum_{i=1}^{m}\left[(x_{ij}^{l})^{2} + (x_{ij}^{m})^{2} + (x_{ij}^{u})^{2}\right]}\text{”}$$

$$\text{“}r_{ij}^{m} = x_{ij}^{m} / \sqrt{\sum_{i=1}^{m}\left[(x_{ij}^{l})^{2} + (x_{ij}^{m})^{2} + (x_{ij}^{u})^{2}\right]}\text{”}$$

$$\text{“}r_{ij}^{u} = x_{ij}^{u} / \sqrt{\sum_{i=1}^{m}\left[(x_{ij}^{l})^{2} + (x_{ij}^{m})^{2} + (x_{ij}^{u})^{2}\right]}\text{”}$$

Step 3 “Obtain the weighted normalized fuzzy decision matrix by multiplying normalized fuzzy decision matrix and diagonal matrix of weights obtained from fuzzy SWARA.”

As $\tilde{v}_{ij} = (v_{ij}^{l}, v_{ij}^{m}, v_{ij}^{u})$ where $\tilde{v}_{ij} = \tilde{r}_{ij} \otimes \tilde{w}_{j}$

Step 4 Compute the normalized performance values by subtracting the cost criteria from the total of benefit criteria

Table 12 Fuzzy weight for main criteria

Criteria	Sj			$Kj = 1 + Sj$			Recalculated weight qj			Weight		
C1: Economic				1.00	1.000	1.000	1.000	1.000	1.000	0.4352	0.4545	0.4838
C2: Environment	0.400	0.500	0.667	1.400	1.500	1.667	0.600	0.667	0.714	0.2611	0.3032	0.3454
C3: Social	0.222	0.250	0.286	1.222	1.250	1.286	0.467	0.533	0.584	0.2032	0.2423	0.2825
Total							2.067	2.200	2.298			

Table 13 Final weight of sub-criteria

Criteria	Weight *wj*			Final weights		
E1: Ordering and logistic cost	0.314011	0.361193	0.41225	0.1366454	0.164179	0.199443
E2: Custom and insurance cost	0.243987	0.288955	0.33722	0.1061735	0.131343	0.163144
E3: Quality management	0.097657	0.144477	0.20200	0.0424967	0.065672	0.097727
E4: On time delivery rate	0.058406	0.096077	0.14429	0.0254160	0.043672	0.069805
E5: Transportation	0.023237	0.048039	0.08657	0.0101118	0.021836	0.041883
E6: Delivery and services	0.011618	0.023839	0.04329	0.0050559	0.010836	0.020942
S1: Occupational health and safety program	0.004396	0.011919	0.02556	0.0008934	0.002888	0.007221
S2: Operations	0.003140	0.009030	0.01979	0.0006381	0.002188	0.005591
S3: Wages	0.001978	0.005779	0.01402	0.0004020	0.001400	0.003960
S4: Prevention and risk control program	0.001539	0.004696	0.01113	0.0003127	0.001138	0.003145
S5: Flexible working facilities	0.0009	0.0028	0.007	0.00018	0.0007	0.00221
EN 1: Air emission	0.000440	0.001445	0.00371	0.0001148	0.000438	0.001282
EN 2: Wastewater	0.000185	0.000722	0.00218	0.0000483	0.000219	0.000755
EN 3: Use of harmful material	0.000110	0.000506	0.00153	0.0000286	0.000153	0.000527
EN4: Use of env. friendly tech. and material	0.000053	0.000253	0.00074	0.0000139	0.000077	0.000256
EN5: Recycle	0.000022	0.000108	0.00045	0.0000057	0.000033	0.000157
EN 6: Reuse	0.000013	0.000072	0.00029	0.0000033	0.000022	0.000100

$$\tilde{Yi} = \sum_{j=1}^{g} \tilde{v}_{ij} - \sum_{j=g+1}^{n} \tilde{v}_{ij}$$

"Here,
$\sum_{j=1}^{g} \tilde{v}_{ij}$ Benefit criteria for 1, …, g
$\sum_{j=g+1}^{n} \tilde{v}_{ij}$ Cost criteria for $g + 1, \ldots, n$
g, maximum number of criteria to be done.
$(n - g)$, minimum number of criteria to be done."

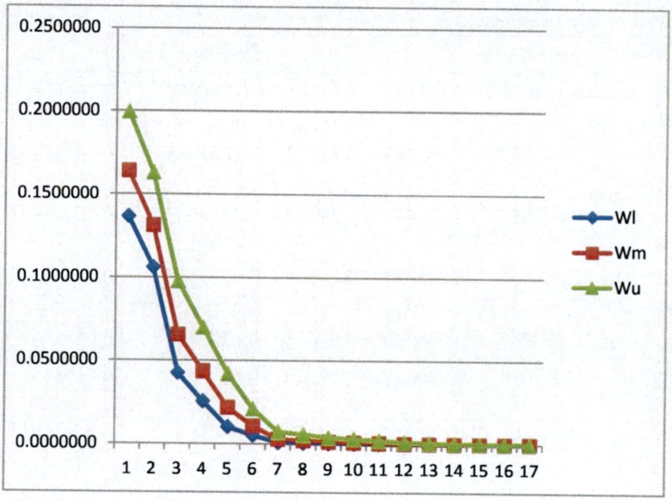

Fig. 4 Results by fuzzy SWARA

Table 14 Transformation for fuzzy membership functions

Linguistic scale	Triangular fuzzy numbers	Attribute grade
"Very low (VL)"	"(0, 0, 0.25)"	"1"
"Low (L)"	"(0, 0.25, 0.5)"	"2"
"Medium (M)"	"(0.25, 0.5, 0.75)"	"3"
"High (H)"	"(0.5, 0.75, 1)"	"4"
"Very high (VH)"	"(0.75, 1.0, 1.0)"	"5"

Ranking shows that alternative I is the best among all alternatives and alternative D is the worst choice (Table 15 and Fig. 5).

4.3 *Evaluation of Reverse Logistic Provider Selection by WASPAS Method*

This "subsection extends WASPAS to the fuzzy atmosphere. The worth of using a fuzzy approach is to allocate the relative importance of attributes using fuzzy numbers instead of accurate numbers."

The WASPAS method consists of two aggregated parts:

1. Weighted Sum Model (WSM);
2. Weighted Product Model (WPM).

Based on the briefly summarized fuzzy theory above, Fuzzy WASPAS steps can be outlined as follows:

Table 15 Best non-fuzzy performance value and ranking

3PRLP	Y_i			BNP(Yi)	Ranking
	Y_i^l	Y_i^m	Y_i^u		
A	0.0204	−0.0174	−0.0208	−0.0077	6
B	0.0171	−0.0295	−0.0315	−0.0146	2
C	0.0257	−0.0262	−0.0270	−0.0092	5
D	0.0269	−0.0155	−0.0213	−0.0033	10
E	0.0346	−0.0244	−0.0268	−0.0055	8
F	0.0286	−0.0221	−0.0227	−0.0054	9
G	0.0315	−0.0265	−0.0269	−0.0073	7
H	0.0170	−0.0215	−0.0262	−0.0102	4
I	0.0236	−0.0350	−0.0373	−0.0162	1
J	0.0235	−0.0289	−0.0362	−0.0138	3

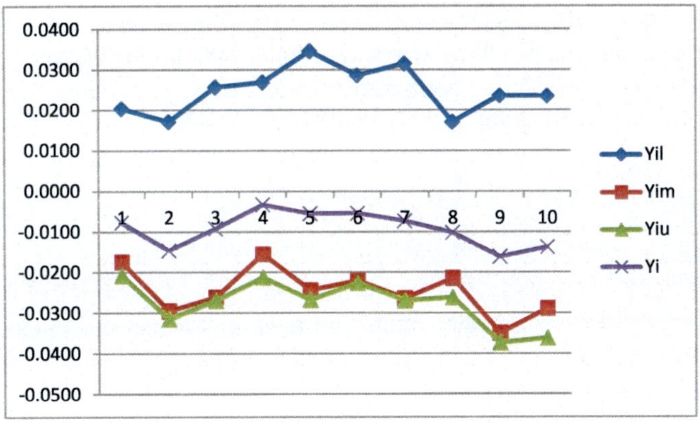

Fig. 5 Results by MOORA method

Step 1 Creating of fuzzy DM matrix (FDMM). The performance values $\tilde{xij}$ and the attributes weights $\tilde{wj}$ are entries of a DMM. Choose the linguistic ratings. "The structure of attributes as well as the values and initial weights of attributes are determined by decision-makers' experts. The distinct optimization problem is characterized by the partialities for m reasonable alternatives (rows) rated on n attributes (columns)":

$$\tilde{X} = \begin{bmatrix} \widetilde{X_{11}} \dots \widetilde{X_{1j}} \dots \widetilde{X_{1n}} \\ \widetilde{X_{i1}} \dots \widetilde{X_{ij}} \dots \widetilde{X_{in}} \\ \widetilde{X_{m1}} \dots \widetilde{X_{m1}} \dots \widetilde{X_{mn}} \end{bmatrix}; i = 1, m, j = 1, n,$$

"where $\widetilde{X}_{ij}$– fuzzy value representing the performance value of the i alternative in terms of the j attribute. A tilde ˜ is placed above a symbol if the symbol represents a fuzzy set. Then the purpose of the priorities of alternatives is conceded out in several steps."

Step 2 "The initial values of all the attributes $\widetilde{X}_{ij}$ are normalized—defining values $\overline{X}_{ij}$ of normalized decision-making matrix $\overline{X}= \left[\overline{X}_{ij}\right] m \times n$."

$$\overline{\overline{X_{ij}}} = \begin{cases} \dfrac{\widetilde{X_{ij}}}{\max\limits_{i} \widetilde{X_{ij}}} \text{ if } \max\limits_{i} \widetilde{X_{ij}} \text{ is preferable} \\ \dfrac{\min\limits_{i} \widetilde{X_{ij}}}{\widetilde{X_{ij}}} \text{ if } \min\limits_{i} \widetilde{X_{ij}} \text{ is preferable} \end{cases} \quad i = 1, m, \; j = 1, n.$$

Step 3a Compute the weighted normalized fuzzy DM Xq for WSM.
Step 3b Compute the weighted normalized fuzzy DM Xp for WPM.
Step 4 Compute the value of optimality function:
"According to the WSM for each alternative":
$\widetilde{Q}i= \sum_{j=1}^{n} \widetilde{x}_{ij}, i = 1, m,$
According to the WPM for each alternative:
$\widetilde{P}i= \prod_{j=1}^{n} \widetilde{x}_{ij}, i = 1, m.$
The result of fuzzy performance measurement for each alternative is fuzzy numbers $\widetilde{Q}i$ and $\widetilde{P}i$. The center-of-area is the most practical and simple to apply for defuzzification:
$Qi = \frac{1}{3}(Qi\alpha + Qi\beta + Qi\gamma)$
$Pi = \frac{1}{3}(Pi\alpha + Pi\beta + Pi\gamma)$
Step 5 "The integrated utility function value of the WASPAS-F method for an alternative could be determined as follows":
$Ki = \lambda \sum_{j=1}^{m} Qi + (1 - \lambda) \sum_{j=1}^{m} Pi, \lambda = 0, \ldots, 0.1, 0 \leq Ki \leq 1.$
Step 6 "Rank preference orders. Choose an alternative with maximal *Ki* value".

It shows that the service provider mentioned as A is the best choice and service provider mentioned as D is the worst choice among alternatives as shown in Fig. 6 (Tables 16 and 17).

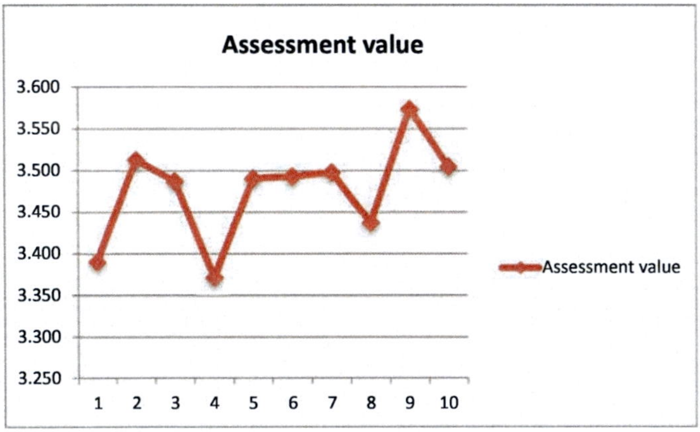

Fig. 6 Results by WASPAS method

5 Discussion and Proposed Solution

Comparative analysis ranking of case alternatives as per implementation of WASPAS and MOORA is given in Table 18. According to MOORA method "alternatives are ranked as I > B > J > H > C > A > G > E > F > D in the decreasing order of preference," and "according to WASPAS method alternatives are ranked as I > B > J > G > F > E > C > H > A > D in the decreasing order of preference." It is clear from the ranking results of the two MCDM approaches MOORA and WASPAS method that 3PRLSPS designated as I is the best choice for the given automobile industry operation under the given conditions while RLSP designated as D is the worst choice.

RLSP I is ideal according to the criteria transportation (E5), wastewater (EN2), use of harmful material (EN3), reuse (EN6) and closer to ideal according to the criteria ordering and logistic cost (E1), quality management (E3), on time delivery rate (E4), use of environment friendly material (EN4) and farthest to the ideal according to the criteria recycle (EN5).

RLSP B is ideal according to the criteria custom and insurance cost (E2), reuse (EN6) and closer to ideal conferring to the criteria transportation (E5), wages (S3), use of harmful material (EN3) and farthest to the ideal according to criteria Ordering and logistic cost (E1), wastewater (EN2), use of environment friendly material (EN4).

As an alternative for a final solution, service provider designated as I could be considered the best compromise from the ranking results of MOORA and WASPAS method.

Finally, study concludes that "the ranking depends on the judgments of relative importance made by the user. The ranking may change if the user assigns different relative importance values to the criteria's. The same is true with all these MCDM approaches (Table 19)."

Table 16 *Qi* value for WSM

3PRLP (*Qi*)	E1	E2	E3	E4	E5	E6	S1	S2	S3	S4	S5	EN1	EN2	EN3	EN4	EN5	EN6
A	0.1668	0.1103	0.0463	0.0242	0.0243	0.0105	0.0029	0.0022	0.0017	0.0010	0.0007	0.0004	0.0003	0.0003	0.0001	0.0001	0.0001
B	0.1096	0.0908	0.0579	0.0342	0.0239	0.0115	0.0032	0.0023	0.0013	0.0014	0.0009	0.0007	0.0003	0.0003	0.0002	0.0001	0.0001
C	0.1154	0.1001	0.0629	0.0335	0.0155	0.0083	0.0037	0.0028	0.0016	0.0010	0.0007	0.0004	0.0004	0.0002	0.0001	0.0001	0.0001
D	0.1308	0.1336	0.0537	0.0392	0.0178	0.0123	0.0027	0.0022	0.0014	0.0013	0.0011	0.0005	0.0003	0.0002	0.0002	0.0001	0.0001
E	0.1274	0.0967	0.0612	0.0388	0.0206	0.0105	0.0036	0.0029	0.0011	0.0014	0.0008	0.0004	0.0003	0.0002	0.0002	0.0001	0.0001
F	0.1106	0.1118	0.0662	0.0331	0.0226	0.0110	0.0034	0.0025	0.0014	0.0013	0.0008	0.0005	0.0003	0.0002	0.0001	0.0001	0.0001
G	0.1195	0.0923	0.0703	0.0257	0.0242	0.0087	0.0029	0.0017	0.0017	0.0016	0.0009	0.0005	0.0003	0.0002	0.0001	0.0001	0.0001
H	0.1274	0.1213	0.0496	0.0351	0.0216	0.0109	0.0021	0.0021	0.0020	0.0011	0.0009	0.0005	0.0003	0.0002	0.0001	0.0001	0.0001
I	0.0881	0.0844	0.0678	0.0463	0.0216	0.0107	0.0026	0.0023	0.0015	0.0011	0.0009	0.0006	0.0002	0.0002	0.0002	0.0001	0.0001
J	0.0971	0.1051	0.0653	0.0304	0.0174	0.0089	0.0031	0.0027	0.0017	0.0013	0.0009	0.0004	0.0003	0.0002	0.0001	0.0001	0.0001

Table 17 *Pi* value for WPM

3PRLP (Pi)	E1	E2	E3	E4	E5	E6	S1	S2	S3	S4	S5	EN1	EN2	EN3	EN4	EN5	EN6
A	1.0000	0.9691	0.9685	0.9527	1.0000	0.9972	0.9988	0.9987	0.9996	0.9988	0.9992	0.9994	0.9999	0.9999	1.0000	1.0000	1.0000
B	0.9295	0.9393	0.9855	0.9806	0.9989	0.9985	0.9989	0.9989	0.9989	0.9996	0.9996	1.0000	0.9998	1.0000	1.0000	1.0000	1.0000
C	0.9339	0.9525	0.9944	0.9785	0.9843	0.9929	1.0000	0.9998	0.9995	0.9988	0.9993	0.9993	1.0000	0.9998	1.0000	1.0000	1.0000
D	0.9607	1.0000	0.9822	0.9878	0.9883	1.0000	0.9979	0.9987	0.9987	0.9994	1.0000	0.9996	0.9997	0.9999	1.0000	1.0000	1.0000
E	0.9588	0.9464	0.9946	0.9873	0.9946	0.9966	0.9996	1.0000	0.9980	0.9997	0.9995	0.9993	0.9997	0.9999	1.0000	1.0000	1.0000
F	0.9288	0.9702	0.9969	0.9779	0.9967	0.9977	0.9993	0.9994	0.9989	0.9994	0.9995	0.9996	0.9999	0.9999	1.0000	1.0000	1.0000
G	0.9436	0.9385	1.0027	0.9616	0.9990	0.9938	0.9986	0.9978	0.9995	1.0000	0.9998	0.9996	0.9997	0.9999	1.0000	1.0000	1.0000
H	0.9588	0.9838	0.9719	0.9823	0.9951	0.9973	0.9967	0.9986	1.0000	0.9993	0.9998	0.9997	0.9999	0.9999	1.0000	1.0000	1.0000
I	0.8933	0.9269	0.9997	1.0000	0.9932	0.9972	0.9978	0.9993	0.9994	0.9991	0.9997	0.9998	0.9996	0.9998	1.0000	1.0000	1.0000
J	0.9077	0.9596	0.9965	0.9737	0.9859	0.9940	0.9988	0.9999	0.9996	0.9997	0.9998	0.9994	0.9997	0.9999	1.0000	1.0000	1.0000

Table 18 Integrated utility functions value

3PRLP	Final *Qi*	Final *Pi*	Assessment value	Ranking
A	−0.164	6.945	3.390	9
B	−0.065	7.090	3.513	2
C	−0.086	7.062	3.488	7
D	−0.134	6.993	3.372	10
E	−0.084	7.066	3.491	6
F	−0.081	7.067	3.493	5
G	−0.075	7.071	3.498	4
H	−0.124	6.999	3.437	8
I	−0.019	7.166	3.573	1
J	−0.072	7.081	3.505	3

Table 19 Comparative ranking of alternatives for reverse logistics providers

3PRLSPs	MOORA	WASPAS
A	6	9
B	2	2
C	5	7
D	10	10
E	8	6
F	9	5
G	7	4
H	4	8
I	1	1
J	3	3

6 Conclusions

MOORA and WASPAS method ranking give service provider I as the best alternative for automobile industry in management of sustainable supply chain. MOORA method ranks alternative by comparing each of the criteria with other alternative and finally finds the assessment value based on normalize data. It also considers into account the weightage given by SWARA method. WASPAS method works on a pair-wise comparison method of alternatives in each single criterion in order to regulate partial relations denoting the preference of an alternative over the other. Ranking by MOORA considering a linear preference function gives the same results as ranking by WASPAS. The similar results are obtained using decision-making framework model. MOORA and WASPAS method are consistent with the discussion. The effect of parameter λ on the ranking enactment of WASPAS method is also studied, revealing the fact that better results are attained at higher value of λ values.

When the value of λ is set at 0, WASPAS method works like a WPM method, and when λ is 1, it is transformed into WSM method. The main advantage of this method is identified as its strong resistance against rank reversal of the considered alternatives. It is also found that this method has the unique capability of dealing with both single and multi-response optimization problems in various machining operations.

References

1. Jayant A, Dhillon MS (2015) Use of analytic hierarchy process (AHP) to select welding process in high pressure vessel manufacturing environment. Int J Appl Eng Res 10(8):586–595
2. Jayant A, Gupta P, Garg SK (2011) Design and simulation of reverse logistics network: a case study. In: Proceedings of the world congress on engineering, vol 1, pp 1–5
3. Jayant A, Gupta P, Garg SK (2014) Reverse logistics network design for spent batteries: a simulation study. Int J Logist Syst Manag 18(3):343–365
4. Mavi RK, Goh M, Zarbakhshnia N (2017) Sustainable third-party reverse logistic provider selection with fuzzy SWARA and fuzzy MOORA in plastic industry. Int J Adv Manuf Technol 91(5–8):2401–2418
5. Sari K (2017) A novel multi-criteria decision framework for evaluating green supply chain management practices. Comput Ind Eng 105:338–347
6. Yazdani M, Zarate P, Coulibaly A, Zavadskas EK (2017) A group decision making support system in logistics and supply chain management. Expert Syst Appl 88:376–392
7. Rezaei J (2015) A systematic review of multi-criteria decision-making applications in reverse logistics. Transp Res Procedia 10:766–776
8. Prakash C, Barua MK (2016) A combined MCDM approach for evaluation and selection of third-party reverse logistics partner for Indian electronics industry. Sustain Prod Consum 7:66–78
9. Jayant A, Paul V, Kumar U (2014) Application of analytic network process (ANP) in business environment: a comprehensive literature review. Int J Res Mech Eng Technol 4(3):29–37
10. Jayant A, Singh A, Patel V (2011) An AHP based approach for supplier evaluation and selection in supply chain management. Int J Adv Manuf Syst 2(1):1–6
11. Jayant A (2013) Evaluation of 3PL service provider in supply chain management: an analytic network process approach. Int J Bus Insights Transform 6(2)

Effect of Power Level on the Processing of Ni-Based Casting Through Microwave Heating

Gurjot Singh, Dinesh, Sarbjeet Kaushal, and Satnam Singh

Abstract In the present research, Ni-based powder casting was developed using a domestic microwave oven with a variable power level of 180–900 W at 2.45 GHz frequency. The power level and exposure time were optimized using a set of experiments. It was found that with increase in power level the exposure time for the development of casting was reduced. The casting was developed within 25 min of exposure time at 900 W of power. Further, the microstructural results revealed the formation of equiaxed grains throughout the casting. The XRD analysis revealed the intermetallic compounds formation during processing. The intense heating led to the formation of nickel silicides and some chromium carbides. The formed carbides precipitated along the grain boundaries as indicated by EDS analysis. The microhardness of developed casting was observed as 404 ± 52 HV and was due to the presence of hard intermetallic compounds.

Keywords Microwave heating · Metal powder casting · Nickel · Microstructure · XRD

1 Introduction

Material processing industries are looking for green, eco-friendly, and sustainable techniques of manufacturing to reduce the global warming effect. The environmental friendly manufacturing techniques are currently focused research areas for academicians, scientists, and researchers. Casting is one of the universally adopted manufacturing techniques throughout the world and is used to manufacture a variety of material components. However, the conventional casting methods are characterized

G. Singh · Dinesh · S. Kaushal (✉)
Gulzar Group of Institutes, Ludhiana, Punjab, India
e-mail: sarbjeet.kaushal@ggi.ac.in

S. Singh
The NorthCap University, Gurugram 122017, India

© Springer Nature Singapore Pte Ltd. 2021
P. M. Pandey et al. (eds.), *Advances in Production and Industrial Engineering*, Lecture Notes in Mechanical Engineering,
https://doi.org/10.1007/978-981-15-5519-0_33

by higher processing time, lower efficiency, high energy consumptions, and pollution generation. Many government organizations are restricting and penalizing such industries to reduce the pollution and energy conservations. In recent decade, material processing through microwave energy has evolved as an environmental friendly method [1, 2]. Microwave energy is associated with the property of volumetric heating, which allows the heating at the molecular levels and also reduces the thermal gradient which commonly occurs in case of conventional heating processes. Apart from this volumetric heating, the microwave processing of materials results in the reduction in processing time, less energy consumption, and excellent structure–property relations of the material being processed [3, 4]. Earlier, the microwave heating applications were only limited to the processing of ceramic materials due to the misconception that metallic materials do not absorb the microwaves; however, Roy et al. [5] carried out the successful sintering of metallic powders through microwave heating. This investigation further encouraged the researchers to explore the possibility of microwave heating for the processing of metallic-based materials. Sharma et al. [6] filed a patent on the joining of metallic materials using microwave irradiation. Sharma and Gupta [7] extended the utilization of microwave energy in the domain of microwave cladding on bulk metallic materials. Kaushal et al. [8–10] successfully produced the metal-ceramic-based composite cladding on stainless steel substrates through microwave heating. Based upon high-temperature applications, Singh et al. [11] successfully attempted the development of metal-ceramic-based composite castings through microwave hybrid heating.

Present work focuses on the study of various power level and exposure time effects on the development of Ni-based metal casting through microwave heating. The Ni metal powder was selected as the metallic material in the present investigation owing to its excellent toughness and corrosion resistance properties.

2 Materials and Methods

Nickel (Ni) is a high corrosion resistance material and is widely used in aircraft turbine components, springs, switches, diaphragms, and valves. In the present work commercially available Ni-based powder of 35–40 μm particle size was used as the raw material for casting. The SEM and EDS micrographs of Ni-based powder are shown in Fig. 1a, b. Figure 1a depicts the spherical morphology of nickel powder particles, while EDS analysis confirms the presence of nickel as major elements and some peaks of iron in powder.

The raw powder was preheated in microwave convection mode at 180 °C. The preheated powder was then placed inside the mold cavity which is made of graphite material. This mold cavity was placed on the refractory brick. To provide the initial conventional heating, charcoal powder was used as susceptor material in the present study. A graphite sheet of approximately 1 mm thickness was used as a separator material to avoid the mixing of charcoal particles with Ni powder. The whole setup was placed inside a multimode type microwave oven which works at 2450 MHz

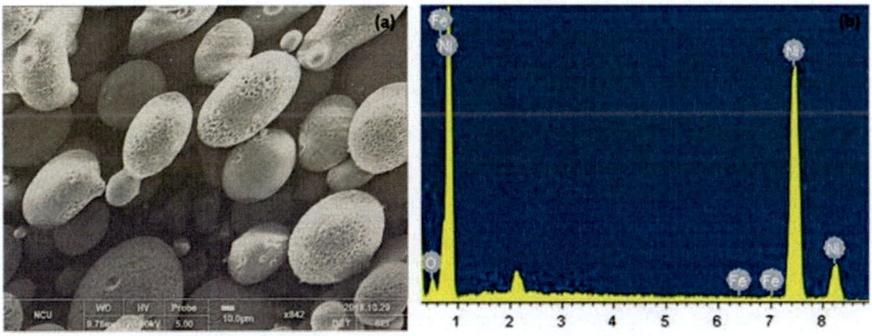

Fig. 1 **a** Spherical morphology and **b** EDS analysis of Ni-based powder

frequency and variable power level of 180–900 W. Principle of microwave hybrid heating (MHH) was utilized for the development of composite casting. Mishra and Sharma [12] explained in detail about the principle of MHH. The schematic arrangement of MHH is illustrated in Fig. 2. Various process parameters utilized in microwave casting are shown in Table 1. The microwave castings trials were carried out at various power levels starting from 360 to 900 W with the step size of 180 W and simultaneously the exposure time was varied. It was observed from

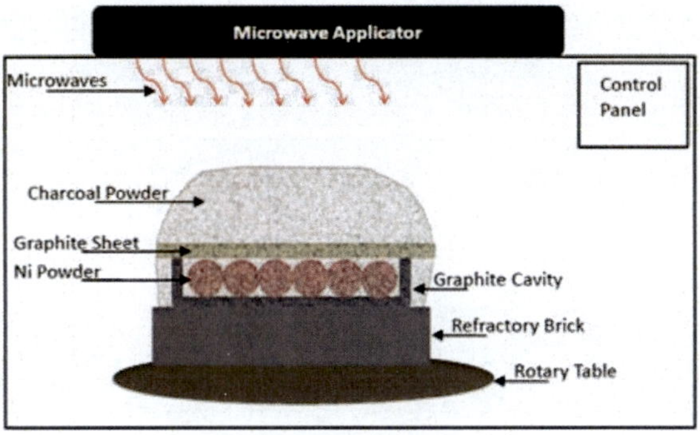

Fig. 2 Mechanism of microwave hybrid heating for development of casting

Table 1 Various process parameters for the development of microwave casting

Parameter	Detail
Applicator	Multimode microwave oven
Power level and frequency	180–900 W, 2.45 GHz
Exposure time	20–30 min
Powder used	Ni powder, average particle size (35–40 μm)
Susceptor	Charcoal powder
Separator	Graphite sheet (~1 mm thickness)

experimental trials that at 900 W power the casting was developed within 25 min of exposure time. The developed casting was characterized through microstructure analysis using XRD, SEM, and EDS. Prior to the characterization, the developed casting was polished using standard metallographic techniques.

2.1 *Mechanism of Casting Through Microwave Heating*

The processing of metal-based powder is a difficult task through microwave heating process. The major role in processing of metallic-based powder is of skin depth associated with the materials. Metals have low value of skin depth which causes the microwaves to reflect from surface and prevents the heating. However, if skin depth can be altered, the metallic powders can be easily processed through microwaves. Authors [13] have reported that some of the parameters are temperature dependent which can affect the skin depth of materials and hence can make microwaves to directly interact with metal powders. The mechanism of microwave casting of metallic powder is shown in Fig. 3.

Initially microwaves are reflected from the metallic powder surfaces (Fig. 3a) and hence cannot be directly processed. However, susceptor/hybrid heating allows initial heating of powder particles (through charcoal and graphite cavity) which helps in altering (increasing) the skin depths such that microwaves gets absorbed within the particles (Fig. 3b). The absorption of microwaves gets converted into heat

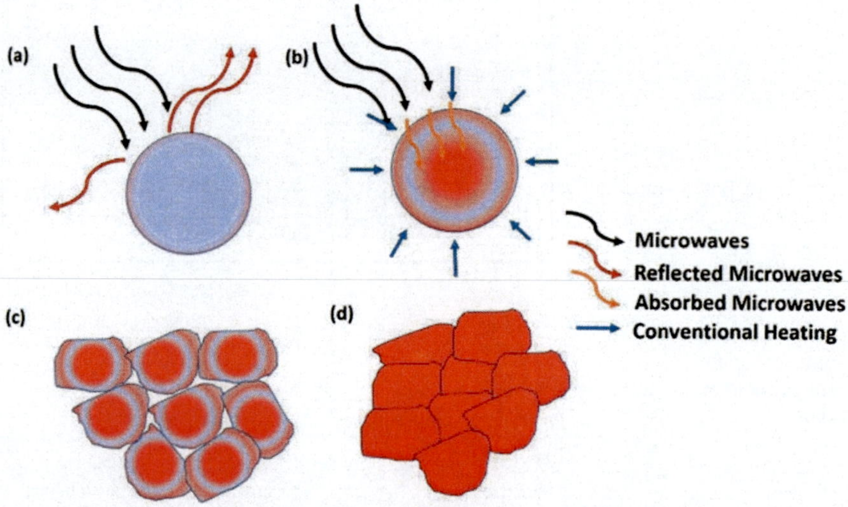

Fig. 3 **a** Reflection of microwaves due to lower skin depths of metallic powder, **b** increase in skin depth due to conventional/susceptor heating, **c** bi-directional heating causing partial melting and diffusion of powder particles, and **d** complete melting and solidification of casting

through various mechanisms as discussed by authors [11] and allows heat to flow from inner to outer surface; while conventional heating continues heating from outwards. This causes bi-directional heating within the powder and allows partial melting and diffusion of particles (Fig. 3c). This intense bi-directional heating causes semi molten state of particles and suddenly reduces the skin depth as single particles do not exists anymore and diffused melt behaves as bulk metal. At this stage, the microwaves again start reflecting and do not allow any further heating. The formation of melt due to bi-directional heating gets solidified on cooling and cavity shaped casting is obtained.

3 Results and Discussion

3.1 Effect of Power Level on the Development of Casting

On increasing the microwave power level, greater amount of microwave energy is injected inside the material. In the current study, the experimental trials were conducted at various power level from 360 to 900 W with step size of 180 W. Effect of various power level on development of microwave casting is illustrated in Table 2. It was observed that at lower power levels of 360 and 540 W, no casting was formed even after 35 min of microwave exposure time. However, at 720 W, the casting was achieved after the exposure time of 45 min. On further experimentations it was

Table 2 Effect of microwave power level and exposure time on the development of castings

Trial	Power level (W)	Exposure time (min)	Observation
1	360	25	No casting was formed, raw heated powder was present in the cavity
2	360	45	Cavity was heated, some layers of powder were sintered and loosely bound
3	540	25	Sintering of side layers was observed
4	540	45	Powder gets sintered with adjacent layers but no diffusion was observed. Layers get separated on applying load
5	720	25	Sintered powder was observed with loosely bound particles at the center
6	720	45	Diffused layers were observed but strength was not achieved due to partial melting at the center
7	900	25	The complete diffusion of powder layers was observed throughout the casting
8	900	35	Over melting of casting which caused distortion of casting. Hot spots were observed at the center of casting

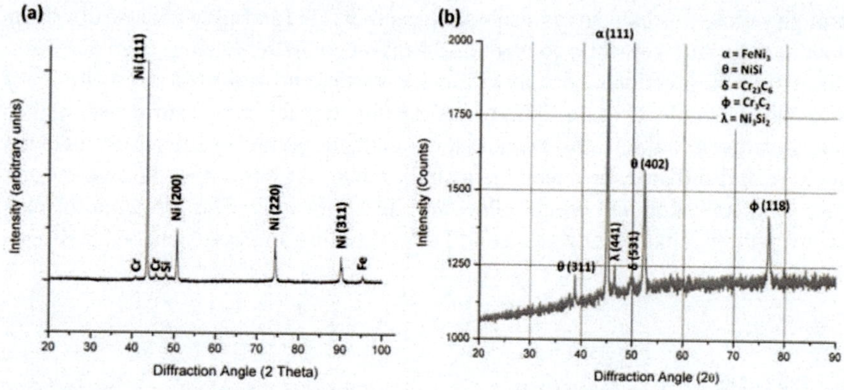

Fig. 4 XRD patterns of **a** nickel-based powder [11] and **b** microwave processed nickel-based casting

observed that at a maximum power level of 900 W, the casting was successfully achieved within 25 min of microwave exposure time.

3.2 *X-ray Diffraction Analysis of Casting*

To study the various phase formation in the casting, XRD analysis was carried out on nickel powder [11] and microwave processed nickel powder casting. The typical pattern obtained by using Cu Kα rays is shown in Fig. 4a, b.

The XRD analysis of raw powder shows the major peaks of nickel and some peaks of iron, chromium, and silicon. The microwave processed casting revealed the formation of some intermetallic as represented by phases. The highest peak of $FeNi_3$ was observed at diffraction angle of 44.12°. At elevated temperatures, the chromium reacts with carbon (from cavity walls) to form chromium carbides, and their peaks are represented corresponding to the diffraction angle of 50.62° ($Cr_{23}C_6$) and 76.49° (Cr_3C_2). Nickel reacts with silicon to form nickel silicides corresponding to 2θ of 46.73 (Ni_3Si_2) and NiSi at 52.23°.

3.3 *Microstructure Analysis*

The cavity shaped microwave processed casting is shown in Fig. 5a. Casting obtained through MHH process is sound and without any visible defects such as cracks and porosity. The microstructure of the developed casting was analyzed through scanning electron microscope and is shown in Fig. 5b, which clearly depicts the presence of equiaxed grains throughout the casting. The formation of these equiaxed grains is the

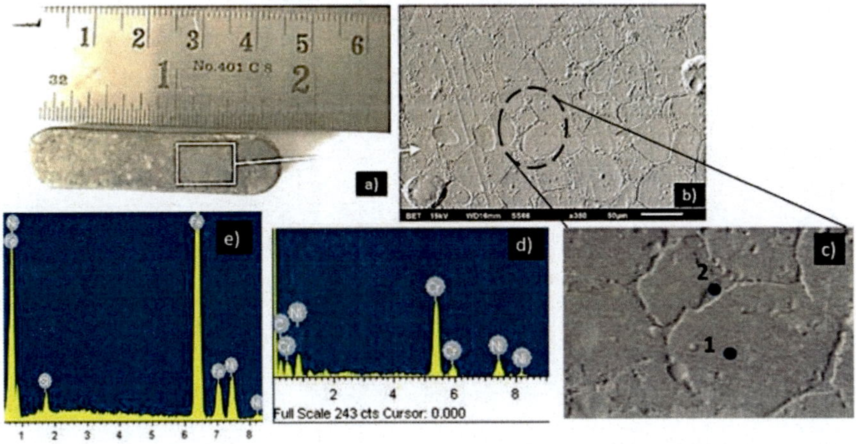

Fig. 5 **a** Microwave processed cavity shaped casting, **b** SEM image showing the equiaxed grains, **c** enlarged image of microstructure, **d** EDS analysis on grain boundaries i.e., location 2, and **e** EDS analysis on grain surface i.e., location 1

result of volumetric heating nature of microwaves, which prevents the formation of dendrites due to absence of differential heating. The enlarged view of microstructure is shown in Fig. 5c, which also shows the locations of EDS analysis on grain surface (1) and grain boundary (2).

The EDS analysis depicted that carbon and chromium content increases along the grain boundary (Fig. 5d). This is due to the precipitation of hard carbides during solidification process along the grain boundaries.

The grain surface represents the abundance presence of nickel, iron, and silicon (Fig. 5e) which is due to the formation of nickel-based intermetallic compounds. The formation of such compounds is shown by XRD analysis of casting. The high temperature during the processing allows the formation of such compounds.

3.4 Microhardness Study

The microhardness of the developed casting was evaluated using Vicker's microhardness tester. The mean value of microhardness of the microwave processed nickel casting was observed as 404 ± 52 HV. Microhardness was observed to be higher when indent falls on the grain boundaries whereas, lower values were observed when indent falls on grain surface. This causes higher variations in the microhardness values. The indent morphology on the grain surface is shown in Fig. 6.

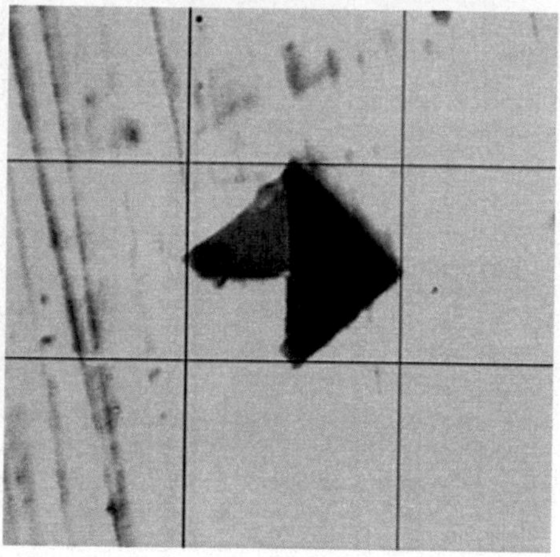

Fig. 6 Microhardness indent morphology on nickel powder-based casting

4 Conclusions

In the present research work, the Ni-based metallic powder casting was successfully achieved through microwave heating. The effect of various power level on development of casting was studied. The casting was achieved within 25 min of exposure time at 900 W of power level. The developed casting was free from any type of pores and voids. XRD analysis revealed the development of intermetallic compounds and formation of silicides and carbides were observed. EDS analysis revealed that carbides were precipitated along grain boundaries and silicides were present within the grain surface. The microstructure of the casting revealed the presence of equiaxed grains. The average microhardness of the developed casting was observed as 404 $\pm$ 52 HV. This method can further be optimized for obtaining large castings of composite materials with energy efficient processing.

References

1. Clark DE, Folz DC, West JK (2000) Processing materials with microwave energy. Mater Sci Eng: A 287(2):153–158
2. Gupta D, Sharma AK (2014) Microwave cladding: a new approach in surface engineering. J Manuf Process 16(2):176–182
3. Singh S, Gupta D, Jain V (2018) Microwave melting and processing of metal–ceramic composite castings. Proc Inst Mech Eng Part B: J Eng Manuf 232(7):1235–1243
4. Singh S, Gupta D, Jain V, Sharma AK (2015) Microwave processing of materials and applications in manufacturing industries: a review. Mater Manuf Process 30(1):1–29
5. Roy R, Agrawal D, Cheng J, Gedevanishvili D (1999) Full sintering of powdered metal bodies in a microwave field. Nature 399:668–670
6. Sharma AK, Srinath MS, Kumar P (2009) Microwave joining of metallic materials. Indian Patent application no. 1994/Del/2009

7. Sharma AK, Gupta D (2012) On microstructure and flexural strength of metal–ceramic composite cladding developed through microwave heating. Appl Surf Sci 258(15):5583–5592
8. Kaushal S, Gupta D, Bhowmick HL (2018) On processing of Ni-WC based functionally graded composite clads through microwave heating. Mater Manuf Process 33(8):822–828
9. Kaushal S, Gupta D, Bhowmick HL (2018) An approach for functionally graded cladding of composite material on austenitic stainless steel substrate through microwave heating. J Compos Mater 52(3):301–312
10. Kaushal S, Gupta D, Bhowmick HL (2018) On surface modification of austenitic stainless steel using microwave processed Ni/Cr_3C_2 composite cladding. Surf Eng 34(11):809–817
11. Singh S, Gupta D, Jain V (2016) Novel microwave composite casting process: theory, feasibility and characterization. Mater Des 111:51–59
12. Mishra RR, Sharma AK (2016) A review of research trends in microwave processing of metal-based materials and opportunities in microwave metal casting. Crit Rev Solid State Mater Sci 41(3):217–255
13. Singh S, Gupta D, Jain V (2018) Processing of Ni-WC-8Co MMC casting through microwave melting. Mater Manuf Process 33(1):26–34

Optimization of Input Parameters for CNC Turning of SS304: A Grey Relational Analysis and Response Surface Methodology Approach

Anmol Bhatia, Mayank Juneja, and Nikhil Juneja

Abstract The present study deals with the investigation of the effect of input parameters on material removal rate and surface roughness during CNC turning of SS304. The turning of the specimen was done using TNMG tool inserts. The paper presents multi-response optimization with the Response surface models. The input parameters chosen for the study are speed, feed, and depth of cut. L-27 orthogonal array was used for the experimentation and Taguchi with Grey relational analysis were used for analysis. The significant parameters were obtained using analysis of variance. Optimal and significant levels were obtained and their regression models were made which are further verified by response surface plots. The relational model demonstrates the relationship between the response variables and input variables. The optimized parameters were further confirmed using a confirmatory experiment.

Keywords Turning · Input parameters · Grey relational analysis · Surface roughness · Multi-response optimization

1 Introduction

In today's fierce and spirited market environment, both large-scale and small-scale industries, have allotted a high primacy to an economically viable machining for producing quality products. In order to obtain high-quality products at the minimal cost, adopting the right cutting parameters becomes a very critical task [1]. One of the most common stainless steel, SS-304, is chosen in this study and because of its excellent corrosion resistance and ease to morph into various shapes, it finds its employment in various spheres ranging from manufacturing to architecture to heat exchangers, braces, marine, and even in medical applications [2]. Asilturk et al. [3] investigated the effects of process parameters on surface roughness and concluded that the most critical factor is feed. Moshat et al. [4] used PCA-based Taguchi method for optimizing cutting parameters of the CNC milling process in order to obtain

A. Bhatia (✉) · M. Juneja · N. Juneja
Department of Mechanical Engineering, The NorthCap University, Gurugram, India
e-mail: anmolbhatia@ncuindia.edu

© Springer Nature Singapore Pte Ltd. 2021

P. M. Pandey et al. (eds.), *Advances in Production and Industrial Engineering*, Lecture Notes in Mechanical Engineering,
https://doi.org/10.1007/978-981-15-5519-0_34

an optimal value of surface roughness and material removal rate. Shetty et al. [5] employed Taguchi and response surface methodologies for turning aluminum alloy 6061 to study effect of cutting parameters in order to determine the minimum surface roughness possible. Neseli et al. [6] presented a model to demonstrate the effect of geometry of tool on surface roughness while turning AISI 1040. The authors applied RSM and concluded that the nose radius was the most significant parameter. Gupta et al. [7] have done grey relation to enhance the material removal rate and surface roughness in unidirectional glass fiber reinforced plastic composite while rough cutting operation.

The selection of input parameters for different performance characteristics is a challenging task and researcher's acumen determines weightage of responses [8]. The optimization studies have been conducted in the recent past by many researchers [9–11]. It is commonly observed that the two performance parameters, surface roughness, and material removal rate, that are discussed above are often found to be conflicting. Optimizing one could adversely affect the other important parameter so it is necessary to reach a win–win situation where none of the parameters gets compromised. Thus, the objective of this study is to conduct multi-output optimization which provides a perfect balance between the pair of performance characteristics presented above and gives a more optimized and better outcome of turning operation of SS-304.

2 Experimentation

In the present study, the experimentation was carried on CNC Lathe. The machine will guide the movements of tools and machiner by a Program which is generated using a computer. Turning operation of SS-304 was executed using TNMG insert which have 6 cutting fringes. In order to obtain accurate results and relationship between the output and input parameters, not only individual parameters but all other possibilities, L27 orthogonal array design was followed during experimentation. Grey relational analysis is best done by incorporating the full factorial method. The levels obtained after the pilot experimentation is shown in Table 1. The experimentation was conducted on a round bar of diameter 28 mm.

The response characteristics chosen for the study were surface roughness and metal removal rate. Mitutoyo surface roughness tester (Surftest SJ-210P) was used to measure the surface roughness, whereas, MRR was calculated using Eq. 1.

Table 1 Input parameters and their levels

Level	v (m/min)	f (mm/rev)	a (mm)
1	150	0.1	0.1
2	175	0.15	0.2
3	200	0.2	0.3

$$\mathrm{MRR} = \pi v f a D_{\mathrm{a}} \tag{1}$$

where, v is the cutting speed, f represents the feed, a represents the depth of cut, and D_{a} represents the average diameter. The turning of the specimens was done at the suitable combination of the levels and the obtained values of R_a and MRR are shown in Table 2.

Table 2 Factor information

Run	V (m/min)	f (mm/rev)	a (mm)	R_a (μm)	MRR (mm^3/min)
1	150	0.1	0.1	1.425	1490.057
2	150	0.1	0.2	1.296	2969.433
3	150	0.1	0.3	1.49	4438.128
4	150	0.15	0.1	2.297	2235.086
5	150	0.15	0.2	2.175	4454.150
6	150	0.15	0.3	2.229	6657.192
7	150	0.2	0.1	3.522	2980.115
8	150	0.2	0.2	3.545	5938.867
9	150	0.2	0.3	3.452	8876.256
10	175	0.1	0.1	1.393	1753.009
11	175	0.1	0.2	1.857	3493.451
12	175	0.1	0.3	1.594	5221.327
13	175	0.15	0.1	2.512	2629.513
14	175	0.15	0.2	2.315	5240.177
15	175	0.15	0.3	2.353	7831.990
16	175	0.2	0.1	3.352	3506.017
17	175	0.2	0.2	3.522	6986.902
18	175	0.2	0.3	3.84	10442.654
19	200	0.1	0.1	1.47	2015.960
20	200	0.1	0.2	1.312	4017.469
21	200	0.1	0.3	1.693	6004.526
22	200	0.15	0.1	2.563	3023.940
23	200	0.15	0.2	2.407	6026.203
24	200	0.15	0.3	2.446	9006.789
25	200	0.2	0.1	3.677	4031.920
26	200	0.2	0.2	3.697	8034.937
27	200	0.2	0.3	3.614	12009.052

3 Optimization Methodology

3.1 Grey Relational Analysis

Grey Relation Analysis (GRA) is implemented to convert single-objective optimization into a multi-output optimization process and solve it effectively [12, 13]. Following steps are used in Grey Relational Analysis.

Normalizing the data

Data needs to be pre-processed. The data has to be normalized first and this normalized data lies in the range 0 and 1. This is done to convert the original (experimental) data into a form in which these values can be compared. Normalizing the data was done according to the parameter being optimized. Higher the better approach was used for normalizing of MRR and lower the better for surface roughness. The equations are as follows:

$$x_i^*(k) = \frac{x_i(k) - x_{i\min}(k)}{x_{i\max}(k) - x_{i\min}(k)} \quad (2)$$

$$x_i^*(k) = \frac{x_{i\max}(k) - x_i(k)}{x_{i\max}(k) - x_{i\min}(k)} \quad (3)$$

where $x_i^*(k)$ and $x_i(k)$ indicate the normalized data and observed data, respectively,

Calculating Grey Relation Coefficient

This is calculated to compare the actual values with the ideal value. After the data is pre-processed, $\zeta_i(k)$ which is the grey relational can be calculated for the kth response characteristics using Eq. 4.

$$\zeta_i(k) = \frac{\Delta_{\min} + \zeta\,\Delta_{\max}}{\Delta_i(k) + \zeta\,\Delta_{\max}} \quad (4)$$

where $x_o^*(k) =$ denotes reference sequence.
$x_j^*(k) =$ denotes comparability sequence.
$\Delta_i =$ difference in absolute value between $x_o^*(k)$ and $x_j^*(k)$
$\Delta_{\min} = \min_{(\forall j \epsilon i)} \min_{(\epsilon k)} \left| x_o^*(k) - x_j^*(k) \right| =$ smallest value of Δ_i
$\Delta_{\max} = \max_{(\forall j \epsilon i)} \max_{(\epsilon k)} \left| x_o^*(k) - x_j^*(k) \right| =$ smallest value of Δ_i

Grey Relation Grade

Grey Relation Grade (GRG) is calculated by the values of grey relation coefficients. This also ranges from 0 to 1. Higher the GRG i.e., the value closest to 1 depicts that the corresponding factors would give the result closest to the optimal one.

$$\Upsilon_i = \frac{1}{m}\sum_{k=1}^{n} wX\zeta_i(k) \tag{5}$$

where,
Υ_i = Grey relation grade.
m = number of run.
w = weight factor ($w = 1$, while finding the optimal values).

4 Response Surface Methodology

The main idea of response surface methodology approach is to establish a relationship between the one or many output parameters over the input parameters, the relationship can be explained using the second order polynomial which is represented by Eq. 6 [14, 15]. The relationship can be very easily computed using this technique as compared to other techniques. Least square is the method which is used to compute the regression coefficients and is illustrated using 3D surface graphs or counter maps.

$$\eta = \beta_o + \sum_{i=1}^{k} \beta_i X_i + \sum_{i=1}^{k} \beta_{ii} X_i^2 + \sum_{i}\sum_{j} \beta_{ij} X_i X_j + \varepsilon \tag{6}$$

5 Results and Discussions

5.1 *Significant Parameters Using ANOVA*

Tables 3 and 4 represent the ANOVA table for Roughness values and MRR respectively, The parameters were analyzed for 95% confidence intervals. From the ANOVA

Table 3 ANOVA table for surface roughness

Source	DF	Adj SS	Adj MS	F-value	P-value
v (m/min)	2	0.1417	0.07083	3.51	0.049
f (mm/rev)	2	19.5931	9.79655	485.26	0
a (mm)	2	0.0222	0.0111	0.55	0.586
Error	20	0.4038	0.02019		
Total	26	20.1607			

Table 4 ANOVA table for MRR

SOURCE	DF	Adj SS	Adj MS	F-Value	P-Value
v (m/min)	2	11094424	5547212	9.4	0.001
f (mm/rev)	2	54787279	27393640	46.41	0
a (mm)	2	121796505	60898252	103.16	0
Error	20	11806050	0		
Total	26	199484258			

table, it can be observed that the major significant parameter for the surface roughness was feed and depth of cut for MRR whereas, feed and cutting velocity have the least contribution to surface roughness and MRR respectively,

5.2 *Grey Relational Analysis*

Using Grey Relation Analysis, the two performance parameters, R_a and MRR, are coalesced into single response, GRG. The cutting parameters that are considered are cutting speed (v), Feed (f), and Depth of cut (a). The experimental values in Table 5 are used to first normalize the data followed by calculating grey relation coefficients. Subsequently, grey relation grade is calculated. According to GRG rule, for MRR, "higher the better" equation was used, whereas, for R_a, "lower the better" equation was used. The GRG 0.6822 is closest to 1 and hence this level is chosen as the optimal one. The initial input parameters were 175 m/min, 0.1 mm/rev, and 0.2 mm as cutting speed, feed, and depth of cut. The new conditions, that when employed would yield the best result were 200 m/min, 0.1 mm/rev, and 0.2 mm. The performance of the process was enhanced when the new conditions were employed. Both the performance parameters are refined. MRR is increased by 15% and surface roughness is improved by 29.35%.

5.3 *Response Surface Methodology*

The full quadratic response surface model is constructed in MINITAB depicting the roughness as a function of process parameters f, v, a. The R^2 value is 0.9803 and the Adj. R^2 is 0.9698 which reveals that all explanatory variables account for about 98.03% variation in the surface roughness, which is fairly high. The R^2 value in this case is high and close to 1, which is desirable [16]. This technique helps to demonstrate that the model fits the original data and also inhibits the model overfitting. A model with overfitted data indicates that excess items are included in the model. It is to be noted that if the $(\text{R-square})_{\text{predicted}}$ is smaller than R^2, then this will act as a warning sign that the data is overfitted. The predicted R^2 value is 0.9560 which

Table 5 Grey relational analysis

Run	v (m/min)	f (mm/rev)	a (mm)	R_a(μm)	MRR (mm^3/min)	Normalized		GRC,$\zeta_i(k)$		GRG (Υ)
						R_a	MRR	R_a	MRR	
1	150	0.1	0.1	1.425	1490.057	0.9393	0	0.8918	0.3333	0.6126
2	150	0.1	0.2	1.296	2969.433	0.9895	0.1406	0.9794	0.3678	0.6736
3	150	0.1	0.3	1.49	4438.128	0.9140	0.2803	0.8533	0.4099	0.6316
4	150	0.15	0.1	2.297	2235.086	0.6002	0.0708	0.5557	0.3499	0.4528
5	150	0.15	0.2	2.175	4454.150	0.6476	0.2818	0.5866	0.4104	0.4985
6	150	0.15	0.3	2.229	6657.192	0.6266	0.4912	0.5725	0.4956	0.5341
7	150	0.2	0.1	3.522	2980.115	0.1237	0.1417	0.3633	0.3681	0.3657
8	150	0.2	0.2	3.545	5938.867	0.1147	0.4229	0.3609	0.4642	0.4126
9	150	0.2	0.3	3.452	8876.256	0.1509	0.7022	0.3706	0.6267	0.4987
10	175	0.1	0.1	1.393	1753.009	0.9518	0.0250	0.9120	0.3390	0.6255
11	175	0.1	0.2	1.857	3493.451	0.7713	0.1905	0.6861	0.3818	0.5340
12	175	0.1	0.3	1.594	5221.327	0.8736	0.3547	0.7982	0.4366	0.6174
13	175	0.15	0.1	2.512	2629.513	0.5165	0.1083	0.5084	0.3593	0.4338
14	175	0.15	0.2	2.315	5240.177	0.5932	0.3565	0.5514	0.4373	0.4943
15	175	0.15	0.3	2.353	7831.990	0.5784	0.6029	0.5425	0.5574	0.5499
16	175	0.2	0.1	3.352	3506.017	0.1898	0.1916	0.3816	0.3822	0.3819
17	175	0.2	0.2	3.522	6986.902	0.1237	0.5226	0.3633	0.5115	0.4374
18	175	0.2	0.3	3.84	10442.654	0.0000	0.8511	0.3333	0.7705	0.5519
19	200	0.1	0.1	1.47	2015.960	0.9218	0.0500	0.8648	0.3448	0.6048
20	200	0.1	0.2	1.312	4017.469	0.9833	0.2403	0.9676	0.3969	0.6823

(continued)

Table 5 (continued)

Run	v (m/min)	f (mm/rev)	a (mm)	R_a (μm)	MRR (mm^3/min)	Normalized		GRC, $\zeta_i(k)$		GRG (Υ)
						R_a	MRR	R_a	MRR	
21	200	0.1	0.3	1.693	6004.526	0.8351	0.4292	0.7520	0.4669	0.6095
22	200	0.15	0.1	2.563	3023.940	0.4967	0.1458	0.4984	0.3692	0.4338
23	200	0.15	0.2	2.407	6026.203	0.5574	0.4312	0.5304	0.4678	0.4991
24	200	0.15	0.3	2.446	9006.789	0.5422	0.7146	0.5220	0.6366	0.5793
25	200	0.2	0.1	3.677	4031.920	0.0634	0.2416	0.3480	0.3973	0.3727
26	200	0.2	0.2	3.697	8034.937	0.0556	0.6222	0.3462	0.5696	0.4579
27	200	0.2	0.3	3.614	12009.052	0.0879	1	0.3541	1	0.6770

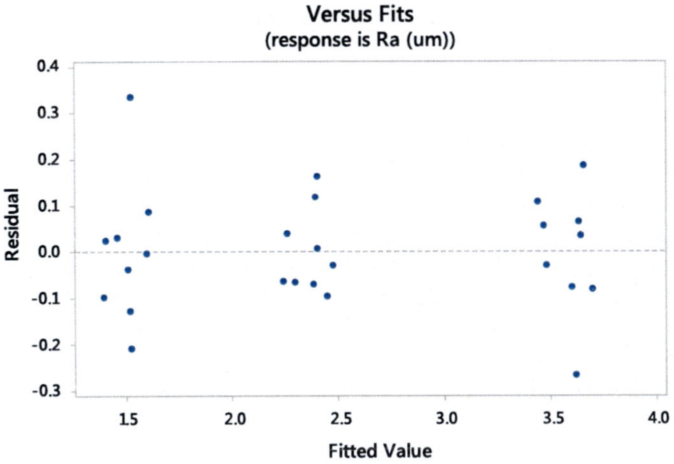

Fig. 1 Residual plot

indicates that the model is not overfit. The trustworthiness of the regression analysis is determined by analyzing the plots of the residuals versus the predicted response. Adequacy is determined whether the residuals are consistent with random error. It is observed in Fig. 1 that the residuals are randomly scattered around zero for the entire range of fitted values. As there are no patterns observed in the residuals, therefore the model is trustworthy and do not require any adjustments [17].

In order to verify the assumption that the residuals are normally distributed, the normal probability plots were examined. It is to be noted that the points plotted on the normal probability plots of the residuals should preferably follow a straight line. If the residuals do not follow a normal distribution, the confidence intervals and p-values can be inaccurate. Figure 2 indicates the normal probability plot for the residuals in which the residuals generally are on a straight line which indicates that the errors are normally distributed.

RSM approach provides a relationship between output parameters and input parameters which is illustrated using the regression model.

Response Surface Regression Model: R_aVersus v, f, a

$$R_a = -2.37 + 0.0367v - 2.22f - 1.55a - 0.000104v * v + 70.2f * f \\ + 3.72a * a + 0.0137v * f + 0.0039v * a - 2.23f * a \quad (7)$$

Equation 7 is used to forecast the value of R_a at any specific design points. The comparison between actual and forecasted values of regression equation is illustrated in Fig. 3.

The 3-D surface graphs for roughness, R_a, is given in Figs. 4, 5, and 6. It is observed that in Fig. 4, R_a (at hold value $v = 175$ m/min) value elevates with increase in feed rate but is not affected severely when depth of cut changes. In Fig. 5, it is seen that

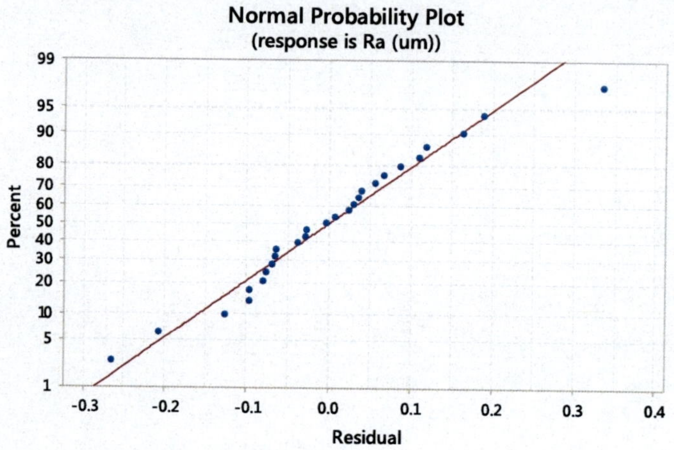

Fig. 2 Normal probability plot

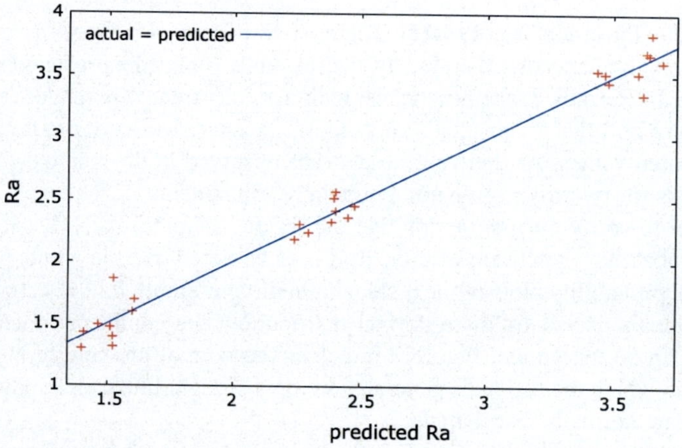

Fig. 3 Actual and forecasted values of R_a

R_a value decreases as cutting velocity and depth of cut decreases at a constant value of feed rate ($f = 0.15$ mm/rev). R_a value in Fig. 6 is altered cogently when feed rate increases at a fixed value of depth of cut but does not depict any observable relationship with the cutting speed.

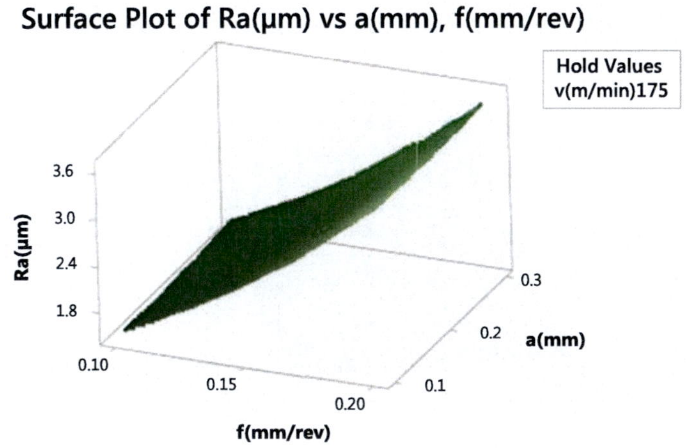

Fig. 4 Surface plot of R_a versus a, f

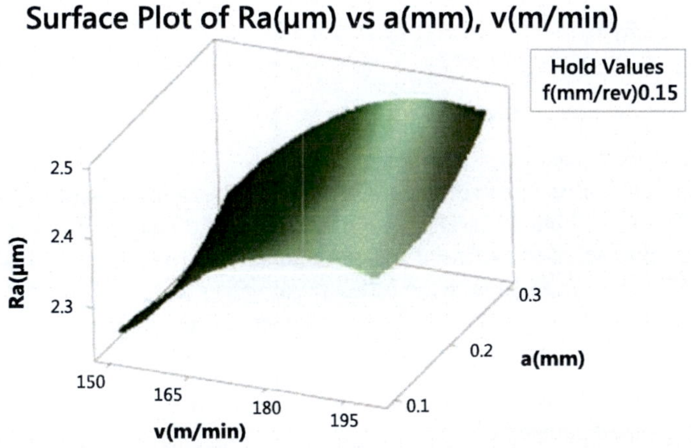

Fig. 5 Surface plot of R_a versus a, v

6 Confirmation Experiment

In order to validate the results, a confirmatory test was done using the combination of optimal parameters. A workpiece is cut by keeping the values of parameter as obtained by the above techniques. The confirmation experiment reveals the surface roughness to be 1.319, which is close to the result obtained and shows that MRR was increased by 15% and surface roughness was improved by 29.35%.

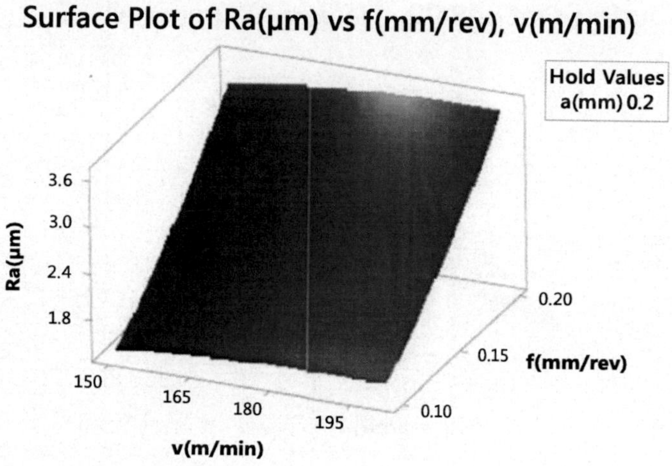

Fig. 6 Surface plot of R_a versus f, v

7 Conclusions

Following are the drawn conclusions:

1. The most significant parameter was feed for R_a.
2. Depth of cut was the most influential parameter affecting the MRR.
3. Using GRA, the optimal levels are found to be $v = 200$ m/min, $f = 0.1$ mm/rev, $a = 0.2$ mm. MRR is increased by 15% and surface roughness is improved by 29.35% when compared with the initial set of values.
4. Regression equations are developed and surface models are drawn which shows relation between the response and independent parameters.

Hence, it could be concluded that the quality traits could be increased significantly by employing these methods.

References

1. Yang WH, Tarng YS (1998) Design optimization of cutting parameters for turning operations based on the Taguchi method. J Mater Process Technol 84:122–129
2. Korkut I, Kasap M (2004) Determination of optimum cutting parameters during machining of AISI 304 austenitic stainless steel. Mater Des 23:303–305
3. Asiltürk I, Akkuş H (2011) Determining the effect of cutting parameters on surface roughness in hard turning using the Taguchi method. Measurement 44:1697–1704
4. Moshat S, Data S, Bandopaddhayay A, Pal PK (2008) Optimization of CNC milling process parameters using PCA based Taguchi method. Int J Eng Sci Technol 2:92–102
5. Shetty R, Pai R, Kamath V, Rao SS (2008) Study on surface roughness minimization in turning of DRACs using surface roughness methodology and Taguchi under pressured steam jet approach. ARPN J Eng Appl Sci 3:59–67

6. Neşeli S, Yaldiz S, Turkeş Y (2011) Optimization of tool geometry parameters for turning operations based on the response surface methodology. Measurement 44:580–587
7. Gupta M, Kumar S (2013) Multi-objective optimization of cutting parameters in turning using grey relation analysis. Int J Ind Eng Comput 4:547–558
8. Gupta A, Singh H, Aggarwal A (2011) Taguchi-fuzzy multi output optimization (MOO) in high speed CNC turning of AISI P-20 tool steel. Expert Syst Appl 38:6822–6828
9. Juneja M, Juneja N, Bhatia A (2018) Optimization of machining characteristics using CNC lathe: a critical review. Int J Adv Eng Res Appl 4(1):36–42
10. Mukherjee S, Kamal A, Kumar K (2014) Optimization of material removal rate during turning of SAE 1020 material in CNC lathe using Taguchi technique. Procedia Eng 97:29–35
11. Bhatia A, Kumar S, Kumar P (2014) A study to achieve minimum surface roughness in wire EDM. Procedia Mater Sci 5:2560–2566
12. Nayak SK, Patro JK, Dewangan S, Gangopadhyay S (2014) Multi-output optimization of machining parameters during dry turning of AISI 304 austenitic stainless steel using Grey relation analysis. Procedia Mater Sci 6:701–708
13. Sahoo AK, Baral AK, Rout AK, Routra BC (2012) Multi-objective optimization and predictive modeling of surface roughness and material removal rate in turning using Grey relation and regression analysis. Procedia Eng 38:1606–1627
14. Asiltürk I, Neşeli S (2012) Multi response optimization of CNC turning parameters via Taguchi method-based response surface analysis. Measurement 45:785–794
15. Kathleen MC, Natalia YK, Jeff YK (2004) Response surface methodology. Center for computational analysis of social and organizational systems (CASOS), Technical Report
16. Makadia AJ, Nanavati JI (2013) Optimisation of machining parameters for turning operations based on response surface methodology. Measurement 46:1521–1529
17. Noordin MY, Venkatesh VC, Sharif S, Elting S, Abdullah A (2004) Application of response surface methodology in describing the performance of coated carbide tools when turning AISI 1045 steel. J Mater Process Technol 145:46–58